MATHEMATICS
FOR
MANAGERIAL
DECISIONS

MATHEMATICS
FOR
MANAGERIAL
DECISIONS

Robert L. Childress

School of Business Administration
University of Southern California

Prentice-Hall, Inc., Englewood Cliffs, New Jersey

Library of Congress Cataloging in Publication Data

CHILDRESS, ROBERT L
    Mathematics for managerial decisions.

    (Quantitative analysis for business)
    Includes bibliographies.
    1. Mathematics—1961– I. Title.
QA37.2.C34   510   73-17352
ISBN 0-13-562231-X

Printed in the United States of America

10 9 8 7 6 5 4 3

PRENTICE-HALL INTERNATIONAL, INC., *London*
PRENTICE-HALL OF AUSTRALIA, PTY. LTD., *Sydney*
PRENTICE-HALL OF CANADA, LTD., *Toronto*
PRENTICE-HALL OF INDIA PRIVATE LIMITED, *New Delhi*
PRENTICE-HALL OF JAPAN, INC., *Tokyo*

To my parents
MARTHA AND LEON CHILDRESS

Contents

Preface

The importance of modern quantitative techniques in the business, economics, and social science curricula is well established. The primary reason for this importance comes from the fact that the techniques are extremely useful in making the managerial decisions required in a business or governmental enterprise. Quantitative methods are used not only in the traditional areas of production and engineering but are also widely employed in the functional areas of finance, marketing, and accounting. For this reason, the competent administrator must understand certain quantitative tools and the associated language. This text provides an introduction to several of the more important of these tools—sets, matrices, linear programming, calculus, and probability.

The objectives of this text are twofold. The first is to provide an introduction to the important quantitative tools of sets, matrices, linear programming, calculus, and probability in a manner that permits the nonmathematically inclined student to easily grasp the basic mathematical concepts. This introduction should prove quite useful both to the student pursuing a general course of study as well as to the student who is interested in studying additional mathematical techniques. A second, and equally important objective, is to provide numerous examples of the applicability of quantitative techniques in the administration of an enterprise. On the basis of these examples, the reader will hopefully be able to recognize problems in his job that can be solved using quantitative techniques. He will then be in the enviable position of being able to apply the techniques or, alternatively, of directing the work of the specialist assigned to the problem.

Mathematics for Managerial Decisions is designed for a one-semester course. Unless the students have an excellent mathematical background, however, most instructors will wish to omit certain chapters. For instance, certain instructors may wish to omit Chapter 4, entitled "Additional Elements of Matrix Algebra." If the time available for the study of linear programming is limited, it may be desirable to concentrate on the formulation of linear programming problems, Chapter 5, and the simplex algorithm, Chapter 6, and return to the topics of "Duality and Sensitivity," Chapter 7, and the "Transportation and Assignment Problems," Chapter 8, in a later course. Similarly, Chapter 16, entitled "Probability Functions," could be deferred to a course in statistics.

Numerous students at the University of Southern California and the University of Missouri-St. Louis have contributed helpful suggestions on ways of improving the text. Mr. Robert Hochhalter, a M.B.A. candidate at the University of Southern California, was especially helpful in reviewing the manuscript. I am also indebted to Dr. Warren Erikson of the University of Southern California and Dr. Robert Markland of the University of Missouri-St. Louis for their helpful suggestions and critique. I am especially grateful to Mrs. Glenell Smoot, Miss Chris Kutschinski, and Mrs. Toni Graham, who worked tirelessly on the difficult task of typing and retyping the manuscript.

ROBERT L. CHILDRESS
Los Angeles, California

Chapter 1

Sets

The concept of sets and the algebra of sets is enjoying increasing popularity and usage in business and the sciences. One reason for this popularity is that an understanding of the basic concepts of sets and set algebra provides a form of language through which the business specialist or practitioner can communicate important concepts and ideas to his associates. Specialists in electrical engineering, for example, use set algebra in the design of circuits. At the opposite extreme, specialists in written communications use sets in the analysis of statements and preparation of reports. Students of business administration use sets in the study of probability, statistics, programming. optimization, etc. This chapter provides an introduction to sets and set algebra along with selected applications. The reader can expect additional applications later in this text as well as in his studies of marketing, finance, economics, production, and statistics.

1.1 Sets Defined

A *set* is defined as a collection or aggregate of objects. To illustrate this definition, the members of the reader's quantitative business methods course are a set of students. Similarly, the members of the American Economic Association are a set of individuals, the faculty of the school of business is a set of instructors, and the *Encyclopedia Britannica* is a set of books.

From these examples the student can see that the concept of a set is relatively straightforward. There are, however, certain requirements for the collection or aggregate of objects that constitute the set. These requirements are:

1. The collection or aggregate of objects must be well defined, i.e., we must be able to determine unequivocally whether or not any object belongs to the set.
2. The objects of a set must be distinct; i.e., we must be able to distinguish between the objects and no object may appear twice.
3. The order of the objects within the set must be immaterial, i.e., *a, b, c* is the same set as *c, b, a.*

On the basis of these requirements for a set, suppose we are asked to verify that the students in a quantitative analysis class are a set. To determine if the students are a set, we ask three questions. First, can we determine if a student is registered for the course? Second, is it possible to distinguish between the students? Third, is the order of individuals in the class immaterial, i.e., is the class the same set whether arranged alphabetically by student or by social security number? If the answer to all three questions is yes, we conclude that the class is a set.

Example: Verify that the letters in the word Mississippi satisfy the requirement for a set. The letters in this word are m, i, s, and p. These letters are well-defined, distinct, and the order of the letters is immaterial. Therefore, the letters are a set.

Example: A coin is tossed three times. Denoting a head by H and a tail by T, determine the set that represents all possible outcomes of the three tosses.

The possible outcomes are {HHH, HHT, HTH, THH, HTT, THT, TTH, TTT}, where HHH represents a head on the first toss, a head on the second toss, and a head on the third toss. HHT represents a head on the first toss, a head on the second toss, and a tail on the third toss, etc.

Example: The set of digits is defined as the numbers {0, 1, 2, 3, 4, 5, 6, 7, 8, 9}. Determine which of the following numbers are members of this set: 3, 7.2, IV, 137, $\frac{4}{3}$.

The digit 3 is the only member of the set. The reader should be able to verify that the remaining numbers do not belong to the set of digits.

1.2 Specifying Sets and Membership in Sets

It is customary to designate a set by a capital letter. For instance, the set of digits defined in the above example could be designated as D.

The objects that belong to the set are termed the *elements* of the set or *members* of the set. The elements or the members of the set are designated by one of two methods: (1) the roster method, or (2) the descriptive method. The roster method involves listing within braces all members of the set. The descriptive method (also called the defining property method) involves describing the membership in a manner such that one can determine if an object belongs in the set.

To illustrate the specification of sets and membership, consider again the set of digits. The set of digits is designated by the capital letter D. The elements in the set (or alternatively the members of the set) are shown either by listing all the elements in the set within braces or by describing within braces the membership. If the roster method is used, the set of digits would appear as

$$D = \{0, 1, 2, 3, 4, 5, 6, 7, 8, 9\}$$

This is read as "D is equal to that set of elements 0, 1, 2, 3, 4, 5, 6, 7, 8, 9 " If the descriptive method of specifying the set is used, the set would be

$$D = \{x \mid x = 0, 1, 2, 3, \ldots, 9\}$$

This is read as "D is equal to that set of elements x such that x equals 0, 1, 2, 3, . . . , 9." In interpreting the symbolism used in set notation, it is useful to think of the left brace as shorthand for "that set of elements" and the vertical line as shorthand for "such that." Commas are used to separate the elements, and the raised periods mean "continuing in the established pattern." The 9 in the above set is interpreted as the final number of the set. The right brace designates set completion.

Example: The positive integers or "natural numbers" are the numbers 1, 2, 3, 4, 5, ... Show the set of natural numbers.

Assume that the set of natural numbers is represented by N. We cannot, of course, list all of the members of the set. We therefore use the descriptive method of specifying set membership and write

$$N = \{x \mid x = 1, 2, 3, 4, 5, \ldots\}$$

Example: Develop the set notation for the English alphabet.

We can use either the roster method or the descriptive method of specifying set membership. Representing the set of letters in the English alphabet by A, the set is

$$A = \{a, b, c, d, e, f, g, h, i, j, k, l, m, n, o, p, q, r, s, t, u, v, w, x, y, z\}$$

The descriptive method would conserve some space,

$$A = \{a, b, c, \ldots, y, z\}$$

Either method is acceptable. When one is using the descriptive method, however, it is important to remember that the description must be sufficient for one to determine the membership of the set. The elements must be well-defined, distinct, and order must be immaterial.

Example: The possible convention sites for the Western Farm Equipment Association are Los Angeles, San Francisco, Phoenix and Las Vegas. The set of convention sites is

$$S = \{\text{Los Angeles, San Francisco, Phoenix, Las Vegas}\}$$

1.2.1 SET MEMBERSHIP

The Greek letter ϵ (epsilon) is customarily used to indicate that an object belongs to a set. If A again represents the set of letters in the English alphabet, then $a \in A$ means that a is an element of the alphabet. The symbol $\notin$ (epsilon with a slashed line) represents nonmembership. We could thus write $\alpha \notin A$, meaning that alpha is not a member of the English alphabet. Similarly, referring to the convention site set S, we can write Los Angeles $\in S$ and San Diego $\notin S$.

1.2.2 FINITE AND INFINITE SETS

A set is termed *finite* or *infinite*, depending upon the number of elements in the set. The set A defined above is finite since it has 26 members, the letters in the English alphabet. The set D is also finite, since it has only the ten digits. The set N of positive integers or natural numbers is infinite, since the process of counting continues infinitely.

Example: Rational numbers are defined as that set of numbers a/b, where a represents all integers, both positive and negative including 0, and b represents all positive and negative integers, excluding 0. Develop the set R of rational numbers and specify whether the set is finite or infinite.

Letting R represent the set of rational numbers, we have

$$R = \{a/b \mid a = \text{all integers including 0}, b = \text{all integers excluding 0}\}$$

The set of rational numbers is infinite. Examples of rational numbers include any number that can be expressed as the ratio of two positive or negative whole numbers, such as $\frac{3}{2}$, $-\frac{6}{2}$, 7, etc.

Example: The individuals who are members of the American Economic

Association comprise a set. Assuming that a list of the membership is available, we could write the set as

$$S = \{\text{all members of the American Economic Association}\}$$

This set is finite, although quite large. It is a set because the elements are well-defined, distinct, and of inconsequential order.

1.3 Set Equality and Subsets

Two sets P and Q are said to be equal, written $P = Q$, if every element in P is in Q and every element in Q is in P. Set equality thus requires all elements of the first set to be in a second set and all elements in the second set to be in the first set. As an example, consider the set

$$P = \{0, 1, 2, 3, 4, 5\} \quad \text{and} \quad Q = \{2, 0, 1, 3, 5, 4\}$$

These sets are equal, since every element in P is in Q and every element in Q is in P.

The student will often have occasion to consider only certain elements of a set. These elements form a *subset* of the original set. As an example, assume that S represents the stockholders of company XYZ. Those stockholders who are employees of the company represent a subset of S. Thus, if Mr. Jones is a stockholder of XYZ and is employed by XYZ, he is a member of the subset of stockholders who are employed by the company.

Subset can be defined as follows. A set R is a subset of another set S if every element in R is in S. For example, if $S = \{0, 1, 2, 3, 4\}$ and $R = \{0, 1, 2\}$, then every element in R is in S, and R is a subset of S. The symbol for subset is $\subseteq$. R is a subset of S is written $R \subseteq S$.

Example: Let A represent the letters in the English alphabet and C represent the letters in the word "corporation." Verify that C is a subset of A.

Since $A = \{a, b, c, d, e, f, g, h, i, j, k, l, m, n, o, p, q, r, s, t, u, v, w, x, y, z\}$ and $C = \{c, o, r, p, a, t, i, n\}$, we see that $C \subseteq A$.

Example: Let D represent the set of digits and I represent the set of all integers including 0. Since $D = \{0, 1, 2, 3, 4, 5, 6, 7, 8, 9\}$ and $I = \{\ldots, -5, -4, -3, -2, -1, 0, 1, 2, 3, 4, 5, \ldots\}$ the finite set D is a subset of the infinite set I, i.e., $D \subseteq I$.

In the preceding example D was defined as the set of all digits including 0. We previously defined N as the set of all positive integers excluding 0. Assume that one is interested in determining if D is a subset of N. By inspection of the sets, D includes the element 0, whereas N does not. Consequently, D is not a subset of N. This is written as $D \not\subseteq N$. The notation for subset $\subseteq$ and not subset $\not\subseteq$ parallels the notation for member ϵ and not member $\not\epsilon$.

1.3.1 PROPER SUBSETS

The term *subset* is often differentiated from that of *proper subset*. A proper subset is designated by the symbol $\subset$. A proper subset P is a subset of another set U, written $P \subset U$, if all elements in P are in U but all elements in U are not in P. This simply means that for P to be a proper subset of U, then U must have all elements that are in P plus at least one element that is not in P. As an example, if

$$S = \{0, 1, 2, 3, 4\}$$

and

$$R = \{0, 1, 2\}$$

then R is a proper subset of S, i.e., $R \subset S$.

Example: Verify that the set C, the letters in corporation, is a proper subset of A, the letters in the English alphabet. Since every letter in C is in A, but every letter in A is not in C, we conclude that $C \subset A$.

1.3.2 UNIVERSAL SET

In discussing sets and subsets, the term *universal set* is often encountered. The term "universal set" is applied to the set that contains all the elements the analyst will wish to consider. If, for example, we are interested in categorizing the stockholders of the XYZ Company, the universal set would be all stockholders of the XYZ Company. The various categories of stockholders would then be subsets of the universal set. Similarly, if the analyst were interested in certain combinations of letters, the universal set would be defined as A, the letters of the English alphabet. It would then be possible to specify various subsets of the univeral set A, such as C, the letters in the word corporation.

The universal set contains all elements under consideration. In contrast, the *null* set is defined as a set that has no elements or members. To illustrate, assume that three students are to take an exam. We define the universal set as consisting of the students who score an A on the exam. If we refer to the students as S_1, S_2, and S_3, the results of the test could be $\{S_1, S_2, S_3\}$, $\{S_1, S_2\}$, $\{S_1, S_3\}$, $\{S_2, S_3\}$, $\{S_1\}$, $\{S_2\}$, $\{S_3\}$, $\{\ \}$. Note that there are eight possible outcomes and these are shown as subsets. One of the subsets is the null set $\{\ \}$. The null set, also referred to as the *empty* set, is designated by the Greek letter ϕ (phi). In this example of eight possible subsets, one of the possibilities is the universal set $\{S_1, S_2, S_3\}$ and another is the null set ϕ. This illustrates that both the universal set and the null sets are included as subsets of the universal set. This concept is also illustrated by the following examples.

Example: A stock market analyst is concerned with the price movement of the stock of the three large automobile manufacturers. Using the ticker symbols of Chrysler, Ford, and General Motors, determine the universal set representing upward movements in price of the stocks and specify all possible subsets.

The universal set representing upward price movement in all three stocks is $U = \{C, F, GM\}$. The possible subsets are $\{C, F, GM\}$, $\{C, F\}$, $\{C, GM\}$, $\{F, GM\}$, $\{C\}$, $\{F\}$, $\{GM\}$, $\{ \ \}$.

We again note that there are eight possible subsets. One of the subsets is the universal set $U = \{C, F, GM\}$ and another is the null set $\phi = \{ \ \}$.

Example: A businessman, vacationing at Santa Anita Racetrack, is considering betting on the first three races. If we denote betting on race 1 as R_1, betting on race 2 as R_2, etc., the universal set is $U = \{R_1, R_2, R_3\}$. The possible subsets are $\{R_1, R_2, R_3\}$, $\{R_1, R_2\}$, $\{R_1, R_3\}$, $\{R_2, R_3\}$, $\{R_1\}$, $\{R_2\}$, $\{R_3\}$, $\{ \ \}$.

1.3.3 COUNTING SUBSETS

The number of possible subsets of the universal set can be calculated through the use of a straightforward formula. The number of possible subsets is given by the formula

$$N = 2^n \tag{1.1}$$

where n represents the number of elements in the universal set and N is the number of possible subsets. If the universal set contains three members, there are eight possible subsets. Similarly, for a universal set with four members there are $N = 2^4 = 16$ possible subsets.

1.4 Set Algebra

Set algebra consists of certain operations on sets whereby the sets are combined to produce other sets. As an example, consider a group of students who are enrolled in introductory quantitative methods and another group who are enrolled in introductory finance. From these two sets we can specify, using the algebra of sets, a third set containing as members those individuals who are enrolled in both courses. These operations are most easily illustrated through use of the Venn diagram.

1.4.1 VENN DIAGRAM

The *Venn diagram*, named after the English logician John Venn (1834–83), consists of a rectangle that conceptually represents the universal set. Subsets of the universal set are represented by circles drawn within the rectangle or universal set. In Fig. 1.1 the universal set U is represented by the rectangle

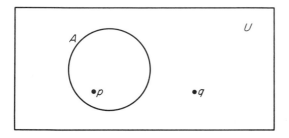

Figure 1.1

and the subset A by the circle. The Venn diagram shows that p is a member of A and that q is not a member of A. Both p and q are members of the universal set.

1.4.2 COMPLEMENTATION

The first set operation we consider is that of *complementation*. Let P be any subset of a universal set U. The complement of P, denoted by P' (read "P complement"), is the subset of elements of U that are not members of P. The complement of P is indicated by the shaded portion of the Venn diagram in Fig. 1.2.

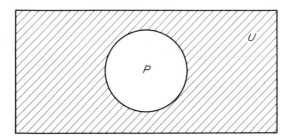

Figure 1.2

Example: For the universal set $D = \{0, 1, 2, 3, 4, 5, 6, 7, 8, 9\}$ with the subset $P = \{0, 1, 3, 5, 7, 9\}$, determine P'.

The complement of P contains all elements in D that are not members of P. Thus, $P' = \{2, 4, 6, 8\}$.

1.4.3 INTERSECTION

A second set operation is *intersection*. Again, let P and Q be any subsets of a universal set U. The intersection of P and Q, denoted by $P \cap Q$ (read "P intersect Q"), is the subset of elements of U that are members of both P and Q. $P \cap Q$ is shown by the shaded area of Fig. 1.3.

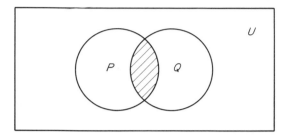

Figure 1.3

Example: For the universal set $D = \{0, 1, 2, 3, 4, 5, 6, 7, 8, 9\}$ with subsets $P = \{0, 1, 3, 5, 7, 9\}$ and $Q = \{0, 2, 3, 5, 9\}$, determine the intersection of P and Q.

The intersection of P and Q is the subset that contains the elements in D that are simultaneously in P and Q. Thus,

$$P \cap Q = \{0, 3, 5, 9\}$$

With a little thought the reader can also recognize that

$$P \cap U = P$$

and

$$P \cap P' = \phi$$

Sets such as $P \cap P' = \phi$ which have no common members are termed *disjoint* or *mutually exclusive*.

1.4.4 UNION

A third set operation is *union*. If we again let P and Q be any subsets of a universal set U, then the union of P and Q, denoted by $P \cup Q$ (read "P union Q"), is the set of elements of U that are members of either P or Q. $P \cup Q$ is shown by the shaded portion of the Venn diagram in Fig. 1.4.

Example: For the universal set $D = \{0, 1, 2, 3, 4, 5, 6, 7, 8, 9\}$ with subsets $P = \{0, 1, 3, 5, 7, 9\}$ and $Q = \{0, 2, 3, 5, 9\}$, determine the union of P and Q.

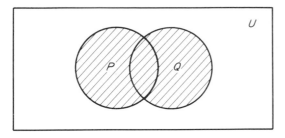

Figure 1.4

The union of P and Q is the subset that contains the elements in D that are in P or in Q. Thus,

$$P \cup Q = \{0, 1, 2, 3, 5, 7, 9\}$$

From the definitions given it also follows that

$$P \cup U = U$$
$$P \cup P' = U$$

1.4.5   OTHER SET OPERATIONS

Two additional set operations are sometimes included in the algebra of sets. The two operations are *difference* and *exclusive union*. Since subsets formed by either of these set operations can also be formed by use of complementation, intersection, and union, these operations are often excluded in discussing set algebra.

Let P and Q be any subsets of the universal set U. The difference of P and Q, denoted by $P - Q$ (read "P minus Q"), is the subset that consists of those elements that are members of P but are not members of Q. This subset is shown in the Venn diagram in Fig. 1.5. The difference of $P - Q$ can also be expressed as $P \cap Q'$.

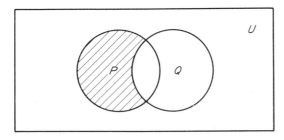

Figure 1.5

The exclusive union of P and Q, denoted by $P \underline{\cup} Q$ (read "P exclusive union Q"), is the set of elements of U that are members of P or of Q but not of both. The subset is shown in the Venn diagram. The subset can also be expressed as $(P \cap Q)' \cap (P \cup Q)$ or $(Q' \cap P) \cup (P' \cap Q)$.

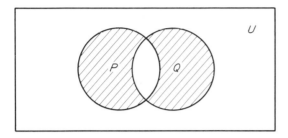

Figure 1.6

1.4.6 COMBINING SET OPERATIONS

Part of the utility of the algebra of sets occurs because of the ability to combine two or more sets into new sets through the use of the set operations. As an example, consider a national firm with district offices. The accounting system of this firm is computerized, and the various accounts are stored in a data bank accessible to the computer. The collection department manager at the national office is concerned with accounts that are sixty days past due. Accounts over sixty days past due are subsets of the set of accounts receivable. In analyzing the accounts due in both the western and southwestern districts, the manager would request the union of the two subsets of past due accounts. If he were interested in customers who simultaneously had accounts past due in both districts, the manager would request the intersection of the subsets.

The use of set algebra to form sets can be illustrated by Fig. 1.7. This

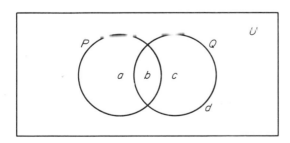

Figure 1.7

figure consists of a Venn diagram, representing the universal set, and two subsets P and Q. Areas in the Venn diagram are shown by the letters a, b, c, and d. The set that consists of areas a, b, c, and d is the universal set U, while the set P consists of a and b, the set Q consists of b and c, etc. Using this notation, we can use the algebra of sets to construct all possible subsets. The student should carefully study Fig. 1.7 and the construction of subsets using the set operations describing the areas a, b, and c presented in Table 1.1

Table 1.1

Area	*Set*
a, b, c, d	U
a, b	P
b, c	Q
a, d	Q'
c, d	P'
b	$P \cap Q$
a, b, c	$P \cup Q$
d	$(P \cup Q)'$
a	$P \cap Q'$ or $P - Q$
c	$P' \cap Q$ or $Q - P$
a, c	$(P \cap Q') \cup (P' \cap Q)$ or $(P \cap Q)' \cap (P \cup Q)$, or $(P \cup Q)$, or $(P - Q) \cup (Q - P)$
a, c, d	$(P \cap Q)'$

before proceeding.

The construction of subsets using the algebra of sets and Venn diagrams is further illustrated by the following examples.

Example: Business Month is developing a profile of its subscribers. Of the information requests returned by the subscribers to the business magazine, the following data were obtained: 30 percent were in a service industry, 40 percent were self-employed, 20 percent sold through retail channels. Of those who were self-employed, 40 percent were in the service industry, 20 percent were in a retail business, and 10 percent were in a retail service business. The response also indicated that 50 percent of the retail businesses were service-oriented. From these data, develop a Venn diagram that shows the reader profile as subsets of the response set.

The circles shown in the Venn diagram in Fig. 1.8 represent the service industry S, self-employed E, and retail business R. From the data we know that $S = 30$, $E = 40$, and $R = 20$. Since 40 percent of the self-employed are in the service industry, it follows that $E \cap S = 16$. Similarly, $E \cap R = 8$ and $E \cap R \cap S = 4$. We also know from the data that $R \cap S = 10$.

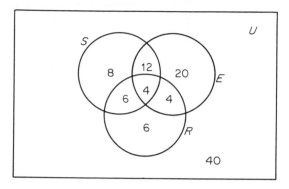

Figure 1.8

The example can be expanded to give the reader additional practice in the use of set operations to specify subsets. This is done by developing set notation for statements describing subsets of the *Business Month* Venn diagram. The statements and subsets are given in Table 1.2 and refer to Fig. 1.8. This table and figure also deserve careful study.

Table 1.2

Statement	Set	Number
1. Self-employed in retail service	$E \cap R \cap S$	4
2. Self-employed in retail nonservice	$E \cap R \cap S'$	4
3. Self-employed in nonretail nonservice	$E \cap R' \cap S'$	20
4. Self-employed in nonretail service	$E \cap R' \cap S$	12
5. Non-self-employed in nonretail service	$E' \cap R' \cap S$	8
6. Non-self-employed in retail service	$E' \cap R \cap S$	6
7. Non-self-employed in retail nonservice	$E' \cap R \cap S'$	6
8. Non-self-employed, nonservice, nonretail	$E' \cap R' \cap S'$	40
9. Service or self-employed	$S \cup E$	54
10. Nonretail service or nonretail self-employed	$(R' \cap S) \cup (R' \cap E)$	40
11. Non-self-employed in service	$E' \cap S$	14

Example. The Venn diagram shown in Fig. 1.9 is subdivided into areas $a_1, a_2, a_3, a_4, a_5,$ and a_6. The areas are described by the subsets given as follows.

Area	Set
a_1, a_2, a_3, a_4	$P \cup Q$
a_5	$R \cap (P \cup Q)'$
a_3, a_4	$(P \cap R) \cup (Q \cap R)$
a_1, a_2, a_5	$(R' \cap P) \cup (R' \cap Q) \cup ((P \cup Q)' \cap R)$

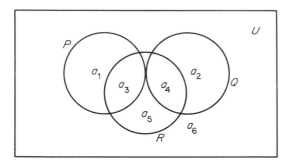

Figure 1.9

1.4.7 LAWS OF THE ALGEBRA OF SETS

The algebra of sets is composed of certain laws. In some cases these laws are similar to the algebra of numbers. We shall state these laws in the form of postulates, thus relieving the reader of the burden of studying mathematical proofs of the laws. The less obvious postulates are illustrated with Venn diagrams.

Postulate 1: $P \cup Q = Q \cup P$
Postulate 2: $P \cap Q = Q \cap P$

Postulates 1 and 2 are termed the *commutative law*. These postulates state that the order in which we combine the union or intersection of two sets is immaterial. This corresponds to the commutative law in ordinary algebra, which states that $p + q = q + p$ and that $p \cdot q = q \cdot p$.

Postulate 3: $P \cup (Q \cup R) = (P \cup Q) \cup R$
Postulate 4: $P \cap (Q \cap R) = (P \cap Q) \cap R$

Postulates 3 and 4 are termed the *associative law*. These postulates state that the selection of two of three sets for grouping in a union or intersection is immaterial. Thus, the order in which the sets are combined is immaterial. These postulates correspond to the associative law in ordinary algebra, which enables us to state that $p + (q + r) = (p + q) + r$ and that $p \cdot (q \cdot r) = (p \cdot q) \cdot r$.

Postulate 5: $P \cup (Q \cap R) = (P \cup Q) \cap (P \cup R)$
Postulate 6: $P \cap (Q \cup R) = (P \cap Q) \cup (P \cap R)$

Postulates 5 and 6 are called the *distributive law*. Postulate 5 has no analogous postulate in ordinary algebra. Postulate 6 corresponds to the distributive law in ordinary algebra that enables us to state that $p(q + r) = p \cdot q + p \cdot r$.

Since Postulates 5 and 6 are not as obvious as Postulates 1 through 4, we shall verify one of them through Venn diagrams. To verify Postulate 5, we must show that the area represented by $P \cup (Q \cap R)$ is the same as that represented by $(P \cup Q) \cap (P \cup R)$. The area representing the set $P \cup (Q \cap R)$ is shown in Fig. 1.10(a). In Fig. 1.10(a) the intersection of Q

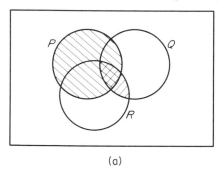 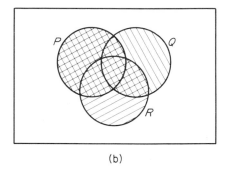

(a)   (b)

Figure 1.10

and R is shown by the shading from lower left to upper right, and P is shown by shading from upper left to lower right. The union of P with $Q \cap R$, represented by both the shaded and crosshatched area in the figure, gives $P \cup (Q \cap R)$.

The area representing the set $(P \cup Q) \cap (P \cup R)$ is shown in Fig. 1.10(b). In Fig. 1.10(b) the union of P and Q is shown by shading from upper left to lower right, and the union of P and R is shown by shading from lower left to upper right. The intersection of the two shaded areas gives the crosshatched area described as the set $(P \cup Q) \cap (P \cup R)$. Since the shaded and crosshatched area in Fig. 1.10(a) equals the crosshatched area in Fig. 1.10(b), we conclude that Postulate 5 is true. Postulate 6 is verified in the same manner as Postulate 5.

Postulate 7: $P \cap P = P$
Postulate 8: $P \cup P = P$
Postulate 9: $P \cup \phi = P$

Postulates 7 through 9 follow directly from the operations of union and intersection. They can be verified by the Venn diagram in Fig. 1.11.

Postulate 10: $P \cap U = P$
Postulate 11: $P \cup P' = U$
Postulate 12: $P \cap P' = \phi$

Postulates 10 through 12 are based upon the definition of the null set, the universal set, and the complement. These postulates are obvious and can also easily be verified by the Venn diagram shown in Fig. 1.11.

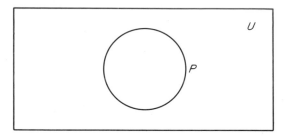

Figure 1.11

Postulate 13: $(P \cup Q)' = P' \cap Q'$
Postulate 14: $(P \cap Q)' = P' \cup Q'$

Postulates 13 and 14 are termed De Morgan's law. Postulate 13 states that the complement of the union of two sets is equal to the intersection of the complement of each set. Postulate 14 states that the complement of the intersection of two sets is equal to the union of the complement of each set. We shall verify Postulate 13 through the use of Venn diagrams. The student is asked to verify Postulate 14.

Figure 1.12(a) shows the set $(P \cup Q)'$ as the area shaded from upper left

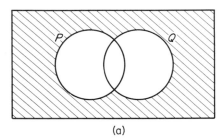 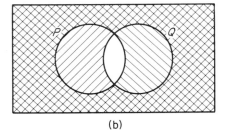

(a)   (b)

Figure 1.12

to lower right. In Fig. 1.12(b), P' is shown by shading from upper left to lower right and Q' is shown by shading from lower left to upper right. The intersection of P' with Q' is shown by the crosshatched area. Since the shaded area in Fig. 1.12(a) equals the crosshatched area in Fig. 1.12(b), it follows that $(P \cup Q)' = P' \cap Q'$.

The laws of set algebra are used analagously with the laws of ordinary algebra. Just as the laws of ordinary algebra can be used to simplify algebraic expressions, the laws of set algebra can be used to simplify sets. This is illustrated by the following examples.

Example: Simplify the set $(A \cup B) \cup (A \cap B)$.

    1. Let $P = (A \cup B)$: $P \cup (A \cap B)$
    2. Postulate 5: $(P \cup A) \cap (P \cup B)$
    3. Substitute for P: $((A \cup B) \cup A) \cap ((A \cup B) \cup B)$
    4. Postulate 1: $(A \cup A \cup B) \cap (A \cup B \cup B)$
    5. Postulate 7: $(A \cup B) \cap (A \cup B)$
    6. Postulate 7: $A \cup B$

In this example, it is important for the student to use parentheses to enclose $P = (A \cup B)$. This is further illustrated by the following example.

Example: On page 11 and again on page 12 we stated that $(P \cap Q)' \cap (P \cup Q)$ was equal to $(Q' \cap P) \cup (P' \cap Q)$. Using the laws of set algebra, verify this relationship.

Given: $(P \cap Q)' \cap (P \cup Q)$
    1. Postulate 14: $(P' \cup Q') \cap (P \cup Q)$
    2. Let $R = (P' \cup Q')$ $R \cap (P \cup Q)$
    3. Postulate 6: $(R \cap P) \cup (R \cap Q)$
    4. $(P' \cup Q') = R$ $((P' \cup Q') \cap P) \cup ((P' \cup Q') \cap Q)$
    5. Postulate 6: $(P \cap P') \cup (P \cap Q') \cup (P' \cap Q) \cup$
                       $(Q \cap Q')$
    6. Postulate 12: $(P \cap Q') \cup (P' \cap Q)$

Thus

$$(P \cap Q)' \cap (P \cup Q) = (P \cap Q') \cup (P' \cap Q)$$

Example: Simplify the expression $P \cup (P' \cap Q)$.

Given: $P \cup (P' \cap Q)$
    1. Postulate 5: $(P \cup P') \cap (P \cup Q)$
    2. Postulate 11: $U \cap (P \cup Q)$
    3. Postulate 10: $P \cup Q$

The expression simplifies to $P \cup Q$.

Example: Show that the set $(P \cup Q') \cap (P \cap Q)'$ equals Q'.

Given: $(P \cup Q') \cap (P \cap Q)'$
    1. Postulate 14: $(P \cup Q') \cap (P' \cup Q')$
    2. Postulate 5: $Q' \cup (P \cap P')$
    3. Postulate 12: Q'

The primary difficulty in understanding this example is in converting $(P \cup Q') \cap (P' \cup Q')$ to $Q' \cup (P \cap P')$. This step involves rewriting (Postulate 1) the expression as $(Q' \cup P) \cap (Q' \cup P')$. Postulate 5, the distributive law, is applied to write the expression as $Q' \cup (P \cap P')$. Since $P \cap P'$ is equal to ϕ, the expression reduces to Q'.

1.5   Cartesian Product

Suppose that one is asked to list the possible outcomes of two tosses of a coin. Since either a head or a tail occurs on a single toss, the possible outcomes are described by the set $O = \{(H, H), (H, T), (T, H), (T, T)\}$. There are four elements in this set and, corresponding to the requirements given for a set, the order of the elements is immaterial. Are each of the elements, however, distinct? To answer this question we must ask if the element (H, T) differs from the element (T, H). The answer is that these elements do differ, since the order of the occurrence is important. If, for instance, an individual bet \$1 that a head would occur on the first toss of the coin and \$1 that a tail would occur on the second toss, he would be \$2 richer if element (H, T) occurred and \$2 poorer if (T, H) occurred.

The elements (H, H), (H, T), (T, H) and (T, T) are examples of *ordered pairs*. One of the two components of each ordered pair is designated as the first element of the pair, and the other, which need not be different from the first, is designated as the second element. If the first element is designated as a and the second element is designated as b, we have the ordered pair (a, b). This ordered pair differs from the ordered pair (b, a), and both ordered pairs differ from the set $\{a, b\}$.

Ordered pairs are formed by the *Cartesian product* of two sets. If A and B are two sets, the Cartesian product of the sets, designated by $A \times B$, is the set containing all possible ordered pairs (a, b) such that $a \in A$ and $b \in B$. If the set A contains the elements a_1, a_2, a_3 and the set B contains the elements b_1 and b_2, the Cartesian product $A \times B$ is the set $A \times B = \{(a_1, b_1), (a_1, b_2), (a_2, b_1), (a_2, b_2), (a_3, b_1), (a_3, b_2)\}$. All possible ordered pairs are included in the set.

The concept of the Cartesian product is quite useful in many decision problems. The student of probability and statistics will often be asked to consider the possible outcomes of an experiment. As an example, consider again the problem of determining all possible outcomes of two tosses of a coin. If we define the outcome of the first toss as the set $O_1 = \{H, T\}$ and the outcome of the second toss as the set $O_2 = \{H, T\}$, then the Cartesian product $O_1 \times O_2$ gives all possible outcomes of the two tosses. As we have seen, these outcomes are $O_1 \times O_2 = \{(H, H), (H, T), (T, H), (T, T)\}$.

The Cartesian product of two sets can be determined quite easily with the aid of a box diagram. Figure 1.13 shows the box diagram for the Cartesian

$$O_2$$

	H	T
H	H, H	H, T
T	T, H	T, T

O_1 labels the rows.

Figure 1.13

product of O_1 and O_2. The method of constructing the diagram involves listing the elements of O_1 to the left of the box and O_2 above the box. The blanks in the box are then filled in with the ordered pairs. The Cartesian product $O_1 \times O_2$ consists of the elements in the box, i.e., (H, H), (H, T), (T, H), (T, T).

The Cartesian product can be expanded to combine more than two sets. This means that the concepts discussed for ordered pairs can, for example, be applied to *ordered triplets*. To illustrate, assume that we are asked to list all possible outcomes of three tosses of a coin. Denoting $O_i = \{H, T\}$, where O_i represents the possible outcomes on the ith toss (i.e., $i = 1$ represents the first toss, $i = 2$ represents the second toss, etc.), the possible outcomes would be given by the Cartesian product $O_1 \times O_2 \times O_3$. This Cartesian product is determined by finding $O_1 \times O_2$ and then $(O_1 \times O_2) \times O_3$. From Fig. 1.13, we know that $O_1 \times O_2 = \{(H, H), (H, T), (T, H), (T, T)\}$. $(O_1 \times O_2) \times O_3$ is shown in Fig. 1.14. The Cartesian product of $O_1 \times O_2 \times$

$$O_1 \times O_2$$

	H, H	H, T	T, H	T, T
H	H, H, H	H, H, T	H, T, H	H, T, T
T	T, H, H	T, H, T	T, T, H	T, T, T

O_3 labels the rows.

Figure 1.14

O_3 is $\{(H, H, H), (H, T, H), (T, H, H), (T, T, H), (H, H, T), (H, T, T), (T, H, T), (T, T, T)\}$.

The box diagrams of Figs. 1.13 and 1.14 provide a straightforward method of determining of the Cartesian products of sets. With some practice, the student can apply this concept without difficulty. Before illustrating the concept with examples, however, let us carry the Cartesian product one additional step. Assume that we are asked to list all possible outcomes of four tosses of a coin. There are 16 such outcomes. Most individuals would be extremely hard pressed to think of all sixteen. If we use the concept of the Cartesian product, the task becomes routine. The possible outcomes are given by the set $O_1 \times O_2 \times O_3 \times O_4$. The box diagram can be expressed in terms of $(O_1 \times O_2) \times (O_3 \times O_4)$ or by any other grouping of the parentheses. Figure 1.15 shows the Cartesian product of $(O_1 \times O_2) \times (O_3 \times O_4)$.

$O_3 \times O_4$

		H, H	H, T	T, H	T, T
	H, H	H, H, H, H	H, H, H, T	H, H, T, H	H, H, T, T
	H, T	H, T, H, H	H, T, H, T	H, T, T, H	H, T, T, T
$O_1 \times O_2$	T, H	T, H, H, H	T, H, H, T	T, H, T, H	T, H, T, T
	T, T	T, T, H, H	T, T, H, T	T, T, T, H	T, T, T, T

Figure 1.15

The elements in the box diagram such as (H, H, H, T) contain four members. Mathematicians call such an element an ordered "4-tuple." Using this term, an ordered pair could be referred to as an ordered 2-tuple, an ordered triplet as an ordered 3-tuple, etc. In general, then, an element containing n members is referred to as an ordered *n-tuple*.

Example: A retailer specializes in three products: color television, black and white television, and stereos. He offers a service contract with the sale of each of the products, which the customer may or may not elect to purchase. Determine the possible combination of sales options.

Let the products be represented by the set $P = \{C, B, S\}$ and the sales contract by $R = \{E, E'\}$. The Cartesian product of $P \times R$ gives the combination of sales options. The box diagram shows the elements of $P \times R$.

P

		C	B	S
	E	E, C	E, B	E, S
R	E'	E', C	E', B	E', S

Example: A builder has three basic floor plans: single story, two story, and trilevel. Each of these plans can have either a shake roof or a wood shingle roof. In addition, the plans are available with or without fireplaces. Determine the number of combinations of plans and show these plans in a box diagram.

Let the basic floor plans be represented by the set $F = \{1, 2, 3\}$, the roofing material by the set $R = \{S, W\}$, and the fireplace option by the set $O = \{f, f'\}$. The combination of plans is represented by the set $\{F \times R \times O\}$. The box diagram for the set is shown below.

F

		1	2	3
R	S	S, 1	S, 2	S, 3
	W	W, 1	W, 2	W, 3

$F \times R$

O	f	f, S, 1	f, S, 2	f, S, 3	f, W, 1	f, W, 2	f, W, 3
	f'	f', S, 1	f', S, 2	f', S, 3	f', W, 1	f', W, 2	f', W, 3

There are twelve combinations of plans.

Example: An advertising agency is placing ads for three products. The media available for advertising are radio, television, and newspaper. The ads will be written by either the agency or the sponsor. Develop the possible combinations with the aid of a box diagram.

Let the products be represented by the set $P = \{1, 2, 3\}$, the media by the set $M = \{R, T, N\}$, and the source of the advertisement by the set $S = \{A, A'\}$. The box diagram for $P \times M \times S$ is constructed as follows:

M

		R	T	N
	1	1, R	1, T	1, N
P	2	2, R	2, T	2, N
	3	3, R	3, T	3, N

		$M \times P$								
		1, R	1, T	1, N	2, R	2, T	2, N	3, R	3, T	3, N
S	A	A, 1, R	A, 1, T	A, 1, N	A, 2, R	A, 2, T	A, 2, N	A, 3, R	A, 3, T	A, 3, N
	A'	A', 1, R	A', 1, T	A', 1, N	A', 2, R	A', 2, T	A', 2, N	A', 3, R	A', 3, T	A', 3, N

The total number of members formed by the Cartesian product of two sets is given by the product of the number of elements in each set. Thus if set A contains five elements and set B contains four elements, then the Cartesian product $A \times B$ contains $5(4) = 20$ elements. The rule applies to the Cartesian product of more than two sets. For the three sets A, B, and C, containing N_1, N_2 and N_3 elements respectively, the Cartesian product of the sets $A \times B \times C$ contains $N_1 \cdot N_2 \cdot N_3$ elements.

Example: Set A has 10 elements, set B has 6 elements, set C has 12 elements, and set D has 3 elements. Determine the number of elements in the Cartesian product of $A \times B \times C \times D$.

The number of elements is given by product of the number of elements in each set. Thus the Cartesian product contains $10 \cdot 6 \cdot 12 \cdot 3$, or 2160 elements.

1.6 Applications to Logic†

The algebra of sets can be applied to problems of logic. To illustrate, assume that we receive the following guidelines concerning the allocation of expenditures in a firm:

Advertising expenditures are to be directed toward men who are college graduates or are over thirty years old, but not to college graduates under thirty years old.

To simplify this unnecessarily complicated directive, let A represent men who are college graduates and B represent men who are over thirty years old. The set $A \cup B$ then represents men who are college graduates or over thirty years old. Similarly, the set $A \cap B'$ represents college graduates under thirty years old. Since advertising expenditures are not to be directed toward college graduates who are under thirty years old, the set describing the allocation of advertising expenditures is

$$(A \cup B) \cap (A \cap B')'$$

---

† This section can be omitted without loss of continuity.

This set can be simplified by using either the laws of set algebra or a Venn diagram. To simplify the set using the laws of set algebra, we perform the following steps:

Given: $(A \cup B) \cap (A \cap B')'$
1. Postulate 14: $(A \cup B) \cap (A' \cup B)$
2. Postulate 5: $B \cup (A \cap A')$
3. Postulate 12: $B \cup \phi$
4. Postulate 9: B

Interestingly, the analysis shows that the advertising guide lines can be reduced to the simple statement, "Advertising expenditures should be directed to men who are over thirty years old." The reader should verify that the same result can be shown through use of the Venn diagram.

Example: Simplify the following edict made by the president of Amalgamated Industries:

> *Due to recent cutbacks in the major divisions of our firm, employees who are over sixty years of age or have over thirty years of service with the company are eligible for early retirement. This excludes, however, individuals who have thirty years of service but are under sixty.*

To simplify this policy statement, let A represent the set of individuals over sixty years old and B represent the set of individuals who have over thirty years of service with the company. Those individuals who are over sixty years of age or have over thirty years with the company are described by the set $A \cup B$. Individuals under sixty years of age who have over thirty years of service are given by the set $A' \cap B$. Those eligible for early retirement are thus described by the set $(A \cup B) \cap (A' \cap B)'$. This set is simplified as shown in the preceding example to give $(A \cup B) \cap (A' \cap B)' = A$. The analysis shows that regardless of the number of years of service, only individuals over sixty years of age are eligible for early retirement.

Example: The president of Amalgamated Industries was told that his policy on early retirement specifically included all individuals over sixty years of age, regardless of the length of service to the company. Upon hearing this interpretation, he revised the policy as follows:

> *Due to recent cutbacks in the major divisions of our firm, employees who are over sixty years of age and have over thirty years of service with the company are eligible for early retirement. This specifically excludes individuals who are over sixty but have less than thirty years of service with the company.*

Simplify this policy statement, using a Venn diagram.

The Venn diagram describing the president's policy is shown below. Sets

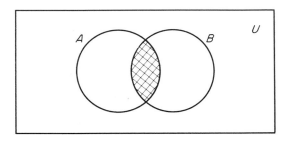

A and *B* are defined as before. Individuals who are over sixty years of age and have over thirty years of service are described by the set $A \cap B$. Those who are over sixty years of age but have less than thirty years of service are described by the set $A \cap B'$. The individuals described by the president as eligible for early retirement are described by the intersection of these sets, $(A \cap B) \cap (A \cap B')'$. This set can be reduced by using set algebra to the set $A \cap B$, the set shown by the crosshatched area in the Venn diagram.

These examples have used the operations of intersection, union, and complementation together with the laws of set algebra to clarify certain complicated statements. It can be seen that these operators are employed in the logical analysis of statements in much the same fashion as they were in the earlier description of sets and set membership.

1.6.1 STATEMENTS AND CONNECTIVES

One of the important concepts in logic is that of the *statement*. A statement is a simple declaration. For instance, "The sun is shining" and "I received an A on the last examination" are simple statements. The distinguishing characteristic of a statement is that the statement makes an assertion. The assertion made by the statement can be either true or false, but not both. It should be emphasized that interrogations (such as "Is the sun shining?" and imperatives (such as "Please turn in your homework.") do not assert anything and, consequently, are not statements.

The set operators are used to combine simple statements to form *compound* statements. The operations of complementation, intersection, and inclusive and exclusive union that were discussed in Sec. 1.4 have analogous counterparts in logic. These counterparts, termed *connectives* in logic, are negation, conjunction, and inclusive and exclusive disjunction.

The *negation* of a statement A is the statement "not A." To illustrate, let

$$A = \text{The sun is shining}$$

The negation of A is

$$A' = \text{The sun is not shining}$$

The *conjunction* of two statements A and B is the statement "A and B." This is denoted by $A \cap B$. For instance, if

$$A = \text{The sun is shining}$$

and

$$B = \text{I will go swimming}$$

then

$$A \cap B = \text{The sun is shining and I will go swimming}$$

Similarly,

$$A' \cap B' = \text{The sun is not shining and I will not go swimming}$$

The *inclusive disjunction* of two statements A and B is the statement A or B or both and is denoted by $A \cup B$. The *exclusive disjunction* of the two statements is A or B but not both and is denoted by $A \underline{\cup} B$. Defining the statement A as

$$A = \text{I will major in engineering}$$

and the statement B as

$$B = \text{I will major in business}$$

then the inclusive disjunction of the two statements is

$$A \cup B = \text{I will major in engineering or business or both}$$

The exclusive disjunction of the two statements is

$$A \underline{\cup} B = \text{I will major in engineering or business but not both}$$

Example: Define the statements A and B as $A =$ "The Dow-Jones stock market averages advanced" and $B =$ "Corporate profits were reported higher during the past quarter." Use the connectives (i.e., operators) negation, conjunction, and inclusive and exclusive disjunction to describe the following compound statements:

1. The Dow-Jones average advanced, and corporate profits were reported higher during the quarter.

Answer: $A \cap B$.

2. The Dow-Jones average fell, but corporate profits were reported higher during the quarter.

Answer: $A' \cap B$.

3. Although not reflected by the Dow-Jones average that continued to advance, corporate profits were reported lower during the past quarter.

Answer: $A \cap B'$.

Example: Define the statements A and B as $A = $ "The labor contract is inflationary" and $B = $ "The price increase is inflationary." Use the logical connectives to describe the following compound statements.

1. The labor contract is inflationary, but the price increase is not.

Answer: $A \cap B'$.

2. Either the labor contract or the price increase is inflationary.

Answer: $A \cup B$.

3. Neither the labor contract nor the price increase is inflationary.

Answer: $(A \cup B)'$.

4. Taken together, the labor contract and the price increase are inflationary.

Answer: $A \cap B$.

5. The labor contract could be considered inflationary or the price increase could be considered inflationary, but certainly both would not be considered inflationary.

Answer: $A \underline{\cup} B$.

1.6.2 LOGICAL ARGUMENTS

The principles of set algebra can easily be extended to the analysis of logical arguments. A *logical argument* consists of a series of statements or *premises* followed by a conclusion. The argument is termed *logically true* if the conclusion is the logical consequence of the premises. Conversely, the argument is *logically false* if the conclusion does not follow from the premises.

The statements or premises upon which the argument is based are termed *factual*. The factual statements may themselves be either true or false. Consequently, an argument that is logically true need not be based on correct facts.

To illustrate the concept of a logical argument, consider the following statement:

Automobile insurance premiums in states such as Massachusetts that have passed "no fault" insurance laws are lower than the premiums in states such as California that have not passed the "no fault" insurance laws. Consequently, insurance premiums in Massachusetts are lower than those in California.

To examine the validity of the argument, let

A = Automobile insurance premiums in states that have passed "no fault" laws are lower than the premiums in states that have not passed the laws.

B = Massachusetts has a "no fault" insurance law.

C = California does not have a "no fault" insurance law.

D = Massachusetts insurance premiums are lower than those in California.

Statements A, B, and C are premises. Statement D is the conclusion based upon the three premises. To determine if D is logically true, consider the Venn diagram in Fig. 1.16. The figure shows that B is a subset of A and that

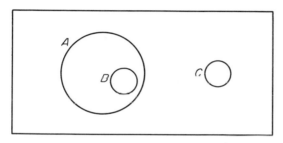

Figure 1.16

C is not a subset of A. Since the insurance premiums in all states that are members of A are lower than those in all states that are not members of A, it logically follows that the insurance premiums in Massachusetts are lower than those in California. The argument, therefore, is logically true.

Example: Determine the validity of the following argument.

Rapid expansion in an economy is often accompanied by inflation. Germany and Japan have experienced rapid economic growth during the past decade. Therefore, Germany and Japan have experienced inflation.

Let:

A = Rapid expansion is often accompanied by inflation

B = Germany and Japan have experienced rapid economic growth

C = Germany and Japan have experienced inflation

To determine the validity of the argument, consider the following Venn diagram. The diagram illustrates that the argument is logically true only if C is a member of the conjunction of A and B. Since there is no factual

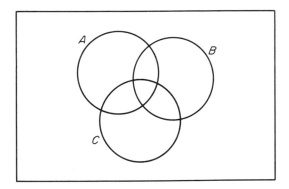

statement to the effect that C is a member of the conjunction of A and B, we conclude that the argument is logically false.

PROBLEMS

1. In a recent editorial, a journalist referred to the Republican voters in an upcoming election. Would it be possible to define this group of voters as a set?

2. In a recent opinion poll, one hundred individuals were questioned concerning a legislative proposal. Both the opinion and the sex of the respondent were noted. Give the subsets that a political analyst might consider important.

3. An economist referred to the set of all family units in the United States whose combined family income is less than $6000. Is the use of the term "set" appropriate in this instance?

4. Let set $A = \{5, 7, 9, 10, 12, 14\}$. List elements in the following sets that are also in A.
   (a) $\{x \mid x = \text{all odd numbers}\}$
   (b) $\{x \mid 2x + 6 = 20\}$
   (c) $\{x \mid x - 3 = \text{an odd integer}\}$

5. List all possible subsets of the set $S = \{a, b, c\}$.

6. A student can take one or more of five courses. How many different course selections are possible?

7. A restaurant offers pickles, ketchup, mustard, onions, lettuce, and tomatoes on its hamburgers. How many different hamburgers are possible?

8. For the sets $U = \{0, 1, 2, 3, 4, 5, 6, 7, 8, 9\}$, $R = \{0, 2, 4, 6, 8\}$, $S = \{2, 3, 4, 5, 6\}$, and $T = \{5, 6, 7, 8, 9\}$, determine the membership of the following:
   (a) $R \cap S$
   (b) $R \cap S \cap T'$
   (c) $R \cap S' \cap T'$
   (d) $R \cup S'$
   (e) $(R \cap S \cap T)'$

9. Simplify the following expressions:
   (a) $(P \cup Q)' \cup (P \cap Q')$
   (b) $(P \cap Q) \cap (P \cup Q)'$
   (c) $(P \cap Q)' \cup P$

10. Kawar Travel Agency handles the winter travel arrangements for all ski clubs in the Los Angeles, San Diego, and Phoenix areas that are affiliated with the United States Ski Association. The billing system for the agency is computerized. From time to time, the agency has need for the information stored in the computer and must call for it by set nomenclature. Following are examples of the type of information which is available.

$$A = \text{ski clubs in Los Angeles}$$
$$B = \text{ski clubs in San Diego}$$
$$C = \text{ski clubs in Phoenix}$$
$$D = \text{set of ski clubs whose billings}$$
$$\text{are more than 45 days old}$$

Develop set nominclature for the following.
   (a) All the ski clubs in California with chapters both in Los Angeles and San Diego that have billings with the agency.
   (b) The billings in Phoenix or San Diego that are more than 45 days old.
   (c) The billings in Phoenix or Los Angeles less than 45 days old.

11. A highway construction crew is composed of men who can operate heavy equipment as follows: bulldozer, 33; crane, 22; cement mixer, 35; mixer and dozer, 14; mixer and crane, 10; dozer and crane, 10; mixer, dozer, and crane, 4. How many men are in the construction crew?

12. An analysis of the membership at a local country club shows that 60 percent of the members are men, 40 percent of the members score below 85, and 50 percent of the individuals belonging to the club are over 30 years old. It was also determined that 50 percent of the men are over

30 years old, 60 percent of the men score below 85, 40 percent of the men over 30 years old score below 85, and that 10 percent of the members who score below 85 are women over 30 years old.

(a) What percentage of the membership are men over 30 years old that score below 85?

(b) What percentage of the membership are men that score below 85?

13. The Johnson Corporation had 500 employees, of which 200 got a raise, 100 got a promotion, and 80 got both.

(a) How many employees got a promotion with no raise?

(b) How many employees got neither a raise nor a promotion?

14. A student surveyed 525 people. He determined that 350 read the newspaper for news, 215 listened to radio, and 140 watched TV. Additionally, 75 read the newspaper and watched television, 40 listened to radio and watched TV, and 100 read the newspaper and listened to radio. If 25 used all three sources of news, how many people utilized none of the three?

15. Define the sets F, C, and S, respectively, as stockholders in Ford, Chrysler, and U.S. Steel. If there are 500 stockholders in Ford, 800 stockholders in Chrysler, 700 stockholders in U.S. Steel, 200 stockholders in both Ford and Chrysler, 100 stockholders in both Ford and U.S. Steel, 100 stockholders in both Chrysler and U.S. Steel, and 50 stockholders in all three, determine numerically the number of stockholders in the following sets:

(a) $F \cup C$

(b) $(F \cap C) \cup (F \cap S)$

(c) $(C \cap S) \cap F'$

(d) $(F \cap C)' \cap (F \cap S)'$

16. A color television set and a black and white television set are being distributed by a company. Both sets offer the option of a remote control unit. Determine the Cartesian product of the two sets.

17. A firm is about to market a new line of hair spray. The product can be called Hair Mist (m), Hair Set (s), or Hair Hold (h). The product can be placed in either the cosmetics section (c) or in the personal care section (p) of a store and can be distributed either through wholesalers (w) or directly to the retailer (r). Determine the possible number of combinations and construct a box diagram to show these combinations.

18. A firm must raise additional capital. Two methods have been suggested, stock or bonds. With either method, the firm has the choice of a public or private offering of the securities. Depending on market conditions, the offering could be undersubscribed or oversubscribed. Determine all possibilities for the offering. How many possibilities must be considered?

19. An automobile manufacturer offers five different models of cars. Each model has nine separate option packages and is available in 14 colors.

If a car dealer were to maintain a complete inventory of cars, what is the minimum number of cars the dealer must stock?

20. Consider the following two statements.

A = "The illegal sale of drugs is rising."
B = "The Department of H. E. W. is concerned with rising drug traffic."

Describe in words the statements represented by the following logical operations.

(a) $A' \cap B$
(b) $A \cap B$
(c) $(A \cup B)'$
(d) $A \cup B$
(e) B'

21. A railroad company executive released the following directive.

All passenger service to cities where revenue is less than $5000 per month or the number of passengers is less than 500 per month will be discontinued. This does not include cities that have less than 500 passengers per month which produce revenue of $5000 per month, or more.

Where will passenger service be discontinued?

SUGGESTED REFERENCES

CANGELOSI, VINCENT E., *Compound Statements and Mathematical Logic* (Columbus, Ohio: Charles E. Merrill Books, Inc., 1967).

FREUND, JOHN E., *College Mathematics with Business Applications* (Englewood Cliffs, N.J.: Prentice-Hall, Inc., 1969) 1.

HANNA, SAMUEL C. and JOHN C. SABER, *Sets and Logic* (Homewood, Ill.: Richard D. Irwin, Inc., 1971).

KEMENY, JOHN G., et al., *Finite Mathematical Structures* (Englewood Cliffs, N.J.: Prentice Hall, Inc., 1950), 1, 2.

KEMENY, JOHN G., et al., *Finite Mathematics with Business Applications*, 2nd ed. (Englewood Cliffs, N.J.: Prentice-Hall, Inc., 1972), 1, 2.

LIPSCHULTZ, SEYMOUR, *Set Theory and Related Topics*, Schaum's Outline Series (New York, N.Y.: McGraw-Hill Book Company, Inc., 1964).

THEODORE, CHRIS A., *Applied Mathematics: an Introduction* (Homewood, Ill.: Richard D. Irwin, Inc., 1965), 1–4.

Chapter 2

Functions
and Systems
of Equations

For mathematics to be of use in business, it is important that relationships that exist between business variables be formally defined. For instance, the retailer knows that profits are related to the number of units of product sold and the cost of the product. If it is possible for the retailer to state these relationships mathematically, the break-even point can be calculated and profit forecast. In this chapter we discuss the properties of mathematical relationships and offer examples of these properties. One of the most important relationships is the function.

2.1 Functions, Sets, and Relations

A *function* is a mathematical relationship in which the values of a single *dependent variable* are determined from the values of one or more *independent variables*. Before we discuss the general definition and properties of functions, it will be helpful to consider one of the most elementary functional forms, the *linear function*. The functional form of the linear function is

$$f(x) = a + bx \tag{2.1}$$

where $f(x)$ is the dependent variable.

x is the independent variable.

a is the value of the dependent variable when x is zero.

b is the coefficient of the independent variable.

The linear function has one dependent and one independent variable. The symbol $f(x)$, read "f of x," represents values of the dependent variable, and x represents values of the independent variable; $f(x)$ varies according to the rule of the function as x varies. For the linear function, the *rule of the function* states that b is to be multiplied by x and this product added to a. This sum determines the value of the dependent variable $f(x)$. Since the value of the dependent variable depends upon the value of the independent variable, $f(x)$ is termed the dependent variable and x is termed the independent variable. The term *variable* refers to a quantity that is allowed to assume different numerical values.

The properties of a linear function can be illustrated by an example. The linear function $f(x) = -4 + 2x$ is graphed in Fig. 2.1.

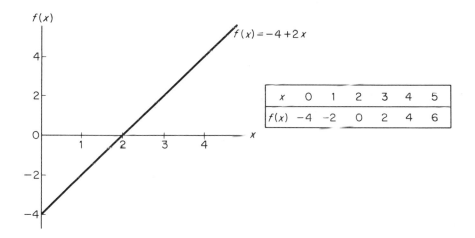

x	0	1	2	3	4	5
$f(x)$	−4	−2	0	2	4	6

Figure 2.1

For each value of the independent variable x, there is one and only one value of the dependent variable $f(x)$. The value of the dependent variable is calculated by the rule of the function. The rule of the function in the example in Fig. 2.1 states that $f(x)$ is equal to -4 plus $2x$.

Both $f(x)$ and x are termed variables, since they are both permitted to take on different numerical values. The table in Fig. 2.1 lists six possible values of x and $f(x)$. For this linear function, however, there are an infinite number of possible values of the variables. For example, between $x = 0$ and $x = 1$ there are an infinite number of possible values of x. Similarly, there are an infinite number of possible values of x between $x = 1$ and $x = 2$. From the infinite number of possible values, we have selected the six values in the table to illustrate the function.

2.1.1 DOMAIN AND RANGE

The permissible values of the independent variable are termed the *domain* of the function. If, for instance, the analyst is interested in all positive values of the independent variable (including zero), he would state that the domain of the function consists of all positive numbers. The function graphed in Fig. 2.1 could thus be expressed as

$$f(x) = -4 + 2x \qquad \text{for } x \geq 0$$

where $x \geq 0$ is read "x greater than or equal to zero." The domain of this function is shown by $x \geq 0$ to consist of all positive numbers.

In many business situations, the analyst will be concerned with only selected values of the independent variable. If he were concerned with only positive integer values of x, then the domain of the function would be values of the independent variable that are positive integers. The function graphed in Fig. 2.1 could be expressed as

$$f(x) = -4 + 2x \qquad \text{for } x = 0, 1, 2, 3, \ldots$$

This function consists of the set of ordered pairs $(x, f(x))$,

$$\{(0, -4), (1, -2), (2, 0), (3, 2), (4, 4), \ldots\}$$

The function is graphed in Fig. 2.2.

The permissible values of the dependent variable are termed the *range* of the function. The range of the function is those numbers that the dependent variable assumes as the independent variable takes on all values in the domain. In the example illustrated in Fig. 2.1, the domain of the function is all values of x along the continuous line between $x = 0$ and $x = \infty$. The range for this function contains all the numbers along the continuum from $f(x) = -4$ to $f(x) = \infty$. Similarly, if the domain of the function consists of positive integer values, the corresponding range consists of the values calculated by $f(x) = -4 + 2x$ for $x = 0, 1, 2, 3, \ldots$.

To summarize, a linear function is a mathematical relationship in which the values of the dependent variable $f(x)$ are determined from the values of the independent variable according to the rule that $f(x) = a + bx$. For the linear function, as for all functions, one and only one value of the dependent variable is possible for each value of the independent variable. The permissible values of the independent variable are termed the *domain* of the function, and the permissible values of the dependent variable are termed the *range* of the

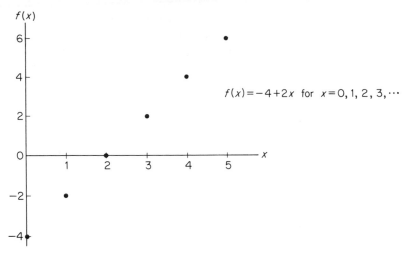

$f(x) = -4 + 2x$ for $x = 0, 1, 2, 3, \cdots$

Figure 2.2

function. The concepts of function, domain, and range are further illustrated by the following examples.

Example: Harvey West, an analyst for Pacific Soft-Drink Company, is attempting to develop a cost function for a diet cola. From the accounting department he has learned that fixed costs are $10,000 and variable costs are $0.03 per bottle. Present capacity limitations are 50,000 bottles per month, or 600,000 bottles per year. Mr. West wishes to develop the cost function and to specify the domain and range of the function.

We shall represent the number of bottles of the diet cola by x and the total cost by $f(x)$. Since total cost is composed of fixed and variable costs, the cost function on a yearly basis is

$$f(x) = 10,000 + 0.03x$$

The domain of the function is $x = 0$ to $x = 600,000$. The corresponding range is $f(x) = \$10,000$ to $f(x) = \$28,000$.

Example: Bill Short owns a small auto repair shop. Mr. Short acts as supervisor and employs three auto repairmen. His shop is operated on a 40-hour per week basis, and his employees are guaranteed 40 hours of pay each week. Weekly labor costs consist of a fixed component and a sum that varies with the amount of overtime during the week. The weekly salaries of Mr. Short and the three repairmen total $750. The overtime rate is $5.00 per hour for the employees. Mr. Short does not draw overtime pay. The maximum amount of overtime per employee is 20 hours per week. Determine the cost function and the domain and range of the function.

The cost function consists of the sum of the weekly salaries and the overtime pay. If we represent the overtime hours per week by x and the total weekly salary by $f(x)$, then the cost function is

$$f(x) = 750 + 5x$$

The domain of the function is $x = 0$ to $x = 60$, and the corresponding range of the function is $f(x) = \$750$ to $f(x) = \$1,050$.

2.1.2 FUNCTIONS AND SET TERMINOLOGY

The language of sets can be usefully employed to define and identify the essential properties of a function. Expressed in set notation, a linear function is the set of ordered pairs of numbers $(x, f(x))$ given by

$$\{(x, f(x)) \mid f(x) = a + bx\} \tag{2.2}$$

The expression is read "that set of ordered pairs $(x, f(x))$ such that $f(x) = a + bx$."

The functional relationship between x and $f(x)$ has been described by the defining property method of specifying sets. This method consists of stating a rule within the braces that allows one to determine set membership. The rule must conform to the requirement that for each value of x there can be only one value of $f(x)$. The members of the set consist of the ordered pairs $(x, f(x))$. The permissible values of x are termed the *domain* of the function, and the corresponding values of $f(x)$, determined from the rule of the function, are termed the *range* of the function.

An alternative method of specifying set membership is the roster method. This method consists of listing within the braces all members of the set. A set of ordered pairs

$$\{(x_1, f(x_1)), (x_2, f(x_2)), (x_3, f(x_3))\} \tag{2.3}$$

is termed a function provided that for each value of the independent variable there exists only one value of the dependent variable. The values of the first component of each ordered pair make up the domain of the function, while the values of the second component of each ordered pair make up the range of the function.

Example: Use set terminology to describe the linear function

$$f(x) = 6 - 3x, \quad \text{for } -10 \le x \le 10$$

The function consists of the set of ordered pairs

$$\{(x, f(x)) \mid f(x) = 6 - 3x \text{ for } -10 \le x \le 10\}$$

For each value of x in the domain $-10 \leq x \leq 10$, the value of $f(x)$ is determined by the rule of the function, $f(x) = 6 - 3x$. The domain of the function is the continuum $-10 \leq x \leq 10$, and the range of the function is the continuum $-24 \leq f(x) \leq 36$.

Example: A probability function is a functional relationship between the value of a variable and the probability of occurrence of that value. If we define the variable as the number of heads occurring in two tosses of a fair coin, then the probability function for the experiment of tossing the coin is given by the set of ordered pairs $(x, p(x))$,

$$\{(0, \tfrac{1}{4}), (1, \tfrac{1}{2}), (2, \tfrac{1}{4})\}$$

where x represents the number of heads and $p(x)$ represents the probability of this number occurring. The domain of the function (the possible number of heads) is $x = 0, 1, 2$, and the range of the function (the associated probabilities) is $p(x) = \tfrac{1}{4}$ and $\tfrac{1}{2}$. Notice that we have used the roster method of specifying set membership to list the elements of the set. This method of specifying a functional relationship can be employed only when a limited number of ordered pairs are included in the function.

Example: Verify that the following set is a function. Give the domain and range of the function.

$$\{(0, 3), (1, 2), (3, 3), (4, 2), (5, -1)\}$$

The roster method is used to specify the ordered pairs $(x, f(x))$ that comprise the set. For each value of x there is only one value of $f(x)$; consequently, the set is a function. The domain of the function is $x = 0, 1, 3, 4, 5$ and the range of the function is $f(x) = -1, 2, 3$. This example illustrates that for each value of x there is only one value of $f(x)$. This value of $f(x)$ need not be unique, however. In this function $f(x) = 3$ when $x = 0$ and $f(x) = 3$ when $x = 3$.

Example: State why the following set is not a function.

$$\{(0, 0), (0, 1), (2, 1), (3, 2)\}$$

This set does not qualify as a function, since the first two ordered pairs $(0, 0)$ and $(0, 1)$ indicate that $f(x)$ is 0 or 1 when $x = 0$.

The Cartesian product of two or more sets can be used in defining and illustrating functions. The Cartesian product of two sets A and B is the set of ordered pairs (a, b), where $a \in A$ and $b \in B$. Utilizing set notation, we find that the Cartesian product $A \times B$ is the set

$$A \times B = \{(a, b) \mid a \in A, b \in B\} \qquad (2.4)$$

The elements of the set A are termed the *domain of the Cartesian product* and the elements of the set B are termed the *range of the Cartesian product.*

For the set $A \times B$, a subset in which elements of A occur only once is a function. The domain of the function consists of elements in A and the range of the function of elements in B. Although the elements in A may occur only once in the function, there is no requirement that all elements in A be included in the function. Elements in B can occur zero, once, or repeatedly in the function.

Example: Given the set $A = \{a, b, c\}$ and the set $B = \{1, 2\}$, the Cartesian product of A and B is the set

$$A \times B = \{(a, 1), (a, 2), (b, 1), (b, 2), (c, 1), (c, 2)\}$$

Examples of subsets that are functions include $\{(a, 1), (b, 1), (c, 1)\}$, $\{(a, 1), (b, 2), (c, 1)\}$, and $\{(a, 2), (b, 1), (c, 1)\}$. The subset $\{(a, 1), (a, 2), (b, 1)\}$ is, however, not a function.

Example: Set $A = \{$boy, girl$\}$ and $B = \{$Yale, Princeton, Vassar, Smith$\}$. The Cartesian product of A and B is

$$A \times B = \{(\text{boy, Yale}), (\text{boy, Princeton}), (\text{boy, Vassar}), (\text{boy, Smith}),$$
$$(\text{girl, Yale}), (\text{girl, Princeton}), (\text{girl, Vassar}), (\text{girl, Smith})\}$$

Two subsets of the Cartesian product that satisfy the requirements of a function are $\{(\text{boy, Yale}), (\text{girl, Vassar})\}$ and $\{(\text{boy, Princeton}), (\text{girl, Smith})\}$.

2.1.3 RELATIONS

A subset of the Cartesian product in the preceding example which is not a function but nevertheless is of interest is the subset

$$\{(\text{boy, Yale}), (\text{boy, Princeton}), (\text{girl, Vassar}), (\text{girl, Smith})\}$$

This subset fails to satisfy the requirements for a function since elements in the domain are associated with more than one element in the range. The subset of a Cartesian product in which elements in the domain and range may both appear repeatedly is called a *relation*. A relation in $A \times B$ is thus a subset of $A \times B$. The difference between a relation and a function is that a relation may have elements in the domain associated with more than one element in the range. A function consists of ordered pairs in which each element in the domain is associated with one and only one element in the range. These concepts are illustrated by the following examples.

Example: Let $S = \{$all stocks traded on the New York Stock Exchange on a Friday$\}$ and let $A = \{+, 0, -\}$ represent the stock closing up, closing unchanged, or closing down. The Cartesian product $S \times A$ is the set of all possible movements in the stock prices. From the Cartesian product, the relation that describes the possible stock movements in a portfolio that contains Ford and IBM is

$$\{(\text{Ford}, +), (\text{Ford}, 0), (\text{Ford}, -), (\text{IBM}, +), (\text{IBM}, 0), (\text{IBM}, -)\}.$$

A subset of the relation is the function which describes the actual price movement for a given Friday. This function is

$$\{(\text{Ford}, +), (\text{IBM}, -)\}.$$

Notice that in the relation elements in the domain may appear more than once and may be associated with more than one element in the range. In the function, elements in the domain are associated with only one element in the range.

Example: Develop the relation between x and y such that x is greater than y and both x and y are positive integers.

The set I consists of all positive integers, i.e.,

$$I = \{1, 2, 3, 4, \ldots\}$$

The Cartesian product, $I \times I$, represents all integer pairs (x, y) in the first quadrant. The relation

$$R = \{(x, y) \mid x > y \text{ and } x, y \in I\}$$

is a subset of the Cartesian product $I \times I$. This relation is shown in Fig. 2.3.

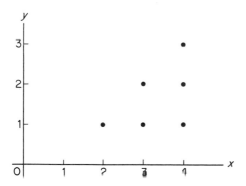

Figure 2.3

Only a few of the infinite number of ordered pairs that comprise the relation are shown in this figure.

A relation that will be of major importance in the chapters on linear programming is the linear inequality. A linear inequality is a relation of the form

$$y \geq kx + c \qquad\qquad (2.5)$$

or alternatively of the form

$$y \leq kx + c \qquad\qquad (2.6)$$

An inequality is a relation rather than a function, since for each value of x in the domain there is more than one value of y in the range. In fact, for any value of x in the linear inequality $y \geq kx + c$ there are an infinite number of values of y that satisfy the requirement that $y \geq kx + c$.

As an example of a linear inequality, consider $y \geq 2x - 3$. This relation is shown by the shaded area in Fig. 2.4. Using set notation, we can express

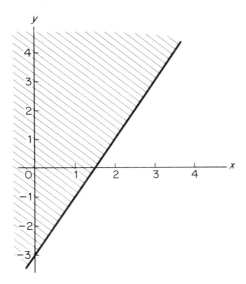

Figure 2.4

the linear inequality as $\{(x, y) \mid y \geq 2x - 3\}$.

Example: A university limits class enrollment to sixty-four students per teacher and cancels any class in which enrollment is six or less. Develop relations that describe the limitations.

If we represent the faculty size as x and the student enrollment as y, then the relation which describes the limitation is

$$\{y \leq 64x \text{ and } y \geq 7x \text{ for } x, y \in I\}$$

The relation is a subset of the Cartesian product of $I \times I$, where I is the set of positive integers.

Example: Show the inequality $y \le x - 1$ on a graph.

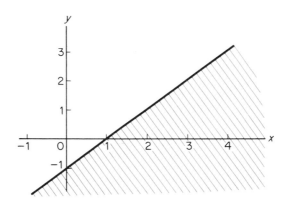

The inequality is shown by the shaded area.

Both functions and relations have been shown to be subsets of Cartesian products. In examples such as on p. 38 concerning the set $A = \{\text{boy, girl}\}$ and $B = \{\text{Yale, Princeton, Vassar, Smith}\}$, the Cartesian product was obvious. We have, however, not indicated the Cartesian product of which the function $f(x) = 2x + 3$ or the relation $y \le 6x - 4$ are subsets. If x is unrestricted, the domain for both the function and the relation is $-\infty \le x \le \infty$.

The continuum $-\infty \le x \le \infty$ is described by the set of real numbers $R = \{-\infty \le x \le \infty\}$. The range of both the function and the relation is similarly $R = \{-\infty \le f(x) \le \infty\}$. The Cartesian product $R \times R$ is the set of all ordered pairs of real numbers, i.e.,

$$R \times R = \{(x, f(x)) \mid x \in R \text{ and } f(x) \in R\}$$

or alternatively

$$R \times R = \{(x, y) \mid x \in R \text{ and } y \in R\}$$

Both the relation and the function are subsets of the Cartesian product of real numbers.

Example: Verify that the mathematical relationship

$$y^2 = x$$

is a relation rather than a function.

If $y^2 = x$, then $y = \pm\sqrt{x}$ and for each value of x in the domain $0 \le x \le \infty$, there are two values of y. This violates the requirement that there exist only one value of the dependent variable for each value of the independent variable. Consequently, $y = \pm\sqrt{x}$ is a relation rather than a function. This relation is graphed in Fig. 2.5.

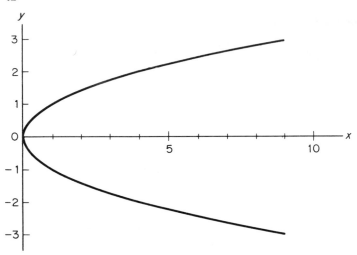

Figure 2.5

2.2 Establishing and Recognizing Functions

Numerous types of functions are used to describe functional relationships between business variables. The analyst should be able to recognize various types of functions and, when appropriate, establish functions that model the relationships that are observed to exist between business variables. Our purpose in this section is to describe the linear and quadratic functions and to illustrate selected applications of these functions. Methods of establishing these functions are presented and illustrated.

2.2.1 LINEAR FUNCTIONS

The linear function was defined earlier as

$$f(x) = a + bx \qquad (2.1)$$

where $f(x)$ is the dependent variable.

- x is the independent variable.
- a is the value of the dependent variable where x is zero.
- b is the coefficient of the independent variable.

As an example of the linear function, the function

$$f(x) = 8 - 2x \qquad \text{for } -2 \le x \le 8$$

is graphed in Fig. 2.6. Values of $f(x)$ are given on the vertical axis and values

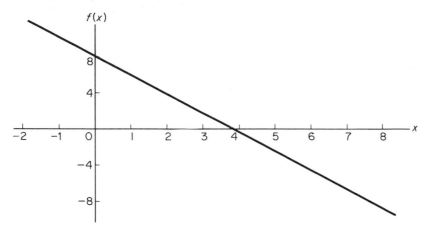

Figure 2.6

of x on the horizontal axis. The linear function plots as a straight line that intersects the vertical axis at $(0, 8)$, i.e., at $x = 0$ and $f(x) = 8$.

The slope of a linear function is defined as the change in the value of the dependent variable divided by the change in the value of the independent variable. For a linear function, the slope is given by b, the coefficient of x. It can also be calculated by determining the change in the dependent variable for a one-unit change in the independent variable. The slope of the function $f(x) = 8 - 2x$ is $b = -2$.

Example: Graph the linear function $f(x) = -3 + 1.5x$ for $-5 \le x \le 6$. Give the intercept and slope of the function.

A linear function can be graphed from two data points. One obvious data point is the ordered pair $(0, -3)$. This data point is the intercept of the function, i.e., the value of the dependent variable when the independent variable is 0. Since there are an infinite number of values of x in the domain $-5 \le x \le 6$, there are an infinite number of ordered pairs $(x, f(x))$ that could be used for a second data point. One of these ordered pairs is the data point $(4, 3)$. The graph of the function is shown on p. 44.

The intercept of the function is $f(x) = -3$ and the slope is $b = 1.5$.

One of the many uses of the linear function is in describing linear revenue and cost models. For certain products, the revenue generated from the sales of the product is a linear function of the number of units of the product sold. If, for instance, a firm is capable of supplying up to 1000 units of a product at a price of $10 per unit, the revenue function for this product would be

$$R = 10x, \quad \text{for } 0 \le x \le 1000$$

If the cost of producing the product is composed of a fixed cost and a variable

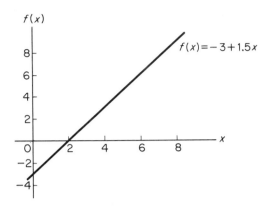

cost, the total cost of production is $TC = FC + VC$. Provided that variable cost is a linear function of the number of units manufactured and the fixed cost is constant for the specified domain of the cost function, the total cost function becomes $TC = FC + cx$, where c is the variable cost per unit of production. If fixed costs are \$1000 and variable costs are \$8, the total cost function is

$$TC = 1000 + 8x, \qquad \text{for } 0 \leq x \leq 1000$$

The revenue and cost functions for the product are graphed in Fig. 2.7.

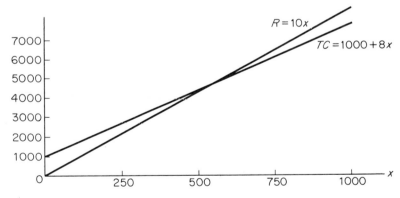

Figure 2.7

The breakeven level of production is the number of units for which revenue equals total cost. This quantity is equal to the value of x at which the revenue and total cost curves intersect. The quantity is calculated by equating R with TC.

$$R = TC$$
$$10x = 1000 + 8x$$
$$x = 500$$

The breakeven level of production is $x = 500$.

Example: The anticipated price of a product is $15. The fixed cost of manufacturing the product is $10,000, and the variable cost is $10 per unit. Plant limitations restrict production to 5000 units. Develop revenue and total cost functions and calculate breakeven production.

The revenue function is $R = 15x$ for $0 \leq x \leq 5000$, and the total cost function is $TC = 10,000 + 10x$ for $0 \leq x \leq 5000$. Breakeven is found by equating the two functions.

$$R = TC$$
$$15x = 10,000 + 10x$$
$$5x = 10,000$$
$$x = 2000$$

Example: Profit is defined as revenue minus total cost. Develop the profit function for the product in the preceding example and verify that profit is 0 at breakeven sales of $x = 2000$.

Revenue is $R = 15x$, and total cost is $TC = 10,000 + 10x$. Profit is thus

$$P = R - TC$$
$$P = 15x \quad (10,000 + 10x)$$
$$P = 5x - 10,000$$

Substituting 2000 for x shows that profit is 0 at breakeven.

It is often necessary to determine the function that passes through data points. In these instances, the analyst must determine the appropriate type of function and the parameters of the function (e.g., the coefficients a and b in the linear function $f(x) = a + bx$). To illustrate the method for determining the parameters of the linear function, consider the example of Bud Brewer.

Example: Bud Brewer is attempting to predict sales for his company for the coming year. Sales for the past five years are plotted on the graph shown on p. 46. On the basis of his knowledge of the market and the historical data, Bud believes that a linear function that passes through the data points for year 1 and year 5 will provide the best forecast for sales in the coming year (year 6).

The procedure for establishing functions through data points requires substituting the data points into the general form of the function and solving

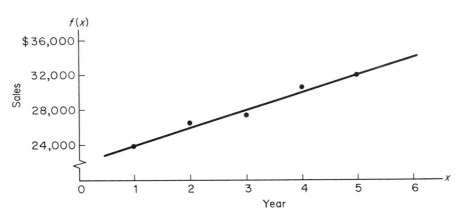

the resulting equations simultaneously for the parameters of the function.
Two equations are required for a linear function. From Bud's statements,
the two data points are the ordered pairs $(x = 1, f(x) = \$24{,}000)$ and $(x = 5,$
$f(x) = \$32{,}000)$. These data points are each substituted into the general form
of the linear function

$$f(x) = a + bx \qquad (2.1)$$

where x is the independent variable, time, and $f(x)$ is the dependent variable,
sales. The substitution gives the two equations

$$32{,}000 = a + b(5)$$
$$24{,}000 = a + b(1)$$

These two equations are solved simultaneously by the method of substitution.† Substituting $a = 32{,}000 - b(5)$ from the first equation into the second
equation gives

$$24{,}000 = 32{,}000 - b(5) + b(1)$$

or

$$4b = 8000$$

and

$$b = 2000$$

From the first equation we obtain

$$a = 32{,}000 - 5b = 22{,}000$$
$$f(x) = 22{,}000 + 2000x$$

Sales in year 6 are determined by evaluating the function for $x = 6$.

$$f(6) = 22{,}000 + 2000(6) = \$34{,}000$$

---

† The method of subtraction is discussed in Appendix A.

Since Bud Brewer is interested in predicting sales based upon this function for year 6 only, he has implied that the domain of the function is $x = 1$ to 6. The corresponding range of the function is $f(x) = \$24,000$ to $f(x) = \$34,000$. The forecast for year 6 is $f(6) = \$34,000$.

Example: Joe Findley is a stockholder in Electronic Products, Inc. In studying the annual report to stockholders, Joe notes that Electronic Products' computer is being depreciated on a straight-line basis. This means that the depreciation charge is constant each year. If the book value of the computer was $1 million in year 2 and $700,000 in year 5, determine the year in which the computer is fully depreciated, i.e., in which the book value will equal the scrap value. The scrap value of the computer is $200,000.

The function that describes the depreciation and book value can be determined by substituting the data points (2, $1,000,000) and (5, $700,000) into the linear function

$$B(t) = a + bt$$

where $B(t)$ is the book value in year t, a is the original book value, and b is the yearly straight-line depreciation. The two equations are

$$1,000,000 = a + b(2)$$
$$700,000 = a + b(5)$$

These equations are solved simultaneously to give $a = \$1,200,000$, $b = -\$100,000$ and

$$B(t) = 1,200,000 - 100,000t$$

The year in which the computer is fully depreciated is determined by equating the function with $200,000 and solving for t.

$$200,000 = 1,200,000 - 100,000t$$

The solution is $t = 10$ years. The depreciable life of the computer is ten years.

2.2.2 QUADRATIC FUNCTIONS

The *quadratic* function is a second function which has important applications in business and economics. The general form of the quadratic function is

$$f(x) = a + bx + cx^2 \tag{2.7}$$

where $f(x)$ is the dependent variable, x is the independent variable, and a, b, and c are the parameters of the function. The shape of the quadratic function is determined by the magnitude and signs of the parameters a, b, and c. Common forms of the quadratic function are shown in Fig. 2.8.

From the diagrams in Fig. 2.8, it can be seen that the value of a positions the function vertically. In Fig. 2.8(a), for instance, if a were negative rather than positive, the function would intercept the vertical axis below the hori-

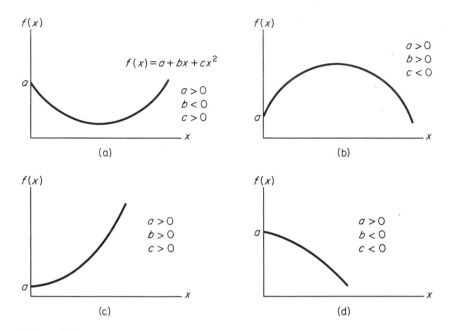

Figure 2.8

zontal axis. This would have the effect of moving the entire function downward. The values of b and c determine the general shape of the function. For
b negative and c positive, the function will have the concave shape illustrated
by Fig. 2.8(a). For b positive and c negative, the function will be convex as
shown in Fig. 2.8(b). If b and c are both positive, the function has the shape
illustrated by Fig. 2.8(c). Conversely, if b and c are both negative, the function
appears as illustrated by Fig. 2.8(d).

Three data points are necessary to establish a quadratic function. The procedure in establishing a quadratic function is to substitute the data points into
the general form of the quadratic function and solve the resulting three equations simultaneously for the parameters a, b, and c. This procedure is illustrated in Appendix A and by the following two examples.

Example: Bill Cook owns a small machine shop. Because of the shop's
limited capacity, average costs are related to the volume of output. Cost
studies undertaken by Bill show that the average cost of producing 90 units
of output is $850 per unit, the average cost of producing 100 units is $800
per unit, and the average cost per unit of producing 120 units is $825 per
unit. Determine the functional relationship between average cost and output.

The three data points, when plotted, show that a function that decreases and then increases describes the cost-output data. This type of functional relationship is described by the quadratic function

$$f(x) = a + bx + cx^2 \qquad (2.7)$$

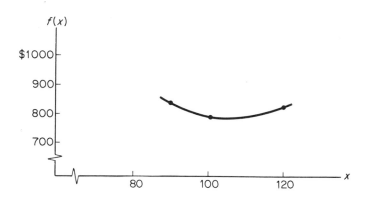

The procedure for determining the function is to substitute the data points into the general form of the quadratic function, Eq. (2.7), and solve the resulting three equations simultaneously for the parameters a, b, and c.† The three equations are

$$850 = a + b(90) + c(90)^2$$
$$800 = a + b(100) + c(100)^2$$
$$825 = a + b(120) + c(120)^2$$

The solution of the three simultaneous equations is $a = 3175$, $b = -44.58$, and $c = 0.2083$. The average cost function is

$$f(x) = 3175 - 44.58x + 0.2083x^2$$

Example: James Mason is concerned with the effect of fatigue on the productivity of skilled machinists. He has observed that productivity, which he defines as standard units of output per man-hour, increases as the number of hours that an individual works increases up to approximately 40 hours per week and then begins to decline as fatigue begins to become a factor. Mason believes that three representative observations of the effect of man-hours on productivity are 9.3 units of output during the 35th hour of work, 9.5 units during the 40th hour, and 8.5 units during the 45th hour.

---

† These equations can be solved simultaneously by the method of substitution, the method of subtraction, or by using matrix algebra. The method of subtraction is discussed in Appendix A, and matrix algebra is discussed in Chapter 3.

The observations are plotted below. Productivity as a function of the hour of work increases and then decreases. The type of a relationship shown in the illustration is described by a quadratic function.

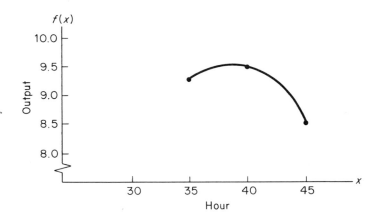

We substitute the data points in the general form of the quadratic function, Eq. (2.7), and obtain

$$9.3 = a + b(35) + c(35)^2$$
$$9.5 = a + b(40) + c(40)^2$$
$$8.5 = a + b(45) + c(45)^2$$

Solving the three equations simultaneously for a, b, and c gives $a = -25.7$, $b = 1.84$, and $c = -0.024$. The productivity function is

$$f(x) = -25.7 + 1.84x - 0.024x^2$$

2.2.3 POLYNOMIAL FUNCTIONS

Linear and quadratic functions belong to the class of functions termed *polynomial* functions. The general form of the polynomial function is

$$f(x) = a_0 + a_1x + a_2x^2 + a_3x^3 + \ldots + a_nx^n \qquad (2.8)$$

where $a_0, a_1, a_2, a_3, \ldots, a_n$ are parameters and n is a positive integer. The parameters may be positive, negative, or zero. The polynomial function is linear if $n = 1$ and quadratic if $n = 2$. This can be verified by comparing Eq. (2.8) for $n = 1$ with the general form of the linear function (2.1), and by comparing Eq. (2.8) for $n = 2$ with the general form of the quadratic function (2.7).

A polynomial function in which the largest exponent is $n = 3$ is termed a *cubic* function. The cubic function has the general form

$$f(x) = a_0 + a_1x + a_2x^2 + a_3x^3 \qquad (2.9)$$

The general shape of the cubic function is shown in Fig. 2.9.

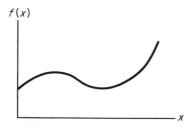

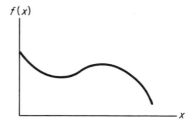

Figure 2.9

Although the two illustrations in Fig. 2.9 do not represent all possible shapes for the cubic function, they do illustrate a distinguishing characteristic of the cubic function. The reader will remember that the quadratic function was described as convex or concave. Referring to Fig. 2.9, we see that the cubic function changes from convex to concave or from concave to convex. An example of a cubic function that changes from convex to concave as the value of the independent variable x increases is

$$f(x) = 5 + 2.5x - x^2 + 0.1x^3$$

This function is plotted in Fig. 2.10.

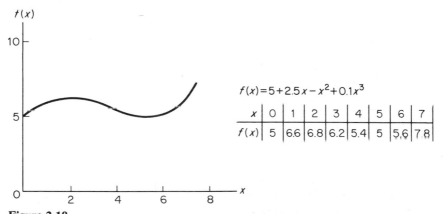

$f(x) = 5 + 2.5x - x^2 + 0.1x^3$

x	0	1	2	3	4	5	6	7
$f(x)$	5	6.6	6.8	6.2	5.4	5	5.6	7.8

Figure 2.10

The number of data points required to establish a polynomial function is equal to $n + 1$, where n is the largest exponent in the function. Thus, two

data points are required for the linear function, three data points are required for the quadratic function, and four data points are required for the
cubic function. The procedure in establishing the function is to substitute
each data point into the general form of the polynomial function, and solve
the resulting $n + 1$ equations simultaneously for the $n + 1$ parameters.
This procedure has been illustrated for the linear and quadratic functions.†

2.3 Multivariate Functions

We have stated that a function has only one dependent variable but may
have one or more independent variables. Functions in which the single dependent variable is related to more than one independent variable are termed
multivariate functions. An example of a multivariate function is

$$f(x_1, x_2) = 2x_1 + 3x_1x_2 + 6x_2$$

where $f(x_1, x_2)$ is the dependent variable, x_1 is an independent variable, and
x_2 is a second independent variable. Multivariate functions are difficult to
plot, since they require one axis (or dimension) for each variable. A function
with two independent variables and one dependent variable, for instance,
requires three dimensions for plotting. Multivariate functions with three
independent variables cannot be plotted, since four dimensions are required.
In spite of the fact that multivariate functions with more than two independent variables cannot be plotted, they are quite necessary to describe many
business and economic models. Multivariate functions with two independent
variables are illustrated by the following examples.

Example: Sales of the Southern Distributing Company consist of sales
made through retail and wholesale outlets. Profit per dollar of sales at retail
is \$0.15 and profit per dollar of sales at wholesale is \$0.05. Determine the
profit function.

If retail sales in dollars are represented by x_1, wholesale sales in dollars
by x_2, and profit by $P(x_1, x_2)$, the profit function is

$$P(x_1, x_2) = 0.15x_1 + 0.05x_2$$

Example: Assume that the Neshay Candy Company makes net profit of
\$0.05 per almond bar and \$0.07 per chocolate bar. If net profit P is represented as a function of monthly sales in units of the almond bar x_1 and the
chocolate bar x_2, determine the appropriate functional relationship and
calculate net profit for a month in which $x_1 = 10,000$ and $x_2 = 20,000$.

---

† An example of establishing a cubic function is given in Appendix A.

$$P = 0.05x_1 + 0.07x_2$$
$$P = 0.05(10,000) + 0.07(20,000)$$
$$P = \$1,900$$

Multivariate functions can be established through data points by using the same technique described for polynomial functions. This technique involves substituting the appropriate data points into the specified general form of the multivariate function and solving the resulting equations simultaneously for the parameters of the function. The technique is illustrated by the following example.

Example: Research has demonstrated that the output Y of a firm is related to labor L and capital K. The general form of the function is

$$Y = aL + bK + c(LK)$$

It has been determined that: $Y = 100,000$ when $L = 9.0$ and $K = 4.0$; $Y = 120,000$ when $L = 10.0$ and $K = 5.0$; and $Y = 150,000$ when $L = 11.5$ and $K = 7.0$. The domains of L and K are $9.0 \leq L \leq 11.5$ and $4.0 \leq K \leq 7.0$. Determine the values of a, b, and c.

The three data points are substituted into the general form of the function to give

$$100,000 = a(9.0) + b(4.0) + c(36.0)$$
$$120,000 = a(10.0) + b(5.0) + c(50.0)$$
$$150,000 = a(11.5) + b(7.0) + c(80.5)$$

These equations are solved simultaneously to give $a = -60,000$, $b = 304,000$, and $c = -16,000$. The function is

$$Y = -60,000L + 304,000K - 16,000(LK) \qquad \text{for } 9.0 \leq L \leq 11.5$$
$$4.0 \leq K \leq 7.0.$$

2.4 Systems of Equations

A function was defined earlier in this chapter as a mathematical relationship in which the value of a single dependent variable is determined from the values of one or more independent variables. In certain instances, it is not appropriate to consider one variable as "dependent" upon other variables. Although the variables are explicitly related and the requirements for a function are satisfied, the relationship is referred to as an *equation* rather than a function. Expressed in set notation, a linear equation with two variables is the set of ordered pairs (x, y) given by

$$\{(x, y) \mid ax + by = c\} \qquad (2.10)$$

In comparing the definition of a linear equation given by Eq. (2.10) with that of a linear function given by Eq. (2.2), the primary difference is that neither variable is distinguished in Eq. (2.10) as the dependent or the independent variable.

A *system of equations* consists of one or more equations with one or more variables. As an example, a system of two equations with two variables is

$$2x + 4y = 28$$
$$3x - 2y = 10$$

Similarly, a system of two equations with three variables is

$$20x + 15y - 10z = 20$$
$$30x - 25y + 35z = 85$$

The concept of a system of equations is illustrated by the following examples.

Example: A firm produces two products, A and B. Each unit of product A requires two man-hours of labor and each unit of product B requires three man-hours of labor. Develop the equation describing the relationship between the two products, assuming that 110 man-hours of labor will be used.

The number of units of product A is represented by x and the number of units of product B is represented by y. The equation describing the relationship between the two variables is

$$2x + 3y = 110$$

Example: Products A and B discussed in the preceding example require machine time as well as labor time. Each unit of product A requires one hour of machine time, and each unit of product B requires 0.5 hour of machine time. Develop the equation describing the relationship between the two products, assuming that thirty-five hours of machine time will be used.

If we again represent the number of units of product A by x and the number of units of product B by y, the equation is

$$x + 0.5y = 35$$

2.4.1 SOLUTION SETS FOR SYSTEMS OF EQUATIONS

The preceding examples illustrate a system of two equations with two variables. The equation that relates units of products A and B with man-hours can be described by the set

$$L = \{(x, y) \mid 2x + 3y = 110\}$$

Similarly, the equation that relates units of products A and B with machine time can be described by the set

$$M = \{(x, y) \mid x + 0.5y = 35\}$$

There are an infinite number of ordered pairs (x, y) that are members of set L. For instance, (0, 36.67), (1, 36), (2, 35.33), (3, 34.67) are four ordered pairs which are solutions to the equation $2x + 3y = 110$. There are also an infinite number of ordered pairs (x, y) that are members of set M. These include (0, 70), (1, 68), (2, 66), (3, 64), ... The ordered pairs (x, y) that are members of the set L are termed *solutions* to the equation $2x + 3y = 110$. Similarly, the ordered pairs that are members of the set M are solutions to the equation $x + 0.5y = 35$. The sets L and M are referred to as *solution sets*.

The sets L and M describe the labor and machine constraints on the production of products A and B. The intersection of sets L and M contains those elements (x, y) that are members of both L and M. That is,

$$L \cap M = \{(x, y) \mid 2x + 3y = 110 \text{ and } x + 0.5y = 35\}$$

The elements (x, y) represent the quantities of products A and B that are solutions for both the labor and the machine constraints.

The method of determining the solution set for a system of two equations is to determine the values of the variables that are solutions to both equations. These values are determined by solving the equations simultaneously. One of the methods commonly used in algebra for solving equations simultaneously is the method of substitution. This method involves solving for one variable in terms of the remaining variables and substituting this expression for the variable in the remaining equations. The two equations are

$$2x + 3y = 110$$
$$x + 0.5y = 35$$

Using the method of substitution, we solve for x in the second equation and obtain $x = 35 - 0.5y$. The expression $35 - 0.5y$ is substituted for x in the first equation to give

$$2(35 - 0.5y) + 3y = 110$$
$$2y = 40$$
$$y = 20$$

Substituting $y = 20$ in the second equation gives $x = 25$. The values of x and y that are solutions to both equations are $L \cap M = \{(25, 20)\}$, or 25 units of product A and 20 units of product B.

Example: A firm uses three processes to produce three products. Product 1 requires 2 hours of process A time, 4 hours of process B time, and 6 hours of process C time. Product 2 requires 1 hour of process A time and 2 hours of process B and C time. Product 3 requires 3 hours of process A time and 4 hours of process B and C time. One hundred hours of process A time, 160 hours of process B time, and 190 hours of process C time have been allocated to the production of the products. Determine the quantities of products 1, 2, and 3 that can be manufactured.

We let x_1, x_2, and x_3 represent the number of units of products 1, 2, and 3. The process A equation is given by

$$A = \{(x_1, x_2, x_3) \mid 2x_1 + x_2 + 3x_3 = 100\}$$

The process B equation is

$$B = \{(x_1, x_2, x_3) \mid 4x_1 + 2x_2 + 4x_3 = 160\}$$

The process C equation is

$$C = \{(x_1, x_2, x_3) \mid 6x_1 + 2x_2 + 4x_3 = 190\}$$

There are an infinite number of ordered triplets (x_1, x_2, x_3) that are solutions to each equation. We require, however, the set of ordered triplets (x_1, x_2, x_3) that are solutions to all three equations. This is the solution set for the system of simultaneous equations. The solution set is determined by solving the three equations simultaneously for x_1, x_2, and x_3. The solution set is $A \cap B \cap C = \{(15, 10, 20)\}$, or $x_1 = 15$ units, $x_2 = 10$ units, and $x_3 = 20$ units.

The solution set for $A \cap B \cap C$ can be determined by using any one of the methods introduced in algebra. Using the method of substitution, we solve the first equation for x_1 to obtain $x_1 = -0.5x_2 - 1.5x_3 + 50$. Substituting this quantity for x_1 in the second and third equations gives

$$4(-0.5x_2 - 1.5x_3 + 50) + 2x_2 + 4x_3 = 160$$

and

$$6(-0.5x_2 - 1.5x_3 + 50) + 2x_2 + 4x_3 = 190$$

These two equations reduce to

$$2x_3 = 40$$

and

$$x_2 + 5x_3 = 110$$

These two equations are solved for x_2 and x_3 to give $x_2 = 10$ and $x_3 = 20$. Substituting these values in equation A gives $x_1 = 15$.

2.4.2 CONSISTENT AND INCONSISTENT SYSTEMS OF LINEAR EQUATIONS

The solution set for the system of linear equations in the preceding two examples had only a single element. In the first example the element was $L \cap M = \{(25, 20)\}$ and in the second example the element was $A \cap B \cap C = \{(15, 10, 20)\}$. These examples illustrate the case in which a *unique* solution exists for the system of equations. A unique solution exists when there is only one element in the intersection of the sets describing the individual equations.

A unique solution is one of three possibilities for the solution to a system of linear equations. A system of linear equations may have: (1) no solution; (2) exactly one solution (unique); or (3) an infinite number of solutions. A system that has no solution is termed *inconsistent*. A system that has one or more solutions is said to be *consistent*. These three possibilities are illustrated by Fig. 2.11.

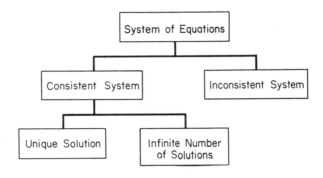

Figure 2.11

The concepts presented in Fig. 2.11 can be explained by using set notation. Assume that $f(x_1, x_2, x_3)$, $g(x_1, x_2, x_3)$, and $h(x_1, x_2, x_3)$ represent three different equations. The set of ordered triplets (x_1, x_2, x_3) that are solutions to each equation is described by the sets

$$F = \{(x_1, x_2, x_3) \mid f(x_1, x_2, x_3)\}$$

$$G = \{(x_1, x_2, x_3) \mid g(x_1, x_2, x_3)\}$$

$$H = \{(x_1, x_2, x_3) \mid h(x_1, x_2, x_3)\}$$

The solution to the system of equations is defined as all ordered triplets (x_1, x_2, x_3) that are common to the three equations. This solution set is the set of ordered triplets contained in the intersection of F, G, and H, i.e., $F \cap G \cap H$. If this intersection is null (empty), the system of equations is inconsistent. If, however, the intersection contains one or more elements, the system is consistent. For those cases in which the intersection contains exactly one element, this element is the unique solution to the system of equations. If the intersection contains more than one element, the system of equations is consistent, but the solution is not unique.

Graphs of equations are quite useful as an aid to understanding the solution sets for systems of equations. We shall show a system of equations that is consistent and has a unique solution, a system of equations that is consistent with an infinite number of solutions, and a system of equations that is inconsistent and consequently has no solution.

2.4.3 CONSISTENT SYSTEM WITH A UNIQUE SOLUTION

As an illustration of a consistent system of equations with a unique solution, consider the equations $2x + y = 8$ and $3x - 2y = -2$. These equations are shown in Fig. 2.12.

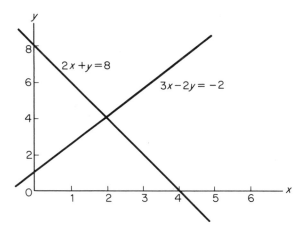

Figure 2.12

The equations can be described by the sets

$$A = \{(x, y) \mid 2x + y = 8\}$$
$$B = \{(x, y) \mid 3x - 2y = -2\}$$

The set A contains all ordered pairs (x, y) that are solutions to the equation $2x + y = 8$. These ordered pairs are shown in Fig. 2.12 by the linear equation $2x + y = 8$. The set B contains all ordered pairs (x, y) that are solutions to the equation $3x - 2y = -2$. These ordered pairs are shown by the linear equation $3x - 2y = -2$. The solution set for the system of equations contains all ordered pairs that are solutions to both A and B. This set is shown in Fig. 2.12 by the intersection of the two equations. The ordered pairs that are common to both equations are

$$A \cap B = \{(x, y) \mid 2x + y = 8 \text{ and } 3x - 2y = -2\}$$

The values of the variables x and y are determined by solving the two equations simultaneously for x and y. Substituting $y = 8 - 2x$ from the first equation for y in the second equation gives

$$3x - 2(8 - 2x) = -2$$
$$7x = 14$$
$$x = 2$$

Substituting $x = 2$ in the first equation gives $y = 4$. The ordered pair $(2, 4)$ is the unique solution to the system of equations.

Example: Determine the unique solution to the following system of three linear equations and three variables.

$$A = \{(x, y, z) \mid 2x + y - 2z = -1\}$$
$$B = \{(x, y, z) \mid 4x - 2y + 3z = 14\}$$
$$C = \{(x, y, z) \mid x - y + 2z = 7\}$$

A unique solution exists for the system of equations if there is an ordered triplet (x, y, z) that is a common solution to all three equations. Solving the three equations simultaneously gives $x = 2$, $y = 3$, and $z = 4$. The solution set for the system of equations is $A \cap B \cap C = \{(2, 3, 4)\}$.

Example: Verify that the following system of three equations with two variables is consistent and has a unique solution.

$$A = \{(x, y) \mid 2x + y = 9\}$$
$$B = \{(x, y) \mid x - y = 3\}$$
$$C = \{(x, y) \mid x + 2y = 6\}$$

The system of three equations with two variables is consistent and has a unique solution if there is a single ordered pair (x, y) that is a common solution to all three equations. Solving the first two equations simultaneously gives $A \cap B = \{(4, 1)\}$ as the unique solution to the first two equations. To determine if $(4, 1)$ is also a solution to the third equation, the ordered pair is substituted into the third equation. Since $4 + 2(1) = 6$, the ordered pair $(4, 1)$ is a solution for the third equation. The solution for the system of equations is $A \cap B \cap C = \{(4, 1)\}$. The system of equations is consistent with a unique solution.

The equations described by the sets A, B, and C are shown in Fig. 2.13. The unique solution $x = 4$, $y = 1$ occurs at the intersection of the three equations.

A system of three equations with two variables has a unique solution when one of the equations can be expressed as a linear combination of the remaining two equations. If we use the sets A, B, and C to represent the equa-

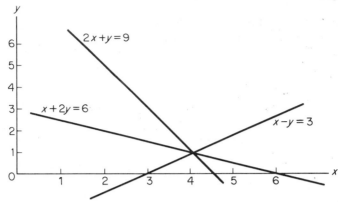

Figure 2.13

tions, a linear combination exists if there are numbers a and b such that $aA + bB = C$.

To determine if a linear combination exists in the preceding example, we first write the equations as $2x + y - 9 = 0$, $x - y - 3 = 0$, and $x + 2y - 6 = 0$. A linear combination exists if there are numbers a and b such that

$$a(2x + y - 9) + b(x - y - 3) = x + 2y - 6$$

In this example equation C can be formed by $1A - 1B$. This is shown by

$$1(2x + y - 9) - 1(x - y - 3) = x + 2y - 6$$

The appropriate numbers are thus $a = 1$ and $b = -1$, and equation C is a linear combination of equations A and B.†

2.4.4 CONSISTENT SYSTEM WITH AN INFINITE NUMBER OF SOLUTIONS

A system of equations has been defined as consistent if one or more solutions exist for the system of equations. As an illustration of a system of equations for which there is more than one solution, consider the equations

$$A = \{(x, y, z) \mid 2x - 4y + 4z = 20\}$$
$$B = \{(x, y, z) \mid 3x + 4y - 2z = 30\}$$

The solution to the system of two equations and three variables consists of

---

† In general, a unique solution to a system of m equations with n variables exists if $m - n$ of the equations can be expressed as a linear combination of the remaining n equations. The technique for determining the existence of a unique solution to a system of equations can be easily demonstrated by the use of matrix algebra. Consequently, we shall postpone further discussion of the requirements for a unique solution until Chapter 4.

all ordered triplets (x, y, z) that are solutions to both equations. From introductory algebra, we remember that it is impossible to determine a unique solution for three variables with two equations. We can, however, determine ordered triplets that are solutions to both equations. The procedure is to arbitrarily specify the value for one of the variables and to solve the two equations simultaneously for the remaining two variables. For instance, if the value of x is specified as 0, the solution set is $x = 0$, $y = 20$, and $z = 25$. These values are determined by solving the two equations

$$2(0) - 4y + 4z = 20$$

$$3(0) + 4y - 2z = 30$$

for y and z. Similarly, if $x = 5$ the equations can be solved simultaneously to obtain $y = 10$ and $z = 12.5$. Since an infinite number of values of x, y, or z could arbitrarily be specified, there are an infinite number of solutions to the system of two equations.

Example: Verify that the following system of two equations with three variables is consistent.

$$A = \{(x, y, z) \mid 2x + 3y + 4z = 20\}$$

$$B = \{(x, y, z) \mid 5x - 4y + 3z = 15\}$$

The system of equations is consistent if one or more solutions exist for the system. Equating $x = 0$ gives $y = 0$ and $z = 5$. Equating $x = 5$ gives $y = 2.8$ and $z = 0.4$. Since an infinite number of values of x, y, or z could be specified, the solution set $A \cap B$ contains an infinite number of triplets (x, y, z), and the system of equations is consistent.

Example: Verify that the following system of equations is consistent and has an infinite number of solutions.

$$A = \{(x, y, z) \mid 2x + 3y - 2z = 40\}$$

$$B = \{(x, y, z) \mid 3x - 2y + z = 50\}$$

$$C = \{(x, y, z) \mid x - 5y + 3z = 10\}$$

The system of three equations with three variables would appear at initial inspection to be consistent with a unique solution. When attempting to determine the solution, however, we find that this is not the case. Substituting the expression for x from the third equation in the first equation gives

$$2(10 - 3z + 5y) + 3y - 2z = 40$$

$$13y - 8z = 20$$

Substituting the same expression for x in the second equation gives

$$3(10 - 3z + 5y) - 2y + z = 50$$

$$13y - 8z = 20$$

Using the method of substitution, we would normally solve the two equations simultaneously for y and z. The equations that resulted from the original substitution are, however, the same. Consequently, there are an infinite number of ordered triplets (x, y, z) that are members of the solution set $A \cap B \cap C$. To obtain a solution to this system of equations, the value of one of the variables must be arbitrarily specified. If z is specified, the complete solution to the system of equations is

$$z = \text{specified}$$

$$y = \frac{20 + 8z}{13}$$

$$x = 10 - 3z + 5y$$

For instance, if $z = 0$, then $y = \frac{20}{13}$ and $x = \frac{230}{13}$. One solution is thus the ordered triplet $(\frac{230}{13}, \frac{20}{13}, 0)$. If $z = 1$, then $y = \frac{28}{13}$ and $x = \frac{231}{13}$. A second solution is $(\frac{230}{13}, \frac{28}{13}, 1)$. It is obvious that an infinite number of values of z could be specified and therefore an infinite number of solutions are possible.

Example: Verify that the equation $x - y = 6$ represents a consistent system of equations with an infinite number of solutions.

A system of equations is termed consistent if there are one or more solutions for the system of equations. The equation $x - y = 6$ is a system of one equation with two variables. The solution set consists of an infinite number of ordered pairs (x, y) that are solutions to the equation. One of the variables must be specified to determine a solution. For instance, if $x = 0$, then $y = -6$ and $(0, -6)$ is a solution. Similarly, if $x = 1$, then $y = -5$ and $(1, -5)$ is another solution. We thus conclude that the equation $x - y = 6$ is a consistent system of equations with an infinite number of solutions.

2.4.5   INCONSISTENT SYSTEM OF EQUATIONS

A system of equations is termed inconsistent when the solution set for the system of equations is null. As an example, consider the two equations

$$A = \{(x, y) \mid x + y = 6\}$$

$$B = \{(x, y) \mid 2x + 2y = 8\}$$

These equations are shown in Fig. 2.14. The solution to the system of two equations consists of the ordered pairs (x, y) that are common to both set A and set B, i.e., $A \cap B$. The equations described by the sets A and B are, however, parallel. Consequently, there is no ordered pair (x, y) that is a

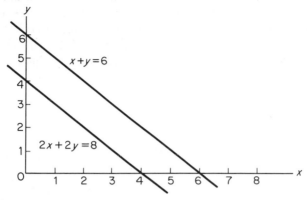

Figure 2.14

common solution to both equations. The solution set for the system of two equations with two variables is thus null, i.e., $A \cap B = \phi$.

Another example of an inconsistent system of equations is

$$A = \{(x, y) \mid x + y = 8\}$$
$$B = \{(x, y) \mid 2x - y = 2\}$$
$$C = \{(x, y) \mid x - 2y = -2\}$$

These equations are shown in Fig. 2.15. The solution to the system of equa-

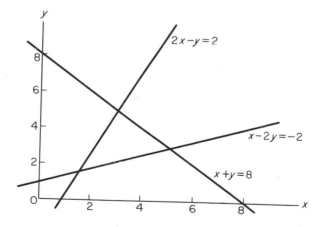

Figure 2.15

tions consists of the ordered pairs (x, y) that are common solutions to all three equations. Figure 2.15 shows that there are no ordered pairs (x, y) that are members of the solution set $A \cap B \cap C$. This can be verified algebraically by determining the solution to any two of the three equations and determining if this ordered pair is a solution to the third equation. For instance, the solution set for equation A and B is $A \cap B = \{(3\frac{1}{3}, 4\frac{2}{3})\}$. The ordered pair $(3\frac{1}{3}, 4\frac{2}{3})$ is, however, not a solution for equation C. Since there is no ordered pair that is a common solution for all three equations, we conclude that the equations are inconsistent.

PROBLEMS

1. Plot the following functions and specify the slope of each function.
   (a) $f(x) = 4$ (b) $f(x) = x + 3$
   (c) $f(x) = 2x + 3$ (d) $f(x) = 4 - 3x$
   (e) $f(x) = -0.5x + 2$ (f) $f(x) = -x$

   (g) $f(x) = -3x - 4$ (h) $f(x) = \dfrac{x}{3} - 2$

2. Establish linear functions through the data points $(x, f(x))$.
   (a) $(3, 4)$ and $(5, 2)$ (b) $(9, 4)$ and $(4, -1)$
   (c) $(8, 2)$ and $(5, 5)$ (d) $(4, -3)$ and $(-5, 6)$
   (e) $(8, 4)$ and $(4, -3)$ (f) $(-4, -5)$ and $(4, 5)$
   (g) $(3, 3)$ with $b = 0.5$ (h) $(-5, 7)$ with $b = 0.75$
   (i) $(\frac{3}{4}, \frac{6}{5})$ with $b = -\frac{3}{2}$ (j) $(\frac{4}{3}, 70)$ with $b = \frac{3}{4}$
   (k) $(0.8, 0.5)$ with $b = 0.5$ (l) $(3, -4)$ with $b = 5$

3. Establish linear functions for the following problems.
   (a) Costs are $3000 when output is 0 and $4000 when output is 1500 units.
   (b) Fixed costs are $75,000 and variable costs are $3.50 per unit.
   (c) Costs are $5000 for 150 units and $6000 for 250 units.
   (d) If 1000 units can be sold at a price of $2.50 per unit and 2000 units can be sold at a price of $1.75 per unit, determine price as a function of the number of units sold.
   (e) If 1500 units can be sold at a price of $10 per unit and 2500 units can be sold at a price of $9 per unit, determine price as a function of the number of units sold.
   (f) A firm has 250 employees and expects to add to their work force by 10 employees per year.
   (g) An individual now earning $10,000 per year expects raises of $500 per year.
   (h) A dealer advertises that a car worth $3000 will be reduced in price by $100 per month until sold.

4. For the sets A and B defined below, determine $A \cap B$.
   $A = \{x \mid x^2 - x - 2 \leq 0\}$
   $B = \{x \mid x + 1 > 0\}$

5. For the sets A and B, find $A \cap B$ and $A \cup B$.

$$A = \{-5, -4, -3, -2, -1, 0, 1, 2, 3, 4\}$$
$$B = \{x \mid x^2 - 9x + 20 = 0\}$$

6. For the set A in problem 5, list the elements in the following sets that are members of A.
   (a) $\{x \mid x^2 = 9\}$
   (b) $\{x \mid -3 < 2x < 4\}$

7. Verify that the following set is a function. Give the domain and range of the function.

$$\{(3, 7), (4, 10), (5, 8), (0, 6)\}$$

8. Is the following set a function or relation?

$$\{(A, 1), (A, 2), (B, 1), (B, 2)\}$$

9. Develop inequalities for the following problems and give the domain of the variables.
   (a) Two products can be manufactured by using the turret lathe. One product requires 2 hours per unit on the lathe and the other product requires 3 hours per unit. The maximum capacity of the lathe is 80 hours per week.
   (b) An airplane has a useful load of 1800 pounds. Gasoline weighs 6 pounds per gallon, and the fuel cells have a maximum capacity of 140 gallons. The standard weight for passengers and crew is 170 pounds, and a maximum of six people can be carried on the airplane.
   (c) A construction equipment leasing company is considering the purchase of dump trucks, road graders, and bulldozers at a cost of $20,000 per truck, $60,000 per grader, and $80,000 per dozer. A maximum of $1.2 million has been authorized for the equipment.
   (d) An individual must have a minimum of 300 units of a special dietary supplement each day. This supplement is available in two commercial products, A and B. Each ounce of A contains 25 units of the supplement and each ounce of B contains 40 units.

10. A firm is interested in determining the breakeven production for a new product. The cost of introducing the product is estimated to be $40,000. This cost includes initial advertising and promotion as well as the fixed cost necessary for one year of production. The variable cost per unit is $35, and the proposed selling price is $60.
    (a) Determine breakeven production.
    (b) Determine the profit on sales of 2000 units.

11. The sole proprietor of a rug-cleaning business has determined that he must work at least as many hours as it takes to cover his fixed cost of $45 per day. He averages $9 per hour and can work no more than 12 hours on any one day. Determine the profit function and the domain and range of the function. Also, determine his breakeven point in hours.

12. The Handley Company plans to produce and sell an item. The cost of production includes fixed cost of $50,000 and variable cost of $10 per item. They plan to sell the item for $25. How many items must be sold to obtain a profit of $100,000?

13. The profit function of Graham Manufacturing Company is described by a quadratic function. The following data points were estimated by the chief accountant: output of zero units results in a loss of $10 million, output of 6000 units results in a profit of $8 million, and output of 8000 units results in a profit of $6 million. Determine the quadratic function that describes profit as a function of the number of units produced.

14. The average per unit cost of a certain assembly manufactured by Milwaukee Electronics Corporation is $400 for the 300th unit, $250 for the 500th unit, and $550 for the 600th unit. Determine the quadratic function that describes the relationship between average cost per unit and output.

15. In a controlled experiment the productivity of a work group was found to be related to the number of workers assigned to the group. Specifically, the analyst found that with a group of six workers the productivity was eight units per man. When the group size was increased to eight workers, the productivity rose to 12 units per man. An additional increase in group size to 10 workers resulted in an increase in productivity to 14 units per man. Determine the quadratic function that relates productivity to the group size.

16. Profit derived from gas sales at a service station depends upon the quantities sold of three types of gas; Regular, Unleaded, and Premium.
    (a) State the profit function for total profit on gallons sold of the three types of gas.
    It was determined from past experience that the profit derived from certain sales combinations is as follows:

    Profit = {($440, 5000 gal. R, 1000 gal. U, 4000 gal. P),
              ($245, 3000 gal. R, 500 gal. U, 2000 gal. P),
              ($ 93, 1000 gal. R, 100 gal. U, 1000 gal. P)}

    (b) Determine the profit per gallon for each type of gas by substituting these data points into the profit function.

17. In a certain company, cost per unit is constant at $100 from 0 to 1000 units produced per month. From 1001 to 3000 units, costs decrease

linearly to $50 per unit at 3000 units. Costs then increase by $0.01 per unit produced up to 4000 units. Determine the functions describing the cost per unit for the product.

18. The total cost of four milk shakes and two coffees is $1.40. The cost of one milk shake and three coffees is $0.60. Find the price of a milk shake and a coffee.

19. A firm uses two machines in the manufacture of two products. Product A requires three hours on machine 1 and five hours on machine 2. Product B requires six hours on machine 1 and four hours on machine 2. If 36 hours of machine 1 time and 42 hours of machine 2 time are to be scheduled, determine the product mix that fully utilizes the machine time.

20. A company packages fruit in fancy boxes for sale as Christmas presents. One box contains six apples, four pears, and four oranges and sells for $1.68. A second box contains five apples, four pears, and six oranges and sells for $1.88. A third box contains eight apples, six pears, and ten oranges and sells for $3.02. Assuming that the cost of fruit is constant, determine the per unit cost of each type of fruit.

21. Determine whether the following consistent system of equations has a unique or an infinite number of solutions.

$$3x + 2y + 7z - 5 = 0$$
$$4x - y + 5z + 4 = 0$$
$$7x + y + 12z - 1 = 0$$

22. Verify by graphing that the following system of linear inequalities has an infinite number of solutions.

$$4x_1 + 5x_2 \leq 20$$
$$6x_1 + 3x_2 \geq 18$$

23. Specify whether the following systems of equations are consistent or inconsistent. For those that are consistent, state whether there is a unique or an infinite number of solutions.

(a) $5x + 7y = 70$
    $x + y = 12$

(b) $6x + 15y = 90$
    $4x + 5y = 40$
    $11x + 8y = 87$

(c) $x + y = 10$
    $8x + 11y = 88$
    $7x + 14y = 98$

(d) $2x_1 + 3x_2 + 8x_3 = 56$
    $4x_1 - 3x_2 + 5x_3 = 48$
    $-2x_1 + 6x_2 + 3x_3 = 8$

24. Determine if the following system of equations has a consistent, unique solution.

$$x + y = 4$$

$$2x + y \geq 6$$
$$2x^2 + 3xy + y^2 = 24$$

SUGGESTED REFERENCES

BURNS, CARL M., et al., *Algebra: An Introduction for College Students* (Menlo Park, Ca.: Cummings Publishing Company, Inc., 1972).

FREUND, JOHN E., *College Mathematics with Business Applications* (Englewood Cliffs, N.J.: Prentice-Hall, Inc., 1969), 3, 4.

MOORE, GERALD E., *Algebra*, The Barnes and Noble College Outline Series (New York, N.Y.: Barnes and Noble, Inc., 1970).

NIELSEN, KAJ L., *Algebra: A Modern Approach*, The Barnes and Noble College Outline Series (New York, N.Y.: Barnes and Noble, Inc., 1969).

NIELSEN, KAJ L., *College Mathematics*, The Barnes and Noble College Outline Series (New York, N.Y.: Barnes and Noble, Inc., 1958), 1–6.

RICH, BARNETT, *Elementary Algebra*, Schaum's Outline Series (New York, N.Y.: McGraw-Hill Book Company, Inc., 1960).

THEODORE, CHRIS A., *Applied Mathematics: an Introduction* (Homewood, Ill.: Richard D. Irwin, Inc., 1965), 5–7.

Chapter 3

Matrices
and
Matrix Algebra

Matrices and matrix algebra have begun to play an increasingly important role in modern techniques for quantitative and statistical analysis of business decisions. These tools have also become quite important in the functional business areas of accounting, production, finance, and marketing. Since the matrix provides a method of representing large quantities of data, it has become an important working tool of the computer programmer and program analyst.

Before beginning a formal discussion of matrices and matrix algebra, let us define a *matrix* as an array or group of numbers. To illustrate, the quantity of inventory by product line at three local car dealers can be described by a matrix. This matrix might have the form shown below.

$$
\begin{array}{ccc}
\text{Ford} & \text{Chevrolet} & \text{Plymouth}
\end{array}
$$

$$
\begin{pmatrix}
6 & 8 & 5 \\
10 & 10 & 8 \\
3 & 4 & 2 \\
4 & 5 & 3
\end{pmatrix}
\begin{array}{l}
\text{Compact} \\
\text{Sedan} \\
\text{Convertible} \\
\text{Station wagon}
\end{array}
$$

The matrix shows that the Ford dealer has an inventory of 6 compacts, 10 sedans, 3 convertibles, and 4 station wagons. Similarly, the inventories of the Chevrolet and Plymouth dealers can be read directly from the matrix.

An important characteristic of a matrix is that the position within the matrix of each number as well as the magnitude of the number is important. For instance, the fact that the local Chevrolet dealer has four convertibles in inventory is described by the entry in the matrix at the intersection of the convertible row and the Chevrolet column. The position of the number is important in that a specific location within the matrix is reserved for the inventory of Chevrolet convertibles. The magnitude of the number is important, since it gives the number of Chevrolet convertibles in inventory.

Certain algebraic operations can be performed on matrices. One such operation might involve determining the value of the inventories of the car dealers. The value of the inventories could be found by multiplying the inventory matrix by a price matrix. This would be an example of matrix multiplication. Other examples of matrix algebra might involve the addition or subtraction of two or more matrices or the multiplication of each of the numbers in the matrix by a common number.

This chapter provides an introduction to matrices, matrix algebra, and selected business applications of matrices. The discussion is extended by the introduction of additional topics in Chapter 4. It should be emphasized that the procedure introduced in this chapter for determining solutions to simultaneous equations will be quite important in the chapters on linear programming.

3.1 Vectors

A *vector* is defined as a row or column of numbers in which the position of each number within the row or column is of importance. A vector is a special case of a matrix in that it has only a single row or column rather than multiple rows and columns. Column vectors are written vertically and row vectors horizontally. Examples of column vectors are $\begin{pmatrix} 2 \\ 1 \end{pmatrix}$, $\begin{pmatrix} -3 \\ 6 \end{pmatrix}$, $\begin{pmatrix} 1 \\ 0 \\ 2 \\ 3 \end{pmatrix}$, and $\begin{pmatrix} 3 \\ -4 \\ 7 \end{pmatrix}$.

Examples of row vectors are $(3, 2)$, $(6, 0, 4)$, and $(-6, 4, \frac{1}{2}, 7)$.

To illustrate that the position of the number within the vector is important, the row vector $(3, 2)$ and the row vector $(2, 3)$ have the same components but are not equal. Vectors are equal only if they are exactly the same. Thus, for two vectors to be equal, each component of the first vector must be exactly the same and occupy the same position as each component of the second vector. Row vectors can, therefore, equal other row vectors but not column vectors. Similarly, column vectors can equal other column vectors but not row vectors.

Example: Are the vectors $(6, 4)$ and $\begin{pmatrix} 6 \\ 4 \end{pmatrix}$ equal?

Since $(6, 4)$ is a row vector and $\begin{pmatrix} 6 \\ 4 \end{pmatrix}$ a column vector, the two vectors are not equal.

Example: Are the vectors $(6, 1, 0)$ and $(6, 1)$ equal?
Since the first vector has the element 0 while the second does not, the two row vectors are not equal.

Both row and column vectors are designated by capital letters. The numbers that comprise the vector are called the *elements* or *components* of the vector. We could thus write $A = (6, 10, 12)$, where A designates the row vector with components or elements 6, 10, and 12.

3.2 Vector Algebra

The basic rules for the addition, subtraction, and multiplication of vectors are analogous to the rules of ordinary algebra. The concept of division is not directly extendable to either vectors or matrices. The matrix counterpart of division will be treated in the discussion of matrix algebra.

3.2.1 VECTOR ADDITION AND SUBTRACTION

Row or column vectors may be added or subtracted to other like row or column vectors by adding or subtracting components. If, for instance, $A = (6, 10, 12)$ and $B = (4, 6, -3)$, then

$$A + B = (6 + 4, 10 + 6, 12 - 3) = (10, 16, 9)$$

and

$$A - B = (6 - 4, 10 - 6, 12 + 3) = (2, 4, 15)$$

Similarly, for the column vectors $G = \begin{pmatrix} 3 \\ 2 \end{pmatrix}$ and $H = \begin{pmatrix} -1 \\ 3 \end{pmatrix}$,

$$G + H = \begin{pmatrix} 3 \\ 2 \end{pmatrix} + \begin{pmatrix} -1 \\ 3 \end{pmatrix} = \begin{pmatrix} 2 \\ 5 \end{pmatrix}$$

It is important to recognize that the addition or subtraction of vectors can take place only if the vectors are of the same type; i.e., row vectors may be added to or subtracted from row vectors and column vectors may be added to or subtracted from column vectors. Furthermore, the vectors to be added or subtracted must have the same number of components. The vector $A = (6, 10, 12)$ cannot be added to the vector $E = (3, 2)$ nor the vector $F = (5, 3, 7, 10)$. Vectors that are the same type and have the same number of components are said to be *conformable* for addition or subtraction.

The commutative and associative laws of addition apply to vector addition. The reader may remember that the *commutative law* of addition states that $a + b = b + a$. If this is applied to vector addition, the commutative law of addition states that $A + B = B + A$. This is easily verified, since for $A = (6, 10, 12)$ and $B = (4, 6, -3)$, $A + B = (10, 16, 9)$ and $B + A = (10, 16, 9)$.

The *associative law* of addition states that $a + (b + c) = (a + b) + c$. Applying this law to the vectors $A = (6, 10, 12)$, $B = (4, 6, -3)$, and $C = (-4, 8, 6)$, we have

$$A + (B + C) = (A + B) + C$$
$$(6, 10, 12) + (4 - 4, 6 + 8, -3 + 6) = (6 + 4, 10 + 6, 12 - 3) + (-4, 8, 6)$$
$$(6, 10, 12) + (0, 14, 3) = (10, 16, 9) + (-4, 8, 6)$$
$$(6 + 0, 10 + 14, 12 + 3) = (10 - 4, 16 + 8, 9 + 6)$$
$$(6, 24, 15) = (6, 24, 15)$$

This verifies that the associative law applies to vectors.

Example: A student has 96 semester credits of which 24 units are A, 36 units are B, 30 units are C, 6 units are D, and 0 units are F. If he receives 6 units of A, 6 units of B, 3 units of C, 0 units of D, and 3 units of F, determine the vector that describes his grades.

Let $G = (24, 36, 30, 6, 0)$ represent grades excluding the current semester and $S = (6, 6, 3, 0, 3)$ represent the current semester's grades. Then $G + S = (30, 42, 33, 6, 3)$ is the vector that describes his grades including the current semester.

3.2.2 VECTOR MULTIPLICATION

Vector multiplication includes the multiplication of a vector by a vector or a vector by a scalar. A *scalar* is any number, such as 3, or -10, or 1.10, etc. A vector may be multiplied by a scalar simply by multiplying each component of the vector by the scalar and listing the resulting products as a new vector. To demonstrate, let $X = \begin{pmatrix} 3 \\ -6 \\ 2 \end{pmatrix}$ and $a = -3$. The product of the vector X and the scalar a is

$$aX = -3 \begin{pmatrix} 3 \\ -6 \\ 2 \end{pmatrix} = \begin{pmatrix} -9 \\ 18 \\ -6 \end{pmatrix}$$

Similarly, if we multiply the row vector $A = (6, 10, 12)$ by the scalar $b = 1.5$, we obtain

$$bA = 1.5(6, 10, 12) = (9, 15, 18)$$

Example: The vector $P = (\$10, \$8, \$6)$ represents the prices of three items. If prices increase by 10 percent, determine the new price vector.

The new price vector is formed by the product of the scalar 1.10 and the vector P. Thus, $1.10P = 1.10(\$10, \$8, \$6) = (\$11, \$8.80, \$6.60)$.

Two vectors may be multiplied together, provided that: (1) one of the vectors is a row vector and the other a column vector; and (2) the row vector and column vector each contain the same number of components. To illustrate, consider the row vector $R = (r_1, r_2, r_3)$ and the column vector $C = \begin{pmatrix} c_1 \\ c_2 \\ c_3 \end{pmatrix}$.

Since R is a row vector and C a column vector, and since each vector has three components, it is possible to determine the product of R and C. The vector multiplication is made by placing the row vector to the left of the column vector. The first component of the row vector is then multiplied by the first component of the column vector, the second component of the row vector is multiplied by the second component of the column vector, etc. These products are summed and the resulting sum represents the product of the two vectors. The product of R and C is

$$RC = (r_1, r_2, r_3) \begin{pmatrix} c_1 \\ c_2 \\ c_3 \end{pmatrix} = r_1 c_1 + r_2 c_2 + r_3 c_3 \tag{3.1}$$

The result is termed the *inner product* of R and C and is a scalar.

Example: Determine the inner product of $A = (6, 10, 12)$ and $X = \begin{pmatrix} 3 \\ -6 \\ 2 \end{pmatrix}$.

$$AX = 6 \cdot 3 + 10(-6) + 12 \cdot 2 = -18$$

Example: At the end of a 64-unit MBA program a student has 30 units of A, 20 units of B, 10 units of C, 4 units of D, and 0 units of F. If an A is worth 4 grade points, a B is worth 3 grade points, a C is worth 2 grade points, a D is worth 1 grade point, and an F is worth 0 grade points, determine the student's total number of grade points.

Let

$$A = (30, 20, 10, 4, 0) \text{ and } G = \begin{pmatrix} 4 \\ 3 \\ 2 \\ 1 \\ 0 \end{pmatrix}$$

The total number of grade points is given by the inner product of A and G. Thus,

$$AG = (30, 20, 10, 4, 0) \begin{pmatrix} 4 \\ 3 \\ 2 \\ 1 \\ 0 \end{pmatrix} = 120 + 60 + 20 + 4 + 0 = 204$$

The *commutative law* of multiplication states that $a \cdot b = b \cdot a$. Although this result is true in ordinary algebra, it is not true that $AB = BA$ in vector algebra. If A is a row vector and B a column vector, then AB is a scalar. As we demonstrate later in this chapter, the product of the column vector B and the row vector A, i.e., BA, is a matrix. It will become obvious that the scalar AB is not equal to the matrix BA; thus, the commutative law of multiplication does not apply to vector multiplication.

The *distributive law* of algebra applies to vector algebra. The distributive law states that $a(b + c) = ab + ac$. Applying this to vector algebra gives the result that $P(B + E) = PB + PE$, where P is a row vector and B and E are column vectors. The distributive law is demonstrated in the following example.

Example: A retailer stocks four brands of a product. The costs of these four brands are given by the row vector $P = (60, 70, 90, 100)$. The beginning inventory of the product is given by the column vector $B = \begin{pmatrix} 30 \\ 20 \\ 15 \\ 10 \end{pmatrix}$, and the ending inventory by $E = \begin{pmatrix} 20 \\ 15 \\ 10 \\ 5 \end{pmatrix}$. Assuming no purchases of inventory, determine the cost of goods sold during the period.

To illustrate the distributive law, determine $PB - PE$ and $P(B - E)$.

$$PB - PE = (60, 70, 90, 100) \begin{pmatrix} 30 \\ 20 \\ 15 \\ 10 \end{pmatrix} - (60, 70, 90, 100) \begin{pmatrix} 20 \\ 15 \\ 10 \\ 5 \end{pmatrix}$$

$$PB - PE = (60 \cdot 30 + 70 \cdot 20 + 90 \cdot 15 + 100 \cdot 10)$$
$$- (60 \cdot 20 + 70 \cdot 15 + 90 \cdot 10 + 100 \cdot 5)$$

$$PB - PE = \$1900$$

and

$$P(B - E) = (60, 70, 90, 100) \begin{pmatrix} 30 - 20 \\ 20 - 15 \\ 15 - 10 \\ 10 - 5 \end{pmatrix}$$

$$P(B - E) = (60, 70, 90, 100) \begin{pmatrix} 10 \\ 5 \\ 5 \\ 5 \end{pmatrix} = 60 \cdot 10 + 70 \cdot 5 + 90 \cdot 5 + 100 \cdot 5$$

$$P(B - E) = \$1900$$

The \$1900 obtained by both methods verifies the distributive law for vector algebra.

To summarize, the commutative and associative laws of addition apply to vector addition; i.e., $A + B = B + A$ and $A + (B + C) = (A + B) + C$. The distributive law of algebra applies to vector algebra; i.e., $A(B + C) = AB + AC$. The commutative law of multiplication, however, does not apply to vector algebra; i.e., $AB \neq BA$.

3.3 Matrices

A matrix is an array of numbers. The array is said to be *ordered*, meaning that the position of each number is important. The ordered array of numbers consists of m rows and n columns that are enclosed in parentheses and designated by a capital letter. The general form of a matrix is

$$A = \begin{pmatrix} a_{11} & a_{12} & a_{13} & \cdots & a_{1n} \\ a_{21} & a_{22} & a_{23} & \cdots & a_{2n} \\ a_{31} & a_{32} & a_{33} & \cdots & a_{3n} \\ \vdots & \vdots & \vdots & & \vdots \\ a_{m1} & a_{m2} & a_{m3} & \cdots & a_{mn} \end{pmatrix} \tag{3.2}$$

The matrix A consists of $m \cdot n$ components or elements a_{ij}. A component a_{ij} is designated by its position in the matrix, i referring to the row and j to the column. Component a_{24} thus refers to the component at the intersection of the second row and fourth column.

The matrix A is said to be of order m by n or, alternatively, is said to have dimensions of m by n, m and n again referring to the number of rows and number of columns.

Example:

$$A = \begin{pmatrix} 3 & 1 \\ 2 & 4 \\ 1 & 6 \end{pmatrix} \text{ is a 3 by 2 matrix.}$$

$$B = \begin{pmatrix} 2 & 4 & 1 \\ 3 & 7 & 0 \end{pmatrix} \text{ is a 2 by 3 matrix.}$$

$$C = \begin{pmatrix} 3 & 2 & 6 \\ 1 & 5 & -2 \\ -1 & 6 & 1 \end{pmatrix} \text{ is a 3 by 3 square matrix.}$$

$D = (3, 5, 2)$ is a 1 by 3 matrix or row vector.

$$E = \begin{pmatrix} 2 \\ 1 \end{pmatrix} \text{ is a 2 by 1 matrix or column vector.}$$

Matrices provide a convenient method of storing large quantities of data. As an example, consider a firm with multiple retail outlets. The inventory of the firm can be represented by the matrix G, where the columns represent the outlets and the rows represent products.

$$\begin{array}{cccc} & \text{Outlets} & & \\ & 1 \quad 2 \quad 3 & & \\ & & & \text{Products} \\ G = & \begin{pmatrix} 50 & 60 & 40 \\ 175 & 200 & 125 \\ 40 & 25 & 30 \\ 205 & 235 & 275 \end{pmatrix} & & \begin{matrix} 1 \\ 2 \\ 3 \\ 4 \end{matrix} \end{array}$$

Component a_{21} of matrix G is 175 units of product 2 at outlet 1.
Component a_{42} is 235 units of product 4 at outlet 2.

3.4 Matrix Algebra

Matrix algebra, also termed *linear algebra*, provides a set of rules for the addition, subtraction, multiplication, and inversion of matrices. The rules for addition and subtraction of matrices are analogous to those for ordinary algebra. The multiplication of matrices requires the introduction of a straightforward rule for matrix multiplication. Division of one matrix by another is not possible in matrix algebra. The process of matrix inversion, however, has many similarities with division in ordinary algebra. Matrix addition, subtraction, and multiplication are discussed in this section. Matrix inversion is presented in Sec. 3.5.

3.4.1 MATRIX ADDITION AND SUBTRACTION

Matrices can be added or subtracted, provided that they are conformable for addition or subtraction. Matrices are conformable for addition or subtraction if they have the same dimensions. A 2 by 3 matrix may be added to or subtracted from another 2 by 3 matrix, a 4 by 2 matrix added to or subtracted from another 4 by 2 matrix, etc. The addition or subtraction is performed by adding or subtracting corresponding components from the two matrices. The result is a new matrix of the same order or dimension as the original matrices.

Example: Let

$$A = \begin{pmatrix} 3 & 4 \\ 6 & 5 \\ 4 & 0 \end{pmatrix} \quad \text{and} \quad B = \begin{pmatrix} 2 & 5 \\ 3 & 6 \\ 4 & 2 \end{pmatrix}.$$

Determine $A + B$ and $A - B$.

$$A + B = \begin{pmatrix} 3+2 & 4+5 \\ 6+3 & 5+6 \\ 4+4 & 0+2 \end{pmatrix} = \begin{pmatrix} 5 & 9 \\ 9 & 11 \\ 8 & 2 \end{pmatrix}$$

$$A - B = \begin{pmatrix} 3-2 & 4-5 \\ 6-3 & 5-6 \\ 4-4 & 0-2 \end{pmatrix} = \begin{pmatrix} 1 & -1 \\ 3 & -1 \\ 0 & -2 \end{pmatrix}$$

Example: Let

$$C = \begin{pmatrix} 1 & 3 & 4 \\ 2 & 6 & 5 \\ 1 & -1 & 6 \end{pmatrix}, \quad D = \begin{pmatrix} 2 & 3 & 5 \\ 4 & 1 & 3 \\ 5 & 2 & 1 \end{pmatrix}, \quad E = \begin{pmatrix} 1 & -2 & 5 \\ 3 & -1 & 4 \\ 6 & 5 & 3 \end{pmatrix}$$

Show that $C + (D + E) = (C + D) + E$.

$$C + (D + E) = \begin{pmatrix} 1 & 3 & 4 \\ 2 & 6 & 5 \\ 1 & -1 & 6 \end{pmatrix} + \begin{pmatrix} 3 & 1 & 10 \\ 7 & 0 & 7 \\ 11 & 7 & 4 \end{pmatrix} = \begin{pmatrix} 4 & 4 & 14 \\ 9 & 6 & 12 \\ 12 & 6 & 10 \end{pmatrix}$$

$$(C + D) + E = \begin{pmatrix} 3 & 6 & 9 \\ 6 & 7 & 8 \\ 6 & 1 & 7 \end{pmatrix} + \begin{pmatrix} 1 & -2 & 5 \\ 3 & -1 & 4 \\ 6 & 5 & 3 \end{pmatrix} = \begin{pmatrix} 4 & 4 & 14 \\ 9 & 6 & 12 \\ 12 & 6 & 10 \end{pmatrix}$$

These examples illustrate the commutative and associative laws of addition. The commutative law states that $A + B = B + A$. The associative law states that $C + (D + E) = (C + D) + E$. The examples show that these laws apply to the addition and subtraction of matrices.

3.4.2 MATRIX MULTIPLICATION

Matrix multiplication is subdivided into two cases: (1) the multiplication of a matrix by a scalar, and (2) the multiplication of a matrix by a matrix. The The simpliest of these cases is the multiplication of a matrix by a scalar. A matrix may be multiplied by a scalar by multiplying each component in the matrix by the scalar. The resulting matrix will be of the same order as the original matrix. To illustrate, the 3 by 3 matrix

$$A = \begin{pmatrix} a_{11} & a_{12} & a_{13} \\ a_{21} & a_{22} & a_{23} \\ a_{31} & a_{32} & a_{33} \end{pmatrix}$$

can be multiplied by the scalar k to give the 3 by 3 matrix

$$kA = \begin{pmatrix} ka_{11} & ka_{12} & ka_{13} \\ ka_{21} & ka_{22} & ka_{23} \\ ka_{31} & ka_{32} & ka_{33} \end{pmatrix}$$

Similarly, if a 2 by 4 matrix were multiplied by a scalar, the resulting matrix would be a 2 by 4 matrix.

Example: Multiply the matrix $Y = \begin{pmatrix} 2 & 3 & 5 \\ 4 & 6 & -2 \end{pmatrix}$ by the scalar $k = 3$.

$$kY = 3\begin{pmatrix} 2 & 3 & 5 \\ 4 & 6 & -2 \end{pmatrix} = \begin{pmatrix} 6 & 9 & 15 \\ 12 & 18 & -6 \end{pmatrix}$$

Example: Multiply the matrix $B = \begin{pmatrix} 2 & 1 \\ 1 & -3 \end{pmatrix}$ by the scalar $k = -2$.

$$kB = -2\begin{pmatrix} 2 & 1 \\ 1 & -3 \end{pmatrix} = \begin{pmatrix} -4 & -2 \\ -2 & 6 \end{pmatrix}$$

Example: Multiply the matrix $A = \begin{pmatrix} 1 & 3 & -2 \\ 4 & 7 & 1 \end{pmatrix}$ by the scalar $k = 0.1$.

$$kA = 0.1\begin{pmatrix} 1 & 3 & -2 \\ 4 & 7 & 1 \end{pmatrix} = \begin{pmatrix} 0.1 & 0.3 & -0.2 \\ 0.4 & 0.7 & 0.1 \end{pmatrix}$$

Matrix multiplication is somewhat more complicated than the multiplication of a matrix by a scalar. Two matrices may be multiplied together only if the number of columns in the first matrix equals the number of rows in the second matrix. If matrix A has dimensions a by b and matrix B has dimensions c by d, the matrix product AB is defined only if b is equal to c. Thus, a 2 by 3 matrix can be multiplied by a 3 by 4 matrix but not, for example, by a 2 by 4 matrix. The matrix resulting from the multiplication has dimensions equivalent to the number of rows in the first matrix and the number of columns in the second matrix, i.e., a by d. To summarize, matrix A with dimensions a by b may be multiplied by B with dimensions c by d to form matrix AB with dimensions a by d, provided that b equals c. If the number of columns in the first matrix equals the number of rows in the second matrix, the matrices are said to be conformable for multiplication.

To illustrate the multiplication of two matrices, multiply

$$A = \begin{pmatrix} a_{11} & a_{12} & a_{13} \\ a_{21} & a_{22} & a_{23} \end{pmatrix} \text{ by } B = \begin{pmatrix} b_{11} & b_{12} & b_{13} \\ b_{21} & b_{22} & b_{23} \\ b_{31} & b_{32} & b_{33} \end{pmatrix}$$

Since A is a 2 by 3 matrix and B is 3 by 3 matrix, the matrices are conformable for multiplication. The resulting matrix AB has dimensions 2 by 3. The multiplication is performed as follows:

$$AB = \begin{pmatrix} a_{11} & a_{12} & a_{13} \\ a_{21} & a_{22} & a_{23} \end{pmatrix} \begin{pmatrix} b_{11} & b_{12} & b_{13} \\ b_{21} & b_{22} & b_{23} \\ b_{31} & b_{32} & b_{33} \end{pmatrix}$$

$$AB = \begin{pmatrix} a_{11}b_{11} + a_{12}b_{21} + a_{13}b_{31} & a_{11}b_{12} + a_{12}b_{22} + a_{13}b_{32} \\ a_{21}b_{11} + a_{22}b_{21} + a_{23}b_{31} & a_{21}b_{12} + a_{22}b_{22} + a_{23}b_{32} \end{pmatrix}$$

$$\begin{matrix} a_{11}b_{13} + a_{12}b_{23} + a_{13}b_{33} \\ a_{21}b_{13} + a_{22}b_{23} + a_{23}b_{33} \end{matrix}$$

Matrix AB in the preceding example is a 2 by 3 matrix. The components of the matrix are formed by the sum of the products of the rows of the first matrix and columns of the second matrix. The sum of the products of the components in the first row of A and the first column of B give the first row, first column of AB. Similarly, the sum of the products of the components in the first row of A and second column of B gives the first row, second column component of AB. This procedure is continued until the matrix AB is formed.

Example: Determine the product of $A = \begin{pmatrix} 6 & 2 & 4 \\ 1 & 2 & 2 \end{pmatrix}$ and $B = \begin{pmatrix} 3 & 2 \\ 2 & 4 \\ 4 & 5 \end{pmatrix}$.

$$AB = \begin{pmatrix} 6 & 2 & 4 \\ 1 & 2 & 2 \end{pmatrix} \begin{pmatrix} 3 & 2 \\ 2 & 4 \\ 4 & 5 \end{pmatrix} = \begin{pmatrix} 6 \cdot 3 + 2 \cdot 2 + 4 \cdot 4 & 6 \cdot 2 + 2 \cdot 4 + 4 \cdot 5 \\ 1 \cdot 3 + 2 \cdot 2 + 2 \cdot 4 & 1 \cdot 2 + 2 \cdot 4 + 2 \cdot 5 \end{pmatrix}$$

$$AB = \begin{pmatrix} 18 + 4 + 16 & 12 + 8 + 20 \\ 3 + 4 + 8 & 2 + 8 + 10 \end{pmatrix} = \begin{pmatrix} 38 & 40 \\ 15 & 20 \end{pmatrix}$$

Example: Determine the product of $A = \begin{pmatrix} 6 & 1 \\ 3 & 7 \end{pmatrix}$ and $B = \begin{pmatrix} 2 & 4 & 7 \\ 3 & 1 & 5 \end{pmatrix}$.

$$AB = \begin{pmatrix} 6 & 1 \\ 3 & 7 \end{pmatrix} \begin{pmatrix} 2 & 4 & 7 \\ 3 & 1 & 5 \end{pmatrix}$$

$$= \begin{pmatrix} 6 \cdot 2 + 1 \cdot 3 & 6 \cdot 4 + 1 \cdot 1 & 6 \cdot 7 + 1 \cdot 5 \\ 3 \cdot 2 + 7 \cdot 3 & 3 \cdot 4 + 7 \cdot 1 & 3 \cdot 7 + 7 \cdot 5 \end{pmatrix} = \begin{pmatrix} 15 & 25 & 47 \\ 27 & 19 & 56 \end{pmatrix}$$

In the first example, a 2 by 3 matrix is multiplied by a 3 by 2 matrix. The matrices are conformable for multiplication, since the number of columns in the first matrix equals the number of rows in the second. The dimensions of the resulting matrix are equal to the number of rows of the first matrix and the number of columns of the second matrix. The second example shows that a 2 by 2 matrix and a 2 by 3 matrix are conformable for multiplication. The product of the matrices is a 2 by 3 matrix.

Example: The inventory of a firm with multiple retail outlets is given by the matrix G

$$
\begin{array}{cc}
\textit{Outlets} & \textit{Products} \\
\begin{array}{ccc} 1 & 2 & 3 \end{array} & \\
G = \begin{pmatrix} 50 & 60 & 40 \\ 175 & 200 & 125 \\ 40 & 25 & 30 \\ 205 & 235 & 275 \end{pmatrix} & \begin{array}{c} 1 \\ 2 \\ 3 \\ 4 \end{array}
\end{array}
$$

If the cost of inventory is given by the vector $C = (20, 10, 30, 40)$, determine the value of inventory at each outlet.

The value of inventory at each outlet is given by the matrix CG. Since C is a 1 by 4 matrix and G is a 4 by 3 matrix, the matrices are conformable for multiplication. The product is

$$
CG = (20, 10, 30, 40) \begin{pmatrix} 50 & 60 & 40 \\ 175 & 200 & 125 \\ 40 & 25 & 30 \\ 205 & 235 & 275 \end{pmatrix}
$$

$$
CG = (12{,}150,\ 13{,}350,\ 13{,}950)
$$

Inventory has a value of \$12,150 at outlet 1, \$13,350 at outlet 2, and \$13,950 at outlet 3.

Example: For $A = (6, 2, 3)$ and $B = \begin{pmatrix} 2 \\ 5 \\ 4 \end{pmatrix}$, find AB and BA.

The row vector A is a 1 by 3 matrix and the column vector B is a 3 by 1 matrix. The inner product AB gives the 1 by 1 matrix or scalar, $AB = 34$.

$$
AB = (6, 2, 3) \begin{pmatrix} 2 \\ 5 \\ 4 \end{pmatrix} = 12 + 10 + 12 = 34
$$

Multiplying B by A is equivalent to multiplying a 3 by 1 matrix by a 1 by 3 matrix. Since the number of columns in the first matrix equal the number of rows in the second matrix, the product is defined. The product of an n by 1 vector and a 1 by n vector is termed the *outer product* and has dimension n by n. To illustrate,

$$
BA = \begin{pmatrix} 2 \\ 5 \\ 4 \end{pmatrix} (6, 2, 3) = \begin{pmatrix} 12 & 4 & 6 \\ 30 & 10 & 15 \\ 24 & 8 & 12 \end{pmatrix}
$$

The outer product of the vectors B and A is the 3 by 3 matrix BA.

Example: For $A = \begin{pmatrix} 3 & -1 & 4 \\ 6 & 8 & 2 \\ 1 & -5 & 4 \end{pmatrix}$ and $B = \begin{pmatrix} 2 & 6 & -1 \\ 3 & -2 & 4 \\ 5 & 3 & -3 \end{pmatrix}$

find AB and BA.

$$AB = \begin{pmatrix} 3 & -1 & 4 \\ 6 & 8 & 2 \\ 1 & -5 & 4 \end{pmatrix} \begin{pmatrix} 2 & 6 & -1 \\ 3 & -2 & 4 \\ 5 & 3 & -3 \end{pmatrix} = \begin{pmatrix} 23 & 32 & -19 \\ 46 & 26 & 20 \\ 7 & 28 & -33 \end{pmatrix}$$

$$BA = \begin{pmatrix} 2 & 6 & -1 \\ 3 & -2 & 4 \\ 5 & 3 & -3 \end{pmatrix} \begin{pmatrix} 3 & -1 & 4 \\ 6 & 8 & 2 \\ 1 & -5 & 4 \end{pmatrix} = \begin{pmatrix} 41 & 51 & 16 \\ 1 & -39 & 24 \\ 30 & 34 & 14 \end{pmatrix}$$

The preceding two examples illustrate that the commutative law of multiplication does not apply to the multiplication of matrices. The commutative law of multiplication states that $a \cdot b = b \cdot a$. Since AB is not equal to BA, the commutative law of multiplication does not apply to matrix multiplication.

The associative and distributive laws of multiplication do, however, apply to matrix multiplication. Given three matrices A, B, and C that are conformable for multiplication, the associative law states that $A(BC) = (AB)C$. The distributive law, again if conformability is assumed, states that $A(B + C) = AB + AC$. These properties are illustrated by the following two examples.

Example: Show that the associative law of multiplication applies to the matrices

$$A = \begin{pmatrix} 2 & 3 \\ 4 & 5 \end{pmatrix}, \quad B = \begin{pmatrix} 3 & 4 \\ 5 & 6 \end{pmatrix}, \quad \text{and } C = \begin{pmatrix} 4 & 5 \\ 6 & 7 \end{pmatrix}$$

The associative law of multiplication states that $A(BC) = (AB)C$. To verify the law, we multiply A by BC.

$$BC = \begin{pmatrix} 3 & 4 \\ 5 & 6 \end{pmatrix} \begin{pmatrix} 4 & 5 \\ 6 & 7 \end{pmatrix} = \begin{pmatrix} 36 & 43 \\ 56 & 67 \end{pmatrix}$$

$$A(BC) = \begin{pmatrix} 2 & 3 \\ 4 & 5 \end{pmatrix} \begin{pmatrix} 36 & 43 \\ 56 & 67 \end{pmatrix} = \begin{pmatrix} 240 & 287 \\ 424 & 507 \end{pmatrix}$$

Next, we multiply AB by the matrix C.

$$AB = \begin{pmatrix} 2 & 3 \\ 4 & 5 \end{pmatrix} \begin{pmatrix} 3 & 4 \\ 5 & 6 \end{pmatrix} = \begin{pmatrix} 21 & 26 \\ 37 & 46 \end{pmatrix}$$

$$(AB)C = \begin{pmatrix} 21 & 26 \\ 37 & 46 \end{pmatrix} \begin{pmatrix} 4 & 5 \\ 6 & 7 \end{pmatrix} = \begin{pmatrix} 240 & 287 \\ 424 & 507 \end{pmatrix}$$

Since $A(BC)$ equals $(AB)C$, we conclude that the associative law applies in matrix multiplication.

Example: Show that the distributive law of algebra applies to the matrices

$$A = \begin{pmatrix} 2 & 3 \\ 4 & 5 \end{pmatrix}, \quad B = \begin{pmatrix} 2 & 3 & 6 \\ 3 & 5 & 1 \end{pmatrix}, \quad \text{and} \quad C = \begin{pmatrix} 1 & 4 & 5 \\ 6 & 3 & 1 \end{pmatrix}$$

The distributive law states that $A(B + C) = AB + AC$, provided that B and C are conformable for addition and that AB and AC are conformable for multiplication. The matrix $A(B + C)$ is

$$A(B + C) = \begin{pmatrix} 2 & 3 \\ 4 & 5 \end{pmatrix} \left[\begin{pmatrix} 2 & 3 & 6 \\ 3 & 5 & 1 \end{pmatrix} + \begin{pmatrix} 1 & 4 & 5 \\ 6 & 3 & 1 \end{pmatrix} \right]$$

$$A(B + C) = \begin{pmatrix} 2 & 3 \\ 4 & 5 \end{pmatrix} \begin{pmatrix} 3 & 7 & 11 \\ 9 & 8 & 2 \end{pmatrix} = \begin{pmatrix} 33 & 38 & 28 \\ 57 & 68 & 54 \end{pmatrix}$$

and the matrix $AB + AC$ is

$$AB + AC = \begin{pmatrix} 2 & 3 \\ 4 & 5 \end{pmatrix} \begin{pmatrix} 2 & 3 & 6 \\ 3 & 5 & 1 \end{pmatrix} + \begin{pmatrix} 2 & 3 \\ 4 & 5 \end{pmatrix} \begin{pmatrix} 1 & 4 & 5 \\ 6 & 3 & 1 \end{pmatrix}$$

$$AB + AC = \begin{pmatrix} 13 & 21 & 15 \\ 23 & 37 & 29 \end{pmatrix} + \begin{pmatrix} 20 & 17 & 13 \\ 34 & 31 & 25 \end{pmatrix} = \begin{pmatrix} 33 & 38 & 28 \\ 57 & 68 & 54 \end{pmatrix}$$

Since $A(B + C)$ equals $AB + AC$, the distributive law applies to matrix algebra.

3.4.3 THE IDENTITY AND NULL MATRICES

The *identity* matrix is a square matrix that has 1's on the upper left to lower right diagonal and 0's elsewhere. The 2×2 identity matrix, commonly designated by I, is

$$I = \begin{pmatrix} 1 & 0 \\ 0 & 1 \end{pmatrix}$$

Similarly, the 3 by 3 and 4 by 4 identity matrices, also designated by I, are

$$I = \begin{pmatrix} 1 & 0 & 0 \\ 0 & 1 & 0 \\ 0 & 0 & 1 \end{pmatrix} \quad \text{and} \quad I = \begin{pmatrix} 1 & 0 & 0 & 0 \\ 0 & 1 & 0 & 0 \\ 0 & 0 & 1 & 0 \\ 0 & 0 & 0 & 1 \end{pmatrix}$$

The identity matrix has properties in matrix algebra similar to those of the number 1 in ordinary algebra. Specifically, the identity matrix has the property that the product of any matrix A with a conformable identity matrix I is equal to the matrix A. Symbolically, this means that $AI = A$, or alternatively, $IA = A$. Thus, although we pointed out that the commutative law of multiplication does not apply in matrix multiplication, it does apply in the case of multiplication of a matrix by an identity matrix.

Example: For $A = \begin{pmatrix} 2 & 3 \\ 4 & 5 \end{pmatrix}$, verify that $AI = A = IA$.

$$AI = \begin{pmatrix} 2 & 3 \\ 4 & 5 \end{pmatrix} \begin{pmatrix} 1 & 0 \\ 0 & 1 \end{pmatrix} = \begin{pmatrix} 2 & 3 \\ 4 & 5 \end{pmatrix}$$

$$IA = \begin{pmatrix} 1 & 0 \\ 0 & 1 \end{pmatrix} \begin{pmatrix} 2 & 3 \\ 4 & 5 \end{pmatrix} = \begin{pmatrix} 2 & 3 \\ 4 & 5 \end{pmatrix}$$

Therefore, $AI = A = IA$.

Example: For $A = \begin{pmatrix} 2 & 3 & 4 \\ 5 & 6 & 7 \\ 8 & 9 & 10 \end{pmatrix}$, verify that $AI = A = IA$.

$$AI = \begin{pmatrix} 2 & 3 & 4 \\ 5 & 6 & 7 \\ 8 & 9 & 10 \end{pmatrix} \begin{pmatrix} 1 & 0 & 0 \\ 0 & 1 & 0 \\ 0 & 0 & 1 \end{pmatrix} = \begin{pmatrix} 2 & 3 & 4 \\ 5 & 6 & 7 \\ 8 & 9 & 10 \end{pmatrix}$$

$$IA = \begin{pmatrix} 1 & 0 & 0 \\ 0 & 1 & 0 \\ 0 & 0 & 1 \end{pmatrix} \begin{pmatrix} 2 & 3 & 4 \\ 5 & 6 & 7 \\ 8 & 9 & 10 \end{pmatrix} - \begin{pmatrix} 2 & 3 & 4 \\ 5 & 6 & 7 \\ 8 & 9 & 10 \end{pmatrix}$$

This again demonstrates that $AI = A = IA$.

The *null* matrix is a matrix in which all elements are 0. Unlike the identity matrix, the null matrix need not be a square matrix. The null matrix is commonly designated by the symbol ϕ (phi). A 2 by 3 null matrix would be

$$\phi = \begin{pmatrix} 0 & 0 & 0 \\ 0 & 0 & 0 \end{pmatrix}$$

Similarly, a 1 by 3 null matrix or null row vector is

$$\phi = (0, \quad 0, \quad 0)$$

and a 3 by 1 null matrix or null column vector is

$$\phi = \begin{pmatrix} 0 \\ 0 \\ 0 \end{pmatrix}$$

The product of any matrix with a conformable null matrix is a null matrix, i.e., $A\phi = \phi$. Thus if $A = \begin{pmatrix} 1 & 2 & 3 \\ 4 & 5 & 6 \end{pmatrix}$ and $\phi = \begin{pmatrix} 0 & 0 \\ 0 & 0 \\ 0 & 0 \end{pmatrix}$,

$$A\phi = \begin{pmatrix} 1 & 2 & 3 \\ 4 & 5 & 6 \end{pmatrix} \begin{pmatrix} 0 & 0 \\ 0 & 0 \\ 0 & 0 \end{pmatrix} = \begin{pmatrix} 0 & 0 \\ 0 & 0 \end{pmatrix}$$

3.5   Matrix Representation of Systems of Linear Equations

One of the important uses of matrices and matrix algebra is in the representation of systems of linear equations. To illustrate, consider the two linear equations

$$a_{11}x_1 + a_{12}x_2 = b_1$$
$$a_{21}x_1 + a_{22}x_2 = b_2$$

$$(3.3)$$

These two equations can be represented by the matrix equation

$$AX = B \qquad (3.4)$$

where A is the *coefficient* matrix $\begin{pmatrix} a_{11}a_{12} \\ a_{21}a_{22} \end{pmatrix}$, X is the *solution* vector $\begin{pmatrix} x_1 \\ x_2 \end{pmatrix}$ and B is the *right-hand-side* vector $\begin{pmatrix} b_1 \\ b_2 \end{pmatrix}$.

To show that the matrix equation $AX = B$ is an alternative way of writing the two linear equations, we write

$$AX = B$$

or alternatively,

$$\begin{pmatrix} a_{11}a_{12} \\ a_{21}a_{22} \end{pmatrix} \begin{pmatrix} x_1 \\ x_2 \end{pmatrix} = \begin{pmatrix} b_1 \\ b_2 \end{pmatrix} \qquad (3.5)$$

Multiplying the coefficient matrix by the solution vector and equating the sum to the right-hand side gives

$$a_{11}x_1 + a_{12}x_2 = b_1$$
$$a_{21}x_1 + a_{22}x_2 = b_2$$

The components of the matrix equation $AX = B$ are the coefficient matrix A, the solution vector X, and the right-hand-side vector B. The coefficient matrix contains the coefficients of the linear equations. The solution vector contains the variables in the system, $x_1, x_2, \ldots, x_n$. The right-hand-side or B vector contains those elements customarily written on the right-hand side of the equals sign.†

Example: Express the following system of two equations in matrix form.

$$3x_1 + x_2 = 9$$
$$5x_1 - 3x_2 = 1$$

---

† The system of equations is termed *homogeneous* in those cases in which $B = \phi$. Systems of equations of the type illustrated in this text in which $B \neq \phi$ are, conversely, *nonhomogeneous*.

The two equations can be written as

$$\begin{pmatrix} 3 & 1 \\ 5 & -3 \end{pmatrix} \begin{pmatrix} x_1 \\ x_2 \end{pmatrix} = \begin{pmatrix} 9 \\ 1 \end{pmatrix}$$

or alternatively as

$$AX = B,$$

where

$$A = \begin{pmatrix} 3 & 1 \\ 5 & -3 \end{pmatrix}, \quad X = \begin{pmatrix} x_1 \\ x_2 \end{pmatrix}, \quad B = \begin{pmatrix} 9 \\ 1 \end{pmatrix}$$

Example: Express the system of three equations in matrix form.

$$\begin{aligned} 2x - 3y + 6z &= -18 \\ 6x + 4y - 2z &= 44 \\ 5x + 8y + 10z &= 56 \end{aligned}$$

The three equations can be written as

$$\begin{pmatrix} 2 & -3 & 6 \\ 6 & 4 & -2 \\ 5 & 8 & 10 \end{pmatrix} \begin{pmatrix} x \\ y \\ z \end{pmatrix} = \begin{pmatrix} -18 \\ 44 \\ 56 \end{pmatrix}$$

or alternatively as

$$AX = B,$$

where

$$A = \begin{pmatrix} 2 & -3 & 6 \\ 6 & 4 & -2 \\ 5 & 8 & 10 \end{pmatrix}, \quad X = \begin{pmatrix} x \\ y \\ z \end{pmatrix}, \quad \text{and} \quad B = \begin{pmatrix} -18 \\ 44 \\ 56 \end{pmatrix}$$

By applying the rules of matrix multiplication, a matrix equation is easily transformed into a system of equations. Consider the following examples.

Example: Write the matrix equation $AX = B$, where $A = \begin{pmatrix} 4 & 2 & 3 \\ 3 & 4 & -1 \\ 5 & 3 & 1 \end{pmatrix}$, $X = \begin{pmatrix} x_1 \\ x_2 \\ x_3 \end{pmatrix}$, and $B = \begin{pmatrix} 16 \\ 13 \\ 16 \end{pmatrix}$ as three equations in three variables.

The matrix equation $AX = B$ is written as

$$\begin{pmatrix} 4 & 2 & 3 \\ 3 & 4 & -1 \\ 5 & 3 & 1 \end{pmatrix} \begin{pmatrix} x_1 \\ x_2 \\ x_3 \end{pmatrix} = \begin{pmatrix} 16 \\ 13 \\ 16 \end{pmatrix}$$

Applying the rules of matrix multiplication gives

$$4x_1 + 2x_2 + 3x_3 = 16$$

$$3x_1 + 4x_2 - 1x_3 = 13$$
$$5x_1 + 3x_2 + 1x_3 = 16$$

Example: Express $\begin{pmatrix} 3 & 2 \\ 6 & -2 \end{pmatrix} \begin{pmatrix} x_1 \\ x_2 \end{pmatrix} = \begin{pmatrix} 14 \\ 4 \end{pmatrix}$ as two equations.

Multiplying the coefficient matrix by the solution vector and equating this sum to the right-hand side gives

$$3x_1 + 2x_2 = 14$$
$$6x_1 - 2x_2 = 4$$

Matrix algebra can be used in conjunction with the matrix representation of the system for solution of the system of linear equations. To illustrate, consider the preceding example of two equations with variables x_1 and x_2. The two equations were

$$3x_1 + 2x_2 = 14$$
$$6x_1 - 2x_2 = 4$$

The solution set for these two equations consists of values of x_1 and x_2 that . satisfy both equalities. We shall demonstrate that these values are $x_1 = 2$ and $x_2 = 4$.

Techniques for the solution of matrix equations are based upon *row operations*. The term row operations means the application of basic algebraic operations to the rows of a matrix. Three operations are defined. These are

1. Any two rows of a matrix may be interchanged.
2. A row may be multiplied by a nonzero constant.
3. A multiple of one row may be added to another row.

The three row operations are applications of fundamental algebraic operations. To illustrate, consider the two equations

$$3x_1 + 2x_2 = 14$$
$$6x_1 - 2x_2 = 4$$

The first row operation states that the rows may be interchanged; i.e.,

$$6x_1 - 2x_2 = 4$$
$$3x_1 + 2x_2 = 14$$

The second row operation states that a row may be multiplied by a nonzero constant. For example, multiplying $6x_1 - 2x_2 = 4$ by $\frac{1}{2}$ gives

$$3x_1 - 1x_2 = 2$$

The third row operation states that a multiple of one row may be added to another row. Multiplying $3x_1 + 2x_2 = 14$ by -2 and adding the result to the equation $6x_1 - 2x_2 = 4$ gives $0x_1 - 6x_2 = -24$. To demonstrate this result, first multiply $3x_1 + 2x_2 = 14$ by -2.

$$-2(3x_1 + 2x_2 = 14) = -6x_1 - 4x_2 = -28$$

This product is then added to the equation $6x_1 - 2x_2 = 4$, i.e.,

$$
\begin{array}{r}
6x_1 - 2x_2 = 4 \\
-6x_1 - 4x_2 = -28 \\
\hline
0x_1 - 6x_2 = -24
\end{array}
$$

The resulting equation is $0x_1 - 6x_2 = -24$.

This row operation is based upon the algebraic principle that "equals may be added to equals." For the equation $a = b$ and the equation $c = d$, the principle states that $a + c = b + d$. Since $6x_1 - 2x_2 = 4$ and $-6x_1 - 4x_2 = -28$, we can add $6x_1 - 2x_2$ to $-6x_1 - 4x_2$ provided that we add -28 to 4. The result of the addition is $0x_1 - 6x_2 = -24$.

To apply row operations in the solution of matrix equations, we *augment* the coefficient matrix A with the right-hand-side vector B. The augmented matrix is designated by the symbol $A \mid B$ and is expressed in matrix form as $\begin{pmatrix} a_{11} & a_{12} & b_1 \\ a_{21} & a_{22} & b_2 \end{pmatrix}$. Referring to the preceding example, we write the augmented matrix as

$$A \mid B = \begin{pmatrix} 3 & 2 & 14 \\ 6 & -2 & 4 \end{pmatrix}$$

The solution to the system of equations is determined by applying row operations to the rows of the augmented matrix. The row operations are applied with the objective of obtaining an identity matrix in the position originally occupied by the coefficient matrix A. Once the identity matrix is obtained, the solution to the system of equations appears in the position originally occupied by the right-hand-side vector.

To demonstrate the solution technique, row operations are performed on both the augmented matrix and the system of two equations in two unknowns. Our purpose is to show that applying row operations to augmented matrices is analogous to the algebraic operations used in the solution of systems of equations and can, therefore, be used to determine the solution for systems of equations. The augmented matrix and system of equations are

$$\begin{pmatrix} 3 & 2 & 14 \\ 6 & -2 & 4 \end{pmatrix} \qquad \begin{array}{l} 3x_1 + 2x_2 = 14 \\ 6x_1 - 2x_2 = 4 \end{array}$$

The first row operation is to add one times the second row or second equation to the first row or equation. This gives

$$\begin{pmatrix} 9 & 0 & 18 \\ 6 & -2 & 4 \end{pmatrix} \qquad \begin{array}{l} 9x_1 + 0x_2 = 18 \\ 6x_1 - 2x_2 = 4 \end{array}$$

Next, multiply the first row or equation by $\frac{1}{9}$.

$$\begin{pmatrix} 1 & 0 & | & 2 \\ 6 & -2 & | & 4 \end{pmatrix}$$
$$\begin{aligned} 1x_1 + 0x_2 &= 2 \\ 6x_1 - 2x_2 &= 4 \end{aligned}$$

Add -6 times the first row or equation to the second row or equation.

$$\begin{pmatrix} 1 & 0 & | & 2 \\ 0 & -2 & | & -8 \end{pmatrix}$$
$$\begin{aligned} 1x_1 + 0x_2 &= 2 \\ 0x_1 - 2x_2 &= -8 \end{aligned}$$

The last operation is to multiply the second row or equation by $-\frac{1}{2}$.

$$\begin{pmatrix} 1 & 0 & | & 2 \\ 0 & 1 & | & 4 \end{pmatrix}$$
$$\begin{aligned} 1x_1 + 0x_2 &= 2 \\ 0x_1 + 1x_2 &= 4 \end{aligned}$$

The solution to the system of equations is read directly as $x_1 = 2$ and $x_2 = 4$. The top element to the right of the vertical line is the solution value of x_1 and the bottom element the solution value of x_2. This can be demonstrated by writing the augmented matrix as the matrix equation,

$$\begin{pmatrix} 1 & 0 \\ 0 & 1 \end{pmatrix} \begin{pmatrix} x_1 \\ x_2 \end{pmatrix} = \begin{pmatrix} 2 \\ 4 \end{pmatrix}$$

and converting the matrix equation to two equations in the two unknowns,

$$1x_1 + 0x_2 = 2$$
$$0x_1 + 1x_2 = 4$$

As a second example of solving matrix equations using row operations, consider the following three equations

$$2x + 1y - 2z = -1$$
$$4x - 2y + 3z = 14$$
$$1x - 1y + 2z = 7$$

Row operations are performed on the augmented matrix and the system of equations. This will again demonstrate the parallel between the algebraic solution and the matrix solution of systems of equations. The augmented matrix and the systems of equations are

$$\begin{pmatrix} 2 & 1 & -2 & | & -1 \\ 4 & -2 & 3 & | & 14 \\ 1 & -1 & 2 & | & 7 \end{pmatrix}$$
$$\begin{aligned} 2x + 1y - 2z &= -1 \\ 4x - 2y + 3z &= 14 \\ 1x - 1y + 2z &= 7 \end{aligned}$$

Add 1 times the third row to the first row.

$$\begin{pmatrix} 3 & 0 & 0 & | & 6 \\ 4 & -2 & 3 & | & 14 \\ 1 & -1 & 2 & | & 7 \end{pmatrix}$$
$$\begin{aligned} 3x + 0y + 0z &= 6 \\ 4x - 2y + 3z &= 14 \\ 1x - 1y + 2z &= 7 \end{aligned}$$

Multiply the first row by $\frac{1}{3}$.

$$\begin{pmatrix} 1 & 0 & 0 & | & 2 \\ 4 & -2 & 3 & | & 14 \\ 1 & -1 & 2 & | & 7 \end{pmatrix}$$

$$1x + 0y + 0z = 2$$
$$4x - 2y + 3z = 14$$
$$1x - 1y + 2z = 7$$

Add -4 times the first row to the second row and -1 times the first row to the third row.

$$\begin{pmatrix} 1 & 0 & 0 & | & 2 \\ 0 & -2 & 3 & | & 6 \\ 0 & -1 & 2 & | & 5 \end{pmatrix}$$

$$1x + 0y + 0z = 2$$
$$0x - 2y + 3z = 6$$
$$0x - 1y + 2z = 5$$

Multiply the second row by $-\frac{1}{2}$.

$$\begin{pmatrix} 1 & 0 & 0 & | & 2 \\ 0 & 1 & -\frac{3}{2} & | & -3 \\ 0 & -1 & 2 & | & 5 \end{pmatrix}$$

$$1x + 0y + 0z = 2$$
$$0x + 1y - \frac{3}{2}z = -3$$
$$0x - 1y + 2z = 5$$

Add 1 times the second row to the third row.

$$\begin{pmatrix} 1 & 0 & 0 & | & 2 \\ 0 & 1 & -\frac{3}{2} & | & -3 \\ 0 & 0 & \frac{1}{2} & | & 2 \end{pmatrix}$$

$$1x + 0y + 0z = 2$$
$$0x + 1y - \frac{3}{2}z = -3$$
$$0x + 0y + \frac{1}{2}z = 2$$

Multiply the third row by 2.

$$\begin{pmatrix} 1 & 0 & 0 & | & 2 \\ 0 & 1 & -\frac{3}{2} & | & -3 \\ 0 & 0 & 1 & | & 4 \end{pmatrix}$$

$$1x + 0y + 0z = 2$$
$$0x + 1y - \frac{3}{2}z = -3$$
$$0x + 0y + 1z = 4$$

Add $\frac{3}{2}$ times the third row to the second row.

$$\begin{pmatrix} 1 & 0 & 0 & | & 2 \\ 0 & 1 & 0 & | & 3 \\ 0 & 0 & 1 & | & 4 \end{pmatrix}$$

$$1x + 0y + 0z = 2$$
$$0x + 1y + 0z = 3$$
$$0x + 0y + 1z = 4$$

The solution to the three equations is $x = 2$, $y = 3$, and $z = 4$.

It has been demonstrated in the preceding two examples that the row operations are applied to the augmented matrix with the objective of obtaining an identity matrix on the left side of the vertical line. Once the identity matrix is obtained, the solution vector can be read from the right side of the vertical line. This is possible since the row operations on the augmented matrix correspond to algebraic operations on the system of equations.

This method for the solution of matrix equations is termed the *Gaussian elimination* method (named for the German mathematician Karl Friedrich Gauss, 1777–1855) or simply the elimination method. To summarize, the technique involves writing the coefficient matrix and the right-hand-side vector as $A \mid B$. Row operations are applied to transform the coefficient matrix A to an identity matrix I. The solution to the system of equations is then read from the right-hand side of the vertical line.

Any combination of row operations is acceptable in obtaining the identity matrix to the left of the vertical line. The procedure many students find useful is to first obtain a 1 in the a_{11} position and to then use multiples of the first row to obtain zeros elsewhere in the first column. The student next obtains a 1 in the a_{22} position and uses multiples of the second row to obtain zeros elsewhere in the second column. This procedure is followed until the identity matrix is obtained. This method is used in the following examples. It is also used in examples later in this chapter.

Example: Determine the solution vector for the following two equations, using the Gaussian elimination method.

$$-2x_1 + 4x_2 = 4$$
$$3x_1 - 2x_2 = 10$$

Write the equations in matrix form as $A \mid B$.

$$\begin{pmatrix} -2 & 4 & \mid & 4 \\ 3 & -2 & \mid & 10 \end{pmatrix}$$

Multiply the first row by $-\frac{1}{2}$.

$$\begin{pmatrix} 1 & -2 & \mid & -2 \\ 3 & -2 & \mid & 10 \end{pmatrix}$$

Add -3 times the first row to the second row.

$$\begin{pmatrix} 1 & -2 & \mid & -2 \\ 0 & 4 & \mid & 16 \end{pmatrix}$$

Multiply the second row by $\frac{1}{4}$.

$$\begin{pmatrix} 1 & -2 & \mid & -2 \\ 0 & 1 & \mid & 4 \end{pmatrix}$$

Add 2 times the second row to the first row.

$$\begin{pmatrix} 1 & 0 & \mid & 6 \\ 0 & 1 & \mid & 4 \end{pmatrix}$$

The solution is $x_1 = 6$ and $x_2 = 4$.

Example: Determine the solution vector to the following system of equations, using the Gaussian elimination method.

$$2x + 3y - 4z = 9$$
$$3x - 2y + 3z = -15$$
$$1x + 4y - 2z = 12$$

Write the equations in matrix form as $A \mid B$.

$$\begin{pmatrix} 2 & 3 & -4 & \Big| & 9 \\ 3 & -2 & 3 & \Big| & -15 \\ 1 & 4 & -2 & \Big| & 12 \end{pmatrix}$$

Multiply the first row by $\frac{1}{2}$.

$$\begin{pmatrix} 1 & \frac{3}{2} & -2 & \Big| & \frac{9}{2} \\ 3 & -2 & 3 & \Big| & -15 \\ 1 & 4 & -2 & \Big| & 12 \end{pmatrix}$$

Add -3 times the first row to the second row and add -1 times the first row to the third row.

$$\begin{pmatrix} 1 & \frac{3}{2} & -2 & \Big| & \frac{9}{2} \\ 0 & -\frac{13}{2} & 9 & \Big| & -\frac{57}{2} \\ 0 & \frac{5}{2} & 0 & \Big| & \frac{15}{2} \end{pmatrix}$$

Interchange the second and third row.

$$\begin{pmatrix} 1 & \frac{3}{2} & -2 & \Big| & \frac{9}{2} \\ 0 & \frac{5}{2} & 0 & \Big| & \frac{15}{2} \\ 0 & -\frac{13}{2} & 9 & \Big| & -\frac{57}{2} \end{pmatrix}$$

Multiply the second row by $\frac{2}{5}$.

$$\begin{pmatrix} 1 & \frac{3}{2} & -2 & \Big| & \frac{9}{2} \\ 0 & 1 & 0 & \Big| & 3 \\ 0 & -\frac{13}{2} & 9 & \Big| & -\frac{57}{2} \end{pmatrix}$$

Add $-\frac{3}{2}$ times the second row to the first row and add $\frac{13}{2}$ times the second row to the third row.

$$\begin{pmatrix} 1 & 0 & -2 & \Big| & 0 \\ 0 & 1 & 0 & \Big| & 3 \\ 0 & 0 & 9 & \Big| & -9 \end{pmatrix}$$

Multiply the third row by $\frac{1}{9}$.

$$\begin{pmatrix} 1 & 0 & -2 & \Big| & 0 \\ 0 & 1 & 0 & \Big| & 3 \\ 0 & 0 & 1 & \Big| & -1 \end{pmatrix}$$

Add 2 times the third row to the first row.

$$\begin{pmatrix} 1 & 0 & 0 & \Big| & -2 \\ 0 & 1 & 0 & \Big| & 3 \\ 0 & 0 & 1 & \Big| & -1 \end{pmatrix}$$

The solution is $x = -2$, $y = 3$, and $z = -1$.

3.6 The Inverse: Gaussian Method

The four basic arithmetic operations of ordinary algebra are addition, subtraction, multiplication, and division. The first three operations apply in matrix algebra and have been discussed. The fourth operation, division, is not defined for matrix algebra. The inverse of a matrix, however, is used in matrix algebra in much the same manner as division in ordinary algebra.

Division is used in ordinary algebra to determine the solution for systems of equations. For instance, in the algebraic equation $ax = b$, x is determined by dividing both sides of the equation by a to give $x = b/a$. An alternative method of solving this algebraic equation is to multiply both sides of the equation $ax = b$ by the reciprocal of a. In ordinary algebra the reciprocal of a is $1/a$. The solution to the equation is thus

$$(1/a)ax = (1/a)b$$

or

$$x = (1/a)b, \qquad \text{provided } a \neq 0$$

An inverse matrix performs a function in matrix algebra similar to that of the reciprocal in ordinary algebra. The product of a number a and its reciprocal $1/a$ in ordinary algebra is 1. Similarly, in matrix algebra the product of a matrix A and its inverse, A^{-1} (read A inverse) is the identity matrix I, i.e.,

$$AA^{-1} = I \qquad (3.6)$$

The matrix equation $AX = B$ can be solved for the solution vector X by use of the inverse. The procedure is to multiply the matrix equation by the inverse,

$$A^{-1}AX = A^{-1}B \qquad (3.7)$$

Since $A^{-1}A = I$, this reduces to

$$IX = A^{-1}B, \qquad (3.8)$$

and since $IX = X$, the solution vector is

$$X = A^{-1}B \qquad (3.9)$$

In order for a matrix to have an inverse, the matrix must be square.† If A is a square matrix, the inverse of A is a square matrix with components such that the matrix product of A^{-1} and A is the identity matrix I, i.e., $AA^{-1} = I$.

As an example, consider the matrix $A = \begin{pmatrix} 3 & 2 \\ 6 & -2 \end{pmatrix}$ and the matrix $A^{-1} =$

---

† Not all square matrices have inverses. See Chapter 4, p. 125 for further explanation.

$\begin{pmatrix} \frac{1}{9} & \frac{1}{9} \\ \frac{1}{3} & -\frac{1}{6} \end{pmatrix}$. A^{-1} is the inverse of A, provided that $AA^{-1} = I$. Since

$$\begin{pmatrix} 3 & 2 \\ 6 & -2 \end{pmatrix}\begin{pmatrix} \frac{1}{9} & \frac{1}{9} \\ \frac{1}{3} & -\frac{1}{6} \end{pmatrix} = \begin{pmatrix} 1 & 0 \\ 0 & 1 \end{pmatrix}$$

we conclude that A^{-1} is the inverse of A. It also follows that the matrix $A = \begin{pmatrix} 3 & 2 \\ 6 & -2 \end{pmatrix}$ is the inverse of the matrix $A^{-1} = \begin{pmatrix} \frac{1}{9} & \frac{1}{9} \\ \frac{1}{3} & -\frac{1}{6} \end{pmatrix}$, since

$$A^{-1}A = \begin{pmatrix} \frac{1}{9} & \frac{1}{9} \\ \frac{1}{3} & -\frac{1}{6} \end{pmatrix}\begin{pmatrix} 3 & 2 \\ 6 & -2 \end{pmatrix} = \begin{pmatrix} 1 & 0 \\ 0 & 1 \end{pmatrix}.$$

In summary, for two square matrices A and A^{-1}, A^{-1} is the inverse of A if $AA^{-1} = I$. Similarly, since $A^{-1}A = I$ it follows that A is the inverse of A^{-1}. The commutative law of multiplication thus applies to inverse matrices, i.e., $AA^{-1} = I = A^{-1}A$. The relationship between a matrix and its inverse is illustrated in the following examples.

Example: Verify that $A^{-1} = \begin{pmatrix} \frac{2}{28} & \frac{3}{28} \\ \frac{6}{28} & -\frac{5}{28} \end{pmatrix}$ is the inverse matrix of $A = \begin{pmatrix} 5 & 3 \\ 6 & -2 \end{pmatrix}$.

The fact that A^{-1} is the inverse matrix of A can be verified by multiplying A by A^{-1}. If the product of the two matrices is the identity matrix, then A^{-1} is the inverse of A.

$$AA^{-1} = \begin{pmatrix} 5 & 3 \\ 6 & -2 \end{pmatrix}\begin{pmatrix} \frac{2}{28} & \frac{3}{28} \\ \frac{6}{28} & -\frac{5}{28} \end{pmatrix} = \begin{pmatrix} 1 & 0 \\ 0 & 1 \end{pmatrix}$$

Example: Verify that the commutative law of multiplication applies for inverse matrices. Use the matrices in the preceding example.

The commutative law of multiplication states that $AB = BA$. We have previously shown that this law does not hold in general in matrix multiplication. It does hold true, however, for inverse matrices. Thus,

$$AA^{-1} = \begin{pmatrix} 5 & 3 \\ 6 & -2 \end{pmatrix}\begin{pmatrix} \frac{2}{28} & \frac{3}{28} \\ \frac{6}{28} & -\frac{5}{28} \end{pmatrix} = \begin{pmatrix} 1 & 0 \\ 0 & 1 \end{pmatrix}$$

and

$$A^{-1}A = \begin{pmatrix} \frac{2}{28} & \frac{3}{28} \\ \frac{6}{28} & -\frac{5}{28} \end{pmatrix}\begin{pmatrix} 5 & 3 \\ 6 & -2 \end{pmatrix} = \begin{pmatrix} 1 & 0 \\ 0 & 1 \end{pmatrix}$$

This demonstrates that $A^{-1}A = I = AA^{-1}$.

Example: Verify that

$$A = \begin{pmatrix} 2 & 4 & -6 \\ 4 & 2 & 2 \\ 3 & -3 & 1 \end{pmatrix} \text{ and } A^{-1} = \begin{pmatrix} \frac{4}{66} & \frac{7}{66} & \frac{10}{66} \\ \frac{1}{66} & \frac{10}{66} & -\frac{14}{66} \\ -\frac{9}{66} & \frac{9}{66} & -\frac{6}{66} \end{pmatrix}$$

are inverse matrices.

The two matrices are inverse matrices if $AA^{-1} = I$.

$$AA^{-1} = \begin{pmatrix} 2 & 4 & -6 \\ 4 & 2 & 2 \\ 3 & -3 & 1 \end{pmatrix} \begin{pmatrix} \frac{4}{66} & \frac{7}{66} & \frac{10}{66} \\ \frac{1}{66} & \frac{10}{66} & -\frac{14}{66} \\ -\frac{9}{66} & \frac{9}{66} & -\frac{6}{66} \end{pmatrix} = \begin{pmatrix} 1 & 0 & 0 \\ 0 & 1 & 0 \\ 0 & 0 & 1 \end{pmatrix}$$

Since $AA^{-1} = 1$, we conclude that A and A^{-1} are inverse matrices.

3.6.1 CALCULATING THE INVERSE BY GAUSSIAN ELIMINATION

The Gaussian elimination method can be used to calculate an inverse matrix. The procedure involves augmenting the square matrix A with the identity matrix I, i.e.

$$A \mid I \tag{3.10}$$

Row operations are then employed to obtain an identity matrix on the left side of the vertical line. Concurrent with the identity matrix on the left of the vertical line, the inverse matrix is obtained on the right of the vertical line. The Gaussian elimination method involves transforming the augmented matrix $A \mid I$ to the inverse matrix $I \mid A^{-1}$ through row operations.

The justification for this procedure can be demonstrated with matrix algebra. First write the matrix A augmented with I.

$$A \mid I$$

If the inverse matrix A^{-1} were known, we could multiply the matrices on both sides of the vertical line by A^{-1}, i.e.,

$$AA^{-1} \mid IA^{-1}$$

This product would give

$$I \mid A^{-1} \tag{3.11}$$

Instead of using the inverse to obtain the identity matrix on the left side of the vertical line, we use row operations. The row operations applied to obtain the identity matrix concurrently yields the inverse matrix on the right side of the vertical line.

The Gaussian elimination method for obtaining the inverse matrix is illustrated by the following examples.

Example: Determine the inverse of $A = \begin{pmatrix} 5 & 3 \\ 6 & -2 \end{pmatrix}$.

To determine the inverse, we augment the matrix A with the identity matrix I and apply row operations to obtain an identity matrix on the left side of the vertical line. The augmented matrix is

$$A \mid I = \begin{pmatrix} 5 & 3 & 1 & 0 \\ 6 & -2 & 0 & 1 \end{pmatrix}$$

Multiply the first row by $\frac{1}{5}$.

$$\begin{pmatrix} 1 & \frac{3}{5} & \frac{1}{5} & 0 \\ 6 & -2 & 0 & 1 \end{pmatrix}$$

Add -6 times the first row to the second row.

$$\begin{pmatrix} 1 & \frac{3}{5} & \frac{1}{5} & 0 \\ 0 & -\frac{28}{5} & -\frac{6}{5} & 1 \end{pmatrix}$$

Multiply the second row by $-\frac{5}{28}$.

$$\begin{pmatrix} 1 & \frac{3}{5} & \frac{1}{5} & 0 \\ 0 & 1 & \frac{6}{28} & -\frac{5}{28} \end{pmatrix}$$

Add $-\frac{3}{5}$ times the second row to the first row.

$$\begin{pmatrix} 1 & 0 & \frac{2}{28} & \frac{3}{28} \\ 0 & 1 & \frac{6}{28} & -\frac{5}{28} \end{pmatrix}$$

This results in an identity matrix on the left side of the vertical line. From Formula (3.11) we know that the inverse of A is given on the right side of the vertical line. The inverse is

$$A^{-1} = \begin{pmatrix} \frac{2}{28} & \frac{3}{28} \\ \frac{6}{28} & -\frac{5}{28} \end{pmatrix}$$

Example: Determine the inverse of $A = \begin{pmatrix} 3 & 2 \\ 6 & -2 \end{pmatrix}$.

Augment the matrix A with the identity matrix I.

$$\begin{pmatrix} 3 & 2 & 1 & 0 \\ 6 & -2 & 0 & 1 \end{pmatrix}$$

Add 1 times the second row to the first row.

$$\begin{pmatrix} 9 & 0 & 1 & 1 \\ 6 & -2 & 0 & 1 \end{pmatrix}$$

Multiply the first row by $\frac{1}{9}$.

$$\begin{pmatrix} 1 & 0 & \frac{1}{9} & \frac{1}{9} \\ 6 & -2 & 0 & 1 \end{pmatrix}$$

Add -6 times the first row to the second row.

$$\begin{pmatrix} 1 & 0 & \frac{1}{9} & \frac{1}{9} \\ 0 & -2 & -\frac{2}{3} & \frac{1}{3} \end{pmatrix}$$

Multiply the second row by $-\frac{1}{2}$.

$$\begin{pmatrix} 1 & 0 & \frac{1}{9} & \frac{1}{9} \\ 0 & 1 & \frac{1}{3} & -\frac{1}{6} \end{pmatrix}$$

The inverse of A is given on the right side of the vertical line and is

$$A^{-1} = \begin{pmatrix} \frac{1}{9} & \frac{1}{9} \\ \frac{1}{3} & -\frac{1}{6} \end{pmatrix}$$

Example: Determine the inverse of $A = \begin{pmatrix} 2 & 3 \\ -4 & 6 \end{pmatrix}$.

Augment the matrix A with the identity matrix I.

$$\begin{pmatrix} 2 & 3 & 1 & 0 \\ -4 & 6 & 0 & 1 \end{pmatrix}$$

Add 2 times the first row to the second row.

$$\begin{pmatrix} 2 & 3 & 1 & 0 \\ 0 & 12 & 2 & 1 \end{pmatrix}$$

Multiply the first row by $\frac{1}{2}$ and the second row by $\frac{1}{12}$.

$$\begin{pmatrix} 1 & \frac{3}{2} & \frac{1}{2} & 0 \\ 0 & 1 & \frac{1}{6} & \frac{1}{12} \end{pmatrix}$$

Add $-\frac{3}{2}$ times the second row to the first row.

$$\begin{pmatrix} 1 & 0 & \frac{1}{4} & -\frac{3}{24} \\ 0 & 1 & \frac{1}{6} & \frac{1}{12} \end{pmatrix}$$

The inverse of A is

$$A^{-1} = \begin{pmatrix} \frac{1}{4} & -\frac{3}{24} \\ \frac{1}{6} & \frac{1}{12} \end{pmatrix}$$

Example: Determine the inverse of $A = \begin{pmatrix} 2 & 4 & -6 \\ 4 & 2 & 2 \\ 3 & -3 & 1 \end{pmatrix}$.

Augment the matrix A with I.

$$A \mid I = \begin{pmatrix} 2 & 4 & -6 & 1 & 0 & 0 \\ 4 & 2 & 2 & 0 & 1 & 0 \\ 3 & -3 & 1 & 0 & 0 & 1 \end{pmatrix}$$

Multiply the first row by $\frac{1}{2}$.

$$\begin{pmatrix} 1 & 2 & -3 & \frac{1}{2} & 0 & 0 \\ 4 & 2 & 2 & 0 & 1 & 0 \\ 3 & -3 & 1 & 0 & 0 & 1 \end{pmatrix}$$

Add -4 times the first row to the second row and add -3 times the first row to the third row.

$$\begin{pmatrix} 1 & 2 & -3 & \frac{1}{2} & 0 & 0 \\ 0 & -6 & 14 & -2 & 1 & 0 \\ 0 & -9 & 10 & -\frac{3}{2} & 0 & 1 \end{pmatrix}$$

Multiply the second row by $-\frac{1}{6}$.

$$
\begin{pmatrix}
1 & 2 & -3 & \frac{1}{2} & 0 & 0 \\
0 & 1 & -\frac{7}{3} & \frac{1}{3} & -\frac{1}{6} & 0 \\
0 & -9 & 10 & -\frac{3}{2} & 0 & 1
\end{pmatrix}
$$

Add -2 times the second row to the first row and add 9 times the second row to the third row. Multiply the new third row by $-\frac{1}{11}$.

$$
\begin{pmatrix}
1 & 0 & \frac{5}{3} & -\frac{1}{6} & \frac{1}{3} & 0 \\
0 & 1 & -\frac{7}{3} & \frac{1}{3} & -\frac{1}{6} & 0 \\
0 & 0 & 1 & -\frac{3}{22} & \frac{3}{22} & -\frac{1}{11}
\end{pmatrix}
$$

Add $\frac{7}{3}$ times the third row to the second row and add $-\frac{5}{3}$ times the third row to the first row.

$$
\begin{pmatrix}
1 & 0 & 0 & \frac{4}{66} & \frac{7}{66} & \frac{10}{66} \\
0 & 1 & 0 & \frac{1}{66} & \frac{10}{66} & -\frac{14}{66} \\
0 & 0 & 1 & -\frac{9}{66} & \frac{9}{66} & -\frac{6}{66}
\end{pmatrix}
$$

The inverse of A is

$$
A^{-1} = \begin{pmatrix}
\frac{4}{66} & \frac{7}{66} & \frac{10}{66} \\
\frac{1}{66} & \frac{10}{66} & -\frac{14}{66} \\
-\frac{9}{66} & \frac{9}{66} & -\frac{6}{66}
\end{pmatrix}
$$

3.6.2 USING THE INVERSE TO SOLVE MATRIX EQUATIONS

The solution to a matrix equation is relatively straightforward once the inverse has been determined. Given the matrix equation $AX = B$, the solution vector is $X = A^{-1}B$. This result comes from the fact that both sides of the matrix equation can be multiplied by the inverse. That is,

$$A^{-1}AX = A^{-1}B$$

$$IX = A^{-1}B$$

and

$$X = A^{-1}B$$

To illustrate, consider again the example of two simultaneous equations presented on p. 86. The two equations are written in matrix form as

$$
\begin{pmatrix} 3 & 2 \\ 6 & -2 \end{pmatrix} \begin{pmatrix} x_1 \\ x_2 \end{pmatrix} = \begin{pmatrix} 14 \\ 4 \end{pmatrix}
$$

The inverse of the coefficient matrix was calculated on p. 95 and is

$$
A^{-1} = \begin{pmatrix} \frac{1}{9} & \frac{1}{9} \\ \frac{1}{3} & -\frac{1}{6} \end{pmatrix}
$$

Multiplying A^{-1} by B gives

$$X = A^{-1}B = \begin{pmatrix} \frac{1}{9} & \frac{1}{9} \\ \frac{1}{3} & -\frac{1}{6} \end{pmatrix}\begin{pmatrix} 14 \\ 4 \end{pmatrix} = \begin{pmatrix} 2 \\ 4 \end{pmatrix}$$

The solution vector is $x_1 = 2$ and $x_2 = 4$. This is, of course, the solution that was obtained by the Gaussian elimination method.†

Example: Determine the solution vector X for the matrix equation

$$\begin{pmatrix} 5 & 3 \\ 6 & -2 \end{pmatrix}\begin{pmatrix} x_1 \\ x_2 \end{pmatrix} = \begin{pmatrix} 6 \\ 10 \end{pmatrix}$$

The inverse of the coefficient matrix was found on p. 94 to be

$$A^{-1} = \begin{pmatrix} \frac{2}{28} & \frac{3}{28} \\ \frac{6}{28} & -\frac{5}{28} \end{pmatrix}.$$

The solution vector is

$$X = A^{-1}B = \begin{pmatrix} \frac{2}{28} & \frac{3}{28} \\ \frac{6}{28} & -\frac{5}{28} \end{pmatrix}\begin{pmatrix} 6 \\ 10 \end{pmatrix} = \begin{pmatrix} \frac{3}{2} \\ -\frac{1}{2} \end{pmatrix}$$

Example: Determine the solution vector X for the matrix equation

$$\begin{pmatrix} 2 & 4 & -6 \\ 4 & 2 & 2 \\ 3 & -3 & 1 \end{pmatrix}\begin{pmatrix} x_1 \\ x_2 \\ x_3 \end{pmatrix} = \begin{pmatrix} 6 \\ 10 \\ 4 \end{pmatrix}$$

The inverse matrix was given on p. 96 as

$$A^{-1} = \begin{pmatrix} \frac{4}{66} & \frac{7}{66} & \frac{10}{66} \\ \frac{1}{66} & \frac{10}{66} & -\frac{14}{66} \\ -\frac{9}{66} & \frac{9}{66} & -\frac{6}{66} \end{pmatrix}$$

The solution vector is $X = A^{-1}B$.

$$X = \begin{pmatrix} \frac{4}{66} & \frac{7}{66} & \frac{10}{66} \\ \frac{1}{66} & \frac{10}{66} & -\frac{14}{66} \\ -\frac{9}{66} & \frac{9}{66} & -\frac{6}{66} \end{pmatrix}\begin{pmatrix} 6 \\ 10 \\ 4 \end{pmatrix} = \begin{pmatrix} \frac{67}{33} \\ \frac{25}{33} \\ \frac{6}{33} \end{pmatrix}$$

3.7   Application to Input-Output Analysis‡

One of the interesting applications of matrix algebra is in the Leontief input-output model. To illustrate the input-output model, consider the case of the Amalgamated Steel Company.

---

† The reader will note that the Gaussian elimination method can be used to determine the inverse or can be applied directly to $A|B$ to obtain the solution vector.

‡ This section can be omitted without loss of continuity.

Amalgamated Steel is one of the major producers of steel in the country. Their customers include the large automobile and truck manufacturers, the new construction industry, the metal products fabricating industry, the heating and plumbing products industry, and a multitude of other steel buyers. With minor exceptions, sales of Amalgamated Steel are made to industrial users rather than to the consuming public. Thus, the demand for Amalgamated's products is dependent upon the demand for the finished products of Amalgamated's customers.

The demand structure facing Amalgamated Steel is not at all uncommon in industry. Many firms sell to both industrial buyers and the general public. In the case of Amalgamated Steel, the majority of sales are made to industrial users of steel. The steel is used in finished products such as automobiles, office buildings, etc. The demand for Amalgamated's product is thus dependent upon the demand by the consuming public for the products of Amalgamated's customers. This type of demand is termed *derived* demand.

Products purchased directly by the consuming public represent an *autonomous* demand. In the case of an electronics manufacturer, for example, the demand for television sets is autonomous, whereas the demand for picture tubes used in the manufacture of the television sets is derived. Similarly, in the case of Amalgamated Steel, the demand for steel reinforcing bars for use in the construction of a back yard patio by the home owner is autonomous, whereas the demand for steel for use in the manufacture of an automobile is derived.

Since the majority of sales of Amalgamated Steel come from derived demand, the president of Amalgamated is interested in developing a model that describes the effect on Amalgamated of changes in the autonomous demand for Amalgamated's customer's products. Specifically, the president would like answers to questions such as "How would a 10 percent increase in new car sales affect the sales of Amalgamated Steel?" and "How would a 5 percent decrease in new car sales coupled with a 10 percent increase in new construction affect the sales of Amalgamated Steel?"

Questions such as those raised by the president of Amalgamated Steel can be answered by constructing a Leontief input-output model. The model was originally constructed for an entire economy, rather than for a segment of the economy. In applying the model to the economy, it is assumed that the economy contains a number of interacting industries, each producing products and each purchasing as intermediate products the products of other industries. To illustrate, home builders purchase lumber from the wood products industry and cement from the portland cement industry, while the wood products industry purchases steel from the steel industry and trucks from the motor vehicle industry. An industry, such as steel, can purchase some of its own product as well as purchase from an industry to which it sells (e.g., purchase trucks from the motor vehicle industry). In addition to selling its own product as an intermediate product to other industries, each industry also experiences an autonomous demand for its product from consumers, the government, or foreign governments.

The task of developing an interindustry model for the entire economy requires rather restrictive assumptions and is quite complex. Fortunately, the president of Amalgamated Steel recognized that a model that included the four major purchasers of steel would provide useful insights into the questions he had originally raised.

The first step in developing the model is the construction of an input-output matrix. Each element a_{ij} in the matrix represents the value of product i used in the manufacture of one dollar's worth of product j. The jth column thus gives the various amounts of the intermediate products used in the manufacture of one dollar's worth of product j. Conversely, the ith row describes the distribution of product i among the industries (or products) included in the matrix.

The input-output matrix for Amalgamated Steel can be used to illustrate these concepts. Amalgamated's input-output matrix was developed through cooperative studies with Amalgamated's customers and is given below.

Output (final product)

Input (intermediate product)	Steel	Motor Vehicles	Construction	Metal Products	Iron Ore
Steel	0.20	0.25	0.15	0.30	0.10
Motor vehicles	0.05	0.10	0.05	0	0.20
Construction	0	0.05	0.02	0.05	0.05
Metal products	0.10	0.05	0.05	0.05	0.10
Iron ore	0.25	0	0	0	0

The column for steel shows that for every dollar's worth of steel manufactured, $0.20 of steel, $0.05 of motor vehicles, $0.10 of metal products, and $0.25 of iron ore are used. Similarly, the column for motor vehicles shows that one dollar's worth of motor vehicles requires $0.25 of steel, $0.10 of motor vehicles, $0.05 of construction, and $0.05 of metal products. The columns do not sum to one dollar for two reasons. First, the matrix does not include all industries in the economy. Thus, products such as plastics that are used in motor vehicles are not included in the matrix. Second, the matrix does not include the value added by the manufacturer, i.e., the expenses incurred by the manufacturer in combining the factors of production to make a finished product.

The rows of the matrix show the distribution of the products of each of the industries. The row for steel, for instance, shows that steel is used as an input to the steel, motor vehicles, construction, metal products, and iron ore industries. The row for motor vehicles similarly shows that motor vehicles are used in the steel, motor vehicles, construction, metal products, and iron ore industries. Since the components of each row represent the value of product used in the manufacture of one dollar's worth of product shown by the column, the sum of the components of each row has no economic meaning.

The demand for each of the five products is equal to the derived demand plus the autonomous demand. Suppose that we represent the demand for each of the five products in terms of dollar sales by x_j, for $j = 1, 2, \ldots, 5$. The demand for steel is thus $x_1 = 0.20x_1 + 0.25x_2 + 0.15x_3 + 0.30x_4 + 0.10x_5 + d_1$. The autonomous demand is represented by d_1 and the derived demand by the remainder of the terms in the summation. The first term in the summation, $0.20x_1$, gives the dollar value of steel used in the manufacture of steel. The second term shows the dollar value of steel used in the manufacture of motor vehicles, etc. Similarly, the demand resulting from using steel in construction, metal products, and the mining of iron ore is given by the third, fourth, and fifth terms. The sum of the derived demands plus the autonomous demand d_1 gives the total demand for steel x_1.

The objective in the input-output model is to describe the interindustry relationships and to use these relationships to predict changes in demand caused by changes in the autonomous demand for the final products. The interindustry relationships for Amalgamated Steel are described by the following system of demand equations.

$$x_1 = 0.20x_1 + 0.25x_2 + 0.15x_3 + 0.30x_4 + 0.10x_5 + d_1$$
$$x_2 = 0.05x_1 + 0.10x_2 + 0.05x_3 + 0x_4 + 0.20x_5 + d_2$$
$$x_3 = 0x_1 + 0.05x_2 + 0.02x_3 + 0.05x_4 + 0.05x_5 + d_3$$
$$x_4 = 0.10x_1 + 0.05x_2 + 0.05x_3 + 0.05x_4 + 0.10x_5 + d_4$$
$$x_5 = 0.25x_1 + 0x_2 + 0x_3 + 0x_4 + 0x_5 + d_5$$

The system of equations can be rewritten with the variables x_j on the left of the equal sign and the autonomous demands d_j on the right of the equal sign. This gives

$$(1 - 0.20)x_1 - 0.25x_2 - 0.15x_3 - 0.30x_4 - 0.10x_5 = d_1$$
$$-0.05x_1 + (1 - 0.10)x_2 - 0.05x_3 - 0x_4 - 0.20x_5 = d_2$$
$$-0x_1 - 0.05x_2 + (1 - 0.02)x_3 - 0.05x_4 - 0.05x_5 = d_3$$
$$-0.10x_1 - 0.05x_2 - 0.05x_3 + (1 - 0.05)x_4 - 0.10x_5 = d_4$$
$$-0.25x_1 - 0x_2 - 0x_3 - 0x_4 + (1 - 0)x_5 = d_5$$

The system is alternatively represented in matrix form by

$$\begin{pmatrix} (1-0.20) & -0.25 & -0.15 & -0.30 & -0.10 \\ -0.05 & (1-0.10) & -0.05 & -0 & -0.20 \\ -0 & -0.05 & (1-0.02) & -0.05 & -0.05 \\ -0.10 & -0.05 & -0.05 & (1-0.05) & -0.10 \\ -0.25 & -0 & -0 & -0 & (1-0) \end{pmatrix} \begin{pmatrix} x_1 \\ x_2 \\ x_3 \\ x_4 \\ x_5 \end{pmatrix} = \begin{pmatrix} d_1 \\ d_2 \\ d_3 \\ d_4 \\ d_5 \end{pmatrix}$$

This system of equations describes the interindustry relationships between the five industries included in the input-output matrix. Furthermore, the demands for the products of each of the five industries can be determined for any alternative vector of autonomous demands. This means, of course, that for any vector of autonomous demands the system of equations can be solved for the variables x_j.

The solution procedure for obtaining the values of x_j involves calculating the inverse of the coefficient matrix for the system of interindustry equations. If A is the input-output matrix, I is an identity matrix, X is the demand vector, and D is the vector of autonomous demands, the system of equations is

$$(I - A)X = D$$

Providing that the inverse of $(I - A)$ exists, the demand vector X is given by

$$X = (I - A)^{-1}D$$

To illustrate the input-output model, assume that the autonomous demand for steel, motor vehicles, new construction, metal products, and iron ore is $2 billion, $20 billion, $15 billion, $10 billion, and $1 billion. The matrix $(1 - A)^{-1}$ is calculated by using the procedure explained in Sec. 3.5 and is

$$(I - A)^{-1} = \begin{pmatrix} 1.43 & 0.44 & 0.26 & 0.46 & 0.30 \\ 0.16 & 1.16 & 0.09 & 0.06 & 0.26 \\ 0.04 & 0.07 & 1.03 & 0.07 & 0.08 \\ 0.20 & 0.12 & 0.09 & 1.12 & 0.16 \\ 0.36 & 0.11 & 0.07 & 0.12 & 1.07 \end{pmatrix}$$

The solution vector is

$$X = (I - A)^{-1}D$$

$$X = \begin{pmatrix} 1.43 & 0.44 & 0.26 & 0.46 & 0.30 \\ 0.16 & 1.16 & 0.09 & 0.06 & 0.26 \\ 0.04 & 0.07 & 1.03 & 0.07 & 0.08 \\ 0.20 & 0.12 & 0.09 & 1.12 & 0.16 \\ 0.36 & 0.11 & 0.07 & 0.12 & 1.07 \end{pmatrix} \begin{pmatrix} 2 \\ 20 \\ 15 \\ 10 \\ 1 \end{pmatrix} = \begin{pmatrix} 20.46 \\ 25.73 \\ 17.71 \\ 15.51 \\ 6.24 \end{pmatrix}$$

The demand for steel, motor vehicles, new construction, metal products, and iron ore is $20.46 billion, $25.73 billion, $17.71 billion, $15.51 billion, and $6.24 billion.

After the input-output matrix A and the inverse of $(I - A)$ have been determined, it is quite simple to determine the change in demand caused by changes in the autonomous demand for the final products. For instance, suppose that the president of Amalgamated is interested in predicting the change in demand for steel caused by increases of $5 billion in motor vehicle sales, $3 billion in new construction, and a decrease of $2 billion in sales of metal products. The change in the demands for steel, represented by ΔX, caused by the change in autonomous demand, represented by ΔD, is

$$\Delta X = (I - A)^{-1}\Delta D$$

$$\Delta X = \begin{pmatrix} 1.43 & 0.44 & 0.26 & 0.46 & 0.30 \\ 0.16 & 1.16 & 0.09 & 0.06 & 0.26 \\ 0.04 & 0.07 & 1.03 & 0.07 & 0.08 \\ 0.20 & 0.12 & 0.09 & 1.12 & 0.16 \\ 0.36 & 0.11 & 0.07 & 0.12 & 1.07 \end{pmatrix} \begin{pmatrix} 0 \\ 5 \\ 3 \\ -2 \\ 0 \end{pmatrix} = \begin{pmatrix} 2.06 \\ 5.95 \\ 3.30 \\ -1.37 \\ 0.52 \end{pmatrix}$$

The total demand based upon these changes is $(X + \Delta X)$ or, alternatively, $(I - A)^{-1}(D + \Delta D)$. Thus,

$$X + \Delta X = \begin{pmatrix} 20.46 + 2.06 \\ 25.73 + 5.95 \\ 17.71 + 3.30 \\ 15.51 - 1.37 \\ 6.24 + 0.52 \end{pmatrix} = \begin{pmatrix} 22.52 \\ 31.68 \\ 21.01 \\ 14.14 \\ 6.76 \end{pmatrix}$$

Based upon these changes, the demand for steel, motor vehicles, new construction, metal products, and iron ore is, respectively, $22.52 billion, $31.68 billion, $21.01 billion, $14.14 billion, and $6.76 billion.

Example: Determine the total demand for the products of industries A, B, and C based upon the input-output matrix and autonomous demand vector shown below.

Output Industry

Input industry	A	B	C
A	0.5	0.1	0.1
B	0.2	0.6	0.2
C	0.1	0.2	0.6

$$D = \begin{pmatrix} 20 \\ 30 \\ 50 \end{pmatrix}$$

The matrix $(I - A)$ is

$$(I - A) = \begin{pmatrix} 0.5 & -0.1 & -0.1 \\ -0.2 & 0.4 & -0.2 \\ -0.1 & -0.2 & 0.4 \end{pmatrix}$$

and the inverse of $(I - A)$ is

$$(I - A)^{-1} = \frac{1}{21}\begin{pmatrix} 60 & 30 & 30 \\ 50 & 95 & 60 \\ 40 & 55 & 90 \end{pmatrix}$$

The demand vector is

$$X = (I - A)^{-1}D$$

$$X = \frac{1}{21}\begin{pmatrix} 60 & 30 & 30 \\ 50 & 95 & 60 \\ 40 & 55 & 90 \end{pmatrix}\begin{pmatrix} 20 \\ 30 \\ 50 \end{pmatrix}$$

$$X = \begin{pmatrix} 171.5 \\ 326.2 \\ 331.0 \end{pmatrix}$$

PROBLEMS

1. For the vectors $A = (10, 6, 12)$ and $B = (14, 8, 5)$, perform the operations indicated or state why the operation cannot be done.
   (a) $A - B$ (b) $B - A$
   (c) AB (d) $A + B$

2. Using the vectors $X = (6, 4, 8)$, $Y = (3, 6, -2)$, and $Z = (4, 10, 12)$, verify that the associative law of addition applies to the addition of conformable vectors.

3. For the vectors $A = \begin{pmatrix} 6 \\ 4 \\ 2 \end{pmatrix}$ and $B = \begin{pmatrix} 8 \\ -2 \\ 6 \end{pmatrix}$, determine the following.

   (a) $A + B$ (b) $2A - B$
   (c) $3A - 2B$ (d) $0.5A + 1.5B$

4. Find the inner product of $W = (5, 8, 3)$ and $V = \begin{pmatrix} 2 \\ -2 \\ 4 \end{pmatrix}$.

5. For the vectors $A = (2, 8, 6)$, $B = \begin{pmatrix} 5 \\ 4 \\ 6 \end{pmatrix}$, and $C = \begin{pmatrix} 8 \\ 10 \\ 3 \end{pmatrix}$, verify the distributive law by showing that $A(B + C)$ is equal to $AB + AC$.

6. A television dealer has an inventory of 12 black and white sets, 9 color console sets, and 8 color portable sets. The dealer's cost on the different types of sets is $80, $340, and $260, respectively. Form the inventory

and price vectors and determine the value of the inventory by finding the inner product of the two vectors.

7. For the matrices $A = \begin{pmatrix} 3 & 5 \\ 2 & 6 \\ 1 & 4 \end{pmatrix}$ and $B = \begin{pmatrix} 2 & -5 \\ 3 & 6 \\ -5 & 5 \end{pmatrix}$, determine the following.

(a) $A + B$ (b) $B - A$
(c) $3A - B$ (d) $A - 2B$

8. Find the product of $A = \begin{pmatrix} 6 & 4 & 8 \\ 3 & -3 & 5 \end{pmatrix}$ and $B = \begin{pmatrix} 8 & -3 \\ -2 & 5 \\ 6 & 4 \end{pmatrix}$.

9. Using the matrices $A = \begin{pmatrix} 5 & 2 \\ 3 & -8 \\ 2 & 6 \end{pmatrix}$, $B = \begin{pmatrix} 8 & 5 \\ 4 & 7 \end{pmatrix}$, $C = \begin{pmatrix} 4 & 7 \\ 2 & 5 \end{pmatrix}$, verify that $A(B + C) = AB + AC$.

10. For the matrices $A = \begin{pmatrix} 3 & 8 & 2 \\ 5 & -2 & 4 \\ 6 & 1 & -5 \end{pmatrix}$ and $B = \begin{pmatrix} 6 & 13 & 7 \\ 7 & 11 & 5 \\ 8 & -1 & 10 \end{pmatrix}$, find AB and BA.

11. Determine the outer product of the vectors W and V given in Problem 4.

12. The matrices A, B, and C have the following dimensions: A has 3 rows and 4 columns, B has 4 rows and 4 columns, C has 2 rows and 3 columns. Can the following operations be performed? If so, give the dimensions of the resulting matrix.

(a) AB (b) AC
(c) CA (d) $C(AB)$

13. Although the commutative law of multiplication does not in general apply to the multiplication of matrices, it does apply in the multiplication of a matrix by the identity matrix. Verify this statement, using the matrices

$$A = \begin{pmatrix} 6 & 3 & 8 \\ 4 & 2 & 5 \\ 3 & 1 & 7 \end{pmatrix} \quad \text{and} \quad I = \begin{pmatrix} 1 & 0 & 0 \\ 0 & 1 & 0 \\ 0 & 0 & 1 \end{pmatrix}$$

14. Express the following systems of equations in matrix form.

(a) $2x_1 + 6x_2 = 44$
$3x_1 + 4x_2 = 36$
(c) $3x_1 + 5x_2 + 4x_3 = 50$
$6x_1 + 4x_2 - 3x_3 = 26$
$2x_1 - 5x_2 + 6x_3 = 5$

(b) $14x_1 + 23x_2 = 176$
$9x_1 + 8x_2 = 86$
(d) $2x_1 - 5x_2 + 3x_3 = 14$
$7x_1 + 4x_2 - 6x_3 = 36$
$5x_1 - 6x_2 + 3x_3 = 34$

15. Determine the solutions to the systems of equations in Problem 14 by the use of the Gaussian elimination method.

16. Verify that the matrices $A = \begin{pmatrix} 3 & 2 \\ 1 & 5 \end{pmatrix}$ and $A^{-1} = \begin{pmatrix} \frac{5}{13} & -\frac{2}{13} \\ -\frac{1}{13} & \frac{3}{13} \end{pmatrix}$ are inverse by determining AA^{-1}.

17. Use the Gaussian method to determine the inverses of the following matrices.

(a) $A = \begin{pmatrix} 2 & 3 \\ 4 & 7 \end{pmatrix}$

(b) $A = \begin{pmatrix} 2 & 3 \\ -5 & -2 \end{pmatrix}$

(c) $A = \begin{pmatrix} 1 & 2 & 3 \\ 1 & 3 & 5 \\ 2 & 5 & 9 \end{pmatrix}$

(d) $A = \begin{pmatrix} 5 & -6 & 7 \\ -10 & 11 & -13 \\ 1 & -1 & 1 \end{pmatrix}$

18. Use the inverses found in Problem 17 to solve the following systems of equations.

(a) $2x_1 + 3x_2 = 10$
    $4x_1 + 7x_2 = 6$

(b) $\quad 2x_1 + 3x_2 = 6$
    $-5x_1 - 2x_2 = 7$

(c) $\quad x_1 + 2x_2 + 3x_3 = 8$
    $\quad x_1 + 3x_2 + 5x_3 = 5$
    $2x_1 + 5x_2 + 9x_3 = 10$

(d) $\quad 5x_1 - 6x_2 + 7x_3 = \quad 5$
    $-10x_1 + 11x_2 - 13x_3 = -10$
    $\quad x_1 - \quad x_2 + \quad x_3 = -12$

19. Determine the total demand for the products of industries A, B, C, and D based on the input-output matrix and the autonomous demand vector shown below.

Output Industry

Input industry	A	B	C	D			
A	0.4	0.2	0	0.1			30
B	0.1	0.3	0.2	0.1	$D =$		40
C	0.2	0.3	0.1	0.1			60
D	0.1	0.3	0.2	0.2			40

20. Determine the change in total demand for each of the industries in Problem 19 caused by the following change in autonomous demand: industry A, an increase of 3; industry B, a decrease of 5; industry C, an increase of 8; industry D, no change.

SUGGESTED REFERENCES

AYRES, FRANK JR., *Theory and Problems of Matrices*, Schaum's Outline Series (New York, N.Y.: McGraw-Hill Book Company, Inc., 1962).

DORFMAN, R., et al., *Linear Programming and Economic Analysis* (New York, N.Y.: McGraw-Hill Book Company, Inc., 1958).

FULLER, LEONARD E., *Basic Matrix Theory* (Englewood Cliffs, N.J.: Prentice-Hall, Inc., 1962).

GRAYBILL, FRANKLIN A., *Introduction to Matrices with Applications in Statistics* (Belmont, Ca.: Wadsworth Publishing Company, Inc., 1969).

HADLEY, G., *Linear Algebra* (Reading, Mass.: Addison-Wesley Publishing Company, Inc., 1961).

KEMENY, JOHN G., et al., *Finite Mathematical Structures* (Englewood Cliffs, N.J.: Prentice-Hall, Inc., 1958), 4.

LIPSCHUTZ, SEYMOUR, *Linear Algebra*, Schaum's Outline Series (New York, N.Y.: McGraw-Hill Book Company, Inc., 1968).

STEIN, F. MAX, *An Introduction to Matrices and Determinants* (Belmont, Ca.: Wadsworth Publishing Company, Inc., 1967).

YAMANE, TARO, *Mathematics for Economists*, 2nd ed. (Englewood Cliffs, N.J.: Prentice-Hall, Inc., 1962), 10–12.

Chapter 4

Additional Elements
of
Matrix Algebra

The preceding chapter introduced the concepts of vectors and matrices. This chapter continues the discussion of matrices and matrix algebra by introducing the transpose matrix and matrix partitioning. This is followed by a discussion of the determinant of a matrix, the use of the determinant in establishing inverse matrices, Cramer's rule, and the significance of the rank of a matrix to the solution set of systems of equations.

4.1 Transpose

The *transpose* of the matrix A is the matrix A^t (read A transpose) that is formed by writing the rows of matrix A as columns in matrix A^t. In forming the transpose matrix, the first row in A is the same as the first column in A^t, the second row in A is the same as the second column in A^t, and, in general, the ith row in A is the same as the ith column in A^t. To illustrate for the 2 by 3 matrix A, the 3 by 2 matrix A^t is formed as follows.

$$A = \begin{pmatrix} a_{11} & a_{12} & a_{13} \\ a_{21} & a_{22} & a_{23} \end{pmatrix}$$

Interchanging the rows and columns gives

108

$$A^t = \begin{pmatrix} a_{11} & a_{21} \\ a_{12} & a_{22} \\ a_{13} & a_{23} \end{pmatrix}$$

The matrix A with m rows and n columns is thus transposed to form the matrix A^t with n rows and m columns. The transposition of a matrix is illustrated by the following examples.

Example: Form the transpose of $A = \begin{pmatrix} 2 & 1 & 3 \\ 4 & 3 & 6 \\ 1 & 5 & 4 \end{pmatrix}$. Interchanging the rows

and columns gives

$$A^t = \begin{pmatrix} 2 & 4 & 1 \\ 1 & 3 & 5 \\ 3 & 6 & 4 \end{pmatrix}$$

Example: Form the transpose of $A = \begin{pmatrix} 1 & 4 & 3 & 7 \\ 2 & 5 & 1 & 3 \end{pmatrix}$. Interchanging the

rows and columns gives

$$A^t = \begin{pmatrix} 1 & 2 \\ 4 & 5 \\ 3 & 1 \\ 7 & 3 \end{pmatrix}$$

Example: Form the transpose of the row vector $A = (6, 3, 1, 7)$. The transpose of the 1 by 4 row vector is the 4 by 1 column vector

$$A^t = \begin{pmatrix} 6 \\ 3 \\ 1 \\ 7 \end{pmatrix}$$

4.2 Submatrices and Partitioning

In applications of matrix algebra discussed later in this chapter, it is necessary to delete selected rows or columns from a matrix. The components that remain after the rows or columns are deleted form a *submatrix*. If, for instance, the second row and third column are deleted from the matrix

$$A = \begin{pmatrix} 1 & 3 & 4 & 6 \\ 2 & 1 & 3 & 7 \\ 0 & 4 & 7 & 1 \end{pmatrix}$$

the submatrix $B = \begin{pmatrix} 1 & 3 & 6 \\ 0 & 4 & 1 \end{pmatrix}$ remains.

Example: Determine the submatrix B formed by the deletion of the third row and first column of the matrix A.

$$A = \begin{pmatrix} 1 & 3 & 6 \\ 2 & 4 & 7 \\ 5 & 5 & 8 \\ 4 & 6 & 9 \end{pmatrix} \qquad B = \begin{pmatrix} 3 & 6 \\ 4 & 7 \\ 6 & 9 \end{pmatrix}$$

Example: Determine the submatrices B_1, B_2, and B_3 formed by the deletion of the second and third columns, the first and third columns, and the first and second columns, respectively, of the matrix

$$A = \begin{pmatrix} 1 & 2 & 4 \\ 2 & 3 & 5 \\ 3 & 4 & 6 \end{pmatrix}$$

The submatrices are

$$B_1 = \begin{pmatrix} 1 \\ 2 \\ 3 \end{pmatrix}, \qquad B_2 = \begin{pmatrix} 2 \\ 3 \\ 4 \end{pmatrix}, \quad \text{and} \quad B_3 = \begin{pmatrix} 4 \\ 5 \\ 6 \end{pmatrix}$$

In certain applications it is useful to consider matrices in which the components of the matrix are submatrices. For instance, the matrix $A = \begin{pmatrix} B_1 & B_2 \\ B_3 & B_4 \end{pmatrix}$ with components $B_1 = \begin{pmatrix} 1 & 2 \\ 3 & 4 \end{pmatrix}$, $B_2 = \begin{pmatrix} 6 \\ 7 \end{pmatrix}$, $B_3 = \begin{pmatrix} 4 & 3 \\ 2 & 1 \end{pmatrix}$, and $B_4 = \begin{pmatrix} 4 \\ 5 \end{pmatrix}$ is a matrix whose components are submatrices. The matrix A can be written as

$$A = \begin{pmatrix} 1 & 2 & 6 \\ 3 & 4 & 7 \\ 4 & 3 & 4 \\ 2 & 1 & 5 \end{pmatrix}$$

Matrices such as A whose components are submatrices are said to have been *partitioned*.

One interesting use of partitioning of matrices is in the multiplication of large matrices. Given two matrices A and B that are partitioned into $A = \begin{pmatrix} A_1 & A_2 \\ A_3 & A_4 \end{pmatrix}$ and $B = \begin{pmatrix} B_1 & B_2 \\ B_3 & B_4 \end{pmatrix}$, the product AB (where the matrices are conformable for multiplication) is

$$AB = \begin{pmatrix} A_1 & A_2 \\ A_3 & A_4 \end{pmatrix} \begin{pmatrix} B_1 & B_2 \\ B_3 & B_4 \end{pmatrix} = \begin{pmatrix} A_1B_1 + A_2B_3 & A_1B_2 + A_2B_4 \\ A_3B_1 + A_4B_3 & A_3B_2 + A_4B_4 \end{pmatrix}$$

The only requirement for the multiplication is that all matrices must be conformable for multiplication. Matrices A and B must be conformable and the matrices must be partitioned such that the submatrices are conformable.

Example: Determine the product of A and B, where

$$A = \begin{pmatrix} 2 & 1 & 0 & 4 \\ -1 & 3 & 2 & 1 \\ -3 & 0 & 2 & -4 \\ 0 & -2 & 3 & 1 \end{pmatrix} \quad \text{and} \quad B = \begin{pmatrix} 3 & 0 \\ -2 & 2 \\ 1 & 3 \\ 3 & 4 \end{pmatrix}$$

The matrices are partitioned such that the submatrices are conformable.

$$AB = \begin{pmatrix} A_1 & A_2 \\ A_3 & A_4 \end{pmatrix} \begin{pmatrix} B_1 \\ B_2 \end{pmatrix} = \begin{pmatrix} A_1B_1 + A_2B_2 \\ A_3B_1 + A_4B_2 \end{pmatrix}$$

The components can be calculated by matrix multiplication and addition to give

$$A_1B_1 + A_2B_2 = \begin{pmatrix} 4 & 2 \\ -9 & 6 \end{pmatrix} + \begin{pmatrix} 12 & 16 \\ 5 & 10 \end{pmatrix} = \begin{pmatrix} 16 & 18 \\ -4 & 16 \end{pmatrix}$$

$$A_3B_1 + A_4B_2 = \begin{pmatrix} -9 & 0 \\ 4 & -4 \end{pmatrix} + \begin{pmatrix} -10 & -10 \\ 6 & 13 \end{pmatrix} = \begin{pmatrix} -19 & -10 \\ 10 & 9 \end{pmatrix}$$

The matrix resulting from the product of A and B is

$$AB = \begin{pmatrix} 16 & 18 \\ -4 & 16 \\ -19 & -10 \\ 10 & 9 \end{pmatrix}$$

The student should verify that this result is the same as obtained by multiplication without partitioning.

4.3 Determinant

One of the important concepts in matrix algebra is that of the *determinant* of a matrix. The determinant of a matrix is a number that is associated with a square matrix. This number can be positive, negative, or zero. The number is important in that it is used in determining the solutions to systems of simultaneous equations.

The determinant is used in two alternative solution techniques for calculating the solution set to a system of simultaneous equations. One approach uses the determinant to calculate the inverse of a matrix. The inverse of the matrix is then used to determine the solution to the system of equations. The second approach uses the determinant in a solution technique termed *Cramer's rule*. This technique is discussed in Sec. 4.5.

4.3.1 DETERMINANTS OF 2 BY 2 AND
3 BY 3 MATRICES

The determinant of the matrix A is commonly represented by one of two symbols, det A or $|A|$. These symbols are used interchangeably in many texts. We will, however, use the symbol $|A|$ exclusively in this text to represent the determinant.

The determinants of 2 by 2 and 3 by 3 matrices are easily calculated. The determinant of the 2 by 2 matrix A is given by

$$|A| = \begin{vmatrix} a_{11} & a_{12} \\ a_{21} & a_{22} \end{vmatrix} = a_{11} \cdot a_{22} - a_{21} \cdot a_{12} \tag{4.1}$$

To illustrate the formula, assume that $A = \begin{pmatrix} 1 & 3 \\ -2 & 4 \end{pmatrix}$.

$$|A| = \begin{vmatrix} 1 & 3 \\ -2 & 4 \end{vmatrix} = 1 \cdot 4 - (-2) \cdot 3 = 10$$

The determinant of A is the scalar 10.

Example: Find the determinant of $C = \begin{pmatrix} -2 & 3 \\ 6 & 4 \end{pmatrix}$.

$$|C| = \begin{vmatrix} -2 & 3 \\ 6 & 4 \end{vmatrix} = -8 - 18 = -26$$

Example: Find the determinant of $D = \begin{pmatrix} 1 & 3 \\ 1 & 3 \end{pmatrix}$.

$$|D| = \begin{vmatrix} 1 & 3 \\ 1 & 3 \end{vmatrix} = 3 - 3 = 0$$

The determinant of a 3 by 3 matrix is calculated by a similar formula. The determinant of the 3 by 3 matrix A is given by

$$|A| = \begin{vmatrix} a_{11} & a_{12} & a_{13} \\ a_{21} & a_{22} & a_{23} \\ a_{31} & a_{32} & a_{33} \end{vmatrix} = \begin{aligned} & a_{11}a_{22}a_{33} + a_{12}a_{23}a_{31} + a_{13}a_{21}a_{32} \\ & - a_{31}a_{22}a_{13} - a_{32}a_{23}a_{11} - a_{33}a_{21}a_{12} \end{aligned} \tag{4.2}$$

Some students find the formula easy to remember by employing the following scheme. First, repeat the first two columns of the matrix. Next, draw diagonals through the components as shown below. The products of the components joined by

each diagonal are then determined, e.g., $a_{11} \cdot a_{22} \cdot a_{33}$. The products of the components on the diagonals from upper left to lower right are summed, and the products of the components on the diagonals from lower left to upper right are subtracted from this sum. This results in Eq. (4.2) given above, i.e.,

$$|A| = a_{11}a_{22}a_{33} + a_{12}a_{23}a_{31} + a_{13}a_{21}a_{32} - a_{31}a_{22}a_{13} - a_{32}a_{23}a_{11} - a_{33}a_{21}a_{12}$$

Example: Find the determinant of $A = \begin{pmatrix} 1 & 3 & -4 \\ 2 & -1 & 6 \\ 3 & 0 & -2 \end{pmatrix}$. Use the method of repeating the first two columns.

$$|A| = \begin{matrix} 1 & 3 & -4 & 1 & 3 \\ 2 & -1 & 6 & 2 & -1 \\ 3 & 0 & -2 & 3 & 0 \end{matrix}$$

$$|A| = 1(-1)(-2) + 3\cdot 6 \cdot 3 + (-4)2\cdot 0 - 3(-1)(-4) - (0)6\cdot 1 - (-2)2\cdot 3$$
$$|A| = 2 + 54 + 0 - 12 - 0 + 12 = 56$$

Example: Find the determinant of $A = \begin{pmatrix} 1 & 5 & 4 \\ -3 & 2 & 3 \\ 6 & 1 & 4 \end{pmatrix}$.

$$|A| = 1\cdot 2\cdot 4 + 5\cdot 3\cdot 6 + 4(-3)(-1) - 6\cdot 2\cdot 4 - (-1)3\cdot 1 - 4(-3)5$$
$$|A| = 8 + 90 + 12 - 48 + 3 + 60 = 125$$

Example: Find the determinant of $A = \begin{pmatrix} 2 & -1 & 3 \\ 3 & -2 & 6 \\ 2 & 1 & 4 \end{pmatrix}$.

$$|A| = 2(-2)4 + (-1)6\cdot 2 + 3\cdot 3\cdot 1 - 2(-2)3 - 1\cdot 6\cdot 2 - 4\cdot 3(-1)$$
$$|A| = -16 - 12 + 9 + 12 - 12 + 12 = -7$$

4.3.2 MINOR, COFACTOR, AND COFACTOR EXPANSION

The method discussed in Sec. 4.3.1 applies for calculating the determinant of a 2 by 2 or 3 by 3 matrix. It does not, however, apply to matrices of higher dimensions. This section introduces a procedure for calculating a determinant that applies to any square matrix. This procedure is termed the method of *cofactor expansion*.

Before discussing the method of cofactor expansion, we must define two terms. There are *minor* and *cofactor*. Both of these terms are defined with reference to an element a_{ij} of a square matrix A. We thus speak of the minor of the a_{ij} element or the cofactor of the a_{ij} element.

With reference again to the a_{ij} element, the minor of the a_{ij} element is defined as the determinant of the submatrix formed by deleting the ith row and the jth column of the matrix. The minor of an element is thus a determinant. It is, again, the determinant of the submatrix resulting from deleting the ith row and jth column. For the 3 by 3 matrix A shown below, the minor of a_{11}

$$A = \begin{pmatrix} a_{11} & a_{12} & a_{13} \\ a_{21} & a_{22} & a_{23} \\ a_{31} & a_{32} & a_{33} \end{pmatrix}$$

is the determinant of the submatrix formed by deleting the first row and first column of A, i.e.,

$$\text{minor } a_{11} = \begin{vmatrix} a_{11} & a_{12} & a_{13} \\ a_{21} & a_{22} & a_{23} \\ a_{31} & a_{32} & a_{33} \end{vmatrix} = \begin{vmatrix} a_{22} & a_{23} \\ a_{32} & a_{33} \end{vmatrix}$$

Similarly, the minor of the element a_{21} is $\begin{vmatrix} a_{12} & a_{13} \\ a_{32} & a_{33} \end{vmatrix}$ and the minor of a_{22} is $\begin{vmatrix} a_{11} & a_{13} \\ a_{31} & a_{33} \end{vmatrix}$.

A method of determining the elements included in the submatrix is to line out the row and column to which the a_{ij} element belongs. The minor of a_{32} is thus

$$\text{minor } a_{32} = \begin{vmatrix} a_{11} & a_{12} & a_{13} \\ a_{21} & a_{22} & a_{23} \\ a_{31} & a_{32} & a_{33} \end{vmatrix} = \begin{vmatrix} a_{11} & a_{13} \\ a_{21} & a_{23} \end{vmatrix}$$

Example: Find the minors of the elements of the first row and of the third column of the matrix $A = \begin{pmatrix} 2 & 3 & 1 \\ -1 & 3 & 2 \\ 4 & 2 & 3 \end{pmatrix}$. Since the minor of the a_{ij} element is the determinant of the submatrix formed by eliminating the ith row and jth column, the minors of a_{11}, a_{12}, a_{13}, a_{23}, and a_{33} are

$$\text{minor } a_{11} = \begin{vmatrix} 3 & 2 \\ 2 & 3 \end{vmatrix} = 9 - 4 = 5$$

$$\text{minor } a_{12} = \begin{vmatrix} -1 & 2 \\ 4 & 3 \end{vmatrix} = -3 - 8 = -11$$

$$\text{minor } a_{13} = \begin{vmatrix} -1 & 3 \\ 4 & 2 \end{vmatrix} = -2 - 12 = -14$$

$$\text{minor } a_{23} = \begin{vmatrix} 2 & 3 \\ 4 & 2 \end{vmatrix} = 4 - 12 = -8$$

$$\text{minor } a_{33} = \begin{vmatrix} 2 & 3 \\ -1 & 3 \end{vmatrix} = 6 - (-3) = 9$$

Example: Find the minors of the second row of the matrix

$$A = \begin{pmatrix} 3 & 1 & 6 & 4 \\ 2 & 3 & -1 & 7 \\ 2 & 0 & 3 & 2 \\ 4 & 2 & -3 & 1 \end{pmatrix}$$

$$\text{minor } a_{21} = \begin{vmatrix} 1 & 6 & 4 \\ 0 & 3 & 2 \\ 2 & -3 & 1 \end{vmatrix} = 3 + 24 + 0 - 24 + 6 + 0 = 9$$

$$\text{minor } a_{22} = \begin{vmatrix} 3 & 6 & 4 \\ 2 & 3 & 2 \\ 4 & -3 & 1 \end{vmatrix} = 9 + 48 - 24 - 48 + 18 - 12 = -9$$

$$\text{minor } a_{23} = \begin{vmatrix} 3 & 1 & 4 \\ 2 & 0 & 2 \\ 4 & 2 & 1 \end{vmatrix} = 0 + 8 + 16 - 0 - 12 - 2 = 10$$

$$\text{minor } a_{24} = \begin{vmatrix} 3 & 1 & 6 \\ 2 & 0 & 3 \\ 4 & 2 & -3 \end{vmatrix} = 0 + 12 + 24 - 0 - 18 + 6 = 24$$

We stated earlier that both cofactor and minor were defined with reference to an element a_{ij} of a square matrix. The *cofactor* is the product of the minor of the a_{ij} element and $(-1)^{i+j}$. The cofactor is thus simply the minor multiplied by either $+1$ or -1, the sign depending upon the location of the a_{ij} element. For instance, the cofactor of the a_{12} element would be the product of the minor of this element and $(-1)^{1+2}$, or -1. The cofactor of the a_{22} element would be the product of the minor of the a_{22} element and $(-1)^{2+2}$, or $+1$. The cofactor of the a_{ij} element is designated by the symbol A_{ij}. The general formula is

$$A_{ij} = (-1)^{i+j} \text{ minor } a_{ij} \tag{4.3}$$

Example: Find the cofactors of the elements of the first row and of the third column of the matrix $A = \begin{pmatrix} 2 & 3 & 1 \\ -1 & 3 & 2 \\ 4 & 2 & 3 \end{pmatrix}$. The minors of this matrix were calculated on p 114. Since the cofactor A_{ij} is the product of $(-1)^{i+j}$ and the minor of a_{ij}, the cofactors are

$$A_{11} = (-1)^2 \cdot \text{minor } a_{11} = +1(5) = 5$$
$$A_{12} = (-1)^3 \cdot \text{minor } a_{12} = -1(-11) = 11$$
$$A_{13} = (-1)^4 \cdot \text{minor } a_{13} = +1(-14) = -14$$
$$A_{23} = (-1)^5 \cdot \text{minor } a_{23} = -1(-8) = 8$$
$$A_{33} = (-1)^6 \cdot \text{minor } a_{33} = +1(9) = 9$$

Example: Determine the cofactors of the elements of the second row of the matrix

$$A = \begin{pmatrix} 3 & 1 & 6 & 4 \\ 2 & 3 & -1 & 7 \\ 2 & 0 & 3 & 2 \\ 4 & 2 & -3 & 1 \end{pmatrix}$$

The minors of the elements of the second row were determined on p. 115. The cofactors of the elements of the second row are

$$A_{21} = (-1)^3 \cdot \text{minor } a_{21} = -1(9) = -9$$
$$A_{22} = (-1)^4 \cdot \text{minor } a_{22} = +1(-9) = -9$$
$$A_{23} = (-1)^5 \cdot \text{minor } a_{23} = -1(10) = -10$$
$$A_{24} = (-1)^6 \cdot \text{minor } a_{24} = +1(24) = 24$$

We can now introduce a method for calculating a determinant that applies to any square matrix. To calculate the determinant we multiply the cofactors for any row or column by the elements of that row or column. The sum of these products is the determinant of the matrix. To illustrate, the determinant of the 3 by 3 matrix

$$A = \begin{pmatrix} a_{11} & a_{12} & a_{13} \\ a_{21} & a_{22} & a_{23} \\ a_{31} & a_{32} & a_{33} \end{pmatrix}$$

is given by multiplying the elements in the first row by the cofactors of that row,

$$|A| = a_{11}A_{11} + a_{12}A_{12} + a_{13}A_{13}$$

or alternatively, by multiplying the elements in the third column by the cofactors of the column,

$$|A| = a_{13}A_{13} + a_{23}A_{23} + a_{33}A_{33}$$

or by the sum of the products of the cofactors and the elements of any other row or column. This method of calculating determinants is termed *cofactor expansion.*

Example: Calculate the determinant by cofactor expansion of the matrix $A = \begin{pmatrix} 2 & 3 & 1 \\ -1 & 3 & 2 \\ 4 & 2 & 3 \end{pmatrix}$. Verify that expansion about the first row or the third column yields the same value of the determinant.

The cofactors of the first row and of the third column of the matrix were calculated on p. 115. Cofactor expansion about the first row gives

$$|A| = 2A_{11} + 3A_{12} + 1A_{13}$$

$$|A| = 2(5) + 3(11) + 1(-14) = 29$$

Cofactor expansion about the third column gives

$$|A| = 1A_{13} + 2A_{23} + 3A_{33}$$

$$|A| = 1(-14) + 2(8) + 3(9) = 29$$

The reader can verify that the cofactor expansion about any other row or column gives the same value for the determinant.

Example: Find the determinant of the matrix

$$A = \begin{pmatrix} 3 & 1 & 6 & 4 \\ 2 & 3 & -1 & 7 \\ 2 & 0 & 3 & 2 \\ 4 & 2 & -3 & 1 \end{pmatrix}$$

The cofactors of the elements of the second row were calculated on p. 116. The determinant of A can be easily determined by cofactor expansion about the second row.

$$|A| = 2A_{21} + 3A_{22} - 1A_{23} + 7A_{24}$$

$$|A| = 2(-9) + 3(-9) - 1(-10) + 7(24)$$

$$|A| = 133$$

These examples show that the determinant of a 4 by 4 matrix can be obtained by cofactor expansion. To illustrate for a larger matrix, consider the 5 by 5 matrix

$$A = \begin{pmatrix} a_{11} & a_{12} & a_{13} & a_{14} & a_{15} \\ a_{21} & a_{22} & a_{23} & a_{24} & a_{25} \\ a_{31} & a_{32} & a_{33} & a_{34} & a_{35} \\ a_{41} & a_{42} & a_{43} & a_{44} & a_{45} \\ a_{51} & a_{52} & a_{53} & a_{54} & a_{55} \end{pmatrix}$$

The determinant of A may be calculated by cofactor expansion about any row or column. If it is assumed that the first row is selected for expansion, the determinant of A is

$$|A| = a_{11}A_{11} + a_{12}A_{12} + a_{13}A_{13} + a_{14}A_{14} + a_{15}A_{15}$$

where A_{1j} are cofactors of the j submatrices. Since the cofactor is the product of the minor a_{ij} and $(-1)^{i+j}$, we must calculate the minors of the elements of the first row. In this instance, the minors must be calculated for 4 by 4 submatrices. The calculation of the determinant of a 4 by 4 matrix by cofactor expansion was illustrated by the preceding example.

4.3.3 PROPERTIES OF DETERMINANTS

The reader has undoubtedly recognized that the calculation of determinants of higher-order matrices requires a considerable amount of arithmetic. The arithmetic involved in the calculation of a determinant by cofactor expansion is reduced if the row or column selected for cofactor expansion contains mainly zeros. For instance, the determinant of a 5 by 5 matrix was given as

$$|A| = a_{11}A_{11} + a_{12}A_{12} + a_{13}A_{13} + a_{14}A_{14} + a_{15}A_{15}$$

The arithmetic calculations required to determine the determinant are reduced in proportion to the number of elements a_{ij} that have values of zero.

Certain properties of determinants are useful in obtaining zeros as elements in rows or columns. Many of these properties will also prove useful in more advanced studies of matrix applications. These properties of determinants are offered without proof as follows.

1. Interchanging any two rows or any two columns of a matrix changes the sign of the determinant of the matrix but not the absolute value of the determinant. Thus if

$$A = \begin{pmatrix} 1 & \cdot 3 & 2 \\ 3 & 1 & -3 \\ 2 & 0 & 4 \end{pmatrix} \quad \text{and}$$

$$B = \begin{pmatrix} 2 & 3 & 1 \\ -3 & 1 & 3 \\ 4 & 0 & 2 \end{pmatrix}, \quad \text{then} \quad |A| = -|B|$$

2. The determinant of a matrix and of its transpose are equal, i.e.,

$$|A| = |A^t|$$

3. If all of the elements in any row or column are zero, the determinant of the matrix is zero.

4. If the elements in any two rows or two columns are equal or proportional, the determinant of the matrix is zero.

5. If the elements of any row or column are multiplied by a constant k, the determinant of the matrix is multiplied by the constant k, i.e.,

$$A = \begin{pmatrix} 1 & 3 & 2 \\ 3 & 1 & -3 \\ 2 & 0 & 4 \end{pmatrix} \quad \text{and}$$

$$B = \begin{pmatrix} 1k & 3k & 2k \\ 3 & 1 & -3 \\ 2 & 0 & 4 \end{pmatrix}, \quad \text{then} \quad |B| = k|A|$$

6. Any nonzero multiple of one row may be added to another row or any nonzero multiple of one column may be added to another column without changing the value of the determinant.
7. The determinant of any identity matrix is 1.

These properties can be applied in many ingenious ways to obtain the determinant of a matrix. To illustrate, assume that we must calculate the determinant of the 5 by 5 matrix

$$A = \begin{pmatrix} 3 & 1 & 0 & -2 & 4 \\ 2 & 3 & 1 & 2 & -3 \\ 4 & -5 & 0 & 6 & 2 \\ 0 & 6 & 1 & 3 & -1 \\ 5 & 2 & 3 & 1 & 1 \end{pmatrix}$$

Using the method of cofactor expansion, we examine the rows and columns and select the third column for expansion. This column is selected because it has more zeros than any other column or row.

To reduce the computation required in the expansion, subtract the second row from the fourth row and three times the second row from the fifth row. This gives

$$|A| = \begin{vmatrix} 3 & 1 & 0 & -2 & 4 \\ 2 & 3 & 1 & 2 & -3 \\ 4 & -5 & 0 & 6 & 2 \\ -2 & 3 & 0 & 1 & 2 \\ -1 & -7 & 0 & -5 & 10 \end{vmatrix}.$$

Cofactor expansion about the third column gives

$$|A| = a_{13}A_{13} + a_{23}A_{23} + a_{33}A_{33} + a_{43}A_{43} + a_{53}A_{53}$$

and since all elements in the column with the exception of a_{23} are zero, the expression becomes

$$|A| = a_{23}A_{23} = 1(-1)^5 \cdot \text{minor } a_{23} = -1 \begin{vmatrix} 3 & 1 & -2 & 4 \\ 4 & -5 & 6 & 2 \\ -2 & 3 & 1 & 2 \\ -1 & -7 & -5 & 10 \end{vmatrix}$$

The determinant of A equals -1 times the minor of a_{23}. This minor can be evaluated by applying the properties of determinants. Subtracting twice the fourth row from the third row and adding four times the fourth row to the second row and three times the fourth row to the first row gives

$$|A| = -1 \begin{vmatrix} 0 & -20 & -17 & 34 \\ 0 & -33 & -14 & 42 \\ 0 & 17 & 11 & -18 \\ -1 & -7 & -5 & 10 \end{vmatrix}$$

Multiplying the fourth row by -1 gives

$$|A| = +1 \begin{vmatrix} 0 & -20 & -17 & 34 \\ 0 & -33 & -14 & 42 \\ 0 & 17 & 11 & -18 \\ 1 & 7 & 5 & -10 \end{vmatrix}$$

Expanding about the first column gives

$$|A| = 1(-1)^{4+1} \text{ minor } a_{41} = -1 \begin{vmatrix} -20 & -17 & 34 \\ -33 & -14 & 42 \\ 17 & 11 & -18 \end{vmatrix}$$

The determinant of the 3 by 3 matrix can be calculated directly by the method presented in Sec. 4.3.1.

$$|A| = -1\{(-20)(-14)(-18) + (-17)(42)(17) + (34)(-33)(11) \\ -(17)(-14)(34) - (11)(42)(-20) - (-18)(-33)(-17)\}$$
$$|A| = -1\{-5040 - 12138 - 12342 + 8092 + 9240 + 10098\}$$
$$|A| = -1(-2090) = 2090$$

Example: Verify that expansion about the first row of the 5 by 5 matrix A results in the same determinant as expansion about the third column.

$$|A| = \begin{vmatrix} 3 & 1 & 0 & -2 & 4 \\ 2 & 3 & 1 & 2 & -3 \\ 4 & -5 & 0 & 6 & 2 \\ 0 & 6 & 1 & 3 & -1 \\ 5 & 2 & 3 & 1 & 1 \end{vmatrix}$$

Cofactor expansion about the third column in the preceding example resulted in $|A| = 2090$. To expand about the first row, we subtract three times the second column from the first column, add two times the second column to the fourth column, and subtract four times the second column from the fifth column. This gives

$$|A| = \begin{vmatrix} 0 & 1 & 0 & 0 & 0 \\ -7 & 3 & 1 & 8 & -15 \\ 19 & -5 & 0 & -4 & 22 \\ -18 & 6 & 1 & 15 & -25 \\ -1 & 2 & 3 & 5 & -7 \end{vmatrix}$$

The cofactor expansion about the first row is

$$|A| = a_{11}A_{11} + a_{12}A_{12} + a_{13}A_{13} + a_{14}A_{14} + a_{15}A_{15}$$

which reduces to

$$|A| = a_{12}A_{12} = 1(-1)^3 \text{ minor } a_{12} = -1 \begin{vmatrix} -7 & 1 & 8 & -15 \\ 19 & 0 & -4 & 22 \\ -18 & 1 & 15 & -25 \\ -1 & 3 & 5 & -7 \end{vmatrix}$$

To reduce the number of arithmetic operations, subtract the first row from the third row and three times the first row from the fourth row. This gives

$$|A| = -1 \begin{vmatrix} -7 & 1 & 8 & -15 \\ 19 & 0 & -4 & 22 \\ -11 & 0 & 7 & -10 \\ 20 & 0 & -19 & 38 \end{vmatrix}$$

Expanding about the second column gives

$$|A| = -1(a_{12}A_{12} + a_{22}A_{22} + a_{32}A_{32} + a_{42}A_{42})$$

or alternatively,

$$|A| = -1(1 \cdot (-1)^3 \text{ minor } a_{12}) = +1 \begin{vmatrix} 19 & -4 & 22 \\ -11 & 7 & -10 \\ 20 & -19 & 38 \end{vmatrix}$$

The determinant can now be calculated directly by the method presented in Sec. 4.3.1.

$$|A| = 19 \cdot 7 \cdot 38 + (-4)(-10)(20) + 22(-11)(-19) - 20 \cdot 7 \cdot 22$$
$$- (-19)(-10)(19) - 38(-11)(-4)$$
$$|A| = 5054 + 800 + 4598 - 3080 - 3610 - 1672$$
$$|A| = 2090$$

The determinants of A calculated by expansion about the third column and about the first row are equal.

Example: One of the properties of determinants is that the value of the determinant is zero if any two rows or two columns are equal. Verify this property for the 3 by 3 matrix $A = \begin{pmatrix} 1 & 2 & 1 \\ 2 & 1 & 2 \\ 3 & 4 & 3 \end{pmatrix}$.

The determinant is

$$|A| = 1 \cdot 1 \cdot 3 + 2 \cdot 2 \cdot 3 + 1 \cdot 2 \cdot 4 - 3 \cdot 1 \cdot 1 - 4 \cdot 2 \cdot 1 - 3 \cdot 2 \cdot 2$$
$$|A| = 3 + 12 + 8 - 3 - 8 - 12 = 0$$

Example: Verify that the determinant of any matrix in which two rows or two columns are proportional is zero.

Given the matrix $A = \begin{pmatrix} 1 & 2 & 3 \\ 2 & 4 & 6 \\ 3 & 1 & 7 \end{pmatrix}$, it can be seen that the second row is twice the first row. Subtracting twice the first row from the second row gives

$$|A| = \begin{vmatrix} 1 & 2 & 3 \\ 0 & 0 & 0 \\ 3 & 1 & 7 \end{vmatrix}$$

Since all elements in the second row are zero, the determinant of A is zero.

Example: One of the properties of a determinant is that if the elements of any row or column are multiplied by a constant k, the determinant of the matrix is multiplied by the constant. This property often proves useful in calculating a determinant. For instance, if $A = \begin{pmatrix} 3 & 2 & 8 \\ 1 & 4 & -6 \\ 6 & 6 & 12 \end{pmatrix}$ the determinant of A is

$$|A| = \begin{vmatrix} 3 & 2 & 8 \\ 1 & 4 & -6 \\ 6 & 6 & 12 \end{vmatrix}$$

The second column can be factored to give

$$|A| = 2 \begin{vmatrix} 3 & 1 & 8 \\ 1 & 2 & -6 \\ 6 & 3 & 12 \end{vmatrix}$$

Factoring a 3 from the third row gives

$$|A| = 2 \cdot 3 \begin{vmatrix} 3 & 1 & 8 \\ 1 & 2 & -6 \\ 2 & 1 & 4 \end{vmatrix}$$

The third column is factored to give

$$|A| = 2 \cdot 3 \cdot 4 \begin{vmatrix} 3 & 1 & 2 \\ 1 & 2 & -1.5 \\ 2 & 1 & 1 \end{vmatrix}$$

The determinant is

$$|A| = 24(3 \cdot 2 \cdot 1 + 1(-1.5)(2) + 2 \cdot 1 \cdot 1 - 2 \cdot 2 \cdot 2 - 1(-1.5)(3) - 1 \cdot 1 \cdot 1)$$
$$|A| = 24(6 - 3 + 2 - 8 + 4.5 - 1) = 12$$

4.3.4 SINGULAR AND NONSINGULAR MATRICES

The terms *singular* and *nonsingular* are used with reference to the value of the determinant of a square matrix. A square matrix A is said to be singular if the determinant of A is zero. Conversely, matrices whose determinants are nonzero are termed nonsingular.

Example: Determine if the matrix A is singular or nonsingular.

$$A = \begin{pmatrix} 2 & 3 \\ 4 & 3 \end{pmatrix}$$

Since $|A| = -6$ is nonzero, the matrix is nonsingular.

Example: Determine if the matrix A is singular or nonsingular.

$$A = \begin{pmatrix} 4 & 6 \\ 2 & 3 \end{pmatrix}$$

Since $|A| = 0$ is zero, the matrix is singular.

4.4 The Inverse: Adjoint Method

The inverse of the matrix A was defined in Chapter 3 as the matrix A^{-1} such that

$$AA^{-1} = I$$

where A and A^{-1} are both square matrices and I is the identity matrix. A method of calculating the inverse, termed the Gaussian elimination method, was presented in that chapter.

An alternative method of calculating the inverse is given in this section. Before giving the procedure for this technique, we must define the cofactor and adjoint matrices.

4.4.1 COFACTOR MATRIX

The cofactor of an element a_{ij} was defined in Sec. 4.3.2. The cofactor of the element a_{ij} of matrix A was designated by the symbol A_{ij}, where i refers to the row location and j the column location. The *cofactor matrix*, designated cof A, is the matrix of cofactors. That is,

$$\text{cof } A = \begin{pmatrix} A_{11} & A_{12} & \cdots & A_{1n} \\ A_{21} & A_{22} & \cdots & A_{2n} \\ \vdots & \vdots & & \vdots \\ A_{m1} & A_{m2} & \cdots & A_{mn} \end{pmatrix}, \tag{4.4}$$

where the component in row i and column j is the cofactor of the element a_{ij} of the square matrix A.

Example: Determine the cofactor matrix for the matrix

$$A = \begin{pmatrix} 3 & -2 & 4 \\ 6 & 4 & 5 \\ 3 & -1 & 6 \end{pmatrix}$$

The minors of the element a_{ij} are

minor $a_{11} = $	29	minor $a_{12} = $	21	minor $a_{13} = $	-18
minor $a_{21} = $	-8	minor $a_{22} = $	6	minor $a_{23} = $	3
minor $a_{31} = $	-26	minor $a_{32} = $	-9	minor $a_{33} = $	24

The cofactor of each element is given by the product of the minor and $(-1)^{i+j}$. The cofactor matrix is

$$\text{cof } A = \begin{pmatrix} 29 & -21 & -18 \\ 8 & 6 & -3 \\ -26 & 9 & 24 \end{pmatrix}$$

Example: Determine the cofactor matrix for the matrix

$$A = \begin{pmatrix} 6 & -2 & -3 \\ -1 & 8 & -7 \\ 4 & -3 & 6 \end{pmatrix}$$

The minors of the elements a_{ij} are

minor $a_{11} = \quad 27$	minor $a_{12} = \quad 22$	minor $a_{13} = -29$
minor $a_{21} = -21$	minor $a_{22} = \quad 48$	minor $a_{23} = -10$
minor $a_{31} = \quad 38$	minor $a_{32} = -45$	minor $a_{33} = \quad 46$

The cofactor matrix is

$$\text{cof } A = \begin{pmatrix} 27 & -22 & -29 \\ 21 & 48 & 10 \\ 38 & 45 & 46 \end{pmatrix}$$

4.4.2 ADJOINT MATRIX

The adjoint matrix is the transpose of the cofactor matrix. That is,

$$\text{adj } A = (\text{cof } A)^t \tag{4.5}$$

where adj A designates the adjoint matrix of A. This definition is illustrated by the following two examples.

Example: Determine the adjoint matrix for

$$A = \begin{pmatrix} 3 & -2 & 4 \\ 6 & 4 & 5 \\ 3 & -1 & 6 \end{pmatrix}$$

The cofactor matrix was calculated on p. 123 and is

$$\text{cof } A = \begin{pmatrix} 29 & -21 & -18 \\ 8 & 6 & -3 \\ -26 & 9 & 24 \end{pmatrix}$$

The adjoint matrix is the transpose of the cofactor matrix, i.e.,

$$\text{adj } A = \begin{pmatrix} 29 & 8 & -26 \\ -21 & 6 & 9 \\ -18 & -3 & 24 \end{pmatrix}$$

Example: Determine the adjoint matrix for

$$A = \begin{pmatrix} 6 & -2 & -3 \\ -1 & 8 & -7 \\ 4 & -3 & 6 \end{pmatrix}$$

The cofactor matrix was calculated on p. 124. The adjoint matrix is the transpose of the cofactor matrix, i.e.,

$$\text{adj } A = \begin{pmatrix} 27 & 21 & 38 \\ -22 & 48 & 45 \\ -29 & 10 & 46 \end{pmatrix}$$

The technique for calculating the inverse of the square matrix A can now be given. The inverse matrix, A^{-1}, is given by

$$A^{-1} = \left(\frac{1}{|A|}\right) \text{adj } A \tag{4.6}$$

provided A is nonsingular. This definition states that the inverse of A exists provided that the determinant of A is nonsingular, i.e., nonzero. The inverse is given by the product of the scalar $1/|A|$ and the adjoint matrix. This method of calculating an inverse matrix is illustrated by the following examples.

Example: Calculate the inverse of

$$A = \begin{pmatrix} 3 & -2 & 4 \\ 6 & 4 & 5 \\ 3 & -1 & 6 \end{pmatrix}$$

Applying the properties of determinants, we obtain

$$|A| = 3 \begin{vmatrix} 1 & -2 & 4 \\ 2 & 4 & 5 \\ 1 & -1 & 6 \end{vmatrix} = 3 \cdot 2 \begin{vmatrix} 1 & -2 & 4 \\ 1 & 2 & 2.5 \\ 1 & -1 & 6 \end{vmatrix}$$

$$|A| = 3 \cdot 2 \begin{vmatrix} 1 & -2 & 4 \\ 0 & 4 & -1.5 \\ 0 & 1 & 2 \end{vmatrix} = 3 \cdot 2[(1)(8 + 1.5)]$$

$$|A| = 57$$

The adjoint matrix was calculated on p. 124 and is

$$\text{adj } A = \begin{pmatrix} 29 & 8 & -26 \\ -21 & 6 & 9 \\ -18 & -3 & 24 \end{pmatrix}$$

The inverse is

$$A^{-1} = \tfrac{1}{57} \begin{pmatrix} 29 & 8 & -26 \\ -21 & 6 & 9 \\ -18 & -3 & 24 \end{pmatrix} = \begin{pmatrix} \frac{29}{57} & \frac{8}{57} & -\frac{26}{57} \\ -\frac{21}{57} & \frac{6}{57} & \frac{9}{57} \\ -\frac{18}{57} & -\frac{3}{57} & \frac{24}{57} \end{pmatrix}$$

Example: Determine the inverse of

$$A = \begin{pmatrix} 6 & -2 & -3 \\ -1 & 8 & -7 \\ 4 & -3 & 6 \end{pmatrix}$$

The determinant of A is

$$|A| = 6 \cdot 8 \cdot 6 + (-2)(-7)(4) + (-3)(-1)(-3) - 4 \cdot 8(-3)$$
$$\qquad -(-3)(-7)(6) - 6(-1)(-2)$$
$$|A| = 288 + 56 - 9 + 96 - 126 - 12$$
$$|A| = 293$$

The adjoint matrix was calculated on p. 125 as

$$\text{adj } A = \begin{pmatrix} 27 & 21 & 38 \\ -22 & 48 & 45 \\ -29 & 10 & 46 \end{pmatrix}$$

The inverse matrix is given by $A^{-1} = (1/|A|) \text{ adj } A$.

$$A^{-1} = \begin{pmatrix} 27/293 & 21/293 & 38/293 \\ -22/293 & 48/293 & 45/293 \\ -29/293 & 10/293 & 46/293 \end{pmatrix}$$

Example: Solve the following system of linear equations by determining the inverse.

$$3x_1 + 3x_2 + 4x_3 = 20$$
$$-2x_1 + 4x_2 - 2x_3 = -6$$
$$4x_1 - 2x_2 + 3x_3 = 16$$

In matrix form, the system of equations is $AX = B$.

$$\begin{pmatrix} 3 & 3 & 4 \\ -2 & 4 & -2 \\ 4 & -2 & 3 \end{pmatrix} \begin{pmatrix} x_1 \\ x_2 \\ x_3 \end{pmatrix} - \begin{pmatrix} 20 \\ -6 \\ 16 \end{pmatrix}$$

The inverse of A is calculated by Formula (4.6),

$$A^{-1} = (1/|A|) \text{ adj } A \qquad\qquad (4.6)$$

where

$$|A| = \begin{vmatrix} 3 & 3 & 4 \\ -2 & 4 & -2 \\ 4 & -2 & 3 \end{vmatrix}$$

$$|A| = 3 \cdot 4 \cdot 3 + 3(-2)4 + 4(-2)(-2) - 4 \cdot 4 \cdot 4 - (-2)(-2)3 - 3(-2)(3),$$
$$|A| = -30$$

The cofactors of A are

$$A_{11} = (-1)^2 \begin{vmatrix} 4 & -2 \\ -2 & 3 \end{vmatrix} = +1(12 - 4) = 8$$

$$A_{12} = (-1)^3 \begin{vmatrix} -2 & -2 \\ 4 & 3 \end{vmatrix} = -1(-6 + 8) = -2$$

$$A_{13} = (-1)^4 \begin{vmatrix} -2 & 4 \\ 4 & -2 \end{vmatrix} = +1(4 - 16) = -12$$

$$A_{21} = (-1)^3 \begin{vmatrix} 3 & 4 \\ -2 & 3 \end{vmatrix} = -1(9 + 8) = -17$$

$$A_{22} = (-1)^4 \begin{vmatrix} 3 & 4 \\ 4 & 3 \end{vmatrix} = +1(9 - 16) = -7$$

$$A_{23} = (-1)^5 \begin{vmatrix} 3 & 3 \\ 4 & -2 \end{vmatrix} = -1(-6 - 12) = 18$$

$$A_{31} = (-1)^4 \begin{vmatrix} 3 & 4 \\ 4 & -2 \end{vmatrix} = +1(-6 - 16) = -22$$

$$A_{32} = (-1)^5 \begin{vmatrix} 3 & 4 \\ -2 & -2 \end{vmatrix} = -1(-6 + 8) = -2$$

$$A_{33} = (-1)^6 \begin{vmatrix} 3 & 3 \\ -2 & 4 \end{vmatrix} = +1(12 + 6) = 18$$

The cofactor matrix is

$$\text{cof } A = \begin{pmatrix} 8 & -2 & -12 \\ -17 & -7 & 18 \\ -22 & -2 & 18 \end{pmatrix}$$

The adjoint matrix is given by the transpose of cof A,

$$\text{adj } A = \begin{pmatrix} 8 & -17 & -22 \\ -2 & -7 & -2 \\ -12 & 18 & 18 \end{pmatrix}$$

The inverse matrix is

$$A^{-1} = -\tfrac{1}{30} \begin{pmatrix} 8 & -17 & -22 \\ -2 & -7 & -2 \\ -12 & 18 & 18 \end{pmatrix}$$

The solution vector for the three simultaneous equations is

$$X = A^{-1}B = -\tfrac{1}{30} \begin{pmatrix} 8 & -17 & -22 \\ -2 & -7 & -2 \\ -12 & 18 & 18 \end{pmatrix} \begin{pmatrix} 20 \\ -6 \\ 16 \end{pmatrix}$$

$$X = -\tfrac{1}{30} \begin{pmatrix} -90 \\ -30 \\ -60 \end{pmatrix} = \begin{pmatrix} 3 \\ 1 \\ 2 \end{pmatrix}$$

4.5 Cramer's Rule

A technique called Cramer's rule is available for solving a system of n equations with n variables. This technique bypasses the calculation of the inverse of the coefficient matrix or the reduction of the augmented coefficient matrix through row operations. Instead, Cramer's rule requires the calculation of determinants for $n + 1$ matrices. The rationale underlying Cramer's rule is shown for a 2 by 2 system of simultaneous equations and extended to the general case of n by n simultaneous equations.

Given the 2 by 2 system of equations,

(1)
$$ax_1 + cx_2 = e$$

(2)
$$bx_1 + dx_2 = f$$

Cramer's rule is developed as follows.
Multiply Eq. (1) by $(1/a)$.

(1')
$$x_1 + \left(\frac{c}{a}\right) x_2 = \frac{e}{a}$$

Subtract b times Eq. (1') from Eq. (2).

(2')
$$\left(d - \frac{bc}{a}\right) x_2 = f - \frac{be}{a}$$

Solve for x_2.

$$x_2 = \frac{\dfrac{af - be}{a}}{\dfrac{ad - bc}{a}} = \frac{af - be}{ad - bc}$$

This can also be expressed as

$$x_2 = \frac{\begin{vmatrix} a & e \\ b & f \end{vmatrix}}{\begin{vmatrix} a & c \\ b & d \end{vmatrix}}$$

Solve Eq. (1') for x_1.

$$x_1 = \frac{e}{a} - \left(\frac{c}{a}\right) x_2$$

$$x_1 = \frac{e}{a} - \left(\frac{c}{a}\right)\left(\frac{af - be}{ad - bc}\right)$$

$$x_1 = \frac{e}{a}\left(\frac{ad - bc}{ad - bc}\right) - \frac{c}{a}\left(\frac{af - be}{ad - bc}\right)$$

$$x_1 = \frac{ead - ebc}{a(ad - bc)} - \frac{caf + ebc}{a(ad - bc)}$$

$$x_1 = \frac{ed - cf}{ad - bc}$$

This can also be expressed as

$$x_1 = \frac{\begin{vmatrix} e & c \\ f & d \end{vmatrix}}{\begin{vmatrix} a & c \\ b & d \end{vmatrix}}$$

For the 2 by 2 system of equations, the solution is

$$x_1 = \frac{\begin{vmatrix} e & c \\ f & d \end{vmatrix}}{\begin{vmatrix} a & c \\ b & d \end{vmatrix}} \quad \text{and} \quad x_2 = \frac{\begin{vmatrix} a & e \\ b & f \end{vmatrix}}{\begin{vmatrix} a & c \\ b & d \end{vmatrix}}$$

Cramer's rule can be applied to a system of n by n simultaneous equations. For the general case, define A_i as the matrix formed by interchanging the ith *column* of the coefficient matrix A with the right-hand-side vector B. Provided that A is nonsingular (i.e., $|A| \neq 0$), the solution value for the ith variable x_i is given by

$$x_i = \frac{|A_i|}{|A|} \tag{4.7}$$

To illustrate, consider the 3 by 3 system of equations,

$$a_{11}x_1 + a_{12}x_2 + a_{13}x_3 = b_1$$
$$a_{21}x_1 + a_{22}x_2 + a_{23}x_3 = b_2$$
$$a_{31}x_1 + a_{32}x_2 + a_{33}x_3 = b_3$$

The value of x_1 is given by

$$x_1 = \frac{|A_1|}{|A|} = \frac{\begin{vmatrix} b_1 & a_{12} & a_{13} \\ b_2 & a_{22} & a_{23} \\ b_3 & a_{32} & a_{33} \end{vmatrix}}{\begin{vmatrix} a_{11} & a_{12} & a_{13} \\ a_{21} & a_{22} & a_{23} \\ a_{31} & a_{32} & a_{33} \end{vmatrix}}$$

Similarly, the values of x_2 and x_3 are given by

$$x_2 = \frac{|A_2|}{|A|} = \frac{\begin{vmatrix} a_{11} & b_1 & a_{13} \\ a_{21} & b_2 & a_{23} \\ a_{31} & b_3 & a_{33} \end{vmatrix}}{\begin{vmatrix} a_{11} & a_{12} & a_{13} \\ a_{21} & a_{22} & a_{23} \\ a_{31} & a_{32} & a_{33} \end{vmatrix}}$$

and

$$x_3 = \frac{|A_3|}{|A|} = \frac{\begin{vmatrix} a_{11} & a_{12} & b_1 \\ a_{21} & a_{22} & b_2 \\ a_{31} & a_{32} & b_3 \end{vmatrix}}{\begin{vmatrix} a_{11} & a_{12} & a_{13} \\ a_{21} & a_{22} & a_{23} \\ a_{31} & a_{32} & a_{33} \end{vmatrix}}$$

Example: Solve the 2 by 2 system of equations, using Cramer's rule.

$$3x_1 + 6x_2 = 6$$
$$2x_1 - 2x_2 = 16$$

$$x_1 = \frac{\begin{vmatrix} 6 & 6 \\ 16 & -2 \end{vmatrix}}{\begin{vmatrix} 3 & 6 \\ 2 & -2 \end{vmatrix}} = \frac{-108}{-18} = 6$$

$$x_2 = \frac{\begin{vmatrix} 3 & 6 \\ 2 & 16 \end{vmatrix}}{\begin{vmatrix} 3 & 6 \\ 2 & -2 \end{vmatrix}} = \frac{36}{-18} = -2$$

Example: Solve the 3 by 3 system of equations, using Cramer's rule.

$$3x_1 - 4x_2 + 8x_3 = 26$$
$$6x_1 + 3x_2 - 5x_3 = 1$$
$$-2x_1 + 1x_2 + 3x_3 = 11$$

$$x_1 = \frac{\begin{vmatrix} 26 & -4 & 8 \\ 1 & 3 & -5 \\ 11 & 1 & 3 \end{vmatrix}}{\begin{vmatrix} 3 & -4 & 8 \\ 6 & 3 & -5 \\ -2 & 1 & 3 \end{vmatrix}} = \frac{340}{170} = 2$$

$$x_2 = \frac{\begin{vmatrix} 3 & 26 & 8 \\ 6 & 1 & -5 \\ -2 & 11 & 3 \end{vmatrix}}{\begin{vmatrix} 3 & -4 & 8 \\ 6 & 3 & -5 \\ -2 & 1 & 3 \end{vmatrix}} = \frac{510}{170} = 3$$

$$x_3 = \frac{\begin{vmatrix} 3 & -4 & 26 \\ 6 & 3 & 1 \\ -2 & 1 & 11 \end{vmatrix}}{\begin{vmatrix} 3 & -4 & 8 \\ 6 & 3 & -5 \\ -2 & 1 & 3 \end{vmatrix}} = \frac{680}{170} = 4$$

4.6 Rank

Section 2.4 first introduced the concept of a system of equations. One of the important points made in that section was that a system of equations may be consistent or inconsistent and that a consistent system of equations may have a unique or an infinite number of solutions. We can now introduce a method for determining if a system of equations is consistent, has a unique solution, or has an infinite number of solutions. The method is based upon the concept of the rank of a matrix.

4.6.1 RANK OF A MATRIX

The *rank* of a matrix is the number of linearly independent rows (or columns) in the matrix. The rank of a matrix can be determined in two ways. First, row operations can be applied to the matrix with the objective of obtaining zeros in as many rows as possible. The number of rows in the matrix that cannot be reduced to zero by row operations gives the number of linearly independent rows. This number is the rank of the matrix.

A second method of determining the rank of a matrix is to determine the order of the largest square submatrix whose determinant is not zero. If, for instance, the matrix A is square and has a nonzero determinant, the rank of A is equal to the order (i.e., number of rows or columns) of A. If, however, the matrix A is not square or is square but has a zero determinant, the rank of A is equal to the order of the largest square submatrix in A whose determinant is nonzero. If there is no square submatrix in A with a nonzero determinant, the rank of A is zero.

To illustrate these methods, consider the matrix

$$A - \begin{pmatrix} 2 & 3 & -2 \\ 3 & -2 & 1 \\ 1 & -5 & 3 \end{pmatrix}$$

The rank of the matrix is determined by applying row operations with the objective of reducing the elements in the rows to zero. Multiplying the first row by $\frac{1}{2}$ gives

$$\begin{pmatrix} 1 & \frac{3}{2} & -1 \\ 3 & -2 & 1 \\ 1 & -5 & 3 \end{pmatrix}$$

Adding -3 times the new first row to the second row and -1 times the new first row to the third row gives

$$\begin{pmatrix} 1 & \frac{3}{2} & -1 \\ 0 & -\frac{13}{2} & 4 \\ 0 & -\frac{13}{2} & 4 \end{pmatrix}$$

Adding -1 times the second row to the third row gives

$$\begin{pmatrix} 1 & \frac{3}{2} & -1 \\ 0 & -\frac{13}{2} & 4 \\ 0 & 0 & 0 \end{pmatrix}$$

Since no further combinations of row operations can reduce all components of either the first or second row to zero, the rank of the matrix is 2.

To verify that the order of the largest square submatrix of A whose determinant is nonzero is also 2, the determinant of A is calculated.

$$|A| = \begin{vmatrix} 2 & 3 & -2 \\ 3 & -2 & 1 \\ 1 & -5 & 3 \end{vmatrix}$$

$$|A| = 2(-2)(3) + 3 \cdot 1 \cdot 1 + (-2)(3)(-5) - 1(-2)(-2)$$
$$\qquad - (-5)(1)(2) - 3 \cdot 3 \cdot 3$$

$$|A| = -12 + 3 + 30 - 4 + 10 - 27$$

$$|A| = 0$$

Since the determinant of A is zero, we know that the rank of A is not 3. Next, we must determine if one of the square submatrices of order 2 has a nonzero determinant. The determinant of the submatrix $\begin{pmatrix} 2 & 3 \\ 3 & -2 \end{pmatrix}$ is -13. Since A has at least one square submatrix of order 2, we again conclude that the rank of A is 2.

Example: Determine the rank of the matrix A by determining the number of independent rows.

$$A = \begin{pmatrix} 1 & 4 & 6 & 5 \\ 3 & 2 & 4 & 4 \\ 2 & -2 & 3 & -10 \end{pmatrix}$$

To reduce the elements to zero, add -3 times the first row to the second row and -2 times the first row to the third row. This gives

$$\begin{pmatrix} 1 & 4 & 6 & 5 \\ 0 & -10 & -14 & -11 \\ 0 & -10 & -9 & -20 \end{pmatrix}$$

Next, add -1 times the second row to the third row.

$$\begin{pmatrix} 1 & 4 & 6 & 5 \\ 0 & -10 & -14 & -11 \\ 0 & 0 & 5 & -9 \end{pmatrix}$$

Since there is no further combination of row operations that leads to all zeros in one or more of the rows, we conclude that the rank of the matrix is 3.

Example: Determine the rank of the matrix A in the preceding example by determining the order of the largest square submatrix of A whose determinant is nonzero.

The largest square submatrices of A are matrices with dimensions 3 by 3. To find if one or more of these submatrices has a nonzero determinant, we begin by calculating $|S_1|$.

$$|S_1| = \begin{vmatrix} 1 & 4 & 6 \\ 3 & 2 & 4 \\ 2 & -2 & 3 \end{vmatrix}$$

$$|S_1| = 1 \cdot 2 \cdot 3 + 4 \cdot 4 \cdot 2 + 6 \cdot 3(-2) - 2 \cdot 2 \cdot 6 - (-2)(4)(1) - 3 \cdot 3 \cdot 4$$

$$|S_1| = -50$$

The determinant of a 3 by 3 submatrix of A is nonzero; therefore, the rank of A is 3.

4.6.2   RANK OF A SYSTEM OF EQUATIONS

The concept of rank can be extended to a system of equations. Remembering that the rank of a matrix was defined as the number of linearly independent rows in a matrix, we similarly define the rank of a system of equations as the number of linearly independent equations in the system.

To determine the rank of the system of equations $AX = B$, the coefficient matrix A is augmented with the right-hand-side vector B to form the augmented matrix $A \mid B$. The rank of the system of equations is given by the rank of this augmented matrix. To calculate the rank, the augmented matrix can be reduced by row operations and the number of nonzero rows counted. Alternatively, the rank can be determined by finding the order of the largest square submatrix of $A \mid B$ whose determinant is nonzero.

To illustrate, consider the system of equations

$$6x_1 - 2x_2 - 4x_3 = 16$$

$$2x_1 - 2x_2 - x_3 = 2$$

$$x_1 + x_2 - x_3 = 6$$

Augmenting the coefficient matrix with the right-hand-side vector gives

$$A \mid B = \begin{pmatrix} 6 & -2 & -4 & 16 \\ 2 & -2 & -1 & 2 \\ 1 & 1 & -1 & 6 \end{pmatrix}$$

To reduce the augmented matrix, add -2 times the third row to the second row and -6 times the third row to the first row.

$$A \mid B = \begin{pmatrix} 0 & -8 & 2 & -20 \\ 0 & -4 & 1 & -10 \\ 1 & 1 & -1 & 6 \end{pmatrix}$$

Next, add -2 times the second row to the first row.

$$A \mid B = \begin{pmatrix} 0 & 0 & 0 & 0 \\ 0 & -4 & 1 & -10 \\ 1 & 1 & -1 & 6 \end{pmatrix}$$

Since no further combination of row operations leads to a zero row, the rank of the augmented matrix is 2.

The rank of the system of equations can also by found by calculating the determinants of the square submatrices of $A \mid B$. The determinants of the four submatrices whose dimensions are 3 by 3 are zero, i.e.,

$$|S_1| = \begin{vmatrix} 6 & -2 & -4 \\ 2 & -2 & -1 \\ 1 & 1 & -1 \end{vmatrix} = 0$$

$$|S_2| = \begin{vmatrix} -2 & -4 & 16 \\ -2 & -1 & 2 \\ 1 & -1 & 6 \end{vmatrix} = 0$$

$$|S_3| = \begin{vmatrix} 6 & -4 & 16 \\ 2 & -1 & 2 \\ 1 & -1 & 6 \end{vmatrix} = 0$$

$$|S_4| = \begin{vmatrix} 6 & -2 & 16 \\ 2 & -2 & 2 \\ 1 & 1 & 6 \end{vmatrix} = 0$$

The determinant of at least one of the square submatrices of order 2 is, however, nonzero. Consequently, the rank of the augmented matrix and therefore the system of equations is 2.

Example: Determine the rank of the following system of equations,

$$2x_1 + 3x_2 = 6$$

$$6x_1 - 2x_2 = 3$$

The rank of the system of equations is found by augmenting the coefficient matrix with the right-hand-side vector, i.e.,

$$A \mid B = \begin{pmatrix} 2 & 3 & 6 \\ 6 & -2 & 3 \end{pmatrix}$$

The determinant of at least one of the three submatrices whose dimensions are 2 by 2 is nonzero. Therefore, the rank of the system of equations is 2.

4.6.3 RANK, CONSISTENCY, AND UNIQUE SOLUTIONS

At the beginning of this section, we stated that the concept of rank is used to determine if a system of equations is consistent or inconsistent. We also stated that the rank is used to distinguish between consistent systems of equations with a unique solution and consistent systems with an infinite number of solutions. This relationship among rank, consistency, and the number of solutions is given by the following two rules.

> *Rule 1.* A system of equations is consistent if and only if the rank of the augmented matrix is equal to the rank of the coefficient matrix. This rank is termed the rank of the system.
>
> *Rule 2.* A system of linear equations has a unique solution if and only if the rank of the system equals the number of variables.

To illustrate these rules, consider the following system of equations.

$$2x + 3y - 2z = 40$$
$$3x - 2y + 1z = 50$$
$$1x - 5y + 3z = 10$$

It was shown in Chapter 2, p. 61, that this system is consistent with an infinite number of solutions. This can be verified by augmenting the coefficient matrix with the right-hand-side vector, i.e.,

$$A \mid B = \begin{pmatrix} 2 & 3 & -2 & 40 \\ 3 & -2 & 1 & 50 \\ 1 & -5 & 3 & 10 \end{pmatrix}$$

This system is reduced by adding 1 times the first row to the third row. This gives

$$A \mid B = \begin{pmatrix} 2 & 3 & -2 & 40 \\ 3 & -2 & 1 & 50 \\ 3 & -2 & 1 & 50 \end{pmatrix}$$

Adding -1 times the second row to the third row gives

$$A \mid B = \begin{pmatrix} 2 & 3 & -2 & 40 \\ 3 & -2 & 1 & 50 \\ 0 & 0 & 0 & 0 \end{pmatrix}$$

The example shows that the rank of both the augmented matrix $A \mid B$ and the coefficient matrix A is 2. Based upon Rule 1 the system of equations is consistent. Since the rank of the system is less than the number of variables, we conclude from Rule 2 that the system has an infinite number of solutions.

Example: Verify that the following system of equations is consistent with a unique solution.

$$2x + y = 9$$
$$x - y = 3$$
$$x + 2y = 6$$

The rank of $A \mid B$ and A is found by reducing the augmented matrix by row operations or by calculating the determinants of the square submatrices of A and $A \mid B$. The determinant of $A \mid B$ is zero, i.e.,

$$|A \mid B| = \begin{vmatrix} 2 & 1 & 9 \\ 1 & -1 & 3 \\ 1 & 2 & 6 \end{vmatrix}$$

$$|A \mid B| = 2(-1)(6) + 1 \cdot 3 \cdot 1 + 9 \cdot 1 \cdot 2 - 1(-1)(9) - 2 \cdot 3 \cdot 2 - 6 \cdot 1 \cdot 1$$
$$|A \mid B| = 0$$

The determinant of at least one of the submatrices of dimensions 2 by 2 in the augmented matrix $A \mid B$ is nonzero. Similarly, the determinant of at least one of the square submatrices of dimension 2 by 2 in the coefficient matrix A is nonzero. Consequently, the rank of both the augmented matrix and the coefficient matrix is 2. On the basis of Rule 1 we therefore conclude that the system of equations is consistent. Furthermore, since the rank of the system of equation is the same as the number of variables, we also conclude from Rule 2 that the system has a unique solution.

Example: Verify that the following system of equations is inconsistent.

$$x_1 + 3x_2 + x_3 = 2$$
$$x_1 - 2x_2 - 2x_3 = 3$$
$$2x_1 + x_2 - x_3 = 6$$

To determine if the system of equations is consistent we augment the coefficient matrix with the right-hand-side vector and reduce the resulting matrix using row operations

$$A \mid B = \begin{pmatrix} 1 & 3 & 1 & 2 \\ 1 & -2 & -2 & 3 \\ 2 & 1 & -1 & 6 \end{pmatrix}$$

Add the second row to the first row.

$$\begin{pmatrix} 2 & 1 & -1 & 5 \\ 1 & -2 & -2 & 3 \\ 2 & 1 & -1 & 6 \end{pmatrix}$$

Add -1 times the third row to the first row.

$$\begin{pmatrix} 0 & 0 & 0 & | & -1 \\ 1 & -2 & -2 & | & 3 \\ 2 & 1 & -1 & | & 6 \end{pmatrix}$$

No further combination of row operations results in a row composed entirely of zeros. Consequently, the rank of the augmented matrix $A \mid B$ is 3. The rank of the coefficient matrix A, however, is 2. Since the rank of the coefficient matrix is less than the rank of the augmented matrix, the system of equations is inconsistent.

PROBLEMS

1. Form the transpose of each of the following matrices.

(a) $A = \begin{pmatrix} 3 & 1 \\ 4 & 3 \end{pmatrix}$

(b) $A = (1, 5, 6)$

(c) $A = \begin{pmatrix} 2 & 1 & -3 \\ 3 & -2 & 4 \end{pmatrix}$

(d) $A - \begin{pmatrix} -2 & 1 \\ 1 & 3 \\ 3 & 2 \end{pmatrix}$

2. Determine the submatrix B formed by the deletion of the second row and third column of the following matrices.

(a) $A = \begin{pmatrix} 3 & 2 & -4 \\ 2 & -3 & 3 \\ 5 & 4 & -1 \end{pmatrix}$

(b) $A = \begin{pmatrix} 6 & 4 & 5 & 2 \\ -5 & 3 & 1 & 6 \\ 7 & 2 & 3 & 7 \end{pmatrix}$

3. Given two matrices A and B that are partitioned into $A - \begin{pmatrix} A_1 & A_2 \\ A_3 & A_4 \end{pmatrix}$ and $B = \begin{pmatrix} B_1 \\ B_2 \end{pmatrix}$, determine the product AB by multiplying and adding the submatrices. The submatrices are

$$A_1 = \begin{pmatrix} 3 & 2 \\ 6 & 3 \end{pmatrix}, A_2 = \begin{pmatrix} 4 & 1 \\ 2 & 5 \end{pmatrix}, A_3 = \begin{pmatrix} 3 & 1 \\ 2 & 3 \end{pmatrix}, A_4 = \begin{pmatrix} 5 & 6 \\ 4 & 8 \end{pmatrix},$$

$$B_1 = \begin{pmatrix} 3 & 5 \\ 6 & 3 \end{pmatrix}, \text{ and } B_2 = \begin{pmatrix} 2 & 7 \\ 1 & 5 \end{pmatrix}.$$

4. Find the determinants of the following matrices.

(a) $A = \begin{pmatrix} 6 & 4 \\ 3 & 8 \end{pmatrix}$

(b) $A = \begin{pmatrix} 5 & -3 \\ 2 & 4 \end{pmatrix}$

(c) $A = \begin{pmatrix} 3 & -2 & 2 \\ 1 & 4 & 3 \\ 2 & 5 & -4 \end{pmatrix}$

(d) $A = \begin{pmatrix} 3 & 5 & 3 \\ 6 & 4 & -5 \\ 5 & 8 & 2 \end{pmatrix}$

5. Find the minors of the elements in the matrix A.

$$A = \begin{pmatrix} 6 & 3 & 4 \\ 3 & -5 & 2 \\ 4 & 3 & -3 \end{pmatrix}$$

6. Find the cofactors of each of the elements in matrix A of Problem 5.

7. Use the method of cofactor expansion to find the determinant of the matrix in Problem 5. Verify the solution by expanding about both the first row and the first column.

8. In each of the following matrices, how can we know that the determinant is zero without actually calculating the determinant?

(a) $A = \begin{pmatrix} 3 & 4 & 6 \\ 6 & 8 & 5 \\ 12 & 16 & 2 \end{pmatrix}$
    (b) $A = \begin{pmatrix} 6 & 4 & 7 \\ 0 & 0 & 0 \\ 5 & 9 & 4 \end{pmatrix}$

(c) $A = \begin{pmatrix} 2 & 5 & 21 \\ 3 & 6 & 9 \\ 6 & 15 & 63 \end{pmatrix}$
    (d) $A = \begin{pmatrix} 1 & 3 & 1 \\ 1 & 1 & 3 \\ 1 & 3 & 1 \end{pmatrix}$

9. Find the determinants of the following matrices by the method of cofactor expansion.

(a) $A = \begin{pmatrix} 3 & 7 & 1 & 5 \\ 6 & 2 & 0 & 4 \\ 1 & 6 & 1 & 3 \\ 2 & 5 & 0 & 2 \end{pmatrix}$
    (b) $A = \begin{pmatrix} 7 & -2 & 3 & 1 \\ 2 & 3 & -1 & 0 \\ 5 & 7 & 3 & 1 \\ 4 & -6 & 6 & 2 \end{pmatrix}$

10. Determine the cofactor matrix for the matrix A in Problem 5.

11. Determine the adjoint matrix for the matrix A in Problem 5.

12. Determine the inverse of the matrix A in Problem 5.

13. Use the adjoint method to calculate the inverses of the following matrices.

(a) $A = \begin{pmatrix} 3 & 1 & 1 \\ 2 & -2 & 2 \\ 1 & 0 & 1 \end{pmatrix}$
    (b) $A = \begin{pmatrix} 1 & 2 & 3 \\ 1 & 3 & 5 \\ 2 & 5 & 9 \end{pmatrix}$

(c) $A = \begin{pmatrix} 2 & 6 & 4 \\ 6 & 6 & 4 \\ 3 & 2 & 4 \end{pmatrix}$
    (d) $A = \begin{pmatrix} 2 & 8 & -11 \\ -1 & -5 & 7 \\ 1 & 2 & -3 \end{pmatrix}$

14. Solve the following systems of equations. (*Hint:* Use the results of Problem 13 to simplify the calculations.)

(a) $3x_1 + x_2 + x_3 = 16$
    $2x_1 - 2x_2 + 2x_3 = 24$
    $x_1 \qquad + x_3 = 8$

(b) $x_1 + 2x_2 + 3x_3 = 16$
    $x_1 + 3x_2 + 5x_3 = 20$
    $2x_1 + 5x_2 + 9x_3 = 10$

(c) $2x_1 + 6x_2 + 4x_3 = 320$
 $6x_1 + 6x_2 + 4x_3 = 480$
 $3x_1 + 2x_2 + 4x_3 = 192$

(d) $2x_1 + 8x_2 - 11x_3 = 25$
 $-x_1 - 5x_2 + 7x_3 = 10$
 $x_1 + 2x_2 + -3x_3 = 15$

15. Resolve Problem 14, using Cramer's rule.

16. Determine the solutions to the following systems of equations, using Cramer's rule.

(a) $7x_1 + 13x_2 = 91$
 $13x_1 + 8x_2 = 104$

(b) $14x_1 + 23x_2 = 176$
 $9x_1 + 8x_2 = 86$

(c) $3x_1 + 5x_2 + 4x_3 = 42$
 $6x_1 + 4x_2 - 3x_3 = 26$
 $2x_1 - 5x_2 + 6x_3 = 5$

(d) $2x_1 - 5x_2 + 3x_3 = 14$
 $7x_1 + 4x_2 - 6x_3 = 36$
 $5x_1 - 6x_2 + 3x_3 = 34$

17. Determine the ranks of the following matrices.

(a) $A = \begin{pmatrix} 6 & 4 \\ 4 & 6 \end{pmatrix}$

(b) $A = \begin{pmatrix} 3 & 9 \\ 2 & 6 \end{pmatrix}$

(c) $A = \begin{pmatrix} 1 & 2 & 0 \\ 4 & 8 & 5 \end{pmatrix}$

(d) $I = \begin{pmatrix} 1 & 0 & 0 \\ 0 & 1 & 0 \\ 0 & 0 & 1 \end{pmatrix}$

(e) $A = (3 \quad 4 \quad 1)$

(f) $A = (6)$

18. Use the concept of rank to determine if the following systems of equations are consistent or inconsistent. For those systems that are consistent, state whether the system of equations has a unique solution or has an infinite number of solutions.

(a) $3x_1 + 2x_2 = 8$
 $x_1 + 4x_2 = 5$

(b) $2x_1 + 5x_2 = 8$
 $4x_1 + 10x_2 = 4$

(c) $x_1 + 2x_2 = 4$
 $2x_1 + 4x_2 = 8$

(d) $2x_1 + 4x_2 = 6$
 $x_1 + 2x_2 = 8$

(e) $2x_1 + x_2 + 3x_3 = 2$
 $x_1 + 2x_2 + x_3 = 4$
 $2x_1 + 4x_2 + 2x_3 = 10$

(f) $x_1 + 2x_2 + x_3 = 2$
 $2x_1 + x_2 + 3x_3 = 1$
 $x_1 + 2x_2 + x_3 = 2$

(g) $4x_1 + 2x_2 + 3x_3 + x_4 = 18$
 $2x_1 + 3x_2 - x_3 + 2x_4 = 10$

SUGGESTED REFERENCES

The references for this chapter are listed in Chapter 3.

Chapter 5

Linear
Programming

Linear programming is one of the most frequently and successfully applied mathematical approaches to managerial decisions. The objective in using linear programming is to develop a model to aid the decision maker in determining the optimal allocation of the firm's resources among the alternative products of the firm. Since the resources employed in a firm have an economic value and the products of the firm lead to profits and costs, the linear programming problem becomes that of allocating the scarce resources to the products in a manner such that profits are a maximum, or alternatively, costs are a minimum.

The popularity of linear programming stems directly from its usefulness to the business firm. One of the fundamental tasks of management involves the allocation of resources. For instance, the petroleum firm must allocate sources of crude oil for the production of various petroleum products. Similarly, the wood products firm must allocate timber resources for the production of different lumber and paper products. In both examples, the allocation must be made subject to constraints on the availability of resources and the demand for the products. The objective in allocating the resources to the products is to maximize the profit of the firm.

Our objective in this and the following two chapters is twofold. First, we shall describe the linear programming problem and the solution techniques available for this problem. Our approach will be to introduce the important characteristics of linear programming through the use of very simple two-product linear programming problems and to solve these problems using a graphical solution technique. This, in turn, will lead to a discussion in Chapter 6 of a more general solution technique. This technique, termed the *simplex algorithm*, relies largely upon the use of the elementary row operations of matrix algebra to obtain the solution of systems of simultaneous equations. A second, and equally important objective, will be to explain and offer examples of important applications of linear programming. These examples are in sufficient detail for the student to understand industrial applications of linear programming and, hopefully, to use the technique in his job.

The importance of the computer in the solution of linear programming problems should be emphasized. Although "linear programming" and "computer programming" are not related, the computer is extremely valuable in obtaining the solution to linear programming problems. Many industrial applications of linear programming require more than one hundred equations and variables. Problems of this size can be solved only through the use of modern electronic computers.

5.1 The Linear Programming Problem

The linear programming problem consists of allocating scarce resources among competing products or activities. Two factors give rise to the allocation problem. First, resources available to management have a cost and are limited in supply. Therefore, management must determine how these limited resources will be used. Second, the allocation of these resources must be made in accordance with some overall objective. In the business firm, this objective is normally the maximization of profit or the minimization of cost.

As an illustration of the allocation of resources using linear programming, consider the following problem concerning the scheduling of production in a machine shop. Production must be scheduled for two types of machines, machine 1 and machine 2. One hundred twenty hours of time can be scheduled for machine 1, and 80 hours can be scheduled for machine 2. Production during the scheduling period is limited to two products, A and B. Each unit of product A requires 2 hours of process time on each machine. Each unit of product B requires 3 hours on machine 1 and 1.5 hours on machine 2. The contribution margin is $4.00 per unit of product A and $5.00 per unit of product B.† Both types of products can be readily marketed; consequently, production should be scheduled with the objective of maximizing profit. The

---

† Contribution margin is defined as contribution to fixed expenses and profit.

linear programming problem can be formulated in the following manner. Let:

x_1 = the number of units of product A scheduled for production
x_2 = the number of units of product B scheduled for production
P = contribution to fixed expenses and profit

The linear programming problem is

Maximize: $P = 4x_1 + 5x_2$ (profit)
Subject to: $2x_1 + 3x_2 \leq 120$ (machine 1 resource)
                      $2x_1 + 1.5x_2 \leq 80$ (machine 2 resource)
                      $x_1, x_2 \geq 0$ (nonnegativity)

The mathematical formulation states that production of products A and B should be scheduled with the objective of maximizing profit. Production is limited, however, by the availability of resources, i.e., 120 hours of machine 1 time and 80 hours of machine 2 time. An additional limitation, shown by $x_1, x_2 \geq 0$, is that negative production is not allowed, i.e., the number of units of either product must be zero or positive.

The solution in terms of the production quantities of products A and B that maximizes profit can be determined in this problem by either a graphical approach or the simplex method. The graphical approach for solving two variable problems provides important insights into the nature of linear programming problems. This approach is shown in Sec. 5.2. The simplex method is presented in Chapter 6.

To illustrate a linear programming problem in which costs are minimized, consider the problem facing a speciality metals manufacturer. The firm produces an alloy that is made from steel and scrap metal. The cost per ton of steel is $50 and the cost per ton of scrap is $20. The technological requirements for the alloy are (1) a minimum of one ton of steel is required for every two tons of scrap; (2) one hour of processing time is required for each ton of steel, and four hours of processing time are required for each ton of scrap; (3) the steel and scrap combine linearly to make the alloy. The process loss from the steel is 10 percent and the loss from scrap is 20 percent. Although production may exceed demand, a minimum of 40 tons of the alloy must be manufactured. To maintain efficient plant operation, a minimum of 80 hours of processing time must be used. The supply of both scrap and steel is adequate for production of the alloy.

The objective of the manufacturer is to produce the alloy at a minimum cost. The problem can be formulated as follows.

Let: x_1 = the number of tons of steel
               x_2 = the number of tons of scrap
               C = the cost of the alloy

The linear programming problem is

Minimize: $C = 50x_1 + 20x_2$ (cost)
Subject to: $2x_1 - 1x_2 \geq 0$ (tech. req'm't.)
                   $1x_1 + 4x_2 \geq 80$ (process)
                   $0.90x_1 + 0.80x_2 \geq 40$ (demand)
                   $x_1, x_2 \geq 0$ (nonnegativity)

The problem is to minimize the cost of producing the alloy. The cost function is equal to $50 per ton of steel multiplied by the number of tons of steel plus $20 per ton of scrap multiplied by the number of tons of scrap. The first constraint states that a minimum of one ton of steel must be used for every two tons of scrap. The second constraint concerns process time, and the third constraint shows that the steel and scrap combine linearly to make the alloy with a 10 percent and 20 percent loss, respectively. The solution to the problem is determined using the graphical method in Sec. 5.2.

5.2   Graphical Solution

The graphical solution technique provides a convenient method for solving simple two-variable linear programming problems. It is also quite useful in illustrating basic concepts in linear programming. For problems in which more than two variables are required, the simplex algorithm is used to determine the solution.

5.2.1   MAXIMIZATION

The problem concerning the scheduling of production in a machine shop was expressed mathematically as

Maximize: $P = 4x_1 + 5x_2$
Subject to: $2x_1 + 3x_2 \leq 120$
                   $2x_1 + 1.5x_2 \leq 80$
                   $x_1, x_2 \geq 0$

The function to be optimized, $P = 4x_1 + 5x_2$, is termed the *objective function*. If x_1 and x_2 were unrestricted, there would be no limit on the value of the objective function. There are, however, *constraints* upon the values of the variables.† The constraints are expressed as inequalities and are

$$2x_1 + 3x_2 \leq 120$$

$$2x_1 + 1.5x_2 \leq 80$$

$$x_1, x_2 \geq 0$$

---

† Constraints can take the form $ax_1 + bx_2 < r$, $ax_1 + bx_2 = r$, or $ax_1 + bx_2 \geq r$.

The first two constraints come from the technological requirements specified in the problem. The remaining constraints, $x_1 \geq 0$ and $x_2 \geq 0$, are termed *nonnegativity* constraints. These constraints are necessary to assure that the variables take on only zero or positive values.

The constraints are graphed in Fig. 5.1 by plotting the equations

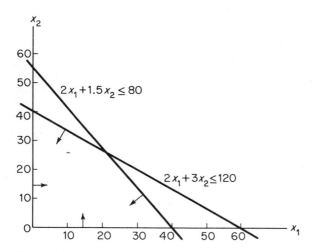

Figure 5.1

Since there are four constraints in the problem, four equations must be plotted. These are

$$2x_1 + 3x_2 = 120$$
$$2x_1 + 1.5x_2 = 80$$
$$x_1 = 0$$
$$x_2 = 0$$

Each of the equations is linear; consequently, each equation can be plotted from two points. The intercept points $(x_1 = 0, x_2 = 40)$ and $(x_1 = 60, x_2 = 0)$ are used to plot the first equation. Similarly, the second equation is plotted by using $(x_1 = 0, x_2 = 53.3)$ and $(x_1 = 40, x_2 = 0)$. The remaining equations, $x_1 = 0$ and $x_2 = 0$, represent the vertical and horizontal axis of the graph.

The constraints in this problem are expressed as inequalities. It was shown in Chapter 2 that the solution to an inequality consisted of the set $\{(x_1, x_2) \mid ax_1 + bx_2 \leq r\}$. The direction of each inequality is shown in Fig. 5.1 by an arrow. For instance, the inequality $2x_1 + 1.5x_2 \leq 80$ can be thought of as consisting of all ordered pairs for which $2x_1 + 1.5x_2 = 80$ and for which $2x_1 + 1.5x_2 < 80$. The line shows the ordered pairs for which $2x_1 + 1.5x_2 =$

80, and the arrow pointing downward and to the left shows the ordered pairs for which $2x_1 + 1.5x_2 < 80$.

The linear programming problem consists of maximizing the objective function, $P = 4x_1 + 5x_2$, subject to the constraints. To define feasible values of (x_1, x_2), we denote the four constraints by c_1, c_2, c_3, and c_4.

$$c_1: \quad 2x_1 + 3x_2 \leq 120$$

$$c_2: \quad 2x_1 + 1.5x_2 \leq 80$$

$$c_3: \quad x_1 \geq 0$$

$$c_4: \quad x_2 \geq 0$$

The values of x_1 and x_2 are limited to the set of ordered pairs that are solutions to the constraints. Using set notation, we express this set of ordered pairs as

$$F = \{(x_1, x_2) \mid c_1 \cap c_2 \cap c_3 \cap c_4\}$$

The set F contains all ordered pairs that are *feasible* solutions to the linear programming problem. These ordered pairs are shown in Fig. 5.2 by the

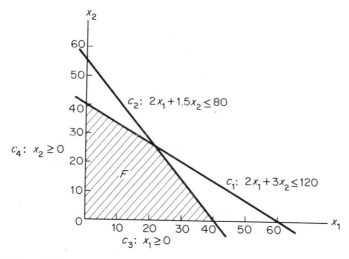

Figure 5.2

shaded area. This area is termed the *feasible region*. The ordered pairs defined by F are members of the *solution set* of the linear programming problem.

The optimal solution to the linear programming problem is the member of F that maximizes the objective function. Even in this simplified example, it is obvious that there are an infinite number of ordered pairs in the feasible region (i.e., shaded area). Since each of these ordered pairs is a member of the solution set, it would not be practical to investigate each ordered pair for optimality. Instead, we can make use of one of the important theorems in linear programming. This theorem, termed the *extreme point theorem*, states that the *optimal value of the objective function occurs at one of the extreme points of the feasible region*.† An extreme point in two-dimensional space is defined by the intersection of two equations. By referring to Fig. 5.2 we see that there are four extreme points. The values of x_1 and x_2 for each of these extreme points are determined by solving the appropriate pairs of intersecting equations. The solutions are

$$E = \{(0, 0), (40, 0), (20, 26.67), (0, 40)\}$$

According to the theorem, the objective function will be a maximum at one of these four ordered pairs. To determine which ordered pair maximizes the objective function, the objective function is evaluated for each extreme point. Table 5.1 shows that the objective function is a maximum when $x_1 = 20$ and

Table 5.1

(x_1, x_2)	P
(0, 0)	$0
(40, 0)	160.00
(20, 26.67)	213.33
(0, 40)	200.00

$x_2 = 26.67$.

The validity of the extreme point theorem can be demonstrated for the two-variable linear programming problem by superimposing the objective function on a graph of the feasible region. The objective function is

$$P = 4x_1 + 5x_2$$

which can also be written as

$$x_2 = 0.2P - 0.8x_1$$

The objective function, superimposed on the feasible region in Fig. 5.3, is shown as a series of parallel lines. Each line shows the combination of ordered pairs (x_1, x_2) that gives a specific value of P. For instance, the set of ordered pairs that are solutions to the equation $x_2 = 0.2(\$100) - 0.8x_1$ represents all combinations of (x_1, x_2) that result in profit of $100. The objective function is also plotted for $P = \$150$, $P = \$200$, and $P = \$213.33$.

Profit increases as the objective function moves upward and to the right

---

† This theorem is true provided that a unique, finite, optimal solution exists for the linear programming problem. There are three exceptions to this theorem. These are discussed in Sec. 5.3.3.

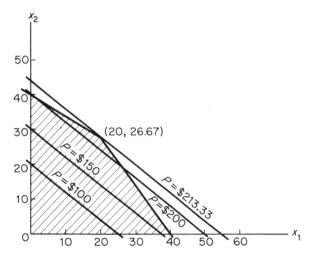

Figure 5.3

in the direction perpendicular to the slope of the function. Figure 5.3 shows that profit will be a maximum when the objective function passes through an extreme point of the feasible region. In this particular problem. the extreme point is (20, 26.67), and profit is $213.33.

Example: Use the graphical solution technique to determine the solution to the following linear programming problem.

$$\text{Maximize:} \quad P = 4x_1 + 3x_2$$
$$\text{Subject to:} \quad 21x_1 + 16x_2 \leq 336$$
$$13x_1 + 25x_2 \leq 325$$
$$15x_1 + 18x_2 \leq 270$$
$$x_1, x_2 \geq 0$$

The feasible region is defined by the intersection of the five linear inequalities. The inequalities $x_1 \geq 0$ and $x_2 \geq 0$ limit the solution to the first quadrant, i.e., nonnegative values of x_1 and x_2. The remaining three inequalities act as an upper bound on the values of x_1 and x_2. The feasible region is shown as the shaded area of Fig. 5.4.

The objective function has an optimal value at an extreme point of the feasible region. If the ordered pair (0, 0) is included as an extreme point, a total of five extreme points must be evaluated. In two-dimensional space, an extreme point is defined by the intersection of two equations. The five extreme points can thus be determined by specifying the five pairs of intersecting equations and solving each pair of equations simultaneously. The sets of ordered pairs specified by the five pairs of intersecting equations are

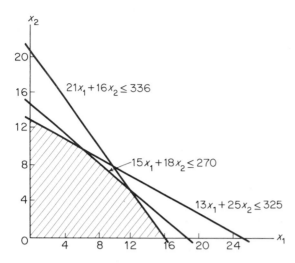

Figure 5.4

1. $\{(x_1, x_2) \mid x_1 = 0 \cap x_2 = 0\}$
2. $\{(x_1, x_2) \mid x_1 = 0 \cap 13x_1 + 25x_2 = 325\}$
3. $\{(x_1, x_2) \mid 15x_1 + 18x_2 = 270 \cap 13x_1 + 25x_2 = 325\}$
4. $\{(x_1, x_2) \mid 21x_1 + 16x_2 = 336 \cap 15x_1 + 18x_2 = 270\}$
5. $\{(x_1, x_2) \mid 21x_1 + 16x_2 = 336 \cap x_2 = 0\}$

Solving each pair of equations simultaneously gives the set of extreme points, $E = \{(0, 0), (0, 13), (6.38, 9.68), (12.45, 4.57), (16,0)\}$.

Not all pairs of intersecting equations in a linear programming problem specify an extreme point. For instance, the equations $21x_1 + 16x_2 = 336$ and $13x_1 + 25x_2 = 325$ intersect at $(9.7, 8.2)$. This point is not in the feasible region. Consequently, the ordered pair $(9.7, 8.2)$ is *nonfeasible* and is not considered in evaluating the objective function. Table 5.2 shows that the

Table 5.2

(x_1, x_2)	P
$(0, 0)$	$0
$(0, 13)$	39.00
$(6.38, 9.68)$	54.56
$(12.45, 4.57)$	63.51
$(16, 0)$	64.00

objective function is a maximum when $x_1 = 16$ and $x_2 = 0$.

5.2.2 MINIMIZATION

The graphical solution for a minimization problem is found by using the technique introduced in the preceding section. The difference between the minimization and maximization problems is that the extreme point that results in a minimum value of the objective function is the solution to the minimization problem rather than that which results in a maximum value of the function. To illustrate, consider the problem of the speciality metals manufacturer previously described in Sec. 5.1. The mathematical formulation of the problem was

$$\text{Minimize:} \quad C = 50x_1 + 20x_2$$
$$\text{Subject to:} \quad 2x_1 - 1x_2 \geq 0$$
$$1x_1 + 4x_2 \geq 80$$
$$0.90x_1 + 0.80x_2 \geq 40$$
$$x_1, x_2 \geq 0$$

All ordered pairs (x_1, x_2) that satisfy the five inequalities are feasible solutions. These are shown by the shaded area in Fig. 5.5.

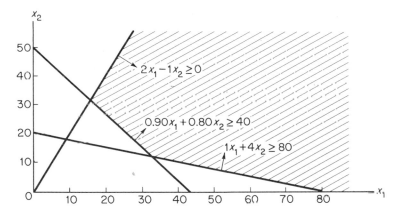

Figure 5.5

The objective function is minimized at an extreme point of the feasible region. Figure 5.5 shows that three extreme points must be evaluated. These points are described by the sets

1. $\{(x_1, x_2) \mid 2x_1 - 1x_2 = 0 \cap 0.90x_1 + 0.80x_2 = 40\}$
2. $\{(x_1, x_2) \mid 0.90x_1 + 0.80x_2 = 40 \cap 1x_1 + 4x_2 = 80\}$
3. $\{(x_1, x_2) \mid 1x_1 + 4x_2 = 80 \cap x_2 = 0\}$

The solution of each of the three pairs of simultaneous equations gives the set of extreme points, $E = \{(16, 32), (34.32, 11.42), (80, 0)\}$. The value of the objective function at each extreme point is shown in Table 5.3. The objective

Table 5.3

(x_1, x_2)	C
(16, 32)	$1440
(34.32, 11.42)	$1944.40
(80, 0)	$4000

function is minimized for $x_1 = 16$ and $x_2 = 32$.

Example: Use the graphical solution technique to determine the solution to the following linear programming problem.

$$\begin{aligned}
\text{Minimize:} \quad & C = 3x_1 + 2x_2 \\
\text{Subject to:} \quad & 2x_1 + x_2 \geq 5 \\
& x_1 + 3x_2 \geq 6 \\
& x_1 + x_2 \geq 4 \\
& x_1, x_2 \geq 0
\end{aligned}$$

The feasible region for the linear programming problem is shown in Fig. 5.6. From the extreme point theorem, we know that the objective function

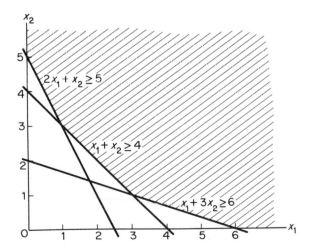

Figure 5.6

is a minimum at one of the extreme points. The extreme points are determined by selecting the pairs of equations that define the extreme points and solving each pair of equations simultaneously. This results in the set of extreme points

$$E = \{(0, 5), (1, 3), (3, 1), (6, 0)\}$$

By evaluating the objective function at each of these extreme points, we find that the objective function has a minimum value of $C = 9$ at the extreme point $x_1 = 1$, $x_2 = 3$.

5.2.3 CHARACTERISTICS ILLUSTRATED BY THE GRAPHIC SOLUTION

The graphical solution illustrates several important characteristics of the linear programming problem. As implied by the name "linear programming," one of these characteristics is that the inequalities and the objective function are linear. This assumption of linearity is made for all linear programming problems, regardless of the number of inequalities or variables.

A second characteristic illustrated by the graphical solution is that an optimal solution occurs at an extreme point of the feasible region. Feasible solutions to the linear programming problem are defined by sets of linear inequalities or equations. The feasible region, as illustrated in Fig. 5.2, contains an infinite number of ordered pairs that are solutions to the system of inequalities. At least one of these solutions will be optimal. The graphical solution shows that this optimal solution occurs at one of the finite number of extreme points rather than at one of the infinite number of ordered pairs that are solutions to the system of inequalities.

A third characteristic important in the linear programming problem is that the extreme points are defined by the intersection of linear equations. With two variables, two equations are required to define an extreme point. Three variables would require three equations to define an extreme point. Similarly, with n variables, n equations are required to define an extreme point. Since an optimal solution occurs at an extreme point, the number of candidates for the optimal solution is limited to the finite number of extreme points.

The simplex solution procedure for the general linear programming problem is presented in Chapter 6. This algorithm provides a series of rules that permits a systematic evaluation of selected extreme points of the feasible region. One important difference between the simplex algorithm and the graphical solution technique is that the algorithm provides a method of reducing the number of extreme points that must be evaluated to a relatively small subset of the total set of extreme points. Since there are often more than one hundred thousand extreme points in a linear programming problem, the importance of the simplex algorithm in reducing the number of extreme points that must be investigated to a manageable number is obvious.

5.3 Assumptions, Terminology, and Special Cases

Certain assumptions and terms are required in the linear programming model. Several of these have been introduced. The purpose of this section is to introduce the additional assumptions of linearity, continuity, and convexity, to show that linear programming implies constant returns to scale, and to illustrate exceptions to the extreme point theorem of linear programming.

5.3.1 LINEARITY, CONTINUITY, AND RETURNS TO SCALE

One of the assumptions in linear programming is that the objective function and constraints are linear. The problem of scheduling production in a machine shop illustrates this assumption. The problem was

$$\text{Maximize:} \quad P = 4x_1 + 5x_2$$
$$\text{Subject to:} \quad 2x_1 + 3x_2 \leq 120$$
$$2x_1 + 1.5x_2 \leq 80$$
$$x_1, x_2 \geq 0$$

The assumption of linearity imposes an important restriction on the structure of the problem. This restriction is that both profits and resource usage are a linear combination of the number of units of each product. In this problem profit is given by $P = 4x_1 + 5x_2$. The linear form of the objective function thus prohibits the competitive and substitutive effects that sometimes exist between products or activities. In certain industries, for instance, the manufacture of one product enables the manufacture of a second product at lesser cost. Such "second-order effects" are not included in the linear programming model.†

The result of the assumption of linearity involves what economists term "constant returns to scale." In the example, two units of both resource one and resource two are required to manufacture one unit of product one. This requirement remains regardless of the number of units of product one manufactured. The assumption of linearity, therefore, implies that the productivity of the resource is independent of the number of units of the product manu-

---

† These types of effects can, however, be included in more advanced programming models. See Harvey M. Wagner, *Principles of Operations Research* (Prentice-Hall, Inc., Englewood Cliffs, New Jersey, 1969), Chapter 14.

factured; i.e., the returns from the manufacturing process are constant regardless of the scale of production.

An additional restriction in the linear programming model comes from the assumption of continuity. The solution to the linear programming problem occurs at an extreme point. Since the equations that define extreme points are continuous, the solutions need not have integer values. This implies that fractional units of products or activities are possible. For instance, the optimal solution to the problem of scheduling production in the machine shop was $x_1 = 20$ and $x_2 = 26.67$.

The restrictions on the linear programming model due to the assumptions of linearity and continuity are not as severe as one might expect. Profit (or cost) and resource usage are often described quite satisfactorily by linear relationships, especially in the relevant range of production. Fractional solutions may present no problem. For those cases in which fractional solutions are impossible, judicial rounding sometimes provides a practical suboptimal solution. If rounding is unsatisfactory, *integer programming* can be employed.†

5.3.2 CONVEX SETS

The solution set or feasible region in a linear programming problem is sometimes referred to as a *convex set*. By definition, a set is *convex* if each point on a straight line that joins two points in the set is also in the set. To illustrate this definition, consider the two sets in Fig. 5.7. The set in Fig. 5.7(a) represents a typical solution set for a two-variable linear programming problem and is convex. It is impossible to select two points in this set such that all points on a straight line that joins the two points are not also in the set. This convex solution set was formed by the intersection of a system of linear inequalities.

The set in Fig. 5.7(b) does not have the property of convexity. This is shown by the fact that points on the line connecting the points $(1, 4)$ and $(3, 3)$ are not in the set.

It can be seen that the set in Fig. 5.7(b) is not formed by the intersection of a system of linear inequalities. If the lines that form the boundaries of the set are extended, we see that there is some area on both sides of at least one line. This means, of course, that at least one of the inequalities that forms the set in Fig. 5.7(b) is not satisfied.

Since the intersection of a system of linear inequalities defines the solution set for a linear programming problem and the set of points generated by this

---

† A good discussion of integer programming is provided by Donald R. Plane and Claude McMillan, Jr., *Discrete Optimization* (Prentice-Hall, Inc., Englewood Cliffs, N.J., 1971).

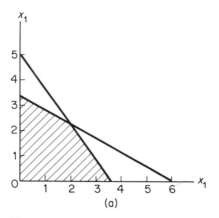

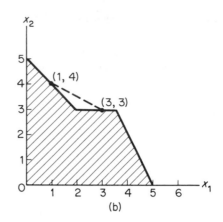

(a)   (b)

Figure 5.7

intersection has the property of convexity, one may ask the reason for introducing the concept of convex sets. The answer, aside from the theoretical
importance of convexity, is that the analyst need be aware that the inequalities in a linear programming problem must be satisfied for all possible values
of the variables rather than only a selected subset of values.

5.3.3 NONFEASIBLE, MULTIPLE OPTIMAL, AND UNBOUNDED SOLUTIONS

The linear programming problems discussed in the preceding sections have
had a single optimal solution. Although problems of this type are common,
other possibilities exist. These include: (1) no feasible solution; (2) multiple
optimal solutions; and (3) unbounded optimal solutions. These are the exceptions referred to in the footnote on p. 146 to the extreme point theorem.

NO FEASIBLE SOLUTION. The solution set for a linear programming problem is formed by the intersection of the system of constraining inequalities.
If this intersection is empty, no feasible solution exists for the problem. As
an example, assume that the machine shop problem is modified by adding
the constraint, $x_1 + x_2 \geq 50$. The problem becomes

$$\begin{aligned}
\text{Maximize:} \quad & P = 4x_1 + 5x_2 \\
\text{Subject to:} \quad & 2x_1 + 3x_2 \leq 120 \\
& 2x_1 + 1.5x_2 \leq 80 \\
& x_1 + x_2 \geq 50 \\
& x_1, x_2 \geq 0
\end{aligned}$$

The constraints are shown in Fig. 5.8. It can be seen from this figure that

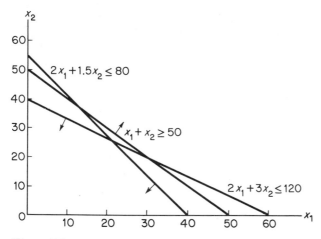

Figure 5.8

there is no set of points that simultaneously satisfies all the inequalities. Consequently, there is no feasible solution to the linear programming problem.

MULTIPLE OPTIMAL SOLUTIONS. We have stressed that an optimal solution occurs at an extreme point of the feasible region. If the objective function is parallel to one of the constraints, however, there will be multiple optimal solutions rather than a unique optimal solution. To illustrate, assume that the objective function for the machine shop problem is changed to $P = 4x_1 + 6x_2$. The problem becomes

$$\text{Maximize:} \quad P = 4x_1 + 6x_2$$
$$\text{Subject to:} \quad 2x_1 + 3x_2 \leq 120$$
$$2x_1 + 1.5x_2 \leq 80$$
$$x_1, x_2 \geq 0$$

The constraints and the objective function are shown in Fig. 5.9.

The figure illustrates that the objective function is parallel to the constraint $2x_1 + 3x_2 \leq 120$. Increases in profit are shown by moving the objective function upward and to the right. Profit is a maximum when the objective function and the constraint are coincident. All ordered pairs described by the constraint between the extreme points $(0, 40)$ and $(20, 26.67)$ are optimal solutions. Profit is \$240 at either extreme point, or at any linear combination of these extreme points.

UNBOUNDED OPTIMAL SOLUTION. In problems with multiple optimal solutions, the objective function has the same value for all optimal solutions. In

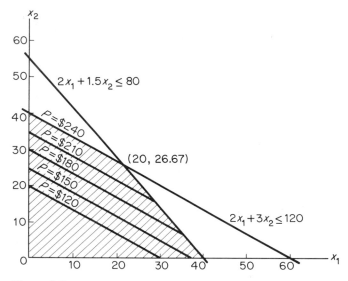

Figure 5.9

the preceding example this value was $P = \$240$. Based upon the constraints, the upper *bound* on the objective function was $240.

Certain linear programming problems do not have a bound on the value of the objective function. In these cases the objective function is *unbounded*. To illustrate, consider the problem

$$
\begin{aligned}
\text{Maximize:} \quad P = \quad & x_1 + 2x_2 \\
\text{Subject to:} \quad & -x_1 + x_2 \le 2 \\
& x_1 + x_2 \ge 4 \\
& x_1, x_2 \ge 0
\end{aligned}
$$

The problem is shown in Fig. 5.10.

The feasible region for this problem is convex. The extreme points of the feasible region are $(4, 0)$ and $(1, 3)$. Neither of these extreme points, however, provides an optimal solution to the problem. This is shown in Fig. 5.10 by the fact that the objective function can increase indefinitely without reaching an upper bound. In fact, given any value of the objective function, there is always a solution that gives a greater value of the objective function. Consequently, the objective function for this problem is unbounded.

These examples illustrate the exception given in the footnote on p. 146 to the extreme point theorem of linear programming. The theorem stated that the optimal value of the objective function occurs at an extreme point of the feasible region. The footnote added that the theorem is true provided that a unique finite optimal solution exists. In the first example, the solution set was

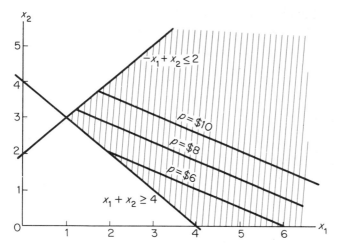

Figure 5.10

empty. There was no feasible solution for the problem. The second example illustrated the case of multiple optimal solutions. Instead of a unique optimal solution, there were an infinite number of optimal solutions, each having the same value. In the final example, the linear programming problem had no finite optimal solution. For any value of the objective function, it was possible to find a solution for which the objective function had a greater value. This third and final exception to the extreme point theorem occurred when the objective function was unbounded.

5.3.4 CHANGING THE SENSE OF OPTIMIZATION

Certain computer programs for solving linear programming problems are written to accept either a maximization problem or a minimization problem. This means that it is sometimes necessary to change the sense of optimization in a linear programming problem, i.e., change a maximization problem to a minimization problem or vice versa.

Changing the sense of optimization in a linear programming problem is straightforward. Any linear maximization problem can be solved as a linear minimization problem by changing the signs of the objective function coefficients and minimizing the objective function subject to the original constraints. Similarly, any linear minimization problem can be solved as a maximization problem by changing the signs of the objective function coefficients and maximizing the objective function subject to the constraints. To illustrate, consider the problem

$$\text{Minimize:} \quad C = x_1 + x_2$$
$$\text{Subject to:} \quad 3x_1 + 8x_2 \geq 24$$
$$3x_1 + 2x_2 \geq 12$$
$$x_1 \quad\quad\; \geq 1$$
$$x_2 \geq 0$$

The minimization problem can be solved as the following maximization problem simply by changing the signs of the objective function coefficients.

$$\text{Maximize:} \quad -C = -x_1 - x_2$$
$$\text{Subject to:} \quad 3x_1 + 8x_2 \geq 24$$
$$3x_1 + 2x_2 \geq 12$$
$$x_1 \quad\quad\; \geq 1$$
$$x_2 \geq 0$$

The graphical solution is shown in Fig. 5.11. The extreme points are $E =$

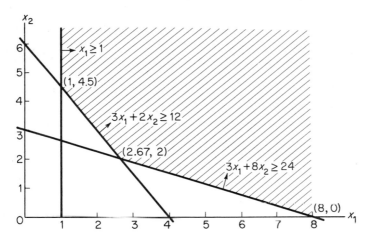

Figure 5.11

$\{(1, 4.5), (2.67, 2), (8, 0)\}$. Table 5.4 gives the value of $C = x_1 + x_2$ and $-C = -x_1 - x_2$.

Table 5.4

(x_1, x_2)	C	$-C$
(1, 4.5)	5.5	−5.5
(2.67, 2)	4.67	−4.67
(8, 0)	8.0	−8.0

The minimum value of C is 4.67 and the maximum value of $-C$ is -4.67.†
The solution obtained by minimizing C is thus equivalent to that obtained
by maximizing $-C$.

5.4 Applications of Linear Programming

One of the objectives of this chapter is to provide examples of important
applications of linear programming. On the basis of these examples, the
reader can begin to express problems in the framework of the linear programming model. These examples involve problems of allocation, such as
product mix, feed mix, fluid blending, portfolio selection, and multiperiod
scheduling. The analytical techniques needed to solve these problems are presented in Chapter 6.‡

5.4.1 PRODUCT MIX

In product mix problems management must determine the quantities of products to be manufactured during a specific time period. To illustrate, consider
the case of the Hall Manufacturing Company. Hall manufactures four types
of electronic subassemblies for use in aircraft avionic equipment. Each subassembly requires assembly labor, test labor, resistors, and capacitors. The
requirements, availability of parts and labor, and the profit per subassembly
is projected for the month of September in Table 5.5. Determine the quanti-

Table 5.5

| Resource | Electronic Subassembly | | | | Resource Availabilities |
	Type 1	Type 2	Type 3	Type 4	
Assembly labor	6.0	5.0	3.5	4.0	600 hours
Test labor	1.0	1.5	1.2	1.2	120 hours
Resistors	4	3	3	3	400 units
Capacitors	2	2	2	3	300 units
Profit	$4.25	$6.25	$5.00	$4.50	

---

† Note that $-4.67 > -5.5$ and $-4.67 > -8.0$. Thus, -4.67 is the maximum.

‡ These applications, while representative, by no means illustrate all uses of linear
programming. Additional examples are given in the references at the end of the chapter.
Problems that utilize the transportation and assignment algorithms of linear programming
are discussed in Chapter 8.

ties of the four products that should be manufactured during September in order to maximize profit. Assume that Hall is contractually required to supply 40 type 1 subassemblies.

To formulate the linear programming problem we let x_j, for $j = 1, 2, 3, 4$, represent the quantity of product j manufactured. The problem can be expressed as

$$\text{Maximize:} \quad P = 4.25x_1 + 6.25x_2 + 5.00x_3 + 4.50x_4$$

$$\begin{aligned}
\text{Subject to:} \quad & 6.0x_1 + 5.0x_2 + 3.5x_3 + 4.0x_4 \leq 600 \\
& 1.0x_1 + 1.5x_2 + 1.2x_3 + 1.2x_4 \leq 120 \\
& 4x_1 + 3x_2 + 3x_3 + 3x_4 \leq 400 \\
& 2x_1 + 2x_2 + 2x_3 + 3x_4 \leq 300 \\
& x_1 \qquad\qquad\qquad\qquad\quad \geq 40 \\
& x_2, x_3, x_4 \geq 0
\end{aligned}$$

Because of the number of variables, the problem cannot be solved graphically. The solution is instead obtained by using the procedure discussed in Chapter 6. The optimal solution is $x_1 = 76.9$ type 1 subassemblies, $x_2 = 20.5$ type 2 subassemblies, $x_3 = 10.3$ type 3 subassemblies, and $x_4 = 0$ type 4 subassemblies. Profit for the month is $506.40.

The following two examples provide additional illustrations of the product mix problem.

Example: A manufacturer can produce three different products during the month of October. Each of these products requires casting, grinding, assembly, and testing. The maximum available hours of capacity of each of the processes during the month, the requirements for each unit of product, and the profit per unit of product are given in Table 5.6. Formulate the linear

Table 5.6

		Product		Available Resource
	A	B	C	Hours
Casting	1.0	1.5	2.0	200
Grinding	0.6	1.0	0.8	120
Assembly	1.5	2.0	1.4	240
Testing	0.3	0.2	0.2	40
Profit	$12.00	$18.00	$16.00	

programming problem. Assume that all products produced during the month can be sold.

Let x_j, for $j = 1, 2, 3$, represent quantities of products A, B, and C. The linear programming problem is

$$
\begin{aligned}
\text{Maximize:} \quad P &= 12x_1 + 18x_2 + 16x_3 \\
\text{Subject to:} \quad 1.0x_1 &+ 1.5x_2 + 2.0x_3 \leq 200 \\
0.6x_1 &+ 1.0x_2 + 0.8x_3 \leq 120 \\
1.5x_1 &+ 2.0x_2 + 1.4x_3 \leq 240 \\
0.3x_1 &+ 0.2x_2 + 0.2x_3 \leq 40 \\
x_j &\geq 0 \quad \text{for } j = 1, 2, 3
\end{aligned}
$$

Example: McGraw Chemical Company uses nitrates, phosphates, potash, and an inert filler material in the manufacture of chemical fertilizers. The firm mixes these ingredients to make three basic fertilizers: 5-10-5, 5-8-8, and 8-12-12 (numbers represent percent by weight of nitrates, phosphates, and potash in each ton of fertilizer). McGraw receives $70, $75, and $80, respectively, for each ton of the 5-10-5, 5-8-8, and 8-12-12 fertilizer. The cost of the ingredients is $200 per ton of nitrates, $60 per ton of phosphates, $100 per ton of potash, and $20 per ton of the inert filler material. The direct costs of mixing, packing, and selling the fertilizer are $25 per ton.

McGraw has 1200 tons of nitrates, 2000 tons of phosphates, 1500 tons of potash, and an unlimited supply of the inert filler on hand. They will not receive additional chemicals until next month. McGraw has a delivery contract for 8000 tons of 5-8-8 during the month. They believe that they can sell or store at negligible cost all fertilizer produced during the month. Formulate the linear programming problem to maximize profits.

Let x_j, for $j - 1, 2, 3$, represent tons of 5-10-5, 5-8-8, and 8-12-12 fertilizers. The coefficients of the objective function are calculated by subtracting the cost of the ingredients and the $25 mixing cost from the sales price. To illustrate, consider the objective function coefficient of x_1 (i.e., 5-10-5 fertilizer). The cost of the nitrates in one ton of 5-10-5 fertilizer is 0.05($200) or $10. The cost of the phosphates is 0.10($60) or $6. Similarly, the cost of the potash is 0.05($100) = $5, and the cost of the inert filler is 0.80($20) = $16. Subtracting these costs plus the cost of mixing, packing, and selling the 5-10-5 fertilizer from the sales price gives an objective function coefficient of

$$
\begin{aligned}
c_1 = \$70 &- 0.05(\$200) - 0.10(\$60) - 0.05(\$100) \\
&- 0.80(\$20) - \$25 = \$8.00
\end{aligned}
$$

The remaining objective function coefficients are calculated in the same manner.

The constraints shown below describe the resource limitations and the contractual requirements. The linear programming problem is

$$\text{Maximize:} \quad P = 8.00x_1 + 11.40x_2 + 6.20x_3$$
$$\text{Subject to:} \quad 0.05x_1 + 0.05x_2 + 0.08x_3 \leq 1200$$
$$0.10x_1 + 0.08x_2 + 0.12x_3 \leq 2000$$
$$0.05x_1 + 0.08x_2 + 0.12x_3 \leq 1500$$
$$x_2 \geq 8000$$
$$x_1, x_3 \geq 0.$$

5.4.2 FEED MIX

The feed mix problem is another form of the allocation problem. This problem involves determining the proportion of several ingredients that, when mixed, satisfy certain criteria and give a minimum cost. To illustrate, consider the problem of mixing feed for a dairy herd. The dairyman may purchase and mix quantities of three types of grain, each containing differing amounts of four nutritional elements. The dairyman has developed nutritional requirements for the feed mixture. His objective is to minimize the cost of the mixture while meeting the nutritional requirements. The data are given in Table 5.7.

Table 5.7

Nutritional Elements	Units of Nutritional Elements Contained in One Pound of			Minimum No. of Units Required
	Grain 1	Grain 2	Grain 3	
1	4	6	8	2500
2	2	3	1	750
3	4	3	1	900
4	0.7	0.3	0.8	250
Cost per pound	$0.025	$0.020	$0.017	
Available	Unlimited	1000 lbs.	Unlimited	

Let x_j, for $j = 1, 2, 3$, represent pounds of grain j in the feed mix. The linear programming problem to minimize cost is

$$\text{Minimize:} \quad C = 0.025x_1 + 0.020x_2 + 0.017x_3$$
$$\text{Subject to:} \quad 4x_1 + 6x_2 + 8x_3 \geq 2500$$
$$2x_1 + 3x_2 + 1x_3 \geq 750$$
$$4x_1 + 3x_2 + 1x_3 \geq 900$$

$$0.7x_1 + 0.3x_2 + 0.8x_3 \geq 250$$
$$x_2 \leq 1000$$
$$x_1, x_2, x_3 \geq 0$$

The optimal solution is found by using the simplex algorithm. The solution is $x_1 = 75.0$ pounds or grain 1, $x_2 = 134.4$ pounds of grain 2, and $x_3 = 196.4$ pounds of grain 3. The cost of this mix is $7.90.

Example: American Foods, Inc., is developing a low-calorie, high-protein diet supplement called Hi-Pro. The specifications for Hi-Pro have been established by a panel of medical experts. These specifications, along with the calorie, protein, and vitamin content of three basic foods, are given in Table 5.8.

Table 5.8

	Units of Nutritional Elements per 8-Ounce Serving of Basic Foods			
	Basic Foods			Hi-Pro
Nutritional Elements	No. 1	No. 2	No. 3	Specifications
Calories	350	250	200	≤ 300
Protein	250	300	150	≥ 200
Vitamin A	100	150	75	≥ 100
Vitamin C	75	125	150	≥ 100
Cost per serving	$0.15	$0.20	$0.12	

Formulate the linear programming model to minimize cost.

Let x_j, for $j = 1, 2, 3$, represent the proportion of basic foods 1, 2, and 3 in an 8-ounce serving of Hi-Pro. The linear programming problem is

Minimize: $C = 0.15x_1 + 0.20x_2 + 0.12x_3$
Subject to: $350x_1 + 250x_2 + 200x_3 \leq 300$
$250x_1 + 300x_2 + 150x_3 \geq 200$
$100x_1 + 150x_2 + 75x_3 \geq 100$
$75x_1 + 125x_2 + 150x_3 \geq 100$
$x_1 + x_2 + x_3 = 1$
$x_j \geq 0$ for $j = 1, 2, 3$

Example: Chico Candy Company, Inc., mixes three types of candies to obtain a one-pound box of candy. The box of candy sells for $0.85 per pound, and the three ingredient candies sell for $1.00, $0.50, and $0.25 per pound, respectively. The mixture must contain at least 0.3 pounds of the first type of candy, and the weight of the first two candies must at least equal the weight of the third. Formulate the linear programming problem to maximize profit.

Let x_j, for $j = 1, 2, 3$, represent the quantity in pounds of each of the three ingredient candies. The objective function coefficients are found by subtracting the cost of each type of candy from the sales price of the box. For instance, the coefficient of x_1 is $\$0.85 - \$1.00 = -\$0.15$. The linear programming problem is

$$\text{Maximize:} \quad P = -0.15x_1 + 0.35x_2 + 0.60x_3$$
$$\text{Subject to:} \quad
\begin{aligned}
x_1 & & & \geq 0.3 \\
x_1 + & x_2 - & x_3 & \geq 0 \\
x_1 + & x_2 + & x_3 & = 1 \\
x_2, & x_3 & & \geq 0
\end{aligned}$$

5.4.3 FLUID BLENDING

Another variation of the allocation problem occurs when fluids, such as chemicals, plastics, molten metals, and oils, must be blended to form a product. Each of the resources included in the blend has certain properties and costs. The fluid blending problem involves blending the different resources to form a final product that meets certain criteria and has a minimum cost or a maximum profit. To illustrate, consider a simplified example of gasoline blending.

The Sigma Oil Company markets three brands of gasoline: Sigma Supreme, Sigma Plus, and Sigma. The three gasolines are made by blending two grades of gasoline, each with a different octane rating, and a special high-octane lead additive. In the gasoline blends, the octane rating of the blend is a linear combination of the octane ratings of the component gasolines and additives. The relevant data are shown in Table 5.9.

Table 5.9

Brands	Minimum Octane	Sales Price to Retailer (per gallon)
Sigma Supreme	100	$0.20
Sigma Plus	94	0.16
Sigma	86	0.14

Blending Component	Octane	Cost (per gallon)	Supply (gallons)
1	110	$0.10	20,000
2	80	0.07	12,000
3	1500	1.00	500

It is assumed that all gasoline blended during the period can be sold or stored at negligible cost. An alternative assumption, not included in this problem, is that forecasts of demand for the three brands of gasoline are available and that gasoline unsold at the end of the period is stored at a cost and sold during the following period.

To formulate the linear programming model, let x_{ij}, for $i = 1, 2, 3$ and $j = 1, 2, 3$, represent the gallons of blending component i used in gasoline blend j. The problem can be expressed as

$$\text{Maximize:} \quad P = 0.10x_{11} + 0.13x_{21} - 0.80x_{31} + 0.06x_{12} + 0.09x_{22}$$
$$- 0.84x_{32} + 0.04x_{13} + 0.07x_{23} - 0.86x_{33}$$

$$\text{Subject to:} \quad 110x_{11} + 80x_{21} + 1500x_{31} \geq 100(x_{11} + x_{21} + x_{31})$$
$$110x_{12} + 80x_{22} + 1500x_{32} \geq 94(x_{12} + x_{22} + x_{32})$$
$$110x_{13} + 80x_{23} + 1500x_{33} \geq 86(x_{13} + x_{23} + x_{33})$$
$$x_{11} + x_{12} + x_{13} \leq 20{,}000$$
$$x_{21} + x_{22} + x_{23} \leq 12{,}000$$
$$x_{31} + x_{32} + x_{33} \leq 500$$
$$x_{ij} \geq 0 \quad \text{for } i = 1, 2, 3 \text{ and } j = 1, 2, 3$$

The objective function shows that there is a $0.10 profit for each gallon of blending component 1 used in Sigma Supreme, a $0.13 profit for each gallon of blending component 2 used in Sigma Supreme, an $0.80 loss for each gallon of blending component 3 used in Sigma Supreme, etc. These numbers are calculated by subtracting the cost of the blending component from the sales price of the gasoline. For instance, each gallon of blending component 1 used in Sigma Supreme costs $0.10 and is sold for $0.20, giving a profit of $0.10.

The first three constraints specify that the blends must have minimum octane ratings. The next three constraints restrict the quantity of each component used for blending to the supply available during the period. The solution to the problem is given in Table 5.10. Profit for the period is $3238.

Table 5.10   Gallons of Blend i Used in Gasoline j

Blend	Supreme	Plus	Sigma	Total
1	20,000	0	0	20,000
2	12,000	0	0	12,000
3	403	0	0	403
Total	32,403	0	0	32,403

Example: American whiskeys are made by blending different types of bourbon whiskeys. Assume that Kentucky Whiskeys, Inc. mixes three bourbon whiskeys, A, B, and C, to obtain two blends, Kentucky Premium and Kentucky Smooth. The recipes used by Kentucky for their two blends are given in Table 5.11.

Table 5.11

Blend	Ingredients	Price (per gallon)
Kentucky Premium	Not less than 60% A Not more than 20% C	$7.00
Kentucky Smooth	Not less than 30% A Not more than 60% C	$5.50

Supplies of the basic bourbon whiskeys and their costs are shown in Table 5.12. Using this data, formulate the linear programming problem to maximize profits.

Table 5.12

Bourbon	Quantity Available (gallons)	Cost (per gallon)
A	1500	$6.75
B	2000	$6.00
C	1000	$4.00

Let x_{ij}, for $i = 1, 2, 3$ and $j = 1, 2$ represent the quantity of bourbon i used in one gallon of blend j. The linear programming problem is

Maximize: $P = 0.25x_{11} - 1.25x_{12} + 1.00x_{21} - 0.50x_{22}$
$+ 3.00x_{31} + 1.50x_{32}$

Subject to: $x_{11} \geq 0.60(x_{11} + x_{21} + x_{31})$
$x_{31} \leq 0.20(x_{11} + x_{21} + x_{31})$
$x_{12} \geq 0.30(x_{12} + x_{22} + x_{32})$
$x_{32} \leq 0.60(x_{12} + x_{22} + x_{32})$

$x_{11} + x_{12} \leq 1500$
$x_{21} + x_{22} \leq 2000$
$x_{31} + x_{32} \leq 1000$
$x_{ij} \geq 0$

The objective function shows that there is a $0.25 profit on each gallon of bourbon A used in Kentucky Premium. This is determined by subtracting the cost per gallon of bourbon A from the price per gallon of Kentucky Premium. Similarly, there is a loss of $1.25 per gallon for each gallon of bourbon A used in Kentucky Smooth.

The first four constraints specify the maximum and minimum quantities of bourbons used in the blended whiskeys. For example, the first constraint shows that the proportion of blend A in Kentucky Premium must be not less than 60 percent. Similarly, the second constraint shows that the proportion of blend C in Kentucky Premium must not exceed 20 percent. The last three constraints limit the supplies of the bourbon whiskeys available for blending during the period.

Example: Plains Coffee Company mixes Brazilian, Colombian, and Mexican coffees to make two brands of coffee, Plains X and Plains XX. The characteristics used in blending the coffees include strength, acidity, and caffeine. Test results of the available supplies of Brazilian, Colombian, and Mexican coffees are shown in Table 5.13.

Table 5.13

Imported Coffee	Price per Pound	Strength Index	Acidity Index	Percent Caffeine	Supply Available
Brazilian	$0.30	6	4.0	2.0	40,000 lb
Colombian	0.40	8	3.0	2.5	20,000 lb
Mexican	0.35	5	3.5	1.5	15,000 lb

The requirements for Plains X and Plains XX coffees are given in Table 5.14. Assume that 35,000 lb of Plains X and 25,000 lb of Plains XX are to be sold. Formulate the linear programming problem to maximize profits.

Table 5.14

Plains Coffee	Price per Pound	Minimum Strength	Maximum Acidity	Maximum Percent Caffeine	Quantity Demanded
X	$0.45	6.5	3.8	2.2	35,000 lb
XX	0.55	6.0	3.5	2.0	25,000 lb

Let x_{ij}, for $i = 1, 2, 3$ and $j = 1, 2$ represent the pounds of imported coffees used in Plains coffees. The linear programming problem is

Maximize: $\quad P = 0.15x_{11} + 0.25x_{12} + 0.05x_{21} + 0.15x_{22}$
$\qquad\qquad + 0.10x_{31} + 0.20x_{32}$

Subject to:
$$6x_{11} + \quad 8x_{21} + \quad 5x_{31} \geq \quad 6.5(x_{11} + x_{21} + x_{31})$$
$$6x_{12} + \quad 8x_{22} + \quad 5x_{32} \geq \quad 6.0(x_{12} + x_{22} + x_{32})$$

$$4.0x_{11} + \quad 3.0x_{21} + \quad 3.5x_{31} \leq \quad 3.8(x_{11} + x_{21} + x_{31})$$
$$4.0x_{12} + \quad 3.0x_{22} + \quad 3.5x_{32} \leq \quad 3.5(x_{12} + x_{22} + x_{32})$$

$$2.0x_{11} + \quad 2.5x_{21} + \quad 1.5x_{31} \leq \quad 2.2(x_{11} + x_{21} + x_{31})$$
$$2.0x_{12} + \quad 2.5x_{22} + \quad 1.5x_{32} \leq \quad 2.0(x_{12} + x_{22} + x_{32})$$

$$x_{11} + x_{12} \leq 40,000$$
$$x_{21} + x_{22} \leq 20,000$$
$$x_{31} + x_{32} \leq 15,000$$

$$x_{11} + x_{21} + x_{31} = 35,000$$
$$x_{12} + x_{22} + x_{32} = 25,000$$
$$x_{ij} \geq 0$$

The coefficients of the objective function represent profits from imported coffee i in mix j. The first six constraints establish restrictions on strength, acidity, and caffeine. Limitations on the supply of imported coffees are given by the next three constraints and the quantity of Plains X and Plains XX coffees demanded during the period is described by the final two constraints.

5.4.4 PORTFOLIO AND MEDIA SELECTION

Another example of the allocation problem occurs in portfolio selection. Various financial institutions such as pension funds, insurance companies, and mutual funds, along with private individuals, are frequently faced with the problem of investing funds among the alternatives available. These alternatives vary in terms of safety, liquidity, income, growth, and other factors. Linear programming can be used to suggest the allocation of funds among the competing alternatives.

To illustrate the use of linear programming in portfolio selection, consider the case of the ABC Mutual Fund. ABC is an income-oriented mutual fund. To aid in its investment decisions, ABC has developed the investment alternatives given in Table 5.15. The return on investment is expressed as an annual rate of return on the invested capital. The risk is a subjective estimate on a scale from 0 to 10 made by the portfolio manager of the safety of the investment. The term of investment is the average length of time required to realize the return on investment indicated in Table 5.15.

ABC's objective is to maximize the return on investment. The guide lines for selecting the portfolio are (1) the average risk should not exceed 2.5; (2)

Table 5.15

Investment Alternative	Yearly Return on Investment	Risk	Term of Investment (yrs)
1. Blue chip stock	12%	2	4
2. Bonds	10	1	8
3. Growth stock	15	3	2
4. Speculation	25	4	10
5. Cash	0	0	0

the average term of investment should not exceed six years; and (3) at least 15 percent of the funds should be retained in the form of cash. To maximize the return on investment, let x_j, for $j = 1, 2, 3, 4, 5$ represent the proportion of funds to be invested in the jth alternative. The objective function and constraints are

$$\text{Maximize:} \quad P = 12x_1 + 10x_2 + 15x_3 + 25x_4 + 0x_5$$
$$\text{Subject to:} \quad x_1 + x_2 + x_3 + x_4 + x_5 = 1$$
$$2x_1 + 1x_2 + 3x_3 + 4x_4 \leq 2.5$$
$$4x_1 + 8x_2 + 2x_3 + 10x_4 \leq 6$$
$$x_5 \geq 0.15$$
$$x_j \geq 0 \quad \text{for } j = 1, 2, 3, 4, 5$$

The first constraint states that the proportions of funds invested in the various alternatives must total 1. The remainder of the constraints describe the risk, term of investment, and cash requirements. The solution to the problem is $x_1 = 40.7$ percent, $x_2 = 2.85$ percent, $x_3 = 0$, $x_4 = 41.4$ percent, and and $x_5 = 15.0$ percent. The overall rate of return on the portfolio is 15.5 percent.

The portfolio selection model is quite similar to the media selection model. Media selection involves the allocation of the advertising budget of a firm among various communication media in an effort to reach appropriate audiences. The objective is to maximize total effective exposures, subject to constraints on the advertising budget and the number of exposures per media.

To illustrate a simplified media selection model, consider the case of the Edwards Advertising Agency. Edwards is formulating an advertising campaign for one of its customers. A careful analysis of the available media has narrowed the allocation problem to that of determining the number of messages to appear in each of three magazines. The cost of advertisements and other relevant data are given in Table 5.16.

The cost per advertisement, characteristics of the media, and the audience size are available from industry sources. Edwards has established the maximum and minimum number of advertisements per medium and the relative

Table 5.16

	Media		
	1	*2*	*3*
Cost per advertisement	$500	$750	$1,000
Maximum number of ads	24	52	24
Minimum number of ads	6	12	0
Reader characteristics			
Age: 21–40	50%	60%	65%
Income: $10,000 or more	40%	50%	80%
Education: college	40%	40%	60%
Audience size	500,000	800,000	1,200,000

importance of the characteristics. The relative importance of the characteristics are: age, 0.5; income, 0.3; education, 0.2.

The linear programming problem is to maximize total effective exposures.
The coefficients of the objective function are the product of the audience size
multiplied by the "effectiveness coefficient" for each medium. The effectiveness coefficient is defined as equalling the sum of the products of the characteristics and their relative importance. The effectiveness coefficient for media
1 is

$$\text{effectiveness coefficient} = 0.50(0.5) + 0.40(0.3) + 0.40(0.2) = 0.450$$

Similarly, the effectiveness coefficients for media 2 and 3 are 0.530 and 0.685,
respectively. The product of the effectiveness coefficients and the audience
size gives the coefficients of the objective function. To illustrate, the coefficient
of the objective function for media 1 is 0.450(500,000) or 225,000.

If it is assumed that an advertising budget of \$35,000 has been established,
the linear problem to maximize total effective exposure is

$$\text{Maximize:} \quad E = 225{,}000x_1 + 424{,}000x_2 + 822{,}000x_3$$

$$
\begin{aligned}
\text{Subject to:} \quad 500x_1 + \quad 750x_2 + \quad 1000x_3 &\leq 35{,}000 \\
x_1 &\geq 6 \\
x_1 &\leq 24 \\
x_2 &\geq 12 \\
x_2 &\leq 52 \\
x_3 &\leq 24 \\
x_3 &\geq 0
\end{aligned}
$$

where x_j, for $j = 1, 2, 3$, represents the number of advertisements in the jth
media.

The solution to the problem is $x_1 = 6$, $x_2 = 12$, and $x_3 = 23$. The total
number of effective exposures is 25,344,000.

5.4.5 MULTIPERIOD SCHEDULING

Linear programming can often be applied to establish production schedules that extend over several time periods. For instance, many manufacturers must plan for seasonal variations in demand for their products. In these cases the costs of overtime must be balanced against the cost of carrying inventories from period to period. The objective is to establish a multiperiod schedule that minimizes all relevant costs of production of the product.

To illustrate multiperiod scheduling, consider the case of Kim Manufacturing, Inc. On October 1, Kim received a contract to supply 6000 units of a specialized product. The terms of the contract require that 1000 units be shipped in October, 3000 in November, and 2000 in December. Kim can manufacture 1500 units per month on regular time and 750 units per month on overtime. The manufacturing cost per item produced during a regular hour is $3, and the cost per item produced during overtime is $5. The monthly storage cost is $1. Formulate the linear programming problem to minimize total cost of production.

Let x_{ijk}, for $i = 1, 2, 3, j = 1, 2$, and $k = 1, 2, 3$, represent the number of units manufactured in month i using shift j and shipped in month k. With this notation, x_{121} represents the number of units manufactured in October using overtime and shipped in October. Similarly, x_{213} represents the number of units manufactured during November using regular time and shipped in December. The linear programming problem is

$$\text{Minimize:} \quad C = 3x_{111} + 5x_{121} + 4x_{112} + 6x_{122} + 5x_{113} + 7x_{123}$$
$$+ 3x_{212} + 5x_{222} + 4x_{213} + 6x_{223} + 3x_{313} + 5x_{323}$$

$$\text{Subject to:} \quad x_{111} + x_{112} + x_{113} \leq 1500$$
$$x_{212} + x_{213} \qquad\quad \leq 1500$$
$$x_{313} \qquad\qquad\qquad \leq 1500$$

$$x_{121} + x_{122} + x_{123} \leq 750$$
$$x_{222} + x_{223} \qquad\quad \leq 750$$
$$x_{323} \qquad\qquad\qquad \leq 750$$

$$x_{111} + x_{121} \qquad\qquad\qquad\qquad\qquad\qquad = 1000$$
$$x_{112} + x_{122} + x_{212} + x_{222} \qquad\qquad\qquad = 3000$$
$$x_{113} + x_{123} + x_{213} + x_{223} + x_{313} + x_{323} = 2000$$
$$x_{ijk} \geq 0$$

The coefficients of the objective function represent manufacturing, overtime, and storage cost. The first three constraints limit production during regular hours to 1500 units per month. The next three constraints limit production during overtime to 750 units per month. The final three constraints assure that the shipping schedule of 1000 units during October, 3000 units during November, and 2000 units during December is met. The solution is $x_{111} = 750$, $x_{121} = 250$, $x_{112} = 750$, $x_{122} = 0$, $x_{113} = 0$, $x_{123} = 0$, $x_{212} = 1500$, $x_{222} = 750$, $x_{213} = 0$, $x_{223} = 0$, $x_{313} = 1500$, and $x_{323} = 500$. The cost of this schedule is \$21,750.

The approach to multiperiod scheduling suggested in the preceding example can easily be adapted to scheduling capital expenditures. To illustrate, assume that the investment alternatives described in Table 5.17 are available

Table 5.17

In- vestment	Amount of Investment	Commitment of Funds	Reinvestment Possible	Total Return on Investment
1	Unlimited	one year*	yes	10%
2	\$30,000 maximum	two years*	no	30
3	\$20,000 maximum	two years†	no	50
4	\$20,000 maximum	one year‡	no	40

* Investment available at the beginning of the first year.
† Investment available at the beginning of the second year.
‡ Investment available at the beginning of the third year.

to a firm. The firm has \$50,000 available to invest at the beginning of the scheduling period. The objective is to schedule the investments during the three years in order to maximize the return on the investment.

Let x_{ij}, for $i = 1, 2, 3, 4$ and $j = 1, 2, 3$, represent the investment in alternative i at the beginning of period j. The linear programming problem is to

$$\text{Maximize:} \quad R = 0.10x_{11} + 0.10x_{12} + 0.10x_{13} + 0.30x_{21} + 0.50x_{32} + 0.40x_{43}$$

The constraint on the available funds at the beginning of the first year is

$$x_{11} + x_{21} \leq 50,000$$

The amount available for investing at the beginning of the second year is $50,000 - x_{21} + 0.10x_{11}$. The constraint on the investments at the beginning of the second year is

$$x_{12} + x_{32} \leq 50,000 - x_{21} + 0.10x_{11}$$

The amount available for investing at the beginning of the third year is

$$50,000 + 0.30x_{21} + 0.10x_{11} + 0.10x_{12} - x_{32}$$

The constraint on the investments at the beginning of the third year is

$$x_{13} + x_{43} \leq 50,000 + 0.30x_{21} + 0.10x_{11} + 0.10x_{12} - x_{32}$$

The constraints on the maximum investment in each of the alternatives are

$$x_{21} \leq 30,000$$
$$x_{32} \leq 20,000$$
$$x_{43} \leq 20,000$$

The linear programming formulation for the multiperiod investment schedule is expressed as

Maximize: $R = 0.10x_{11} + 0.10x_{12} + 0.10x_{13} + 0.30x_{21} + 0.50x_{32} + 0.40x_{43}$

Subject to:
$$x_{11} + \quad x_{21} \leq 50,000$$
$$x_{12} + \quad x_{32} + \quad x_{21} - 0.10x_{11} \leq 50,000$$
$$x_{13} + \quad x_{43} - 0.30x_{21} - 0.10x_{11} - 0.10x_{12}$$
$$+ \quad x_{32} \leq 50,000$$
$$x_{21} \leq 30,000$$
$$x_{32} \leq 20,000$$
$$x_{43} \leq 20,000$$
$$x_{ij} \geq 0$$

The solution is $x_{11} = \$20,000$, $x_{12} = \$2000$, $x_{13} = \$21,200$, $x_{21} = \$30,000$, $x_{32} = \$20,000$, $x_{43} = \$20,000$. The total return on the investment is $31,320.

PROBLEMS

1. Determine the solutions to the following linear programming problems by graphing the linear inequalities and evaluating the objective function at each extreme point of the feasible region.

(a)   Maximize: $Z = 5x_1 + 6x_2$
          Subject to: $3x_1 + 2x_2 \leq 32$
                        $1x_1 + 4x_2 \leq 34$
                          $x_1, x_2 > 0$

(b)   Maximize: $Z = 3.0x_1 + 1.0x_2$
          Subject to: $6.0x_1 + 4.0x_2 \leq 48$
                       $3.5x_1 + 6.0x_2 \leq 42$
                         $x_1, x_2 \geq 0$

(c)   Minimize: $C = 5x_1 + 7x_2$
          Subject to: $10x_1 + 3x_2 \geq 90$
                        $4x_1 + 6x_2 \geq 72$
                        $7x_1 + 5x_2 \geq 105$
                        $x_1, x_2 \geq 0$

(d)   Minimize: $C = \quad 3x_1 + \quad 4x_2$
                   Subject to: $\quad 6x_1 + \quad 5x_2 \geq 45$
                   $\quad 5x_1 + \quad 8x_2 \geq 60$
                   $x_2 \geq 2$
                   $x_1 \geq 0$

2. Solve the following linear programming problems.

(a)   Maximize: $Z = \quad 3x_1 + \quad 5x_2$
                   Subject to: $13x_1 + \quad 9x_2 \leq 234$
                   $9x_1 + 15x_2 \leq 270$
                   $x_1, x_2 \geq 0$

(b)   Maximize: $Z = \quad 2x_1 + \quad 5x_2$
                   Subject to: $4x_1 + \quad 3x_2 \leq 24$
                   $3x_1 + \quad 5x_2 \leq 30$
                   $x_1 + \quad x_2 \geq 8$
                   $x_1, x_2 \geq 0$

(c)   Minimize: $C = \quad x_1 + \quad x_2$
                   Subject to: $4x_1 + \quad 3x_2 \geq 12$
                   $-3x_1 + \quad 4x_2 \leq 12$
                   $x_1, x_2 \geq 0$

(d)   Maximize: $Z = \quad x_1 + \quad x_2$
                   Subject to: $8x_1 + \quad 5x_2 \geq 40$
                   $-4x_1 + \quad 6x_2 \geq 24$
                   $x_1, x_2 \geq 0$

3. Use the graphical solution technique to show that the problem

   Maximize: $Z = \quad x_1 + \quad x_2$
   Subject to: $3x_1 + 2x_2 \leq 24$
   $4x_1 + 7x_2 \leq 56$
   $-5x_1 + 6x_2 \leq 30$
   $x_1, x_2 \geq 0$

   is equivalent to the problem

   Minimize: $-Z = \quad -x_1 - \quad x_2$
   Subject to: $3x_1 + 2x_2 \leq 24$
   $4x_1 + 7x_2 \leq 56$
   $-5x_1 + 6x_2 \leq 30$
   $x_1, x_2 \geq 0$

4. Piper Farms plans to introduce two new gift packages of fruit for the Christmas market. Box A will contain 20 apples and 15 pears. Box B will contain 40 apples and 20 pears. Piper has 18,000 apples and 12,000 pears available for packaging. They believe that all fruit packaged can be sold. Profits are estimated as $0.60 for A and $1.00 for B. Determine the number of boxes of A and B that should be prepared to maximize profits.

5. Radio Manufacturing Company must determine production quantities for this month for two different models, A and B. Data per unit are given in the following table.

Model	Revenue	Subassembly Time (hr)	Final Assembly Time (hr)	Quality Inspection (hr)
A	$10	1.0	0.8	0.5
B	$20	1.2	2.0	0

The maximum time available for these products is 1200 hours for subassembly, 1600 hours for final assembly, and 500 hours for quality inspection. Orders outstanding require that at least 200 units of A and 100 units of B be produced. Determine the quantities of A and B that maximize total revenue.

6. A baseball manufacturer makes two types of baseballs, Major League and Minor League. He has four manufacturing processes that are used in making each type of baseball. These processes are designated as J, K, L, and M. Available time and time required per baseball for each process are listed below.

Process	Major League (time/baseball)	Minor League (time/baseball)	Total Time Available
J	2	3	6,000
K	8	5	20,000
L	1	$3\frac{1}{3}$	5,000
M	3	1	6,000

The contribution to profit is $2 for Major and $1 for Minor League baseballs. Determine the product mix that leads to maximum profit.

7. A company makes two types of leather wallets. Type A is a high-quality wallet and type B is a medium-quality wallet. The profits on the two wallets are $0.80 for type A and $0.60 for type B. The type A wallet requires twice as much time to manufacture as the type B. If all wallets were type B, the company could make 1000 per day. The supply of leather is sufficient for a maximum of 800 wallets per day (both A and B combined). A special process further limits production to a maximum of 450 wallets of type A and 700 wallets of type B. Assuming that all wallets manufactured can be sold, determine the number of each type to maximize profits.

8. Kawar Corporation produces two types of tract homes, the Tempo model and the Trend model. Each type is built on the same size lot and the tract will allow for 120 homes. The building materials used for both types of home are the same except for the framing materials. The Tempo model uses lumber for framing and the Trend model uses aluminum. Fifty thousand board feet of lumber and 75,000 running feet of standard aluminum framing are available during the construction period. Due to architectural styling, the Tempo model requires 600 board feet of lumber and 3000 man-hours, while the Trend model requires 800 running feet of standard aluminum frame and 6000 man hours. Total labor available during the construction period is 500,000 man-hours. The profit margin is $5000 for each Tempo model and $7000 for each Trend model. Determine the product mix for maximum profit.

9. A poultry farmer must supplement the vitamins in the feed he buys. He is considering two supplements, each of which contains the four vitamins required but in differing amounts. He must meet or exceed the minimum vitamin requirements. The vitamin content per ounce of the supplements is given in the following table.

Vitamin	Supplement 1	Supplement 2
1	5 units	25 units
2	25 units	10 units
3	10 units	10 units
4	35 units	20 units

Supplement 1 costs 3 cents per ounce and supplement 2 costs 4 cents per ounce. The feed must contain at least 50 units of vitamin 1, 100 units of vitamin 2, 60 units of vitamin 3, and 180 units of vitamin 4. Determine the mixture that has the minimum cost.

10. The Whoop-Bang Novelty Co. makes three basic types of noisemakers: Toot, Wheet, and Honk. A Toot can be made in 30 minutes and has a feather attached to it. A Wheet requires 20 minutes, has two feathers, and is sprinkled with 0.5 oz of sequin powder. The Honk requires 30 minutes, has three feathers, and 1 oz of sequin powder. The net profit is $0.45 per Toot, $0.55 per Wheet, and $0.70 per Honk. The following resources are available: 80 hours of labor, 90 ounces of sequin powder, and 360 feathers. Set up the linear programming problem to maximize profits.

11. A company manufactures three products: A, B, and C. The products require four operations: grinding, turning, assembly, and testing. The requirements per unit of product in hours for each operation are as follows:

Product	Grinding	Turning	Assembly	Test
A	0.03	0.11	0.30	0.08
B	0.02	0.14	0.20	0.07
C	0.03	0.20	0.26	0.08
Capacity (Hr)	1000	4000	9000	3000

The minimum monthly sales requirements are

A   9000 units
B   9000 units
C   6000 units

The profit per unit sold of each product is $0.15 for A, $0.12 for B, and $0.09 for C. All units produced above the minimum can be sold. What quantities of each product should be produced next month for maximum profit? Set up only.

12. A certain firm has two plants. Orders from four customers have been received. The number of units ordered by each customer and the shipping costs from each plant are shown in the following table.

		Shipping Cost/Unit	
Customer	Units Ordered	From Plant 1	From Plant 2
A	500	$1.50	$4.00
B	300	2.00	3.00
C	1000	3.00	2.50
D	200	3.50	2.00

Each unit of the product must be machined and assembled. These costs, together with the capacities at each plant, are shown below.

	Hours/Unit	Cost/Hour	Hours Available
Plant No. 1:			
Machining	0.10	$4.00	120
Assembling	0.20	3.00	260
Plant No. 2:			
Machining	0.11	4.00	140
Assembling	0.22	3.00	250

Formulate the linear programming problem to minimize cost. Set up only.

13. Crane Feed Company markets two feed mixes for rabbits. The first mix, Fertilex, requires at least twice as much wheat as barley. The second mix, Multiplex, requires at least twice as much barley as wheat. Wheat costs $0.49 per pound, and only 1000 pounds are available this month. Barley costs $0.36 per pound, and 1200 pounds are available. Fertilex sells for $1.59 per pound up to 99 pounds, and each additional pound over 99 sells for $1.43. Multiplex sells at $1.35 per pound up to 99 pounds, and each additional pound over 99 sells for $1.18. Rancho Farms will buy any and all amounts of both mixes Crane Feed Company will mix. Set up the linear programming problem to determine the product mix that results in maximum profits.

14. A farmer must decide how many pounds of each of several types of grain he should purchase in order that his livestock receive the minimum nutrient requirement at the lowest cost possible. The relevant information is as follows:

	One Pound Of				Minimum Nutrient Requirement
	Grain 1	Grain 2	Grain 3	Grain 4	
Nutrient 1	4	5	6	3	1000
Nutrient 2	2	1	0	3	850
Nutrient 3	1	2	3	1	700
Nutrient 4	2	3	1	2	1320
Nutrient 5	0	2	1	1	550
Cost/pound	$0.35	$0.42	$0.45	$0.37	

Set up the linear programming problem to determine the mix of grains that minimizes cost.

15. The Franklin Oil Company produces three oils of different viscosities. Franklin Oil makes its products by blending two grades of oil, each with a different viscosity. The final viscosity is linearly proportional to the blending viscosities. There is an unlimited demand for the three oils.

Blending Component	Viscosity	Cost/Quart	Supply
A	10	$0.20	4000 qt/wk
B	50	$0.30	2000 qt/wk

Brand	Minimum Viscosity	Sales Price/Quart
S	20	$0.43
SS	30	$0.48
SSS	40	$0.53

Formulate the linear programming problem to maximize profits.

16. Western Oil Company makes three brands of gasoline: Super, Low Lead, and Regular. Western makes its gasolines by blending two grades of gasolines and a high-octane lead additive. Each brand of gasoline must have an octane rating of at least the stated minimum. The brands are made by blending the two grades of gasoline and the high-octane lead additive. The relevant data are given below.

Blending Component	Octane	Cost per Gallon	Supply from Refinery (per week)
A	120	$0.12	18,000 gallons
B	80	$0.09	26,000 gallons
L	1200	$1.40	Unlimited

Gasoline Brand	Minimum Octane	Sales Price per Gallon
Super	105	$0.18
Low Lead*	94	$0.16
Regular	96	$0.14

* Low Lead can contain no more than 0.1 percent by volume of blending component L.

Assuming that Western Oil can sell all gasoline produced, set up the linear programming problem to maximize profit.

17. The Hamilton Data Processing Company performs three types of data processing activities: payrolls, accounts receivable, and inventories. The profit and time requirements for key punch, computation, and off-line printing for a "standard job" are shown in the following table.

Job	Profit per Std. Job	Time Requirements (Minutes)		
		Key Punch	Computation	Print
Payroll	$275	1200	20	100
Accounts receivable	$125	1400	15	60
Inventory	$225	800	35	80

Hamilton guarantees overnight completion of the job. Any job scheduled during the day can be completed during the day or during the night. Any job scheduled during the night, however, must be completed during the night. The capacity for both day and night are shown in the following table.

Capacity (minutes)	Key Punch	Operation Computation	Print
Day	4200	150	400
Night	9200	250	650

Set up the linear programming problem to determine the mixture of "standard jobs" that should be accepted during the day and during the night in order to maximize profit.

18. The Trust Department of the Barclay Bank is preparing an investment portfolio for a wealthy customer. The investment alternatives along with their current yields and risk factors are given in the following table.

Investment	Yield	Risk
AAA corporate bonds	6.5%	1.0
Convertible debentures	7.2	1.5
Selected high-grade common stock	8.3	2.0
Preferred stock	6.3	1.8
Growth stock	9.0	3.2

The risk factor is a subjective estimate on a scale from 0 to 10, the lower numbers representing the lower risk investments. On the assumption that the current yields will continue, the Trust Department wishes to design a portfolio with the maximum yield. However, the average risk must not exceed 2.4. In addition, the customer has specified that the investment in growth stock cannot exceed the combined total investment in corporate bonds and convertible debentures. Set up the linear programming problem to determine the proportion of funds in each investment alternative.

19. G and H Advertising, Inc., is preparing a proposal for an advertising campaign for White Chemical Company, manufacturers of Gro-More Lawn products. The advertising copy has been written and approved. The final step in the proposal, therefore, is to recommend an allocation of advertising funds so as to maximize the total number of effective exposures. The characteristics of three alternative publications are shown in the following table.

	House Beautiful	Publication Home & Garden	Lawn Care
Cost per advertisement	$600	$800	$450
Maximum no. of adv.	12	24	12
Minimum no. of adv.	3	6	2
Characteristics:			
Homeowner	80%	70%	20%
Income: $10,000 or more	70%	80%	60%
Occupation: gardener	15%	20%	40%
Audience size	600,000	800,000	300,000

The relative importance of the three characteristics are: homeowner, 0.4; income, 0.2; gardener, 0.4. The advertising budget is $20,000. Formulate the linear programming problem to determine the most effective number of exposures in each magazine.

20. Great Western Airlines, Inc., must determine how many stewardesses to hire and train during the next six months. Their requirements, expressed in terms of flight hours, are 8400 in January, 9100 in February, 9800 in March, 11,200 in April, 11,200 in May, and 11,800 in June.

One month of training is required before a stewardess can be put on a regular flight. Consequently, a girl must be hired at least a month before she is needed. A maximum of 20 girls can be enrolled in the training program during any month.

Each stewardess can work up to 70 flight hours per month. Great Western has 140 stewardesses available for scheduling at the beginning of January. Approximately 10 percent of the girls on flight status either quit or are dismissed each month. Girls on standby normally do not quit. The cost of training a girl is $2500, and the salary of the stewardess is $700. Girls not used on flights are put on standby duty and earn their full salary. Formulate the linear programming problem to determine the hiring schedule during the six-month period.

21. Fast Eddie is a pool shark. To keep sharp he must play at least seven hours a day and not more than twelve hours. Of his playing hours, at least four hours must be straight practicing. Eddie's average income from his three games is: snooker, $30 per hour; pocket billiards, $35 per hour; three-rail billiards, $55 per hour.

Eddie has the following problem: For every hour of "hustling" a given game, he must practice the other two games to remain in top hustling condition. The following table shows the relationships:

Game-1 hour	Minimum Practice Required	
Snooker	$\frac{1}{4}$ hr Pocket Billiards	$\frac{1}{6}$ hr 3-Rail
Pocket billiards	$\frac{1}{6}$ hr Snooker	$\frac{1}{4}$ hr 3-Rail
3-Rail billiards	$\frac{1}{2}$ hr Pocket Billiards	$\frac{1}{3}$ hr Snooker

Eddie also feels that he must practice each of the three games at least one hour apiece. Set up the linear programming problem to determine the mix of hustling and practicing that will maximize Eddie's profit.

22. Next Monday, Gordon Shipley has a term paper due in Marketing and exams in Quantitative Methods and Accounting. Shipley has at most 20 hours of study time available. Shipley has completed the rough draft of the term paper but he has not started the final draft. He estimates that he could turn in the rough draft and receive a grade of 60. With 10 hours of effort on the final draft he could get an 80 and with 20 hours of study effort he could get 100. Without studying he could get a 60 on the Quantitative Methods exam. With 4 hours of study, however, he could get an 80 and 8 hours of study should result in a perfect score. Shipley is a little behind in Accounting. Without any study he would score 40. With 5 hours of study he would score 70, and with 10 hours of study he would expect to score 100 on the exam.

The term paper is 50 percent of the Marketing course grade, the Quantitative Methods exam is 25 percent, and the Accounting exam is 20 percent. Taking into account that Accounting and Quantitative Methods are four-unit courses and Marketing is a two-unit course, how should Shipley allocate his time to maximize his overall GPA (grade point average) and still not receive a grade of less than 70 in any subject? Set up the problem only.

SELECTED REFERENCES

BAUMOL, WILLIAM J., *Economic Theory and Operations Analysis*, 2nd ed. (Englewood Cliffs, N.J.: Prentice-Hall, Inc., 1965).

BOULDING, K. E. AND W. A. SPIVEY, *Linear Programming and the Theory of the Firm* (New York, N.Y.: The Macmillan Company, Inc., 1960).

CHARNES, A. AND W. W. COOPER, *Management Models and Industrial Applications of Linear Programming* (New York, N.Y.: John Wiley and Sons, Inc., 1961).

GARVIN, WALTER W., *Introduction to Linear Programming* (New York, N.Y.: The McGraw-Hill Book Company, Inc., 1960).

GASS, SAUL I., *Linear Programming: Methods and Applications* (New York, N.Y.: The McGraw-Hill Book Company, Inc., 1964).

HADLEY, GEORGE, *Linear Programming* (Reading, Mass.: Addison-Wesley Publishing Company, Inc., 1962).

KIM, CHAIHO, *Introduction to Linear Programming* (New York, N.Y.: Holt, Rinehart, and Winston, Inc., 1971).

LEVIN, R. I. AND RUDY P. LAMONE, *Linear Programming for Management Decisions* (Homewood, Ill.: Richard D. Irwin, Inc., 1969).

NAYLOR, T. H., E. T. BRYNE, AND J. M. VERNON, *Introduction to Linear Programming: Methods and Cases* (Belmont, Ca.: Wadsworth Publishing Company, Inc., 1971).

SPIVEY, W. A. AND R. M. THRALL, *Linear Optimization* (New York, N.Y.: Holt, Rinehart, and Winston, Inc., 1970).

STRUM, JAY E., *Introduction to Linear Programming* (San Francisco, Ca.: Holden-Day, Inc., 1972).

VANDERMEULEN, DANIEL C., *Linear Economic Theory* (Englewood Cliffs, N.J.: Prentice-Hall Inc., 1971).

Chapter 6

The
Simplex Method

The simplex method is a solution algorithm for solving linear programming problems.† The simplex algorithm was developed by George Dantzig in 1947 and made generally available in 1951. The importance of the algorithm is demonstrated by the fact that linear programming has become one of the most widely used quantitative approaches to business decisions.

6.1 System of Equations

The simplex algorithm utilizes the *extreme point theorem* of linear programming. This theorem states that *an optimal solution to a linear programming problem occurs at one of the extreme points of the feasible region*. The graphical solution technique was used to show that extreme points are defined by the intersection of systems of equations. The first step in the simplex algorithm, therefore, is to convert the system of linear inequalities to linear equations. These equations are then used to determine the extreme points.

---

† The term *algorithm* refers to a systematic procedure or series of rules for solving a problem.

6.1.1 CONVERTING INEQUALITIES TO EQUATIONS

A straightforward technique for converting an inequality to an equation involves adding a *slack variable* to the left side of inequalities of the form $ax_1 + bx_2 \leq c$ and subtracting a *surplus variable* from the left side of inequalities of the form $ax_1 + bx_2 \geq c$. To illustrate this technique, consider the system of inequalities

$$3x_1 + 2x_2 \leq 40$$

$$2x_1 + x_2 \geq 10$$

The first inequality is converted to an equation by adding the slack variable x_3. The resulting equation is

$$3x_1 + 2x_2 + x_3 = 40$$

The slack variable x_3 represents the difference between $3x_1 + 2x_2$ and 40. If $3x_1 + 2x_2$ equals 40, then x_3 has the value 0. If, on the other hand, $3x_1 + 2x_2$ is less than 40, x_3 assumes the positive value equal to the difference between $3x_1 + 2x_2$ and 40.

The second inequality is converted to an equation by subtracting the surplus variable x_4. The resulting equation is

$$2x_1 + x_2 - x_4 = 10$$

The surplus variable x_4 represents the difference between the expression $2x_1 + x_2$ and 10. As in the case of slack variables, if $2x_1 + x_2$ equals 10, then the surplus variable x_4 has the value 0. If, however, $2x_1 + x_2$ is greater than 10, then x_4 has a positive value that is equal to the difference between $2x_1 + x_2$ and 10. Since $2x_1 + x_2 \geq 10$, the value of the surplus variable subtracted from the left side of the inequality must be greater than or equal to 0.

Several important characteristics of the new system of equations should be noted. First, both the slack and surplus variables are nonnegative. Second, a slack variable must be added for each "less than or equal to" inequality and a surplus variable subtracted for each "greater than or equal to" inequality. Finally, slack and surplus variables would only coincidently have equivalent values. Consequently, different variables must be used for each inequality.

Example: A manufacturer makes two kinds of bookcases, A and B. Type A requires 2 hours on machine 1 and 4 hours on machine 2. Type B requires 3 hours on machine 1 and 2 hours on machine 2. The machines work no more than 16 hours per day. The profit is \$3 per bookcase A and \$4 per bookcase B. Formulate the linear programming problem, convert the constraining inequalities to equations, and interpret the meaning of the slack variables.

Let x_1 and x_2 represent the number of bookcases of type A and B, respectively. The linear programming problem is

$$\text{Maximize:} \quad P = 3x_1 + 4x_2$$
$$\text{Subject to:} \quad 2x_1 + 3x_2 \leq 16$$
$$4x_1 + 2x_2 \leq 16$$
$$x_1, x_2 \geq 0$$

The constraining inequalities are converted to equations by adding slack variables. The problem becomes

$$\text{Maximize:} \quad P = 3x_1 + 4x_2$$
$$\text{Subject to:} \quad 2x_1 + 3x_2 + x_3 \qquad\quad = 16$$
$$4x_1 + 2x_2 \qquad + x_4 = 16$$
$$x_1, x_2, x_3, x_4 \geq 0$$

The slack variable x_3 represents unused machine 1 time and the slack variable x_4 represents unused machine 2 time. If both machines are fully utilized, x_3 and x_4 would have values of zero.

Example: American Foods, Inc., is developing a low-calorie, high-protein diet supplement. The linear programming formulation of this problem along with the data is given in Chapter 5, p. 163. Convert the system of inequalities to equations by adding slack and subtracting surplus variables. Interpret the meaning of the slack and surplus variables.

After adding the slack and subtracting the surplus variables, the problem becomes

$$\text{Minimize:} \quad C = 0.15x_1 + 0.20x_2 + 0.12x_3$$
$$\text{Subject to:} \quad 350x_1 + 250x_2 + 200x_3 + x_4 \qquad\qquad\qquad = 300$$
$$250x_1 + 300x_2 + 150x_3 \qquad - x_5 \qquad\qquad = 200$$
$$100x_1 + 150x_2 + 75x_3 \qquad\qquad - x_6 \qquad = 100$$
$$75x_1 + 125x_2 + 150x_3 \qquad\qquad\qquad - x_7 = 100$$
$$x_j \geq 0 \text{ for } j = 1, 2, \ldots, 7$$

The slack variable x_4 represents the difference between the calories in the diet supplement and the maximum number of calories allowed by the specifications. The surplus variables x_5, x_6, and x_7 represent, respectively, the number of units of protein, vitamin A, and vitamin C in excess of that required by the specifications.

6.1.2 BASIC FEASIBLE SOLUTIONS

The addition of slack variables and subtraction of surplus variables enables linear inequalities to be expressed in the form of linear equations. As stated earlier, the reason that this transformation from inequalities to equations is important in the simplex algorithm is that it allows one to solve the resulting systems of equations for extreme points. The objective function in a linear programming problem, according to the extreme point theorem of linear programming, reaches an optimal value at one of the extreme points. Since the extreme points are defined by the intersection of sets of linear equations, the transformation from inequalities to equations is necessary.

To illustrate, consider the linear programming problem

$$\text{Maximize:} \quad P = x_1 + x_2$$
$$\text{Subject to:} \quad 3x_1 + 2x_2 \leq 40$$
$$2x_1 + x_2 \geq 10$$
$$x_1, x_2 \geq 0$$

The system of linear inequalities is transformed to a system of equations by adding a slack variable x_3 and subtracting a surplus variable x_4. This gives the system of linear equations

$$3x_1 + 2x_2 + x_3 = 40$$
$$2x_1 + x_2 - x_4 = 10$$

This system of two equations and four variables does not have a unique solution. In Chapter 2 it is shown that a system of m equations and n variables, where $n > m$, has an infinite number of solutions, provided the system of equations is consistent. Since this system of equations is consistent, it has an infinite number of solutions rather than a unique solution.†

Although the system of equations has an infinite number of solutions, it does not have an infinite number of extreme points. In fact, in this system of two equations and four variables there are only six extreme points. The extreme points can be determined by applying the *basis theorem* of linear programming. The *basis theorem* states that *for a system of m equations and n variables, where n > m, a solution in which at least n − m of the variables have values of zero is an extreme point.* Any solution found by setting $n - m$ of the variables equal to zero and solving the resulting system of m equations for the m remaining variables gives an extreme point. This solution is termed a *basic solution.*

The basic solutions for the preceding system of equations are determined by selecting two of the four variables and equating these variables with zero. The resulting system of two equations and two basic variables is solved simultaneously. For instance, if $x_3 = 0$ and $x_4 = 0$, the system of equations is

$$3x_1 + 2x_2 = 40$$
$$2x_1 + x_2 = 10$$

---

† Consistent systems of equations are discussed in Chapter 2, p. 56 and Chapter 4, p. 135.

and the basic solution is $x_1 = -20$ and $x_2 = 50$. Similarly, if $x_1 = 0$ and $x_2 = 0$, the resulting basic solution is $x_3 = 40$ and $x_4 = -10$. The six basic solutions to the system of equations are shown in Table 6.1.†

Table 6.1

x_1	x_2	x_3	x_4	Objective Function
0	0	40	-10	nonfeasible
0	20	0	10	20
0	10	20	0	10
13.3	0	0	16.6	13.3
5	0	25	0	5
-20	50	0	0	nonfeasible

The extreme point theorem of linear programming can be extended to state that *the objective function is optimal at at least one of the basic solutions.* In the example given in Table 6.1 there are only six extreme points or, alternatively, only six basic solutions. Two of the solutions have negative values for variables and are, therefore, *nonfeasible.* The optimal value of the objective function, therefore, occurs at one of the four *basic feasible solutions·* Table 6.1 shows that the maximum value of the objective function occurs when $x_1 = 0$, $x_2 = 20$, $x_3 = 0$, and $x_4 = 10$. The value of the objective function at this basic solution is $P = 20$.

It can be shown that the extreme point and basis theorems of linear programming are related. This relationship is demonstrated through the use of the simple two-variable problem,

$$\text{Maximize:} \quad P = 2x_1 + 3x_2$$
$$\text{Subject to:} \quad 3x_1 + 2x_2 \le 12$$
$$2x_1 + 4x_2 \le 16$$
$$x_1, x_2 \ge 0$$

The feasible region for this problem is shown in Fig. 6.1.

The objective function is a maximum at an extreme point. The four extreme points are $E = \{(0, 0), (0, 4), (2, 3), (4, 0)\}$.

This same set of extreme points is generated from the basis theorem. To

---

† The number of basic solutions is given by

$$\frac{n!}{m!(n - m)!}$$

where $n!$, read *n factorial*, equals $n(n - 1)(n - 2)\ldots 1$. To illustrate, the number of basic solutions when $n = 4$ and $m = 2$ is 6.

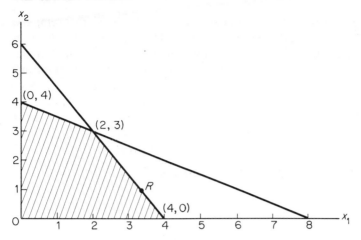

Figure 6.1

apply the basis theorem, the inequalities are first converted to equations by adding the slack variables x_3 and x_4, i.e.,

$$3x_1 + 2x_2 + x_3 = 12$$
$$2x_1 + 4x_2 + x_4 = 16$$

With four variables and two equations, the basic solutions are found by equating two of the four variables to zero. The four basic feasible solutions are given in Table 6.2. These were found by solving the four sets of two simul-

Table 6.2

Solution	x_1	x_2	x_3	x_4
A	0	0	12	16
B	0	4	4	0
C	2	3	0	0
D	4	0	0	8

taneous equations. The basic solutions A, B, C, and D correspond, respectively, to the extreme points $(0, 0)$, $(0, 4)$, $(2, 3)$, and $(4, 0)$.

Nonbasic solutions correspond to points on the boundary of the constraining inequalities. The point R in Fig. 6.1 is an example of a nonbasic solution. The coordinates of R are $x_1 = 3\frac{1}{3}$ and $x_2 = 1$. Substituting these coordinates into the system of equations gives $x_1 = 3\frac{1}{3}$, $x_2 = 1$, $x_3 = 0$, and $x_4 = 5\frac{1}{3}$. This solution is nonbasic, since less than $n - m$ of the variables have values of zero.

This example illustrates the relationship among the basis theorem, the extreme point theorem, and the optimal solution. The extreme point theorem states that the optimal value of the objective function occurs at one of the extreme points of the feasible region. The example shows that basic solutions are extreme points. Consequently, the relationship between the basis theorem, the extreme point theorem, and the optimal solution is that (1) a basic solution occurs for a system of m equations and n variables when at least $n - m$ of the variables have values of zero and (2) the objective function is optimal at one of the basic solutions. We can thus determine the optimal solution by finding all basic solutions and evaluating the objective function at each of these basic solutions.

Example: Determine all basic solutions for the system of equations,

$$2x_1 + x_2 - 2x_3 + x_4 = 300$$
$$3x_1 - 2x_2 + 2x_3 + x_5 = 200$$

Since $m = 2$ equations and $n = 5$ variables, there are

$$\frac{n!}{m!(n-m)!} = \frac{5!}{2!3!} = \frac{5 \cdot 4 \cdot 3 \cdot 2 \cdot 1}{2 \cdot 1 \cdot 3 \cdot 2 \cdot 1} = 10 \text{ basic solutions}$$

The ten basic solutions are found by equating all possible combinations of three of the five variables to zero and solving the resulting systems of two equations and two basic variables. The basic solutions are given in Table 6.3.

Table 6.3

x_1	x_2	x_3	x_4	x_5	Feasible/ Nonfeasible
0	0	0	300	200	feasible
0	0	−150	0	500	nonfeasible
0	0	100	500	0	feasible
0	300	0	0	800	feasible
0	−100	0	400	0	nonfeasible
0	−500	−400	0	0	nonfeasible
150	0	0	0	−250	nonfeasible
66.7	0	0	167.3	0	feasible
100	0	−50	0	0	nonfeasible
114.3	71.4	0	0	0	feasible

The solutions that are feasible are also indicated in the table.

Example: Determine the basic feasible solutions to the following linear programming problem. Show that these solutions correspond to the extreme points in Fig. 5.4.

$$\text{Maximize:} \quad P = 4x_1 + 3x_2$$
$$\text{Subject to:} \quad 21x_1 + 16x_2 \leq 336$$
$$13x_1 + 25x_2 \leq 325$$
$$15x_1 + 18x_2 \leq 270$$
$$x_1, x_2 \geq 0$$

The system of equations is

$$21x_1 + 16x_2 + x_3 \qquad\qquad = 336$$
$$13x_1 + 25x_2 \qquad + x_4 \qquad = 325$$
$$15x_1 + 18x_2 \qquad\qquad + x_5 = 270$$

The basic solutions are given in Table 6.4. The problem was solved graphically in Chapter 5, Fig. 5.4. The extreme points that correspond to the basic feasible solutions are given in the table.

Table 6.4

x_1	x_2	x_3	x_4	x_5	Feasible/ Nonfeasible	Figure 5.4 Extreme Point
0	0	336	325	270	feasible	(0, 0)
0	21	0	−200	−108	nonfeasible	—
0	13	128	0	36	feasible	(0, 13)
0	15	96	−50	0	nonfeasible	—
16	0	0	117	30	feasible	(16, 0)
25	0	−189	0	−105	nonfeasible	—
18	0	−42	91	0	nonfeasible	—
10.08	7.75	0	0	−20.70	nonfeasible	
12.45	4.57	0	48.90	0	feasible	(12.45, 4.57)
6.38	9.68	47.14	0	0	feasible	(6.38, 9.68)

6.1.3 ALGEBRAIC SOLUTIONS

Before we introduce the simplex algorithm, it will be useful formally to introduce the algebraic solution technique. This technique, although inefficient, shares several characteristics with the simplex algorithm. The most important of these is that it uses the same algebraic procedure for determining basic solutions.

The algebraic solution technique requires the following steps: (1) determine all basic solutions; (2) evaluate each basic feasible solution for optimality. To illustrate, consider the linear programming problem

$$\text{Maximize:} \quad P = 3x_1 + 4x_2$$
$$\text{Subject to:} \quad 2x_1 + 3x_2 \leq 16$$
$$4x_1 + 2x_2 \leq 16$$
$$x_1, x_2 \geq 0$$

The inequalities are converted to equalities by adding slack variables. The linear programming problem becomes

$$\text{Maximize:} \quad P = 3x_1 + 4x_2 + 0x_3 + 0x_4$$
$$\text{Subject to:} \quad 2x_1 + 3x_2 + 1x_3 + 0x_4 = 16$$
$$4x_1 + 2x_2 + 0x_3 + 1x_4 = 16$$
$$x_1, x_2, x_3, x_4 \geq 0$$

The slack variables are included in the objective function with zero coefficients. The zero coefficients show that unused resource, or slack, does not contribute to profits.

The basic solutions are determined by selecting all possible combinations of two of the four variables and equating these variables with zero. The solutions of the resulting systems of two equations and two variables are basic solutions. For each basic solution, the variables that are equated to zero are termed *not in solution*, or alternatively, *not in the basis*. Conversely, those variables that are not equated to zero are said to be *in solution*, *in the basis*, or alternatively, *basic variables*.†

As a starting point, assume that the slack variables are in the basis. If the slack variables are in the basis, the variables x_1 and x_2 are equated with zero. Since the coefficients of the slack variables form an identity matrix, the initial basic solution can be read directly from the system of equations. The initial basic solution is $x_1 = 0$, $x_2 = 0$, $x_3 = 16$, and $x_4 = 16$. Since x_1, x_2, x_3, and x_4 are greater than or equal to zero, the solution is feasible. The value of the objective function at this initial basic feasible solution is 0, i.e.,

$$P = 3(0) + 4(0) + 0(16) + 0(16) = 0$$

For a second solution, let x_1 and x_4 be in the basis. According to the basis theorem of linear programming, x_2 and x_3 are not in the basis and are valued at zero. The solution to the system of two equations and two basic variables can be obtained by using the row operations of matrix algebra.‡

The objective is to solve the system of equations for x_1 and x_4. The system could be written as

$$2x_1 + 0x_4 = 16$$

$$4x_1 + 1x_4 = 16$$

Alternatively, the system could be written in matrix form as

$$\begin{pmatrix} 2 & 0 \\ 4 & 1 \end{pmatrix} \begin{pmatrix} x_1 \\ x_4 \end{pmatrix} = \begin{pmatrix} 16 \\ 16 \end{pmatrix}$$

---

† It is possible for the solution value of a basic variable to be zero. When this occurs, the solution is termed *degenerate*. A degenerate solution is treated the same as any other basic solution in the algebraic or simplex algorithms.

‡ Chapter 3, pp. 86–91.

Since x_2 and x_3 have values of zero, both formulations are equivalent to the original system of equations,

$$2x_1 + 3x_2 + 1x_3 + 0x_4 = 16$$
$$4x_1 + 2x_2 + 0x_3 + 1x_4 = 16$$

In solving the system of equations, it is customary to apply row operations to the entire system of equations, rather than to only the coefficients of the basic variables. This has the advantage of reducing the number of arithmetic calculations required in moving from one basic solution to another.

The system of equations can be solved for x_1 and x_4 by using row operations to obtain an identity matrix as the coefficient matrix of x_1 and x_4. Multiplying the first equation by $\frac{1}{2}$ gives

$$1x_1 + \tfrac{3}{2}x_2 + \tfrac{1}{2}x_3 + 0x_4 = 8$$

By multiplying the new equation by -4 and adding this product to the second equation, the second equation becomes

$$0x_1 - 4x_2 - 2x_3 + 1x_4 = -16$$

and the system of equations is

$$1x_1 + \tfrac{3}{2}x_2 + \tfrac{1}{2}x_3 + 0x_4 = 8$$
$$0x_1 - 4x_2 - 2x_3 + 1x_4 = -16$$

The coefficients of the basic variables form an identity matrix. The non-basic variables are valued at zero, and the basic solution can be read directly as $x_1 = 8$, $x_2 = 0$, $x_3 = 0$, and $x_4 = -16$. This solution is nonfeasible.

To determine a third solution, let x_1 and x_2 be basic variables and x_3 and x_4 be nonbasic variables. The values of x_1 and x_2 can be found by solving the original two equations simultaneously. An alternative approach, used in the simplex method, is to use the equations from the preceding iteration in solving for x_1 and x_2. The equations were

$$1x_1 + \tfrac{3}{2}x_2 + \tfrac{1}{2}x_3 + 0x_4 = 8$$
$$0x_1 - 4x_2 - 2x_3 + 1x_4 = -16$$

Multiplying the second equation by $-\frac{1}{4}$ gives

$$0x_1 + 1x_2 + \tfrac{1}{2}x_3 - \tfrac{1}{4}x_4 = 4$$

Multiplying the new second equation by $\frac{3}{2}$ and subtracting it from the first equation gives

$$1x_1 + 0x_2 - \tfrac{1}{4}x_3 + \tfrac{3}{8}x_4 = 2$$

The system of equations is

$$1x_1 + 0x_2 - \tfrac{1}{4}x_3 + \tfrac{3}{8}x_4 = 2$$
$$0x_1 + 1x_2 + \tfrac{1}{2}x_3 - \tfrac{1}{4}x_4 = 4$$

The coefficients of x_1 and x_2 form an identity matrix and x_3 and x_4 are not in the basis. The solution can be read directly as $x_1 = 2$, $x_2 = 4$, $x_3 = 0$, and $x_4 = 0$. The solution is feasible. The value of the objective function is 22, i.e.,

$$P = 3(2) + 4(4) + 0(0) + 0(0) = 22$$

A fourth solution can be found by assuming that x_1 and x_3 are in the basis and x_2 and x_4 are not. Rather than solving the original system of equations, we use the equations from the preceding iteration. The system of equations

$$1x_1 + 0x_2 - \tfrac{1}{4}x_3 + \tfrac{3}{8}x_4 = 2$$
$$0x_1 + 1x_2 + \tfrac{1}{2}x_3 - \tfrac{1}{4}x_4 = 4$$

with basic variables x_1 and x_3 can be solved by multiplying the second equation by 2 and adding the product of $\tfrac{1}{4}$ the new second equation to the first equation. This results in an identity matrix as the coefficient matrix for the basic variables x_1 and x_3.

$$1x_1 + \tfrac{1}{2}x_2 + 0x_3 + \tfrac{1}{4}x_4 = 4$$
$$0x_1 + 2x_2 + 1x_3 - \tfrac{1}{2}x_4 = 8$$

The solution is $x_1 = 4$, $x_2 = 0$, $x_3 = 8$, $x_4 = 0$. The solution is feasible, and the value of the objective function is $P = 12$.

For a fifth solution, let x_2 and x_3 be the basic variables. The system of equations from the preceding iteration,

$$1x_1 + \tfrac{1}{2}x_2 + 0x_3 + \tfrac{1}{4}x_4 = 4$$
$$0x_1 + 2x_2 + 1x_3 - \tfrac{1}{2}x_4 = 8$$

can be solved for x_2 and x_3 by multiplying the first equation by 2 and subtracting the product of 2 times the new first equation from the second equation. This gives

$$2x_1 + 1x_2 + 0x_3 + \tfrac{1}{2}x_4 = 8$$
$$-4x_1 + 0x_2 + 1x_3 - \tfrac{3}{2}x_4 = -8$$

The solution, $x_1 = 0$, $x_2 = 8$, $x_3 = -8$, $x_4 = 0$, is not feasible.

The final solution occurs when x_2 and x_4 are in the basis. The system of equations

$$2x_1 + 1x_2 + 0x_3 + \tfrac{1}{2}x_4 = 8$$
$$-4x_1 + 0x_2 + 1x_3 - \tfrac{3}{2}x_4 = -8$$

can be solved for x_2 and x_4 by multiplying the second equation by $-\tfrac{2}{3}$ and subtracting the product of $\tfrac{1}{2}$ times this equation from the first equation. This gives

$$\tfrac{2}{3}x_1 + 1x_2 + \tfrac{1}{3}x_3 + 0x_4 = \tfrac{16}{3}$$
$$\tfrac{8}{3}x_1 + 0x_2 - \tfrac{2}{3}x_3 + 1x_4 = \tfrac{16}{3}$$

The solution is $x_1 = 0$, $x_2 = \frac{16}{3}$, $x_3 = 0$, $x_4 = \frac{16}{3}$. The value of the objective function is $P = 16$.

The six solutions to the linear programming problem are summarized in Table 6.5. The optimal value of the objective function, $P = 22$, occurs when $x_1 = 2$, $x_2 = 4$, $x_3 = 0$, and $x_4 = 0$.

Table 6.5

Solution	x_1	x_2	x_3	x_4	Objective Function
1	0	0	16	16	0
2	8	0	0	-16	Nonfeasible
3	2	4	0	0	22
4	4	0	8	0	12
5	0	8	-8	0	Nonfeasible
6	0	$\frac{16}{3}$	0	$\frac{16}{3}$	16

The algebraic and simplex algorithms have several common characteristics. The most important of these is that both algorithms utilize the extreme point theorem of linear programming. This theorem states that an optimal value of the objective functions occurs at one of the basic solutions to the system of equations. Basic solutions to a system of m equations and n variables are found in both algorithms by equating $n - m$ of the variables to zero and solving the resulting system of m equations and m basic variables. Consequently, much of the arithmetic required for the algebraic solution to a linear programming problem is also required for a simplex solution.

6.2 The Simplex Algorithm

The simplex algorithm is an iterative procedure for determining basic feasible solutions to a system of equations and testing each solution for optimality. The algorithm involves moving from one basic feasible solution to another, always maintaining or improving the value of the objective function, until an optimal solution is reached.

Several of the inefficiencies of the algebraic method are eliminated by the simplex algorithm. First, feasible solutions found by using the algebraic technique cannot be distinguished from nonfeasible solutions until after the solution has been determined. In contrast, all solutions found by using the simplex algorithm are feasible. Since only feasible solutions are considered, the computational requirements are reduced by the simplex algorithm. Second, the algebraic technique requires that all basic feasible solutions be evaluated. In comparison, each iteration of the simplex algorithm gives a solution that improves or maintains the value of the objective function. This means that only a subset of the set of extreme points need be evaluated. This results in a further reduction in the computational requirements.

6.2.1 SIMPLEX TABLEAU

The first step in applying the simplex algorithm is to transform the inequalities of the linear programming problem to equations by adding slack and subtracting surplus variables. The coefficients of the objective function and constraining equations along with the right-hand-side values are then transferred to the simplex tableau. As an example, consider the linear programming problem

$$\text{Maximize:} \quad P = 3x_1 + 4x_2 + 5x_3 + 4x_4$$
$$\text{Subject to:} \quad 2x_1 + 5x_2 + 4x_3 + 3x_4 \leq 224$$
$$5x_1 + 4x_2 - 5x_3 + 10x_4 \leq 280$$
$$2x_1 + 4x_2 + 4x_3 - 2x_4 \leq 184$$
$$x_j \geq 0 \quad \text{for } j = 1, 2, 3, 4$$

By adding slack variables, the problem becomes

$$\text{Maximize:} \quad P = 3x_1 + 4x_2 + 5x_3 + 4x_4 + 0x_5 + 0x_6 + 0x_7$$
$$\text{Subject to:} \quad 2x_1 + 5x_2 + 4x_3 + 3x_4 + 1x_5 + 0x_6 + 0x_7 = 224$$
$$5x_1 + 4x_2 - 5x_3 + 10x_4 + 0x_5 + 1x_6 + 0x_7 = 280$$
$$2x_1 + 4x_2 + 4x_3 - 2x_4 + 0x_5 + 0x_6 + 1x_7 = 184$$
$$x_j \geq 0 \quad \text{for } j = 1, 2, \ldots, 7$$

The coefficients of the objective function and equations along with the right-hand-side values can now be transferred to the simplex tableau. The initial simplex tableau is shown in Table 6.6.

Table 6.6 Initial Simplex Tableau

c_b	c_j	3	4	5	4	0	0	0	
	Basis	x_1	x_2	x_3	x_4	x_5	x_6	x_7	Solution
0	x_5	2	5	4	3	1	0	0	224
0	x_6	5	4	-5	10	0	1	0	280
0	x_7	2	4	4	-2	0	0	1	184
	z_j	0	0	0	0	0	0	0	0
	$c_j - z_j$	3	4	5	4	0	0	0	

The first two rows of the tableau give the coefficients of the objective function and the column headings. The coefficients of the objective function are copied directly from the objective function. *Both slack and surplus variables have zero coefficients in the objective function.*

The coefficients of the constraining equations are shown in the tableau under the appropriate column heading. The coefficients of the variable x_1 in the constraining equations are given in the column labeled x_1, the coefficients of x_2 are given in the column labeled x_2, etc. The right-hand-side values are listed under the column labeled Solution.

The column labeled Basis contains the basic variables, i.e., the variables that are "in the basis." In the initial tableau, the variables x_5, x_6, and x_7 are in the basis. This means that the variables x_1, x_2, x_3, and x_4 are not in the basis and, according to the basis theorem of linear programming, are valued at zero. By referring to the system of linear equations, it can be seen that if x_1, x_2, x_3, and x_4 equal zero, the system reduces to

$$2(0) + 5(0) + 4(0) + 3(0) + 1x_5 + 0x_6 + 0x_7 = 224$$
$$5(0) + 4(0) - 5(0) + 10(0) + 0x_5 + 1x_6 + 0x_7 = 280$$
$$2(0) + 4(0) + 4(0) - 2(0) + 0x_5 + 0x_6 + 1x_7 = 184$$

The solution to this system of three equations and three basic variables can be read from the tableau. The variables x_1, x_2, x_3, and x_4 are not in the basis and have values of zero. Furthermore, the coefficients of the basic variables x_5, x_6, and x_7 form an identity matrix. The values of the variables in the basis can thus be read directly from the Solution column as $x_5 = 224$, $x_6 = 280$, and $x_7 = 184$.

The column labeled c_b contains the objective function coefficients of the basic variables. Since the basic variables in the initial tableau of our example problem are slack variables, the column labeled c_b contains zeros in the initial tableau.

The row labeled z_j contains the sum of the products of the numbers in the c_b column times the coefficients in the x_j column. In the example,

$$z_1 = 0(2) + 0(5) + 0(2) = 0$$
$$z_2 = 0(5) + 0(4) + 0(4) = 0$$
$$z_3 = 0(4) + 0(-5) + 0(4) = 0, \text{etc.}$$

The values of z_j in this example are zero for each of the j columns in the initial tableau. This, of course, is because of the fact that the objective function coefficients of the slack variables in the c_b column are zero. Many of the z_j values will be nonzero after the initial tableau.

The $c_j - z_j$ row is determined by subtracting the z_j value for the jth column from the objective function coefficient for that column. Since the z_j values are zero in the initial tableau, the $c_j - z_j$ values in this example are the same as the values in the c_j row.

The value of the objective function for the initial basic feasible solution is zero. This value is calculated by summing the products of the objective function coefficients of the basic variables and the solution values of the basic variables. This value is shown in the tableau as the last entry in the Solution column.

6.2.2 CHANGE OF BASIS

The value of the objective function can be increased by including one of the nonbasic variables in the basis. In Table 6.6, the nonbasic variables are x_1, x_2, x_3, and x_4. Since the number of variables in the basis remains constant, adding a nonbasic variable means that one of the current basic variables must be removed. Thus, x_5, x_6, or x_7 must leave the basis. The process of adding a variable and, concurrently, removing a variable is termed a *change of basis*.

The $c_j - z_j$ row is used to determine the variable to include in the basis. The numbers in this row show the change in the objective function resulting from including one unit of variable x_j in the basis. In the initial tableau, the $c_j - z_j$ row shows that the objective function will increase by 3 for each unit of x_1, by 4 for each unit of x_2, by 5 for each unit of x_3, etc. Since the objective is to maximize P, the variable that results in the largest per unit increase in the objective function should be placed in the basis. In this example, the objective function increases by 5 for each unit of x_3. Consequently, variable x_3 enters the basis. The rule for deciding the variable that enters the basis is summarized by Simplex Rule 1.

> Simplex Rule 1. *The selection of the variable to enter the basis is based upon the value of $c_j - z_j$. For maximization, the variable selected should have the largest value of $c_j - z_j$. If all values of $c_j - z_j$ are zero or negative, the current basic solution is optimal. For minimization, the variable selected should have the smallest (most negative) value of $c_j - z_j$. If all values of $c_j - z_j$ are zero or positive, the objective function is optimal.*

On the basis of Simplex Rule 1, the variable x_3 is included in the basis.

In certain cases it is possible that there is no single largest $c_j - z_j$ value. For instance, if the $c_j - z_j$ value for x_3 were changed from 5 to 4, the $c_j - z_j$ values for x_2, x_3, and x_4 would be the same. In these cases of ties, it is acceptable arbitrarily to select one of the variables for entry into the basis.

For x_3 to enter the basis, one of the variables currently included in the basis must be removed. Simplex Rule 2 is used to determine this variable.

Simplex Rule 2. *The selection of the variable to leave the basis is made by dividing the numbers in the Solution column by the coefficients in the x_j column (i.e., the coefficients of the variable entering the basis). Select the row with the minimum ratio (ignore ratios with zero or negative numbers in the denominator). The variable associated with this row leaves the basis.*†

This rule applies for both maximization and minimization problems.

Simplex Rule 2 is illustrated for the example problem in Table 6.7. This table shows the ratios of the Solution column to the coefficients of the variable entering the basis. The row associated with variable x_7 has the minimum ratio, and x_3 therefore replaces x_7 in the basis.

Table 6.7

Basis Variables	Current Solution	÷	Coefficients of Entering Variable		Ratios
x_5	224	÷	4	=	56
x_6	280	÷	−5	=	—
x_7	184	÷	4	=	46

The reason underlying Simplex Rule 2 can be explained by referring to the example problem. This problem was

Maximize: $P = 3x_1 + 4x_2 + 5x_3 + 4x_4$
Subject to: $2x_1 + 5x_2 + 4x_3 + 3x_4 \leq 224$
$5x_1 + 4x_2 - 5x_3 + 10x_4 \leq 280$
$2x_1 + 4x_2 + 4x_3 - 2x_4 \leq 184$
$x_j \geq 0 \quad$ for $j = 1, 2, 3, 4$

To illustrate the rule, assume that this is a product mix problem and that the variables represent quantities of product. By applying Simplex Rule 1 to the initial tableau, variable x_3 is selected to enter the basis. The limitation on the maximum amount of x_3 that can be produced comes from the third constraint. That is, $\frac{184}{4}$ or 46 units of product 3 can be produced without exceeding the third constraint, while $\frac{224}{4}$ or 56 units can be produced without exceeding the first constraint. Notice that 46 units of x_3 is a feasible solution, while 56 units is not (i.e., $x_3 = 56$ violates the third constraint). If it is assumed that 46 units of x_3 are produced, the values of the four product variables are $x_1 = 0$, $x_2 = 0$,

† In case of ties among the ratios, it is acceptable arbitrarily to select one of the tied variables to leave the basis.

$x_3 = 46$, and $x_4 = 0$. The values of the slack variables are $x_5 = 40$, $x_6 = 510$, and $x_7 = 0$. Since $x_7 = 0$, it is removed from the basis and is replaced by x_3.

The second tableau has the variables x_5, x_6, and x_3 in the basis. The system of equations is

$$2x_1 + 5x_2 + 4x_3 + 3x_4 + 1x_5 + 0x_6 + 0x_7 = 224$$
$$5x_1 + 4x_2 - 5x_3 + 10x_4 + 0x_5 + 1x_6 + 0x_7 = 280$$
$$2x_1 + 4x_2 + 4x_3 - 2x_4 + 0x_5 + 0x_6 + 1x_7 = 184$$

The system of three equations can be solved for the basic variables x_5, x_6, and x_3 by using row operations. Multiply the third row by $\frac{1}{4}$:

$$\tfrac{1}{2}x_1 + 1x_2 + 1x_3 - \tfrac{1}{2}x_4 + 0x_5 + 0x_6 + \tfrac{1}{4}x_7 = 46$$

Next, subtract four times the new third row from the first row and add five times the new third row to the second row. The resulting system of equations is

$$0x_1 + 1x_2 + 0x_3 + 5x_4 + 1x_5 + 0x_6 - 1x_7 = 40$$
$$\tfrac{15}{2}x_1 + 9x_2 + 0x_3 + \tfrac{15}{2}x_4 + 0x_5 + 1x_6 + \tfrac{5}{4}x_7 = 510$$
$$\tfrac{1}{2}x_1 + 1x_2 + 1x_3 - \tfrac{1}{2}x_4 + 0x_5 + 0x_6 + \tfrac{1}{4}x_7 = 46$$

Since the coefficient matrix of the basic variables x_5, x_6, and x_3 is an identity matrix and the nonbasic variables have values of zero, the basic solution can be read directly from the system of equations. The coefficients of this system of equations and the solution values of the basic variables are entered in the second tableau. This tableau is shown in Table 6.8.

Table 6.8   Second Simplex Tableau

c_b		c_j	3	4	5	4	0	0	0	
	Basis		x_1	x_2	x_3	x_4	x_5	x_6	x_7	Solution
0	x_5		0	1	0	⑤	1	0	-1	40
0	x_6		$\frac{15}{2}$	9	0	$\frac{15}{2}$	0	1	$\frac{5}{4}$	510
5	x_3		$\frac{1}{2}$	1	1	$-\frac{1}{2}$	0	0	$\frac{1}{4}$	46
	z_j		$\frac{5}{2}$	5	5	$-\frac{5}{2}$	0	0	$\frac{5}{4}$	230
	$c_j - z_j$		$\frac{1}{2}$	-1	0	$\frac{13}{2}$	0	0	$-\frac{5}{4}$	

The use of row operations to solve the systems of equations for the new basic variables is termed *pivoting*. The *pivot element* is the element at the intersection of the column headed by the variable entering the basis and the row associated with the variable leaving the basis. In the example, the pivot element in the initial tableau (Table 6.6) is the element at the intersection of the column labeled x_3 and the row labeled x_7. The pivot element in the second simplex tableau (Table 6.8) is circled and is at the intersection of the column labeled x_4 and the row labeled x_5.

The pivoting process involves obtaining an identity matrix as the coefficient matrix of the basic variables. Since only one new basic variable is entered in the tableau at any iteration, the identity matrix is completed by using row operations to obtain a one as the pivot element and zeros elsewhere in the *pivot column*. The mechanics of pivoting are straightforward. The first row operation is to multiply the *pivot row* by the reciprocal of the pivot element. In the example, the pivot row in the initial tableau is multiplied by $\frac{1}{4}$. This new row is entered in the second tableau, Table 6.8. The remainder of the row operations are made to obtain zeros in the pivot column. The new pivot row is multiplied by -4 and added to the first equation. The resulting equation is entered in the second tableau. The new pivot row is then multiplied by 5 and added to the second equation. This equation is also entered in the second tableau. These operations result in values of 1 for the pivot element and 0 for the remaining elements in the pivot column.

The pivoting process is equivalent to solving the system of equations simultaneously for the new basic variables. Since $x_1 = 0$, $x_2 = 0$, $x_4 = 0$, and $x_7 = 0$, the system of equations shown in the second tableau can be written as

$$0(x_3) + 1(x_5) + 0(x_6) = 40$$
$$0(x_3) + 0(x_5) + 1(x_6) = 510$$
$$1(x_3) + 0(x_5) + 0(x_6) = 46$$

The values of the variables at the new extreme point can be read directly as $x_1 = 0$, $x_2 = 0$, $x_3 = 46$, $x_4 = 0$, $x_5 = 40$, $x_6 = 510$, and $x_7 = 0$.

The solution in the second tableau is optimal if all entries in the $c_j - z_j$ row are less than or equal to zero. The z_j values in Table 6.8 are calculated by summing the products of the entries in the c_b column times the coefficients in the x_j column. In the second tableau the values are

$$z_1 = 0(0) + 0(\tfrac{15}{4}) + 5(\tfrac{1}{4}) \quad - \tfrac{5}{4}$$
$$z_2 = 0(1) + 0(9) + 5(1) = 5$$
$$z_3 = 0(0) + 0(0) + 5(1) = 5$$
$$z_4 = 0(5) + 0(\tfrac{15}{2}) + 5(-\tfrac{1}{2}) = -\tfrac{5}{2}, \text{ etc.}$$

The $c_j - z_j$ values are found by subtracting the z_j values from c_j. The $c_j - z_j$ values for x_1 and x_4 are positive; consequently, at least one more iteration is required.

The selection of the variable to enter the basis is made by using Simplex

Rule 1. For maximization, the variable selected should be that variable that has the largest value of $c_j - z_j$. Since the $c_j - z_j$ value for x_4 is $\frac{13}{2}$ and the $c_j - z_j$ value for x_1 is $\frac{1}{2}$, variable x_4 is selected to enter the basis.

Simplex Rule 2 is used to determine the variable to be removed from the basis. The ratios of the values in the Solution column to the positive entries in the x_4 column are $\frac{40}{5} = 8$ and $510/(15/2) = 68$. For both maximization and minimization problems, the variable leaving the basis is the variable associated with the row that has the smallest ratio. Consequently, x_5 leaves the basis and is replaced by x_4.

The values of the new basic variables are determined by the pivoting calculations. These calculations again involve completing the identity matrix by obtaining a 1 as the pivot element and zeros elsewhere in the pivot column. The pivot element is circled in Table 6.8. The pivot row is multiplied by the reciprocal of the pivot element and entered in the new tableau. The product of $\frac{1}{5}$ times the pivot row is shown as the x_4 row in Table 6.9.

Table 6.9   Third Simplex Tableau

c_b	c_j Basis	3 x_1	4 x_2	5 x_3	4 x_4	0 x_5	0 x_6	0 x_7	Solution
4	x_4	0	$\frac{1}{5}$	0	1	$\frac{1}{5}$	0	$-\frac{1}{5}$	8
0	x_6	$\left(\frac{15}{2}\right)$	$\frac{15}{2}$	0	0	$-\frac{3}{2}$	1	$\frac{11}{4}$	450
5	x_3	$\frac{1}{2}$	$\frac{1}{10}$	1	0	$\frac{1}{10}$	0	$\frac{3}{20}$	50
	z_j	$\frac{5}{2}$	$\frac{63}{10}$	5	4	$\frac{13}{10}$	0	$-\frac{1}{20}$	282
	$c_j - z_j$	$\frac{1}{2}$	$-\frac{23}{10}$	0	0	$-\frac{13}{10}$	0	$\frac{1}{20}$	

The remaining pivot calculations are made to obtain zeros in the pivot column. The new pivot row is multiplied by $-\frac{15}{2}$ and added to the x_6 row. It is then multiplied by $\frac{1}{2}$ and added to the x_3 row. This results in a 1 as the pivot element and zeros elsewhere in the column. The identity matrix associated with the basic variables is now complete. The values of the basic variables at the new extreme point can be read directly from the Solution column in the table. The solution is $x_1 = 0$, $x_2 = 0$, $x_3 = 50$, $x_4 = 8$, $x_5 = 0$, $x_6 = 450$, and $x_7 = 0$.

The $c_j - z_j$ values in Table 6.9 are positive for variables x_1 and x_7. Since the $c_j - z_j$ value for x_1 is the largest, x_1 is entered in the basis. The variable leaving the basis is determined by calculating the ratios of the numbers in the Solution column to the positive numbers in the x_1 column. The ratio for the

x_6 row is the smallest; consequently, x_6 is removed from the basis and replaced by x_1.

The pivot element is circled in Table 6.9. This element is converted to a 1 by multiplying the pivot row by $\frac{2}{15}$. The new pivot row is then used to obtain zeros elsewhere in the pivot column. The results of the pivoting calculations are shown in Table 6.10.

Table 6.10   Fourth Simplex Tableau

	c_j	3	4	5	4	0	0	0	
c_b	Basis	x_1	x_2	x_3	x_4	x_5	x_6	x_7	Solution
4	x_4	0	$\frac{1}{5}$	0	1	$\frac{1}{5}$	0	$-\frac{1}{5}$	8
3	x_1	1	1	0	0	$-\frac{1}{5}$	$\frac{2}{15}$	$\frac{11}{30}$	60
5	x_3	0	$\frac{3}{5}$	1	0	$\frac{1}{5}$	$-\frac{1}{15}$	$-\frac{1}{30}$	20
	z_j	3	$\frac{34}{5}$	5	4	$\frac{6}{5}$	$\frac{1}{15}$	$\frac{2}{15}$	312
	$c_j - z_j$	0	$-\frac{14}{5}$	0	0	$-\frac{6}{5}$	$-\frac{1}{15}$	$-\frac{2}{15}$	

The $c_j - z_j$ values in Table 6.10 are all zero or negative. This indicates that the solution shown in the tableau is optimal. The solution is $x_1 = 60$, $x_2 = 0$, $x_3 = 20$, $x_4 = 8$, $x_5 = 0$, $x_6 = 0$, and $x_7 = 0$. The value of the objective function is $P = 312$.

6.2.3   OPTIMALITY

According to Simplex Rule 1, the solution to a linear maximization problem is optimal if all values of $c_j - z_j$ are zero or negative. Conversely, the solution to the minimization problem is optimal if all values of $c_j - z_j$ are zero or positive. To understand this rule, consider the example problem just completed.

In the initial tableau of the example problem, Table 6.6, slack variables are in the basis. According to Simplex Rule 1, a change of basis should be made if one or more of the $c_j - z_j$ values are positive. Since the $c_j - z_j$ value for x_3 in the initial tableau is greater than the $c_j - z_j$ values for the other variables, x_3 enters the basis. On the basis of Simplex Rule 2, variable x_7 leaves the basis.

The net contribution from including x_3 in the basis and removing x_7 is given by the $c_j - z_j$ row. The $c_j - z_j$ value for x_3 is 5 in the initial tableau. This means that each unit of x_3 will add 5 to the objective function. Since $x_3 = 0$ before the iteration (Table 6.6), and $x_3 = 46$ after the iteration (Table 6.8), a total of 46 units of x_3 has been included in the current solution. The change in the objective function is equal to the net contribution from including x_3 in the basis multiplied by the number of units in the solution, i.e., 5(46) or 230.

The $c_j - z_j$ values in the second tableau, Table 6.8, are positive for x_1 and x_4. Applying the simplex rules, x_4 is entered in the basis and x_5 removed from the basis. The $c_j - z_j$ value for x_4 shows that the net contribution per unit from x_4 is $\frac{13}{2}$. After the pivoting process, the solution is $x_1 = 0$, $x_2 = 0$, $x_3 = 50$, $x_4 = 8$, $x_5 = 0$, $x_6 = 450$, $x_7 = 0$, and $P = 282$. Notice that the increase in the objective function from $P = 230$ to $P = 282$ is equal to the $c_j - z_j$ value of x_4 in the second tableau multiplied by the solution value of x_4 in the third tableau, that is $\frac{13}{2}(8) = 52$.

The $c_j - z_j$ values for x_1 and x_7 are positive in the third tableau. Since the $c_j - z_j$ value of $\frac{1}{2}$ for x_1 is the largest, x_1 is entered in the basis. The solution for the new basic variables is given in Table 6.10 as $x_1 = 60$, $x_2 = 0$, $x_3 = 20$, $x_4 = 8$, $x_5 = 0$, $x_6 = 0$, $x_7 = 0$, and $P = 312$. The increase in the objective function from $P = 282$ to $P = 312$ is again equal to the $c_j - z_j$ value for x_1 multiplied by the solution value of x_1, that is, $\frac{1}{2}(60) = 30$.

The $c_j - z_j$ values in Table 6.10 are less than or equal to zero. This means that introducing any of the nonbasic variables in this tableau into the basis would lead to a decrease in the value of the objective function. For instance, if x_2 is entered in the basis, the objective function will decrease by $\frac{14}{5}$ times the solution value of x_2. Since the objective is to maximize the objective function, and since any change of the basic variables will decrease the objective function, the current solution is optimal.

Example: A firm uses manufacturing labor and assembly labor to produce three different products. There are 120 hours of manufacturing labor and 260 hours of assembly labor available for scheduling. One unit of product 1 requires 0.10 hr of manufacturing labor and 0.20 hr of assembly labor. Product 2 requires 0.25 hr of manufacturing labor and 0.30 hr of assembly labor for each unit produced. One unit of product 3 requires 0.40 hr of assembly labor but no manufacturing labor. The contribution to profit from products 1, 2, and 3 is $3.00, $4.00, and $5.00, respectively. Formulate the linear programming problem and solve, using the simplex algorithm.

Let x_j, for $j = 1, 2, 3$, represent the number of units of products 1, 2, and 3. The linear programming problem is

$$\begin{aligned}
\text{Maximize:} \quad P = \quad & 3x_1 + \quad 4x_2 + \quad 5x_3 \\
\text{Subject to:} \quad & 0.10x_1 + 0.25x_2 + \quad 0x_3 \leq 120 \\
& 0.20x_1 + 0.30x_2 + 0.40x_3 \leq 260 \\
& x_1, x_2, x_3 \geq 0
\end{aligned}$$

The inequalities are converted to equalities by the addition of slack variables. This gives

$$\text{Maximize: } P = 3x_1 + 4x_2 + 5x_3 + 0x_4 + 0x_5$$
$$\text{Subject to: } 0.10x_1 + 0.25x_2 + 0x_3 + 1x_4 + 0x_5 = 120$$
$$0.20x_1 + 0.30x_2 + 0.40x_3 + 0x_4 + 1x_5 = 260$$
$$x_j \geq 0, \quad j = 1, 2, 3, 4, 5$$

The slack variables x_4 and x_5 represent unused manufacturing and assembly labor. Since idle labor hours do not contribute to profit, the coefficients of the slack variables in the objective function are zero.

The initial tableau is shown in Table 6.11. The initial basic feasible solution

Table 6.11

Tableau	c_b	Basis	c_j / x_1	x_2	x_3	x_4	x_5	Solution
			3	4	5	0	0	
Initial	0	x_4	0.10	0.25	0	1	0	120
	0	x_5	0.20	0.30	0.40	0	1	260 ←
		z_j	0	0	0	0	0	0
		$c_j - z_j$	3	4	5	0	0	
Second	0	x_4	0.10	0.25	0	1	0	120 ←
	5	x_3	0.50	0.75	1	0	2.50	650
		z_j	2.50	3.75	5	0	12.50	3250
		$c_j - z_j$	0.50	0.25	0	0	−12.50	
Third	3	x_1	1	2.50	0	10	0	1200
	5	x_3	0	−0.50	1	−5	2.50	50
		z_j	3	5	5	5	12.50	3850
		$c_j - z_j$	0	−1	0	−5	−12.50	

is $x_1 = 0$, $x_2 = 0$, $x_3 = 0$, $x_4 = 120$, $x_5 = 260$, and $P = 0$. By Simplex Rules 1 and 2, variable x_3 enters the basis and x_5 leaves. The solution for the new basic variables is given in the second tableau as $x_1 = 0$, $x_2 = 0$, $x_3 = 650$, $x_4 = 120$, $x_5 = 0$, and $P = \$3250$. Notice that the change in the objective function from $P = 0$ to $P = 3250$ is equal to the $c_j - z_j$ value for x_3 in the initial tableau times the solution value for x_3 after the iteration, that is, $\$5(650) = \$3250$.

We have shown that the $c_j - z_j$ value of variable x_j gives the net contribution to the objective function for each unit of x_j included in the solution. The reason for this relationship can perhaps be better understood by examining the $c_j - z_j$ values for the variables in the second tableau. The $c_j - z_j$ value for x_1 and x_2 in the second tableau are positive. The $c_j - z_j$ value of $\$0.50$ for x_1 is the larger of the two values; therefore, x_1 enters the basis. Notice in the original problem that x_1 and x_3 use assembly hours in the ratio of 1 to 2. Since all assembly hours are being utilized in the second tableau (i.e., the value of the slack variable in the second tableau for assembly hours is $x_5 = 0$), each unit of x_1 included in the solution in the third tableau reduces the production of x_3 by one-half unit. This, of course, results in a decrease in the objective function of $\$2.50$ for each one-half hour of assembly labor diverted from x_3 to x_1. This effect on the objective function of diverting assembly labor from x_3 to x_1 is shown by z_1. The z_1 value in the second tableau of $\$2.50$ means that the production of each unit of x_1 will reduce the contribution to the objective function from the variables currently in the basis by $\$2.50$. Since, however, the per unit contribution of x_1 is $\$3.00$, the net change in the objective function for each unit of x_1 included in the basis in the third tableau is $\$3.00 - \$2.50$ or $\$0.50$. This value is shown in the $c_j - z_j$ row in the second tableau for x_1.

The final tableau contains basic variables x_1 and x_3. All $c_j - z_j$ values are zero or negative; consequently, the solution is optimal. As expected, the increase in the objective function from $P = \$3250$ to $P = \$3850$ is equal to the $c_j - z_j$ value of $\$0.50$ for x_1 times the solution value of x_1, that is, $\$0.50(1200) = \$600$. The solution to the problem is $x_1 = 1200$, $x_2 = 0$, $x_3 = 50$, $x_4 = 0$, $x_5 = 0$, and $P = \$3850$.

6.2.4 MARGINAL VALUE OF A RESOURCE

The linear programming problem has been described as that of allocating scarce resources among competing products or activities. Since these resources are combined to produce a salable product, the resources have a value to the firm. One measure of this value is termed the *marginal value* of the resource.

The marginal value of a resource is given by the change in the objective function resulting from employing one additional unit of the resource. To illustrate, consider the preceding example. In this example the marginal value of manufacturing labor is equal to the change in the objective function resulting from employing an additional hour of manufacturing labor. Similarly, the marginal value of assembly labor is equal to the change in the objective function resulting from employing one additional hour of assembly labor.

The marginal value of a resource can be determined directly from the $c_j - z_j$ row of the final tableau. In the final tableau (Table 6.11) the $c_j - z_j$ values for x_4 and x_5 are -5 and -12.50, respectively. Including x_4 in the basis would, therefore, decrease the objective function by \$5 for each unit of x_4 in the solution. Similarly, including x_5 in the basis would decrease the objective function by \$12.50 for each unit of x_5. Since x_4 represents unused manufacturing labor hours and x_5 represents unused assembly labor hours, the objective function will decrease by \$5.00 for each hour of manufacturing labor withheld from production and by \$12.50 for each hour of assembly labor withheld from production. Conversely, as long as the basic variables do not change, each additional hour of manufacturing labor included in production results in a \$5.00 increase in the objective function, and each additional hour of assembly labor results in a \$12.50 increase. The marginal value of manufacturing and assembly labor is therefore \$5.00 and \$12.50, respectively.

The example illustrates that the marginal value of an additional unit of resource is determined from the $c_j - z_j$ value in the final tableau. The absolute value of $c_j - z_j$ for the slack variable for the resource gives the marginal value of that resource. If the slack variable for a resource is in the basis (i.e., unused resource is available), the marginal value of an additional unit of the resource as shown by $c_j - z_j$ is zero. If, however, the slack variable is not in the basis (i.e., all resource has been allocated to production), the marginal value of an additional unit of resource is positive.

Example: Determine the marginal value of an additional unit of resources 1, 2, and 3 for the following linear programming problem.

$$\text{Maximize:} \quad P = \$30x_1 + \$20x_2$$
$$\text{Subject to:} \quad 2x_1 + \quad x_2 \le 280 \quad \text{(resource 1)}$$
$$3x_1 + \quad 2x_2 \le 500 \quad \text{(resource 2)}$$
$$x_1 + \quad 3x_2 \le 420 \quad \text{(resource 3)}$$
$$x_1, x_2 \ge 0$$

This problem is solved by adding slack variables x_3, x_4, and x_5 to convert the inequalities to equations and then applying the simplex algorithm. The final tableau for this problem is shown in Table 6.12.

Since x_3 represents slack for resource 1 and the $c_j - z_j$ value for x_3 is -14, the marginal value of resource 1 is \$14. The marginal value of resource 2 is given by the $c_j - z_j$ entry for x_4. Since 24 units of resource 2 are in the optimal

Table 6.12

c_b	c_j Basis	30 x_1	20 x_2	0 x_3	0 x_4	0 x_5	Solution
30	x_1	1	0	$\frac{3}{5}$	0	$-\frac{1}{5}$	84
0	x_4	0	0	$-\frac{7}{5}$	1	$-\frac{1}{5}$	24
20	x_2	0	1	$-\frac{1}{5}$	0	$\frac{2}{5}$	112
	z_j	30	20	14	0	2	4760
	$c_j - z_j$	0	0	-14	0	-2	

tableau as slack, resource 2 is currently in excess supply and the marginal value of resource 2 is zero. Similarly, the marginal value of resource 3 is $2.

The value of the resources to the firm comes from the fact that they are used to produce a product that is sold for a profit. Since the products are made from the resources, it follows that the value of the resources in generating profit should be equivalent to the profit made from their utilization. This is, in fact, the case. To illustrate, the marginal values of resources 1, 2, and 3 in Table 6.12 are $14, $0, and $2, respectively. The sum of the marginal values of the resources multiplied by the quantity of resource gives the total value of the resources; i.e.,

$$\$14(280) + \$0(500) + \$2(420) = \$4760$$

The value of the resources is equal to the value of the objective function. This result is true at all iterations.

Example: Show that the value of the resources is equal to the value of the objective function at each iteration for the allocation problem given on p. 204.

The tableaus for this problem are given by Table 6.11. The value of the resources in the initial tableau is

$$\$0(120) + \$0(260) = \$0$$

The value of the resources in the second tableau is

$$\$0(120) + \$12.50(260) = \$3250$$

The value of the resources in the final tableau is

$$\$5(120) + \$12.50(260) = \$3850$$

These values equal the objective function at each iteration.

The marginal value of the resource has a number of uses in decision making. One obvious use is in evaluating the employment of additional resources.

In the example, an additional hour of assembly labor is worth $12.50 to the firm in terms of increased profits. If assembly labor can be employed for less than $12.50 per hour, additional laborers should be hired.

Another important use of marginal value occurs in establishing transfer prices for resources between divisions in a firm. Resources in the form of labor, materials, and intermediate products are often transferred from one profit center to another. Knowledge of the marginal value of the resource is, therefore, quite helpful in establishing the transfer price.

6.3 Minimization

The simplex algorithm applies to both maximization and minimization problems. The only difference in the algorithm involves the selection of the variable to enter the basis. In the maximization problem, the variable with the largest $c_j - z_j$ value is included in the basis. Conversely, the variable with the smallest (i.e., most negative) $c_j - z_j$ value is selected to enter the basis in the minimization problem. The selection of the variable to be removed from the basis and the pivoting calculations are the same for both maximization and minimization problems. The solution is optimal in the minimization problem when all $c_j - z_j$ values are zero or positive.

To illustrate minimization, consider the problem

$$\begin{aligned}
\text{Minimize:} \quad & C = 20x_1 + 10x_2 \\
\text{Subject to:} \quad & x_1 + 2x_2 \leq 40 \\
& 3x_1 + x_2 = 30 \\
& 4x_1 + 3x_2 \geq 60 \\
& x_1, x_2 \geq 0
\end{aligned}$$

To apply the simplex algorithm, the inequalities must be converted to equalities. Adding a slack variable to the first inequality and subtracting a surplus variable from the third inequality, we obtain

$$\begin{aligned}
\text{Minimize:} \quad & C = 20x_1 + 10x_2 + 0x_3 + 0x_4 \\
\text{Subject to:} \quad & 1x_1 + 2x_2 + 1x_3 + 0x_4 = 40 \\
& 3x_1 + 1x_2 + 0x_3 + 0x_4 = 30 \\
& 4x_1 + 3x_2 + 0x_3 - 1x_4 = 60 \\
& x_1, x_2, x_3, x_4 \geq 0
\end{aligned}$$

The simplex algorithm begins with an initial basic feasible solution.† In the examples in the preceding sections, the initial feasible solution was given by including the slack variables in the basis. This procedure does not give a

---

† The exceptions to this requirement are discussed in Chapter 7.

feasible solution for this particular problem. To illustrate, if x_1 and x_2 are equated to zero, the system of equations reduces to

$$1x_3 + 0x_4 = 40$$
$$0x_3 + 0x_4 = 30$$
$$0x_3 - 1x_4 = 60$$

The second equation obviously is not true, i.e., zero is not equal to thirty. Consequently, the solution is not feasible.

There is another problem in obtaining the initial basic feasible solution that is not as obvious as that of the equality. For the moment, assume that the equality $3x_1 + x_2 = 30$ is omitted from the original problem. On the basis of this assumption, the problem reduces to

$$\text{Minimize:} \quad C = 20x_1 + 10x_2$$
$$\text{Subject to:} \quad x_1 + 2x_2 \leq 40$$
$$4x_1 + 3x_2 \geq 60$$
$$x_1, x_2 \geq 0$$

Adding the slack variable and subtracting the surplus variable, we now have the system of equations

$$1x_1 + 2x_2 + 1x_3 + 0x_4 = 40$$
$$4x_1 + 3x_2 + 0x_3 - 1x_4 = 60$$
$$x_1, x_2, x_3, x_4 \geq 0$$

Designating x_3 and x_4 as basic variables in the initial tableau gives the solution $x_1 = 0$, $x_2 = 0$, $x_3 = 40$, and $x_4 = -60$. This solution violates the requirement in the simplex algorithm of $x_j \geq 0$. Thus, even after the original equality is omitted, the solution obtained by including the slack and surplus variables in the basis is not feasible.

6.3.1 ARTIFICIAL VARIABLES

The simplex algorithm requires an initial basic feasible solution. As illustrated by the preceding example, it is not always possible to obtain this initial basic feasible solution by merely adding a slack variable to each "less than or equal to" inequality and subtracting a surplus variable from each "greater than or equal to" inequality. In these cases the problem must be modified by adding *artificial variables*. The initial basic feasible solution to this modified problem is then used as a starting point for applying the simplex algorithm to the original problem.

An artificial variable has no physical interpretation. It is merely a "dummy" variable that is added to constraining equations or inequalities for the purpose of generating an initial basic feasible solution.

The minimization problem discussed above provides an example of the use of artificial variables. The slack and surplus variables in this problem do not provide an initial basic feasible solution. To apply the simplex algorithm, the problem must be modified by adding artificial variables to the second and third constraints. After adding artificial variables, the system of equations is

$$1x_1 + 2x_2 + 1x_3 + 0x_4 + 0A_1 + 0A_2 = 40$$

$$3x_1 + 1x_2 + 0x_3 + 0x_4 + 1A_1 + 0A_2 = 30$$

$$4x_1 + 3x_2 + 0x_3 - 1x_4 + 0A_1 + 1A_2 = 60$$

$$x_1, x_2, x_3, x_4, A_1, A_2 \geq 0$$

This system of equations differs from the original system in that it includes the artificial variables A_1 and A_2. Solutions that include A_1 and A_2 at positive values have no meaning in the linear programming problem. Consequently, the artificial variables cannot have positive values in the final tableau.

The artificial variables merely provide a convenient vehicle for generating an initial basic feasible solution. This solution, obtained by including x_3, A_1, and A_2 in the initial basis, is $x_1 = 0$, $x_2 = 0$, $x_3 = 40$, $x_4 = 0$, $A_1 = 30$, and $A_2 = 60$.

6.3.2   THE "BIG M" METHOD

The basic feasible solution that includes artificial variables is used as a starting point for applying the simplex algorithm. The problem is to minimize the objective function subject to the three constraints. This is accomplished provided the artificial variables have values of zero in the final tableau. If those variables have values of zero, they will have provided a starting point for the simplex algorithm while not affecting the optimal solution.

A simple method is available for assuring that the artificial variables have values of zero in the final tableau. The method is to make the coefficients of the artificial variables in the objective function in a minimization problem extremely large and, conversely, to make the coefficients of the artificial variables in the objective function in a maximization problem extremely small (i.e., large negative numbers). This is analogous in a problem involving minimizing cost to making the artificial variable extremely expensive to produce. Alternatively, in a problem involving maximizing profit, the large negative coefficient has the effect of making the artificial variable extremely costly to produce. Since there are no constraints requiring production of the artificial variable, the simplex algorithm will assure that the artificial variable is not in the basis in the final tableau.

Rather than assigning some arbitrarily large positive or negative numbers as coefficients of the artificial variables in the objective function, it is customary to use the capital letter M. If we adopt this convention, the linear minimization problem becomes

$$\text{Minimize:} \quad C = 20x_1 + 10x_2 + 0x_3 + 0x_4 + MA_1 + MA_2$$

$$\text{Subject to:} \quad
\begin{aligned}
1x_1 + 2x_2 + 1x_3 + 0x_4 + 0A_1 + 0A_2 &= 40 \\
3x_1 + 1x_2 + 0x_3 + 0x_4 + 1A_1 + 0A_2 &= 30 \\
4x_1 + 3x_2 + 0x_3 - 1x_4 + 0A_1 + 1A_2 &= 60 \\
x_1, x_2, x_3, x_4, A_1, A_2 &\geq 0
\end{aligned}$$

The tableaus for this problem are given in Table 6.13. Since this is a minimization problem, the variable with the smallest $c_j - z_j$ value is included in the basis. Notice in the initial tableau that $20 - 7M$ is smaller than either $10 - 4M$ or M. Thus, variable x_1 enters the basis. From Simplex Rule 2, variable A_1 leaves the basis. Row operations are then performed to determine the solution values of the variables in the second tableau. The iteration process continues until entries in the $c_j - z_j$ row are zero or positive. The optimal solution is shown in the final tableau as $x_1 = 6$, $x_2 = 12$, $x_3 = 10$, $x_4 = 0$, and $C = 240$.

The linear minimization problem illustrates the types of constraints for which artificial variables are used. These constraints are of the forms $ax_1 + bx_2 \geq c$ and $ax_1 + bx_2 = c$. One surplus and one artificial variable are needed for each greater than or equal to constraint, whereas the equalities require only the artificial variable.

Artificial variables are used to obtain an initial basic feasible solution in both maximization and minimization problems. With the exception that large negative rather than positive numbers are used as the coefficients of the artificial variables in the objective function of a maximization problem, the solution procedure for both maximization and minimization problems is the same. This is illustrated by the following example.

Example:

$$\text{Maximize: } P = x_1 + 2x_2 + 4x_3$$

$$\text{Subject to:} \quad
\begin{aligned}
x_1 + x_2 + x_3 &\leq 12 \\
2x_1 - x_2 + x_3 &\geq 8 \\
x_1, x_2, x_3 &\geq 0
\end{aligned}$$

A slack variable is required for the first inequality, while both a surplus and an artificial variable are required for the second inequality. Adding the slack, surplus, and artificial variables gives

Maximize: $P = 1x_1 + 2x_2 + 4x_3 + 0x_4 + 0x_5 - MA_1$
Subject to: $1x_1 + 1x_2 + 1x_3 + 1x_4 + 0x_5 + 0A_1 = 12$
$2x_1 - 1x_2 + 1x_3 + 0x_4 - 1x_5 + 1A_1 = 8$
$x_1, x_2, x_3, x_4, x_5, A_1 \geq 0$

The tableaus for the problem are shown in Table 6.14.

An interesting feature of this example is that a variable enters the basis in one of the intermediate tableaus but is not in the basis in the optimal solution. Variable x_1 enters the basis in the second tableau and leaves in the third.

Table 6.13

Tableau	c_b	Basis	c_j 20 x_1	10 x_2	0 x_3	0 x_4	M A_1	M A_2	Solution
Initial	0	x_3	1	2	1	0	0	0	40
	M	A_1	③	1	0	0	1	0	30 ←
	M	A_2	4	3	0	-1	0	1	60
		z_j	$7M$	$4M$	0	$-M$	M	M	$90M$
		$c_j - z_j$	$20 - 7M$	$10 - 4M$	0	M	0	0	
			↑						
Second	0	x_3	0	$\frac{5}{3}$	1	0	$-\frac{1}{3}$	0	30
	20	x_1	1	$\frac{1}{3}$	0	0	$\frac{1}{3}$	0	10
	M	A_2	0	⑤⁄₃	0	-1	$-\frac{4}{3}$	1	20 ←
		z_j	20	$\frac{20}{3} + \frac{5}{3}M$	0	$-M$	$\frac{20}{3} - \frac{4}{3}M$	M	$200 + 20M$
		$c_j - z_j$	0	$\frac{10}{3} - \frac{5}{3}M$	0	M	$\frac{7}{3}M - \frac{20}{3}$	0	
				↑					
Third	0	x_3	0	0	1	1	1	-1	10
	20	x_1	1	0	0	$\frac{1}{5}$	$\frac{3}{5}$	$-\frac{1}{5}$	6
	10	x_2	0	1	0	$-\frac{3}{5}$	$-\frac{4}{5}$	$\frac{3}{5}$	12
		z_j	20	10	0	-2	4	2	240
		$c_j - z_j$	0	0	0	2	$M - 4$	$M - 2$	

Table 6.14

Tableau	c_b	Basis	c_j x_1	x_2	x_3	x_4	x_5	A_1	Solution
			1	2	4	0	0	$-M$	
Initial	0	x_4	1	1	1	1	0	0	12
	$-M$	A_1	②$2$	-1	1	0	-1	1	8 ←
		z_j	$-2M$	$+M$	$-M$	0	$+M$	$-M$	$-8M$
		$c_j - z_j$	$1 + 2M$	$2 - M$	$4 + M$	0	$-M$	0	
			↑						
Second	0	x_4	0	$\frac{3}{2}$	$\frac{1}{2}$	1	$\frac{1}{2}$	$-\frac{1}{2}$	8
	1	x_1	1	$-\frac{1}{2}$	$\left(\frac{1}{2}\right)$	0	$-\frac{1}{2}$	$\frac{1}{2}$	4 ←
		z_j	1	$-\frac{1}{2}$	$\frac{1}{2}$	0	$-\frac{1}{2}$	$\frac{1}{2}$	4
		$c_j - z_j$	0	$\frac{5}{2}$	$\frac{7}{2}$	0	$\frac{1}{2}$	$-M - \frac{1}{2}$	
					↑				
Third	0	x_4	-1	②$2$	0	1	1	-1	4 ←
	4	x_3	2	-1	1	0	-1	1	8
		z_j	8	-4	4	0	-4	4	32
		$c_j - z_j$	-7	6	0	0	4	$-M - 4$	
				↑					
Fourth	2	x_2	$-\frac{1}{2}$	1	0	$\frac{1}{2}$	$\left(\frac{1}{2}\right)$	$-\frac{1}{2}$	2 ←
	4	x_3	$\frac{3}{2}$	0	1	$\frac{1}{2}$	$-\frac{1}{2}$	$\frac{1}{2}$	10
		z_j	5	2	4	3	-1	1	44
		$c_j - z_j$	-4	0	0	-3	1	$-M - 1$	
							↑		

0	x_5	-1	2	0	1	1	-1	4
Fifth 4	x_3	1	1	1	1	0	0	12
	z_j	4	4	4	4	0	0	48
	$c_j - z_j$	-3	-2	0	-4	0	$-M$	

Similarly, x_2 enters the basis in the fourth tableau and is replaced by x_5 in the fifth. Although not illustrated by this example, it is also possible for a variable to enter and leave the basis and then reenter in a later iteration.

6.4 Special Cases

The graphical solution procedure was used in Chapter 5 to define and illustrate the cases of no feasible solution, multiple optimal solutions, and unbounded solutions. The relationship between these cases and the simplex algorithm is shown in this section. The section also introduces a method to allow variables that are not restricted to zero or positive values, i.e., variables unrestricted in sign.

6.4.1 NO FEASIBLE SOLUTION

Realistic linear programming applications often have more than one hundred variables and constraints. In problems of this size, it is impossible to tell by graphing if the problem has feasible solutions. Instead, feasibility must be determined from the simplex tableau.

To illustrate, consider the problem

$$\text{Maximize:} \quad P = x_1 + 2x_2$$
$$\text{Subject to:} \quad x_1 + x_2 \leq 4$$
$$x_1 + x_2 \geq 6$$
$$x_1, x_2 \geq 0$$

It can be seen that the constraints are mutually exclusive and, consequently, there can be no feasible solution. This conclusion can also be made from the simplex tableaus for this problem. These are shown in Table 6.15.

The simplex algorithm is applied to obtain the tableaus in Table 6.15. The elements in the $c_j - z_j$ row of the second tableau are all zero or negative, indicating for the maximization problem that the solution is optimal. The solution, however, contains an artificial variable. This indicates that the

Table 6.15

Tableau		c_j	1	2	0	0	$-M$	
	c_b	Basis	x_1	x_2	x_3	x_4	A_1	Solution
Initial	0	x_3	1	①	1	0	0	4 ←
	$-M$	A_1	1	1	0	-1	1	6
		z_j	$-M$	$-M$	0	M	$-M$	$-6M$
		$c_j - z_j$	$1+M$	$2+M$	0	$-M$	0	
				↑				
Second	2	x_2	1	1	1	0	0	4
	$-M$	A_1	0	0	-1	-1	1	2
		z_j	2	2	$2+M$	M	$-M$	$8-2M$
		$c_j - z_j$	-1	0	$-2-M$	$-M$	0	

solution shown in the final tableau is not a feasible solution to the original linear programming problem.

The example illustrates the characteristic form of linear programming problems that have no feasible solution. This characteristic is that one or more artificial variables are in the basis at a nonzero level in the final tableau. In the example the variable A_1 was in the basis and had the value $A_1 = 2$. Since the $c_j - z_j$ values are all zero or negative, the solution is optimal but not feasible.

6.4.2 MULTIPLE OPTIMAL SOLUTIONS

The existence of multiple optimal solutions to a linear programming problem is determined from the $c_j - z_j$ row of the final tableau. The values of $c_j - z_j$ give the net change in the objective function from including one unit of variable x_j in the basis. As shown by previous examples, the $c_j - z_j$ values for the basic variables in the final tableau are zero.

Multiple optimal solutions to the linear programming problem *exist if the $c_j - z_j$ value for one of the nonbasic variables is zero.* A $c_j - z_j$ value of zero for a nonbasic variable means that the variable can be included in the basis without changing the value of the objective function. If a nonbasic variable can be entered in the basis without changing the value of the objective function, the solution given by including the new variable in the basis is also optimal.

This is illustrated by the problem

$$
\begin{aligned}
\text{Maximize:} \quad P = \quad & 3x_1 + \quad 5x_2 + \quad 5x_3 \\
\text{Subject to:} \quad & 0.10x_1 + 0.25x_2 + \quad 0x_3 \le 120 \\
& 0.20x_1 + 0.30x_2 + 0.40x_3 \le 260 \\
& x_1, x_2, x_3 \ge 0
\end{aligned}
$$

The solution is given in Table 6.16. The $c_j - z_j$ row of the final tableau shows that variable x_1 can enter the basis without changing the value of the objective

Table 6.16

Tableau	c_b	c_j Basis	3 x_1	5 x_2	5 x_3	0 x_4	0 x_5	Solution
Initial	0	x_4	0.10	0.25	0	1	0	120
	0	x_5	0.20	0.30	(0.40)	0	1	260 ←
		z_j	0	0	0	0	0	0
		$c_j - z_j$	3	5	5	0	0	
Second	0	x_4	0.10	(0.25)	0	1	0	120 ←
	5	x_3	0.50	0.75	1	0	2.50	650
		z_j	2.50	3.75	5	0	12.50	3250
		$c_j - z_j$	0.50	1.25	0	0	−12.50	
Third	5	x_2	0.40	1	0	4	0	480
	5	x_3	0.20	0	1	−3	2.50	290
		z_j	3	5	5	5	12.50	3850
		$c_j - z_j$	0	0	0	−5	−12.50	

function. To demonstrate, the tableau that includes variable x_1 in the basis is given by Table 6.17. The value of the objective function is unchanged from the tableau in Table 6.16.

Table 6.17

c_b	c_j Basis	3 x_1	5 x_2	5 x_3	0 x_4	0 x_5	Solution
3	x_1	1	2.50	0	10	0	1200
5	x_3	0	−0.50	1	−5	2.50	50
	z_j	3	5	5	5	12.50	3850
	$c_j - z_j$	0	0	0	−5	−12.50	

The graphical analysis of multiple optimal solutions showed that if more than one optimal solution exists, then an infinite number of optimal solutions exist. These solutions are given by forming a linear combination of the basic solutions.† In the example the basic solutions were $x_1 = 0$, $x_2 = 480$, $x_3 = 290$, $x_4 = 0$, $x_5 = 0$ and $x_1 = 1200$, $x_2 = 0$, $x_3 = 50$, $x_4 = 0$, $x_5 = 0$. The linear combination of these solutions is

$$x_1 = b(0) + (1 - b)1200$$
$$x_2 = b(480) + (1 - b)0$$
$$x_3 = b(290) + (1 - b)50$$
$$x_4 = b(0) + (1 - b)0$$
$$x_5 = b(0) + (1 - b)0$$

where b is a weighting factor with domain $0 \le b \le 1$. If, for instance, $b = 0.4$, the solution is $x_1 = 720$, $x_2 = 192$, $x_3 = 146$, $x_4 = 0$, and $x_5 = 0$. This solution satisfies the original constraints and has the optimal objective function value of $P = 3850$.

6.4.3 UNBOUNDED SOLUTIONS

The concept of an unbounded solution was introduced in Chapter 5. The problem used to illustrate the concept was

---

† In large-scale applications it is possible to have more than two optimal basic solutions.

$$\text{Maximize:} \quad P = \quad x_1 + 2x_2$$
$$\text{Subject to:} \quad -x_1 + \quad x_2 \le 2$$
$$x_1 + \quad x_2 \ge 4$$
$$x_1, x_2 \ge 0$$

The effect of an unbounded solution on the simplex algorithm is shown in Table 6.18. The final tableau of the table shows that variable x_4 should enter the basis. The elements in the x_4 column of this tableau are, however, negative. According to Simplex Rule 2, negative elements are ignored in forming the ratios used in selecting the variable to leave the basis. Since neither of the

Table 6.18

c_b	c_j Basis	1 x_1	2 x_2	0 x_3	0 x_4	$-M$ A_1	Solution
0	x_3	-1	①	1	0	0	2 ←
$-M$	A_1	1	1	0	-1	1	4
	z_j	$-M$	$-M$	0	M	$-M$	$-4M$
	$c_j - z_j$	$1 + M$	$2 + M$	0	$-M$	0	
			↑				
2	x_2	-1	1	1	0	0	2
$-M$	A_1	②	0	-1	-1	1	2 ←
	z_j	$-2 - 2M$	2	$2 + M$	M	$-M$	$4 - 2M$
	$c_j - z_j$	$3 + 2M$	0	$-2 - M$	$-M$	0	
		↑					
2	x_2	0	1	$\frac{1}{2}$	$-\frac{1}{2}$	$\frac{1}{2}$	3
1	x_1	1	0	$-\frac{1}{2}$	$-\frac{1}{2}$	$\frac{1}{2}$	1
	z_j	1	2	$\frac{1}{2}$	$-\frac{3}{2}$	$\frac{3}{2}$	7
	$c_j - z_j$	0	0	$-\frac{1}{2}$	$\frac{3}{2}$	$-M - \frac{3}{2}$	
					↑		

basic variables in the final tableau can leave the basis, the change of basis required by Simplex Rule 1 cannot be made.

The condition illustrated by this example occurs for linear programming problems with unbounded solutions. On the basis of Simplex Rule 1, the final solution in Table 6.18 is not optimal and a change of basis is required. The change of basis cannot be made, however, because all the elements in the column headed by the entering variable are zero or negative. If there are no positive elements in this column, the optimal solution is unbounded.

6.4.4 UNRESTRICTED VARIABLES

In certain special applications of linear programming it is possible for variables to have a negative value. A negative-valued variable may, for instance, represent an increase in the inventory of a product. Alternatively, a negative variable could represent returns from a retail outlet to the factory. Variables that are permitted to assume negative values are termed *unrestricted* or *free variables.*

Several techniques are available for incorporating unrestricted variables in the simplex algorithm. One of the widely used methods involves replacing the unrestricted variable by two nonnegative variables. The unrestricted variable x_j is replaced by $x_j = x_j' - x_j''$ and the simplex method is then used without alteration.

To illustrate this technique, assume that the variable x_1 is unrestricted in sign in the linear programming problem

$$\text{Maximize:} \quad P = 3x_1 + 5x_2$$
$$\text{Subject to:} \quad 2x_1 + 5x_2 \leq 132$$
$$3x_1 + 2x_2 \leq 100$$
$$x_2 \geq 0, x_1 \text{ unrestricted in sign}$$

The variable x_1 is replaced by $x_1 = x_1' - x_1''$. This gives

$$\text{Maximize:} \quad P = 3x_1' - 3x_1'' + 5x_2$$
$$\text{Subject to:} \quad 2x_1' - 2x_1'' + 5x_2 \leq 132$$
$$3x_1' - 3x_1'' + 2x_2 \leq 100$$
$$x_1', x_1'', x_2 \geq 0$$

The simplex tableaus for this problem are shown in Table 6.19.

The optimal solution is $x_1' = 21.44$, $x_1'' = 0$, and $x_2 = 17.82$. Since $x_1 = x_1' - x_1''$, the solution reduces to $x_1 = 21.44$ and $x_2 = 17.82$.

6.5 Matrix Representation

The linear programming problem can be stated in matrix form. The linear programming problem

$$\text{Maximize:} \quad P = c_1x_1 + c_2x_2 + c_3x_3$$
$$\text{Subject to:} \quad a_{11}x_1 + a_{12}x_2 + a_{13}x_3 \leq r_1$$
$$a_{21}x_1 + a_{22}x_2 + a_{23}x_3 \leq r_2$$
$$a_{31}x_1 + a_{32}x_2 + a_{33}x_3 \leq r_3$$
$$x_1, x_2, x_3 \geq 0$$

can be represented in matrix form as

$$\text{Maximize:} \quad CX$$
$$\text{Subject to:} \quad AX \leq R$$
$$X \geq 0$$

Table 6.19

c_b	Basis	c_j 3 x_1'	-3 x_1''	5 x_2	0 x_3	0 x_4	Solution
0	x_3	2	-2	⑤	1	0	132 ←
0	x_4	3	-3	2	0	1	100
	z_j	0	0	0	0	0	0
	$c_j - z_j$	3	-3	5	0	0	
				↑			
5	x_2	0.40	-0.40	1	0.20	0	26.40
0	x_4	(2.20)	-2.20	0	-0.40	1	47.20 ←
	z_j	2	-2	5	1	0	132
	$c_j - z_j$	1	-1	0	-1	0	
		↑					
5	x_2	0	0	1	0.273	-0.182	17.82
3	x_1'	1	-1	0	-0.182	0.454	21.44
	z_j	3	-3	5	0.816	0.404	153.42
	$c_j - z_j$	0	0	0	-0.816	-0.404	

where C is the row vector of objective function coefficients, X is the column vector of variables, A is the matrix of constraint coefficients, and R is the column vector of right-hand-side values.

To illustrate matrix representation of a linear programming problem, consider the following problem.

$$\text{Maximize:} \quad P = 3x_1 + 4x_2 + 5x_3 + 4x_4$$
$$\text{Subject to:} \quad 2x_1 + 5x_2 + 4x_3 + 3x_4 \leq 224$$
$$5x_1 + 4x_2 - 5x_3 + 10x_4 \leq 280$$
$$2x_1 + 4x_2 + 4x_3 - 2x_4 \leq 184$$
$$x_j \geq 0 \quad \text{for } j = 1, 2, 3, 4$$

The initial tableau for this problem was given in Table 6.6 and is repeated in Table 6.20. The matrix format for this tableau is given in Table 6.21.

Table 6.20   Tableau Representation of Linear Programming Problem

c_b	c_j Basis	3 x_1	4 x_2	5 x_3	4 x_4	0 x_5	0 x_6	0 x_7	Solution
0	x_5	2	5	4	3	1	0	0	224
0	x_6	5	4	-5	10	0	1	0	280
0	x_7	2	4	4	-2	0	0	1	184
	z_j	0	0	0	0	0	0	0	0
	$c_j - z_j$	3	4	5	4	0	0	0	

Table 6.21   Matrix Representation of Simplex Tableau

c_b	X_b	C			
		A	I	R	
	z_j	0			
	$c_j - z_j$	C			

The matrix I in Table 6.21 is an m by m identity matrix and 0 is a row vector of zeros. A is, of course, the matrix of constraint coefficients, C is the row vector of objective function coefficients, X_b is the column vector of basic variables, and C_b the column vector of objective function coefficients of the basic variables.

The intermediate and final simplex tableaus can also be represented by matrices. Following any iteration of the simplex algorithm, the simplex tableau is given by Table 6.22. This table shows that the solution vector for the basic variables is $B^{-1}R$, the value of the objective function is $C_b{}^t B^{-1}R$, etc.

Table 6.22

	Basis	C		Solution
C_b	X_b	$B^{-1}A$	B^{-1}	$B^{-1}R$
	z_j	$C_b{}^t B^{-1}A$	$C_b{}^t B^{-1}$	$C_b{}^t B^{-1}R$
	$c_j - z_j$	$C - C_b{}^t B^{-1}A$	$- C_b{}^t B^{-1}$	

The matrices A, R, C_b, and C were defined earlier. The matrix B^{-1} has, however, not yet been mentioned. B^{-1} is the inverse of the matrix of constraining coefficients of the current basic variables. The matrix of constraining coefficients of the current basic variables, B, is an m by m matrix. The first column of B contains the coefficients from the initial constraints of the basic variable shown in the first row of the simplex tableau. The second column of B contains the coefficients from the initial constraints of the basic variable shown in the second row of the simplex tableau, etc. The inverse of the matrix B occupies the position of the matrix I in the original tableau. Since the basic variables change at each iteration, the components of B differ at each iteration of the simplex algorithm. This, of course, also means that B^{-1} changes at each iteration.

The relationships shown in Table 6.22 can be illustrated by using the simplex tableaus for the example problem. The problem was

$$\text{Maximize:} \quad P = 3x_1 + 4x_2 + 5x_3 + 4x_4$$
$$\text{Subject to:} \quad 2x_1 + 5x_2 + 4x_3 + 3x_4 \leq 224$$
$$5x_1 + 4x_2 - 5x_3 + 10x_4 \leq 280$$
$$2x_1 + 4x_2 + 4x_3 - 2x_4 \leq 184$$
$$x_j \geq 0 \qquad \text{for } j = 1, 2, 3, 4$$

The simplex tableaus for this problem were given earlier. The optimal tableau was given in Table 6.10 and is repeated below in Table 6.23.

The basic variables in the final tableau are x_4, x_1, and x_3. The matrix B contains the column vectors from the original tableau corresponding to these basic variables, i.e.,

$$\begin{matrix} & x_4 & x_1 & x_3 \\ B = & \begin{pmatrix} 3 & 2 & 4 \\ 10 & 5 & -5 \\ -2 & 2 & 4 \end{pmatrix} \end{matrix}$$

Table 6.23

c_b	c_j Basis	3 x_1	4 x_2	5 x_3	4 x_4	0 x_5	0 x_6	0 x_7	Solution
4	x_4	0	$\frac{1}{5}$	0	1	$\frac{1}{5}$	0	$-\frac{1}{5}$	8
3	x_1	1	1	0	0	$-\frac{1}{5}$	$\frac{2}{15}$	$\frac{11}{30}$	60
5	x_3	0	$\frac{3}{5}$	1	0	$\frac{1}{5}$	$-\frac{1}{15}$	$-\frac{1}{30}$	20
	z_j	3	$\frac{34}{5}$	5	4	$\frac{6}{5}$	$\frac{1}{15}$	$\frac{2}{15}$	312
	$c_j - z_j$	0	$-\frac{14}{5}$	0	0	$-\frac{6}{5}$	$-\frac{1}{15}$	$-\frac{2}{15}$	

The inverse of B is located in Table 6.23 in the position occupied by the identity matrix in the initial tableau (Table 6.20). Using the methods introduced in Chapter 3, we can verify that

$$B^{-1} = \begin{pmatrix} \frac{1}{5} & 0 & -\frac{1}{5} \\ -\frac{1}{5} & \frac{2}{15} & \frac{11}{30} \\ \frac{1}{5} & -\frac{1}{15} & -\frac{1}{30} \end{pmatrix}$$

The solution values for the basic variables are given by $X_b = B^{-1}R$. Thus,

$$X_b = \begin{pmatrix} \frac{1}{5} & 0 & -\frac{1}{5} \\ -\frac{1}{5} & \frac{2}{15} & \frac{11}{30} \\ \frac{1}{5} & -\frac{1}{15} & -\frac{1}{30} \end{pmatrix} \begin{pmatrix} 224 \\ 280 \\ 184 \end{pmatrix} = \begin{pmatrix} 8 \\ 60 \\ 20 \end{pmatrix}$$

The value of the objective function is given by $P = C_b{}'B^{-1}R$

$$P = (4, 3, 5) \begin{pmatrix} \frac{1}{5} & 0 & -\frac{1}{5} \\ -\frac{1}{5} & \frac{2}{15} & \frac{11}{30} \\ \frac{1}{5} & -\frac{1}{15} & -\frac{1}{30} \end{pmatrix} \begin{pmatrix} 224 \\ 280 \\ 184 \end{pmatrix} = 312$$

The columns of the tableau are given by $B^{-1}A$,

$$B^{-1}A = \begin{pmatrix} \frac{1}{5} & 0 & -\frac{1}{5} \\ -\frac{1}{5} & \frac{2}{15} & \frac{11}{30} \\ \frac{1}{5} & -\frac{1}{15} & -\frac{1}{30} \end{pmatrix} \begin{pmatrix} 2 & 5 & 4 & 3 \\ 5 & 4 & -5 & 10 \\ 2 & 4 & 4 & -2 \end{pmatrix} = \begin{pmatrix} 0 & \frac{1}{5} & 0 & 1 \\ 1 & 1 & 0 & 0 \\ 0 & \frac{3}{5} & 1 & 0 \end{pmatrix}$$

Finally, the z_j values are given by $C_b{}'B^{-1}A$ and $C_b{}'B^{-1}$ and the $c_j - z_j$ values are found by subtracting the z_j row from the c_j row. It is left as an exercise for the student to show that the second and third tableaus can be generated by using the matrix relationships.

PROBLEMS

1. Solve the following linear programming problems by determining the value of the objective function for all basic solutions to the problem.

(a) Maximize: $Z = 3x_1 + 2x_2$
     Subject to: $3x_1 + 2x_2 \leq 24$
                     $7x_1 + 12x_2 \leq 84$
                     $x_1, x_2 \geq 0$

(b) Maximize: $Z = 10x_1 + 14x_2$
     Subject to: $2x_1 + 3x_2 \leq 12$
                     $x_1 + 3x_2 \leq 10$
                     $x_1, x_2 \geq 0$

(c) Maximize: $Z = 7x_1 + 8x_2 + 5x_3$
     Subject to: $2x_1 + x_2 + x_3 \leq 10$
                     $x_1 + 3x_2 + 2x_3 \leq 16$
                     $x_1, x_2, x_3 \geq 0$

(d) Minimize: $Z = 4x_1 + 3x_2$
     Subject to: $x_1 \geq 4$
                     $3x_1 + 2x_2 \geq 18$
                     $x_2 \geq 0$

2. Solve the following linear programming problems, using the simplex algorithm.

(a) Maximize: $Z = 5x_1 + 2x_2$
     Subject to: $4x_1 + x_2 \leq 8$
                     $5x_1 + 2x_2 \leq 12$
                     $x_1, x_2 \geq 0$

(b) Maximize: $Z = 5x_1 + 6x_2$
     Subject to: $x_1 \leq 5$
                     $x_2 \leq 4$
                     $10x_1 + 25x_2 \leq 100$
                     $x_1, x_2 \geq 0$

(c) Maximize: $Z = 3x_1 + 2x_2 + 2.5x_3$
     Subject to: $4x_1 + 2x_2 + 2x_3 \leq 12$
                     $x_2 + 3x_3 \leq 4$
                     $x_1, x_2, x_3 \geq 0$

(d)   Maximize: $Z = 2x_1 + 4x_2 + 3x_3$
                       Subject to: $x_1 + 2x_2 \qquad \leq 80$
                                              $x_1 + 4x_2 + 2x_3 \leq 120$
                                              $x_1, x_2, x_3 \geq 0$

3. Solve the following linear programming problems, using the simplex algorithm.

(a)   Maximize: $Z = 20x_1 + 6x_2 + 8x_3$
               Subject to: $8x_1 + 2x_2 + 3x_3 \leq 200$
                                      $4x_1 + 3x_2 \qquad \leq 100$
                                      $x_3 \leq 20$
                                      $x_1, x_2, x_3 \geq 0$

(b)   Maximize: $Z = 12x_1 + 20x_2 + 16x_3$
               Subject to: $4x_1 + 9x_2 + 6x_3 \leq 6000$
                                      $x_1 + x_2 + 3x_3 \leq 4000$
                                      $6x_1 + 5x_2 + 8x_3 \leq 10{,}000$
                                      $x_1, x_2, x_3 \geq 0$

4. Solve the following linear programming problems, using the simplex algorithm.

(a)   Minimize: $Z = x_1 + x_2$
               Subject to: $5x_1 + 3x_2 \geq 30$
                                      $6x_1 + 17x_2 \geq 103$
                                      $x_1, x_2 \geq 0$

(b)   Minimize: $Z = 2x_1 + 3x_2$
               Subject to: $12x_1 + 7x_2 \geq 47$
                                      $5x_1 + 12x_2 \geq 65$
                                      $x_1, x_2 \geq 0$

(c)   Minimize: $Z = 2x_1 + 5x_2 + 4x_3$
               Subject to: $x_1 + 2x_2 + 4x_3 \geq 61$
                                      $3x_1 + x_2 + 5x_3 \geq 78$
                                      $x_1, x_2, x_3 \geq 0$

(d)   Minimize: $Z = 9x_1 + 11x_2 + 7x_3$
               Subject to: $2x_1 + x_2 + x_3 \geq 50$
                                      $5x_1 + 3x_2 + 6x_3 \geq 153$
                                      $x_1, x_2, x_3 \geq 0$

5. Solve the following linear programming problems, using the simplex algorithm.

(a)   Minimize: $Z = 4x_1 + 12x_2$
               Subject to: $15x_1 + 44x_2 \leq 660$
                                      $-4x_1 + 8x_2 \geq -32$
                                      $9x_1 + 15x_2 = 270$
                                      $x_1, x_2 \geq 0$

(b) Maximize: $Z = 2x_1 + 3x_2$

Subject to: $5x_1 + 4x_2 \leq 40$

$11x_1 - 4x_2 = 0$

$-4x_1 + 5x_2 \geq 10$

$x_1, x_2 \geq 0$

6. The following linear programming problem has multiple optimal solutions. Use the simplex algorithm to find the basic solutions to the problem. Give the linear combinations of the basic solutions that define the multiple optimal solutions.

Maximize: $Z = 4x_1 + 6x_2 + 5x_3$

Subject to: $2x_1 + 4x_2 + 2x_3 \leq 60$

$9x_1 + 4x_2 + 16x_3 \leq 240$

$5x_1 + 4x_2 + 6x_3 \geq 125$

$x_1, x_2, x_3 \geq 0$

7. The following linear programming problems illustrate the special cases of no feasible solution and unbounded solution. Use the simplex algorithm to specify which problems have no feasible solution and which have unbounded solutions.

(a) Minimize: $Z = x_1 + 2x_2$

Subject to: $-10x_1 + 6x_2 \geq 60$

$8x_1 + 15x_2 \leq 120$

$x_1, x_2 \geq 0$

(b) Maximize: $Z = 3x_1 + x_2$

Subject to: $7x_1 + 5x_2 \geq 140$

$-4x_1 + 8x_2 \leq 32$

$x_1, x_2 \geq 0$

(c) Maximize: $Z = 5x_1 + 10x_2$

Subject to: $35x_1 + 30x_2 = 1050$

$25x_1 + 45x_2 \leq 1125$

$40x_1 + 20x_2 \leq 800$

$x_1, x_2 \geq 0$

(d) Minimize: $Z = -2x_1 + 3x_2$

Subject to: $9x_1 + 7x_2 \geq 63$

$5x_1 + 12x_2 \geq 60$

$x_1, x_2 \geq 0$

8. Solve the following linear programming problems, using the simplex algorithm.

(a) Minimize: $Z = 5x_1 + 10x_2$

Subject to: $10x_1 + 7x_2 \geq 36$

$4x_1 - 10x_2 \leq 40$

$x_1 \geq 0, x_2$ unrestricted in sign

(b)   Minimize: $Z = x_1 + 10x_2$

           Subject to: $6x_1 + 16x_2 \geq 94$

                          $x_1 \qquad\quad \leq 21$

                          x_1, x_2 unrestricted in sign

9. A manufacturer makes three types of decorative tensor lamps; model 1200, model 1201, and model 1202. The cost of raw materials for each lamp is the same; however, the cost of production differs. Each model 1200 lamp requires 0.1 hr of assembly time, 0.2 hr of wiring time, and 0.1 hr of packaging time. The model 1201 requires 0.2 hr of assembly time, 0.1 hr of wiring time, and 0.1 hr of packaging time. The model 1202 requires 0.2 hr of assembly time, 0.3 hr of wiring time, and 0.1 hr of packaging time. The manufacturer makes a profit of $1.20 on each model 1200 lamp, $1.90 on each model 1201 lamp, and $2.10 on each model 1202 lamp. The manufacturer can schedule up to 80 hr of assembly labor, 120 hr of wiring labor, and 100 hr of packaging labor. Assuming that all lamps can be sold, determine the optimal quantities of each model and the marginal values of each resource.

10. A clothing manufacturer is scheduling work for the next week. There are three possible products that can be made: sportcoats, topcoats, and raincoats. The following table gives the profit for each product and the time required in each process.

Product	Profit per Unit	Hours Required per Unit in		
		Cutting	Sewing	Detailing
Sportcoat	$ 5	1.0	1.0	0.5
Topcoat	8	2.0	1.5	1.0
Raincoat	12	2.0	2.0	1.5

The maximum number of hours that can be scheduled for each process are: cutting, 80 hr; sewing, 60 hr; and detailing, 50 hr. Assuming that all garments produced can be sold, determine the optimal quantities of each item and the marginal values of the three resources.

11. Use the simplex algorithm to determine the optimal quantities of Toots, Wheets, and Honks referred to in Problem 10, Chapter 5. What is the value of an additional unit of each resource used in the manufacture of the noisemakers?

12. Use the simplex algorithm to determine the optimal investments for the Barclay Bank in Problem 18, Chapter 5. Use the $c_j - z_j$ row to determine the change in the overall yield of the investment portfolio from increasing the risk factor to 2.5.

13. Determine the solution to Problem 13, Chapter 5.

14. Determine the solution to Problem 14, Chapter 5.
15. Determine the solution to Problem 15, Chapter 5.
16. Determine the solution to Problem 16, Chapter 5.
17. Determine the solution to Problem 17, Chapter 5.
18. Determine the solution to Problem 19, Chapter 5.
19. Determine the solution to Problem 21, Chapter 5.
20. Determine the solution to Problem 22, Chapter 5

SUGGESTED REFERENCES

The references for this chapter are listed in Chapter 5.

Chapter 7

Duality
and
Sensitivity Analysis

Two extensions of linear programming are presented in this chapter. The first, the dual formulation of the linear programming problem, provides a method for solving an alternative form of the linear programming problem. This reformulation has the advantage of reducing the computational burden for certain linear programming problems. More importantly, however, the theorems that permit this alternative formulation provide special algorithms and additional applications of linear programming. These applications are discussed both in this chapter and the chapter that follows.

The second extension discussed in this chapter is sensitivity analysis. Sensitivity analysis provides a method for investigating the effect on the optimal solution of changes in certain parameters of the linear programming problem. These parameters include the coefficients of the objective function and the right-hand-side values of the constraining equations. Since both the coefficients of the objective function and the right-hand-side values of the constraining equations are often only estimates, information concerning the sensitivity of the solution to changes in these estimates is quite valuable. To illustrate, sensitivity analysis is used in a product mix problem to investigate the effect of changes in the contribution margin of one of the products. Similarly, sensitivity analysis is used to determine the change in the optimal

230

product mix caused by a change in the available quantity of a resource. Sensitivity analysis is discussed beginning in Sec. 7.4.

7.1 The Dual Theorem of Linear Programming

For every linear programming *maximization* problem there is a closely related linear programming *minimization* problem. Conversely, for every linear programming *minimization* problem there is a closely related linear programming *maximization* problem. These pairs of closely related problems are called *dual linear programming problems*.

The duality relationships are important for a number of reasons. First, they lead to a number of theorems that add substantially to our knowledge of linear programming. Second, the dual formulation can be usefully employed in the solution of linear programming problems. As stated earlier, the dual formulation of the linear programming problem often results in a significant reduction of the computational burden of solving the linear programming problem. Finally, the dual problem often provides important economic information concerning the value of the scarce resources employed in a firm.

Before giving an economic interpretation to the dual formulation of a linear programming problem, we shall show the mathematical relationship between the pair of linear programming problems. It is important to remember that this relationship provides a method for converting a linear maximization problem into a related dual minimization problem. Conversely, the relationship permits one to convert a minimization problem into a maximization problem. The two related problems are called the *primal* problem and the *dual* problem.

To illustrate the duality relationships, consider the following pair of linear programming problems. For a maximization problem,

$$\text{Maximize:} \quad P = c_1 x_1 + c_2 x_2 + c_3 x_3 \tag{7.1}$$
$$\text{Subject to:} \quad a_{11} x_1 + a_{12} x_2 + a_{13} x_3 \leq r_1$$
$$a_{21} x_1 + a_{22} x_2 + a_{23} x_3 \leq r_2$$
$$a_{31} x_1 + a_{32} x_2 + a_{33} x_3 \leq r_3$$
$$x_1, x_2, x_3 \geq 0$$

the related minimization problem is

$$\text{Minimize:} \quad L = r_1 w_1 + r_2 w_2 + r_3 w_3 \tag{7.2}$$
$$\text{Subject to:} \quad a_{11} w_1 + a_{21} w_2 + a_{31} w_3 \geq c_1$$
$$a_{12} w_1 + a_{22} w_2 + a_{32} w_3 \geq c_2$$
$$a_{13} w_1 + a_{23} w_2 + a_{33} w_3 \geq c_3$$
$$w_1, w_2, w_3 \geq 0$$

where w_i represents the dual variables.

Although the choice is arbitrary, we shall refer to the maximization problem in this example as the primal problem and to the minimization problem as the dual problem. Throughout this chapter we follow the custom of denoting the original statement of the problem as the primal and the alternative formulation as the dual.

The dual formulation of the primal problem was obtained by:

1. Replacing the variables x_j in the primal by w_i in the dual.
2. Transposing the rows in the primal coefficient matrix to columns in the dual coefficient matrix.
3. Writing the coefficients of the objective function of the primal as the right-hand-side (resource) values in the dual.
4. Writing the right-hand-side values of the primal as the objective function coefficients in the dual.
5. Reversing the direction of the inequalities from the primal to the dual, i.e., if the primal inequalities are $\leq$, the dual inequalities are $\geq$, and vice versa.
6. Reversing the sense of optimization from the primal to the dual, i.e., a primal maximization problem becomes a dual minimization problem and vice versa.

The dual problem can be solved by using the simplex algorithm. One of the important dual theorems is that the optimal value of the primal objective function is equal to the optimal value of the dual objective function. It is also true that if either the primal or the dual problem has an optimal solution, then the related problem must also have an optimal solution.

To illustrate the relationships between the primal and dual problems, consider the problem

$$
\begin{aligned}
\text{Minimize:} \quad & Z = 30x_1 + 15x_2 \\
\text{Subject to:} \quad & 3x_1 + x_2 \geq 3 \\
& 4x_1 + 3x_2 \geq 6 \\
& x_1 + 2x_2 \geq 2 \\
& x_1, x_2 \geq 0
\end{aligned}
$$

The dual formulation of this minimization problem is

$$
\begin{aligned}
\text{Maximize:} \quad & P = 3w_1 + 6w_2 + 2w_3 \\
\text{Subject to:} \quad & 3w_1 + 4w_2 + w_3 \leq 30 \\
& w_1 + 3w_2 + 2w_3 \leq 15 \\
& w_1, w_2, w_3 \geq 0
\end{aligned}
$$

In the illustration the primal problem has two decision variables and three inequalities. To solve the primal problem using the simplex algorithm requires three surplus variables and three artificial variables. After these variables are added, the problem becomes

Minimize: $Z = 30x_1 + 15x_2 + 0x_3 + 0x_4 + 0x_5 + MA_1 + MA_2 + MA_3$
Subject to: $3x_1 + 1x_2 - 1x_3 + 0x_4 + 0x_5 + 1A_1 + 0A_2 + 0A_3 = 3$
                          $4x_1 + 3x_2 + 0x_3 - 1x_4 + 0x_5 + 0A_1 + 1A_2 + 0A_3 = 6$
                          $1x_1 + 2x_2 + 0x_3 + 0x_4 - 1x_5 + 0A_1 + 0A_2 + 1A_3 = 2$
                          $x_1, x_2, x_3, x_4, x_5, A_1, A_2, A_3 \geq 0$

The dual formulation of the problem has three decision variables but only two inequalities. Furthermore, the inequalities do not require artificial variables. Since only three decision variables and two slack variables are needed for the dual problem, the arithmetic necessary to determine the solution for the dual is considerably less than that required for the primal. The dual formulation is

Maximize: $P = 3w_1 + 6w_2 + 2w_3 + 0w_4 + 0w_5$
Subject to: $3w_1 + 4w_2 + 1w_3 + 1w_4 + 0w_5 = 30$
                          $1w_1 + 3w_2 + 2w_3 + 0w_4 + 1w_5 = 15$
                          $w_i \geq 0$ for $i = 1, 2, \ldots, 5$

The solution to the dual problem is found by using the simplex algorithm. The optimal value of the objective function for the dual problem is $P = 36$. Since the optimal value of the objective function for the dual is equal to the optimal value of the objective function for the primal, the solution to the primal problem is also $Z = 36$. The tableaus for the dual problem are given in Table 7.1.

The tableaus for the primal problem are given in Table 7.2. By comparing Tables 7.1 and 7.2, we see that the optimum values of the objective function for the primal and dual formulations to a linear programming problem are equal. The tables also show that the arithmetic required to obtain the solution in this particular problem for the dual is substantially less than that required for the primal.

7.1.1 OBTAINING THE OPTIMAL PRIMAL
SOLUTION FROM THE DUAL

The solution to a linear programming problem includes both the value of the objective function and the value of the variables. We have shown that the optimal value of the objective function for the primal is equal to the optimal value of the objective function for the dual and is, therefore, given in the final tableau of the dual.

Table 7.1

Tableau	c_b		c_j	3	6	2	0	0	
	c_b	Basis		w_1	w_2	w_3	w_4	w_5	Solution
Initial	0	w_4		3	4	1	1	0	30
	0	w_5		1	③	2	0	1	15
		z_j		0	0	0	0	0	0
		$c_j - z_j$		3	6	2	0	0	
Second	0	w_4		⑤⁄₃	0	$-\frac{5}{3}$	1	$-\frac{4}{3}$	10
	6	w_2		$\frac{1}{3}$	1	$\frac{2}{3}$	0	$\frac{1}{3}$	5
		z_j		2	6	4	0	2	30
		$c_j - z_j$		1	0	-2	0	-2	
Third	3	w_1		1	0	-1	$\frac{3}{5}$	$-\frac{4}{5}$	6
	6	w_2		0	1	1	$-\frac{1}{5}$	$\frac{3}{5}$	3
		z_j		3	6	3	$\frac{3}{5}$	$\frac{6}{5}$	36
		$c_j - z_j$		0	0	-1	$-\frac{3}{5}$	$-\frac{6}{5}$	

The values of the primal variables are also given in the optimal tableau of the dual solution. To illustrate, consider again the primal problem,

$$\text{Minimize:} \quad Z = 6x_1 + 3x_2$$
$$\text{Subject to:} \quad 3x_1 + x_2 \geq 3$$
$$4x_1 + 3x_2 \geq 6$$
$$x_1 + 2x_2 \geq 2$$
$$x_1, x_2 \geq 0$$

The solution to the dual formulation of this problem is shown in Table 7.1. The values of the primal variables are given in the $c_j - z_j$ row of the final tableau of the dual. The decision variables of the primal are equal to

absolute values of the $c_j - z_j$ entries for the dual slack or surplus variables.[†]
In the example the values of x_1 and x_2 are given in the $c_j - z_j$ row of the final
tableau in the columns headed by the slack variables w_4 and w_5. The absolute
value of these entries gives $x_1 = \frac{3}{5}$ and $x_2 = \frac{6}{5}$.

The values of the surplus variables in the primal problem are given by the
absolute value of the $c_j - z_j$ entries for the decision variables in the dual formulation. The values of x_3, x_4, and x_5 are equal to the absolute values of
$c_1 - z_1, c_2 - z_2$, and $c_3 - z_3$ in Table 7.1 and are $x_3 = 0$, $x_4 = 0$, and $x_5 = 1$.

The values of the artificial variables in the primal problem are not shown
in the dual. Since artificial variables are used only to obtain an initial basic
feasible solution and have no physical interpretation in the problem, the value
of the artificial variables in the optimal feasible solution will always be zero.

The values of the dual variables are found in the same manner in the final
tableau of the primal (Table 7.2). In the example problem the dual decision
variables are w_1, w_2, and w_3. The optimal values of these variables are equal
to the $c_j - z_j$ values of the surplus variables x_3, x_4, and x_5. The values are
$w_1 = 6$, $w_2 = 3$, and $w_3 = 0$. The values of the dual slack variables, w_4 and
w_5, are given by the $c_j - z_j$ values of the primal decision variables. From the
$c_j - z_j$ row of the final tableau of Table 7.2 the values of w_4 and w_5 are $w_4 = 0$
and $w_5 = 0$.

Example: Solve the following linear programming problem, using the
dual theorem of linear programming.

$$\text{Minimize:} \quad L = 100x_1 + 80x_2$$
$$\text{Subject to:} \quad 6x_1 + 2x_2 \geq 24$$
$$x_1 + x_2 \geq 8$$
$$3x_1 + 9x_2 \geq 12$$
$$x_1, x_2 \geq 0$$

The arithmetic required to determine the solution can be reduced by
solving the dual problem. The dual problem is

$$\text{Maximize:} \quad P = 24w_1 + 8w_2 + 12w_3$$
$$\text{Subject to:} \quad 6w_1 + w_2 + 3w_3 \leq 100$$
$$2w_1 + w_2 + 9w_3 \leq 80$$
$$w_1, w_2, w_3 \geq 0$$

The tableaus for this problem are shown in Table 7.3.

The solution to the primal linear programming problem is $x_1 = 2$, $x_2 = 6$,
and $L = 680$. The surplus variables associated with the primal problem have
values $x_3 = 0$, $x_4 = 0$, and $x_5 = 48$.

[†] Decision variables refer to the original variables of the problem as opposed to slack,
surplus, or artificial variables.

Table 7.2

Tableau	c_b	Basis	30 x_1	15 x_2	0 x_3	0 x_4	0 x_5	M A_1	M A_2	M A_3	Solution
Initial	M	A_1	③	1	-1	0	0	1	0	0	3
	M	A_2	4	3	0	-1	0	0	1	0	6
	M	A_3	1	2	0	0	-1	0	0	1	2
		z_j	$8M$	$6M$	$-M$	$-M$	$-M$	M	M	M	$11M$
		$c_j - z_j$	$30 - 8M$	$15 - 6M$	M	M	M	0	0	0	
Second	30	x_1	1	$\frac{1}{3}$	$-\frac{1}{3}$	0	0	$\frac{1}{3}$	0	0	1
	M	A_2	0	$\frac{5}{3}$	$\frac{4}{3}$	-1	0	$-\frac{4}{3}$	1	0	2
	M	A_3	0	$\left(\frac{5}{3}\right)$	$\frac{1}{3}$	0	-1	$-\frac{1}{3}$	0	1	1
		z_j	30	$10 + \frac{10M}{3}$	$-10 + \frac{5M}{3}$	$-M$	$-M$	$10 - \frac{5M}{3}$	M	M	$30 + 3M$
		$c_j - z_j$	0	$5 - \frac{10M}{3}$	$10 - \frac{5M}{3}$	M	M	$-10 + \frac{8M}{3}$	0	0	

Third

c_i	Basis									
30	x_1	1	0	$-\frac{2}{5}$	0	$\frac{1}{5}$	$\frac{2}{5}$	0	$-\frac{1}{5}$	$\frac{4}{5}$
M	A_2	0	0	1	-1	⨀ 1	-1	1	-1	1
15	x_2	0	1	$\frac{1}{5}$	0	$-\frac{3}{5}$	$-\frac{1}{5}$	0	$\frac{3}{5}$	$\frac{3}{5}$
	z_j	30	15	$-9+M$	$-M$	$-3+M$	$9-M$	M	$3-M$	$33+M$
	c_j-z_j	0	0	$9-M$	M	$3-M$	$-9+2M$	0	$-3+2M$	

Fourth

c_i	Basis									
30	x_1	1	0	$-\frac{3}{5}$	$\frac{1}{5}$	0	$\frac{3}{5}$	$-\frac{1}{5}$	0	$\frac{3}{5}$
0	x_5	0	0	1	-1	1	-1	1	-1	1
15	x_2	0	1	$\frac{4}{5}$	$-\frac{3}{5}$	0	$-\frac{4}{5}$	$\frac{3}{5}$	0	$\frac{6}{5}$
	z_j	30	15	-6	-3	0	6	3	0	36
	c_j-z_j	0	0	6	3	0	$M-6$	$M-3$	M	

Table 7.3

Tableau	c_b	Basis	c_j w_1	w_2	w_3	w_4	w_5	Solution
			24	**8**	**12**	**0**	**0**	
Initial	0	w_4	⑥	1	3	1	0	100
	0	w_5	2	1	9	0	1	80
		z_j	0	0	0	0	0	0
		$c_j - z_j$	24	8	12	0	0	
Second	24	w_1	1	$\frac{1}{6}$	$\frac{1}{2}$	$\frac{1}{6}$	0	$\frac{100}{6}$
	0	w_5	0	⓶⁄₃	8	$-\frac{1}{3}$	1	$\frac{140}{3}$
		z_j	24	4	12	4	0	400
		$c_j - z_j$	0	4	0	-4	0	
Third	24	w_1	1	0	$-\frac{3}{2}$	$\frac{1}{4}$	$-\frac{1}{4}$	5
	8	w_2	0	1	12	$-\frac{1}{2}$	$\frac{3}{2}$	70
		z_j	24	8	60	2	6	680
		$c_j - z_j$	0	0	-48	-2	-6	

In solving a dual linear programming problem, it is necessary that all inequalities in the primal problem have the same form. This means that the inequalities in a primal maximization problem must all be $\leq$ and, conversely, the inequalities in a primal minimization problem must all be $\geq$. To illustrate, consider the following linear programming problem.

$$\text{Minimize:} \quad C = 30x_1 + 24x_2$$
$$\text{Subject to:} \quad x_1 + 2x_2 \geq 80$$
$$-2x_1 + x_2 \leq 10$$
$$x_1 + x_2 \geq 30$$
$$x_1, x_2 \geq 0$$

In order to obtain the dual, we must reverse the sense of the second inequality. This is done by multiplying the second inequality by -1. The primal problem now becomes

$$\text{Minimize:} \quad C = 30x_1 + 24x_2$$
$$\text{Subject to:} \quad x_1 + 2x_2 \geq 80$$
$$2x_1 - x_2 \geq -10$$
$$x_1 + x_2 \geq 30$$
$$x_1, x_2 \geq 0$$

The dual linear programming problem is

$$\text{Maximize:} \quad P = 80w_1 - 10w_2 + 30w_3$$
$$\text{Subject to:} \quad w_1 + 2w_2 + w_3 \leq 30$$
$$2w_1 - w_2 + w_3 \leq 24$$
$$w_1, w_2, w_3 \geq 0$$

The tableaus for the dual linear programming problem are given in Table 7.4. The solution to the primal linear programming problem is $x_1 = 12$, $x_2 = 34$ and $C = 1176$. The surplus variables in the primal problem are $x_3 = 0$, $x_4 = 0$, and $x_5 = 16$.

7.1.2 EQUALITIES AND UNRESTRICTED VARIABLES

In our discussion of the relationship between the primal and dual formulations of a linear programming problem, we have considered only those problems in which the constraints are inequalities. To complete the discussion, we must include the case of constraining equations. Assume that the primal problem has the form

$$\text{Maximize:} \quad P = c_1x_1 + c_2x_2 + c_3x_3$$
$$\text{Subject to:} \quad a_{11}x_1 + a_{12}x_2 + a_{13}x_3 \leq r_1$$
$$a_{21}x_1 - a_{22}x_2 + a_{23}x_3 \geq r_2$$
$$a_{31}x_1 + a_{32}x_2 + a_{33}x_3 = r_3$$
$$x_1, x_2, x_3 \geq 0$$

The primal maximization problem includes one "less than" inequality, one "greater than" inequality, and one equation. Before the primal problem can be written in the dual format, the sense of the second inequality must be changed to "less than or equal to." This is done by multiplying the inequality by -1. The dual linear programming problem is

$$\text{Minimize:} \quad L = r_1w_1 - r_2w_2 + r_3w_3$$
$$\text{Subject to:} \quad a_{11}w_1 - a_{21}w_2 + a_{31}w_3 \geq c_1$$
$$a_{12}w_1 + a_{22}w_2 + a_{32}w_3 \geq c_2$$
$$a_{13}w_1 - a_{23}w_2 + a_{33}w_3 \geq c_3$$
$$w_1, w_2 \geq 0$$
$$w_3 \text{ unrestricted in sign}$$

Table 7.4

Tableau		c_j	80	−10	30	0	0	
	c_b	Basis	w_1	w_2	w_3	w_4	w_5	Solution
Initial	0	w_4	1	2	1	1	0	30
	0	w_5	②	−1	1	0	1	24
		z_j	0	0	0	0	0	0
		$c_j - z_j$	80	−10	30	0	0	

			w_1	w_2	w_3	w_4	w_5	
	0	w_4	0	$\frac{5}{2}$	$\frac{1}{2}$	1	$-\frac{1}{2}$	18
Second	80	w_1	1	$-\frac{1}{2}$	$\frac{1}{2}$	0	$\frac{1}{2}$	12
		z_j	80	−40	40	0	40	960
		$c_j - z_j$	0	30	−10	0	−40	

			w_1	w_2	w_3	w_4	w_5	
	−10	w_2	0	1	$\frac{1}{5}$	$\frac{2}{5}$	$-\frac{1}{5}$	$\frac{36}{5}$
Third	80	w_1	1	0	$\frac{3}{5}$	$\frac{1}{5}$	$\frac{2}{5}$	$\frac{78}{5}$
		z_j	80	−10	46	12	34	1176
		$c_j - z_j$	0	0	−16	−12	−34	

The formulation shows that if the ith constraint in the primal is an equation, then the ith dual variable is unrestricted in sign. In the preceding formulation, the third constraint is an equation. Consequently, the third dual variable is unrestricted in sign. The duality relationships are summarized in Table 7.5.

Example: Solve the following linear programming problem, using the duality relationships of linear programming.

$$\text{Minimize:} \quad C = 20x_1 + 10x_2$$
$$\text{Subject to:} \quad x_1 + 2x_2 \le 40$$
$$3x_1 + x_2 = 30$$
$$4x_1 + 3x_2 \ge 60$$
$$x_1, x_2 \ge 0$$

Table 7.5

Primal (maximize)	Dual (minimize)
Objective function	Right-hand side
Right-hand side	Objective function
jth column of coefficients	jth row of coefficients
ith row of coefficients	ith column of coefficients
jth variable nonnegative	jth constraint an inequality ($\geq$)
jth variable unrestricted in sign	jth constraint an equality
ith constraint an inequality ($\leq$)	ith variable nonnegative
ith constraint an equality	ith variable unrestricted in sign

The problem was introduced in Chapter 6, p. 209, and the tableaus for the primal problem were given in Table 6.13. To solve the problem, using the dual theorem, we must first reverse the sense of the inequality in the first constraint. This gives

$$\text{Minimize:} \quad C = 20x_1 + 10x_2$$
$$\text{Subject to:} \quad -x_1 - 2x_2 \geq -40$$
$$3x_1 + x_2 = 30$$
$$4x_1 + 3x_2 \geq 60$$
$$x_1, x_2 \geq 0$$

The dual formulation for the problem is

$$\text{Maximize:} \quad P = -40w_1 + 30w_2 + 60w_3$$
$$\text{Subject to:} \quad -w_1 + 3w_2 + 4w_3 \leq 20$$
$$-2w_1 + w_2 + 3w_3 \leq 10$$
$$w_1, w_3 \geq 0$$
$$w_2 \text{ unrestricted in sign}$$

Using the method introduced in Sec. 6.4.4 for solving linear programming problems with unrestricted variables, we replace w_2 by $w_2' - w_2''$ and the problem becomes

$$\text{Maximize:} \quad P = -40w_1 + 30w_2' - 30w_2'' + 60w_3$$
$$\text{Subject to:} \quad -w_1 + 3w_2' - 3w_2'' + 4w_3 \leq 20$$
$$-2w_1 + w_2' - w_2'' + 3w_3 \leq 10$$
$$w_1, w_2', w_2'', w_3 \geq 0$$

The tableaus for this problem are given in Table 7.6. The solution to the primal problem is $x_1 = 6$, $x_2 = 12$, and $C = 240$. The slack variable associated with the first constraint of the primal is given by the $c_j - z_j$ entry for w_1 and is $x_3 = 10$. The surplus variable associated with the third constraint is given by the $c_j - z_j$ entry for w_3 and $x_4 = 0$. The $c_j - z_j$ values for w_2' and w_2'' sum to zero in each tableau, indicating that the second constraint in the primal problem is an equation rather than an inequality.

Table 7.6

Tableau		c_j	-40	30	-30	60	0	0	
	c_b	Basis	w_1	w_2'	w_2''	w_3	w_4	w_5	Solution
Initial	0	w_4	-1	3	-3	4	1	0	20
	0	w_5	-2	1	-1	$③$	0	1	10
		z_j	0	0	0	0	0	0	0
		$c_j - z_j$	-40	30	-30	60	0	0	
	0	w_4	$\frac{5}{3}$	$⑤\!\!/\!\!③$	$-\frac{5}{3}$	0	1	$-\frac{4}{3}$	$\frac{20}{3}$
Second	60	w_3	$-\frac{2}{3}$	$\frac{1}{3}$	$-\frac{1}{3}$	1	0	$\frac{1}{3}$	$\frac{10}{3}$
		z_j	-40	20	-20	60	0	20	200
		$c_j - z_j$	0	10	-10	0	0	-20	
	30	w_2'	1	1	-1	0	$\frac{3}{5}$	$-\frac{4}{5}$	4
Third	60	w_3	-1	0	0	1	$-\frac{1}{5}$	$\frac{3}{5}$	2
		z_j	-30	30	-30	60	6	12	240
		$c_j - z_j$	-10	0	0	0	-6	-12	

7.2   Economic Interpretation of the Dual Problem

The dual problem has an important economic interpretation in certain applications of linear programming. One of these applications is the product mix problem. To illustrate this application, we shall assume that the primal problem is a standard product mix problem and that the objective is to determine the profit-maximizing output levels for each of the firm's several products. Profit and output are, of course, limited by the availability of resources. For a problem involving two resources and three products, the typical primal formulation is

$$\text{Maximize:} \quad P = c_1 x_1 + c_2 x_2 + c_3 x_3$$
$$\text{Subject to:} \quad a_{11} x_1 + a_{12} x_2 + a_{13} x_3 \leq r_1$$
$$a_{21} x_1 + a_{22} x_2 + a_{23} x_3 \leq r_2$$
$$x_1, x_2, x_3 \geq 0$$

In manufacturing the three products in our typical problem, two resources are required. Since the profit resulting from the sale of the products could not have been earned without using the resources, the resources have a certain value to the firm. Suppose, then, that the businessman attempts to determine this value. One approach in attempting to determine the value of the resources would be to calculate their cost. This approach would involve depreciation charges, labor rates, overhead allocations, etc. Another approach would be to recognize that without the resources the firm could not have earned its profits. On the basis of this premise, it follows that the businessman should be able to *impute* a certain portion of the profit to each resource. The dual formulation of the product mix problem involves determining the imputed values of the resources.

In the dual formulation of the product mix problem, we impute an artificial accounting price or value w_i to each of the i resources.[†] The values are calculated so that the total value of all resources used is exactly equal to the total profit resulting from the sale of the products. It should be emphasized that the artificial accounting price or value w_i of each resource is not equal to the cost of the resource. Instead, it is an imputed value that comes from the fact that the resources are used to generate the profit of the firm.

The variables in the dual formulation are the artificial accounting prices or values w_i of the i resources. In order to understand the dual formulation, it is important to recognize that the artificial accounting price w_i of the ith resource is equal to the marginal value of the ith resource. The marginal value of a resource, introduced in Sec. 6.2.4, represents the change in the objective function of the primal problem from employing an additional unit of resource. The artificial accounting price represents the value of the resource in generating profits. It is easy to demonstrate that for a particular set of basic variables in the primal problem, the value of each additional unit of resource is constant. This means that the marginal value of the first unit of resource used in the optimal product mix is equal to the marginal value of all other units of that resource used in the same product mix. It also means that the marginal value of the resource and the artificial accounting price (or imputed value) are the same.

The resources have value to the firm because they are used in the manufacture of products. In order to maximize profits, the resources must be allocated to the most profitable combinations of products. This requires that

---

† These variables are also termed *shadow prices*.

we utilize the resources so that the marginal value of additional units of the resources is a minimum. If this were not the case, additional units of a resource would be of greater value than those already employed, a condition that could exist only when we have failed to employ our input resources optimally. The objective in the dual formulation of the product mix problem is, therefore, to allocate the resources to the products so that the total value of the resources is minimized. The objective function is thus

$$\text{Minimize:} \quad L = r_1w_1 + r_2w_2$$

where w_1 and w_2 represent the marginal or imputed values of resources 1 and 2.

The constraints in the dual formulation of the product-mix problem are

$$a_{11}w_1 + a_{21}w_2 \geq c_1$$
$$a_{12}w_1 + a_{22}w_2 \geq c_2$$
$$a_{13}w_1 + a_{23}w_2 \geq c_3$$
$$w_1, w_2 \geq 0$$

The left side of the jth constraint gives the value to the firm of the resources required for the manufacture of one unit of the jth product. For instance, the left side of the first constraint gives the value of the resources used in the manufacture of one unit of product one. Similarly, the left side of the second constraint gives the value of the resources used in the manufacture of one unit of product two. The constraints show that the value to the firm from allocating the scarce resources to manufacture one unit of a product must at least equal the contribution margin from that product. If we use product one as an example, this means that

$$a_{11}w_1 + a_{21}w_2 \geq c_1$$

To understand this constraint, we must remember that the resources have value to the firm because they are used in the manufacture of products. The direction of the inequality guarantees that the imputed value of the resources consumed in the manufacture of one unit of product one must at least equal the contribution to profits from product one. If the value to the firm from employing the resources to manufacture the product exactly equals the contribution to profits resulting from the sale of that product, the inequality is strictly satisfied, and the product will be included in the optimal tableau of the primal formulation.

A second possibility with regard to the constraint is that the value of the resources required to produce one unit of product one exceeds the contribution margin of that product. This occurs if the firm can more profitably employ the resources in the manufacture of the remaining products. If this is the case, the surplus variable associated with the constraint (i.e., w_3) will be included as a basic variable in the optimal tableau of the dual solution. This means, of course, that the $c_j - z_j$ entry for the surplus variable w_3 is zero.

Since the $c_j - z_j$ value of w_3 is also equal to the value of the primal variable x_1, the primal variable x_1 is zero. To illustrate these concepts, consider the following example.

Example: A firm uses manufacturing labor and assembly labor to produce three different products. There are 120 hours of manufacturing labor and 260 hours of assembly labor available for scheduling. One unit of product 1 requires 0.10 hr of manufacturing labor and 0.20 hr of assembly labor. Product 2 requires 0.25 hr of manufacturing labor and 0.30 hr of assembly labor for each unit produced. One unit of product 3 requires 0.40 hr of assembly labor but no manufacturing labor. The contribution to profit from products 1, 2, and 3 is $3, $4, and $5, respectively. Determine the imputed or marginal values of the resources by solving the dual problem.

The primal formulation for the problem was given in Chapter 6. The dual formulation is

$$\text{Minimize:} \quad V = 120w_1 + 260w_2$$
$$\text{Subject to:} \quad 0.10w_1 + 0.20w_2 \geq 3$$
$$0.25w_1 + 0.30w_2 \geq 4$$
$$0w_1 + 0.40w_2 \geq 5$$
$$w_1, w_2 \geq 0$$

where w_1 and w_2 represent the marginal values of manufacturing and assembly labor, respectively. The tableaus for the primal problem are given in Table 6.11 and those for the dual formulation in Table 7.7.

The optimal tableau of the dual formulation shows that $w_1 = \$5.00$, $w_2 = \$12.50$, $w_3 = \$0$, $w_4 = \$1.00$, $w_5 = \$0$, and $V - \$3850$. In terms of the product mix example, the marginal values of resources 1 and 2 are $5.00 and $12.50, respectively.

These resources have value because they are used to manufacture products. The primal solution (Table 6.11) shows that the profit from the manufacture of products 1 and 3 is $3850. If the resources can be used to generate $3850 in profits, it follows that the value of the resources to the firm is $3850. This is confirmed by $V = \$3850$ for the dual solution; i.e.,

$$120(\$5.00) + 260(\$12.50) = \$3850$$

We explained earlier that the constraints in the dual specify that the value of the resources used to manufacture a product must at least equal the contribution margin from the product. The dual constraint for product 1 in the example is

$$0.10w_1 + 0.20w_2 \geq 3$$

The value of the resources used to manufacture one unit of product 1 is given by $0.10w_1 + 0.20w_2$. Since $w_1 = \$5.00$ and $w_2 = \$12.50$, the value to the firm of the resources required to manufacture one unit of product 1 is

$$0.10(\$5.00) + 0.20(\$12.50) = \$3.00$$

Table 7.7

Tableau	c_b	c_j → Basis	120 w_1	260 w_2	0 w_3	0 w_4	0 w_5	M A_1	M A_2	M A_3	Solution
Initial	M	A_1	0.10	0.20	−1	0	0	1	0	0	3
	M	A_2	0.25	0.30	0	−1	0	0	1	0	4
	M	A_3	0	(0.40)	0	0	−1	0	0	1	5
		z_j	0.35M	0.90M	−M	−M	−M	M	M	M	12M
		$c_j - z_j$	120 − 0.35M	260 − 0.90M	M	M	M	0	0	0	
Second	M	A_1	0.10	0	−1	1	0.50	1	0	−0.50	0.50
	M	A_2	0.25	0	0	−1	(0.75)	0	1	−0.75	0.25
	260	w_2	0	1	0	0	−2.50	0	0	2.50	12.50
		z_j	0.35M	260	−M	−M	−650 + 1.25M	M	M	650 − 1.25M	3250 + 0.75M
		$c_j - z_j$	120 − 0.35M	0	M	M	650 − 1.25M	0	0	−650 + 2.25M	

Third

c_B	Basis								
M	A_1	-0.065	0	-1	0.67 ⃝	0	-0.67	0	0.33
0	w_5	0.33	0	0	-1.33	1	1.33	-1	0.33
260	w_2	0.83	1	0	-3.33	0	3.33	0	13.33
	z_j	$214 + 0.065M$	120	$-M$	$-865 + 0.67M$	M	$865 - 0.67M$	0	$3470 + 0.33M$
	$c_j - z_j$	$94 - 0.065M$	0	M	$865 - 0.67M$	0	$-865 + 1.67M$	M	

Fourth

c_B	Basis								
0	w_4	-0.10	0	-1.50	1	0	1.50	0	0.50
0	w_5	0.20 ⃝	0	-2.00	0	1	2.00	-1	1.00
260	w_2	0.50	1	-5.00	0	0	5.00	0	15.00
	z_j	130	120	-1300	0	0	1300	0	3900
	$c_j - z_j$	-10	140	1300	0	0	$M - 1300$	M	

Fifth

c_B	Basis								
0	w_4	0	0	-2.5	1	0.5	2.50	-0.5	1
120	x_1	1	0	-10	0	5	10	-5	5
260	x_2	0	1	0	0	-2.5	0	2.5	12.50
	z_j	120	260	-1200	0	-50	1200	50	3850
	$c_j - z_j$	0	0	1200	M	50	$M - 1200$	$M - 50$	

The imputed value of the resources used to manufacture product 1 is thus equal to the contribution margin from that product, i.e., \$3.00. Similarly, the third constraint shows that the value of the resources used to manufacture one unit of product 3 is equal to the contribution margin from product 3. Consequently, products 1 and 3 will be manufactured by the firm.

The second constraint describes à product that is not included in the optimal product mix. The constraint for product 2 is

$$0.25w_1 + 0.30w_2 \geq 4$$

The value of the resources required for the manufacture of one unit of product 2 is \$5.00, i.e.,

$$0.25(\$5.00) + 0.30(\$12.50) = \$5.00$$

The contribution margin from manufacturing one unit of product 2 is, however, only \$4.00. Each unit of product 2 manufactured, therefore, reduces profits from the optimal product mix by \$1.00. Consequently, the firm should use the resources to manufacture products 1 and 3 instead of product 2.

The surplus variable for the second constraint is $w_4 = \$1.00$. This verifies that the value of the resources required to manufacture one unit of product 2 exceeds the contribution to profits from that product by $w_4 = \$1.00$. Expressed in a somewhat different manner, the fact that w_4 is positive means that the resources can be more profitably employed to manufacture other products. Furthermore, since w_4 is included as a basis variable in the optimal tableau of the dual, the $c_j - z_j$ entry for w_4 is zero. Remembering that the $c_j - z_j$ entry for w_4 gives the value of x_2, we again conclude that product 2 is not included in the optimal production schedule; i.e., $x_2 = 0$.

Example: McGraw Chemical Company uses nitrates, phosphates, potash, and an inert filler material in the manufacture of three types of chemical fertilizers. McGraw has 1200 tons of nitrates, 2000 tons of phosphates, 1500 tons of potash, and an unlimited supply of the inert filler material on hand. They will not receive additional chemicals during the current production period.

The primal formulation of this problem was given in Chapter 5, p. 161. The objective function and resource constraints for the problem were

$$
\begin{aligned}
\text{Maximize:} \quad & P = 8.00x_1 + 11.40x_2 + 6.20x_3 \\
\text{Subject to:} \quad & 0.05x_1 + 0.05x_2 + 0.08x_3 \leq 1200 \\
& 0.10x_1 + 0.08x_2 + 0.12x_3 \leq 2000 \\
& 0.05x_1 + 0.08x_2 + 0.12x_3 \leq 1500 \\
& \qquad\qquad\quad x_2 \qquad\qquad \geq 8000 \\
& x_1, x_3 \geq 0
\end{aligned}
$$

Management of McGraw Chemical has just learned of a critical shortage of nitrates, phosphates, and potash in the Pharmaceutical Division of the

Company. This shortage means that these chemicals must be transferred from the Fertilizer Division to the Pharmaceutical Division. The cost per ton of nitrates, phosphates, and potash is, respectively, $200, $60, and $100. Since each division is a separate profit center, management realizes that it would be unfair merely to credit the Fertilizer Division with the cost of the chemicals transferred. Rather, they want to establish transfer prices that reflect the value of the resources to the Fertilizer Division. Use the economic interpretation of the dual to establish these transfer prices.

The dual variables represent the marginal value of the resources to the Fertilizer Division. The values of these variables are given in the $c_j - z_j$ row of the optimal primal tableau. This tableau is shown in Table 7.8.

Table 7.8

	c_j	8.00	11.40	6.20	0	0	0	0	$-M$	
c_b	Basis	x_1	x_2	x_3	x_4	x_5	x_6	x_7	A_1	Solution
0	x_4	0	0	0.005	1	-0.375	-0.25	0	0	75
8.00	x_1	1	0	0	0	20	-20	0	0	10,000
0	x_7	0	0	1.5	0	-12.5	25	1	-1	4,500
11.40	x_2	0	1	1.5	0	-12.5	25	0	0	12,500
	z_j	8.00	11.40	17.10	0	17.5	125	0	0	222,500
	$c_j - z_j$	0	0	-10.90	0	-17.5	-125	0	$-M$	

The marginal values of the resources are given by the $c_j - z_j$ entries for x_4, x_5, and x_6. These entries show that $w_1 = 0$, $w_2 = \$17.50$, and $w_3 = \$125.00$. To determine the transfer price of each chemical, we add the cost of the chemical to the imputed value of the chemical. Since there is an excess supply of nitrates, the transfer price of nitrate is merely the $200 cost of the chemical. The transfer price of phosphates is given by the sum of the cost per ton of the chemical and the marginal value of the chemical. The transfer price of phosphate is, therefore, $77.50 (i.e., $60.00 + $17.50). Similarly, the transfer price of potash is $225.00 (i.e., $100 + $125).

7.3 The Dual Simplex Algorithm

The *dual simplex* algorithm is based on the fact that *the solution to the linear programming problem is optimal when both the primal and dual solutions are feasible*. The algorithm applies directly to the primal problem and should not be confused with the dual formulation of the primal problem. To understand the algorithm, we must first discuss the concepts of primal and dual feasibility.

7.3.1 PRIMAL AND DUAL FEASIBILITY

A solution, it will be remembered, is feasible when all constraints in the linear programming problem are satisfied. The term *primal feasibility* thus refers to a solution in which the primal variables have feasible solution values. Similarly, the term *dual feasible* is used to describe a solution in which the dual variables have feasible values. For a linear programming problem, the solution is primal feasible if the solution values of the primal variables are nonnegative. For a linear programming *maximization* problem, the solution is dual feasible if all $c_j - z_j$ entries are nonpositive. Conversely, for a linear programming *minimization* problem, the solution is dual feasible if the $c_j - z_j$ entries are nonnegative.

To illustrate primal and dual feasibility, consider the following problem.

$$\begin{aligned}
\text{Minimize:} \quad & C = 2x_1 + x_2 \\
\text{Subject to:} \quad & 3x_1 + x_2 \geq 6 \\
& 4x_1 + 3x_2 \geq 12 \\
& x_1 + 2x_2 \geq 4 \\
& x_1, x_2 \geq 0
\end{aligned}$$

Assume, for the moment, that we ignore the requirement of nonnegativity for the values of the variables. The initial tableau of the linear programming problem under this assumption is given in Table 7.9. The table was obtained

Table 7.9

c_b	c_j Basis	2 x_1	1 x_2	0 x_3	0 x_4	0 x_5	Solution
0	x_3	-3	-1	1	0	0	-6
0	x_4	-4	-3	0	1	0	-12
0	x_5	-1	-2	0	0	1	-4
	z_j	0	0	0	0	0	0
	$c_j - z_j$	2	1	0	0	0	

by subtracting a surplus variable from each inequality and multiplying the resulting equations by -1. The values of the basic variables in the tableau are $x_3 = -6$, $x_4 = -12$, and $x_5 = -4$.

The initial solution to the primal problem in Table 7.9 is nonfeasible. Interestingly, however, the $c_j - z_j$ entries for the primal problem are nonnegative. Since the primal problem is a minimization problem, the nonnegative $c_j - z_j$ entries indicate that the dual solution is feasible. Remembering that the solution to the dual problem is given by the entries in the $c_j - z_j$ row, we note that the initial basic feasible solution for the dual problem is $w_1 = 0$, $w_2 = 0$, $w_3 = 0$, $w_4 = 2$, and $w_5 = 1$. Table 7.9 illustrates a tableau in which the initial solution is dual feasible but not primal feasible.

One of the duality relationships is that a solution that is both primal and dual feasible is optimal. This relationship provides the basis for the dual simplex algorithm. The idea underlying the algorithm is to restore primal feasibility while retaining dual feasibility. This simply means that in a primal minimization problem we must obtain nonnegative solution values for the primal variables while retaining nonnegative entries in the $c_j - z_j$ row. Conversely, for a primal maximization problem we would eliminate the primal infeasibility by obtaining nonnegative solution values for the primal variables while retaining dual feasibility, i.e., nonpositive entries in the $c_j - z_j$ row. In both maximization and minimization problems, the dual simplex iterations are, therefore, made with the objective of removing the primal infeasibilities while retaining a dual feasible solution.

7.3.2 ADVANTAGES OF THE DUAL SIMPLEX ALGORITHM

There are several advantages to the dual simplex algorithm. One of the advantages is that it eliminates the need for artificial variables for inequalities. This reduces the computational effort required to obtain the solution. A second advantage is that the dual simplex algorithm enables one to add a new constraint directly to the final tableau. This will be shown in Sec. 7.4.

7.3.3 DUAL SIMPLEX RULES

The dual simplex rules for both maximization and minimization problems are the same. The rule for determining which variable leaves the basis is given by Dual Simplex Rule 1.

> Dual Simplex Rule 1. *The primal variable whose solution value is most negative leaves the basis. If the solution value for all variables is zero or positive, the current solution is optimal.*

In the example given in Table 7.9, the solution value of x_4 is the most negative. Variable x_4, therefore, leaves the basis. The rule for selecting the variable to enter the basis is given by Dual Simplex Rule 2.

Dual Simplex Rule 2. *Determine the ratio of the numbers in the* $c_j - z_j$ *row (numerator) with the numbers in the row of the variable leaving the basis (denominator). Ignore ratios with zero or positive numbers in the denominator. The column variable whose ratio has the minimum absolute value enters the basis.*

From Table 7.9, the variable selected to enter the basis is determined by forming the ratios of the $c_j - z_j$ row and the x_4 row. The ratios are $2/-4$, $1/-3$, $0/0$, $0/1$, and $0/0$. The ratios with zero or positive components in the denominator are ignored. From the ratios, the column variable whose ratio has the minimum absolute value enters the basis. Since the absolute value of the ratio of the first column is 0.5 and that of the second column is 0.33, variable x_2 enters the basis. Row operations are used to determine the solution values for the new basic variables. The rules are repeated until the optimal solution is obtained.

The dual simplex algorithm is applied to the example problem in Table 7.10. The primal solution is given in the third tableau as $x_1 = \frac{6}{5}$, $x_2 = \frac{12}{5}$, $x_3 = 0$, $x_4 = 0$, and $x_5 = 2$. The dual solution is $w_1 = \frac{2}{5}$, $w_2 = \frac{1}{5}$, $w_3 = 0$, $w_4 = 0$, and $w_5 = 0$. Since both the primal and dual solutions are feasible, the solution is optimal. The value of the objective function is $C = \frac{24}{5}$.

Example: Solve the following problem, using the dual simplex algorithm.

$$\begin{aligned}
\text{Minimize:} \quad & C = 20x_1 + 10x_2 \\
\text{Subject to:} \quad & x_1 + 2x_2 \le 40 \\
& 3x_1 + x_2 = 30 \\
& 4x_1 + 3x_2 \ge 60 \\
& x_1, x_2 \ge 0
\end{aligned}$$

The problem is converted to the simplex format by adding a slack variable to the first inequality, adding an artificial variable to the equation, and subtracting a surplus variable from the final inequality. The solution resulting from this conversion is given by the initial tableau of Table 7.11. Notice that the value of x_4 in the initial tableau is $x_4 = -60$. This, of course, is not a feasible solution for the primal variable. Furthermore, the solution is dual feasible in a primal minimization problem only if all $c_j - z_j$ entries are nonnegative. Since the $c_j - z_j$ entries in the x_1 and x_2 columns are both negative, the initial solution is not dual feasible. Consequently, to apply the dual simplex algorithm, we must first restore dual feasibility. This is accom-

Table 7.10

Tableau	c_b	c_j	2	1	0	0	0	
		Basis	x_1	x_2	x_3	x_4	x_5	Solution
Initial	0	x_3	-3	-1	1	0	0	-6
	0	x_4	-4	-3	0	1	0	-12
	0	x_5	-1	-2	0	0	1	-4
		z_j	0	0	0	0	0	0
		$c_j - z_j$	2	1	0	0	0	
Second	0	x_3	$-\frac{5}{3}$	0	1	$-\frac{1}{3}$	0	-2
	1	x_2	$\frac{4}{3}$	1	0	$-\frac{1}{3}$	0	4
	0	x_5	$\frac{5}{3}$	0	0	$-\frac{2}{3}$	1	4
		z_j	$\frac{4}{3}$	1	0	$-\frac{1}{3}$	0	4
		$c_j - z_j$	$\frac{2}{3}$	0	0	$\frac{1}{3}$	0	
Third	2	x_1	1	0	$-\frac{3}{5}$	$\frac{1}{5}$	0	$\frac{6}{5}$
	1	x_2	0	1	$\frac{4}{5}$	$-\frac{3}{5}$	0	$\frac{12}{5}$
	0	x_5	0	0	1	-1	1	2
		z_j	2	1	$-\frac{2}{5}$	$-\frac{1}{5}$	0	$\frac{24}{5}$
		$c_j - z_j$	0	0	$\frac{2}{5}$	$\frac{1}{5}$	0	

plished by applying the simplex algorithm to the initial tableau. Our objective in applying the simplex algorithm is to obtain nonnegative entries in the $c_j - z_j$ row. The dual simplex algorithm is then used to obtain the optimal solution. For purposes of comparison, the problem was solved by using only the simplex algorithm in Chapter 6, Table 6.13. The problem was again solved in Table 7.6 by applying the simplex algorithm to the dual formulation of the problem.

Table 7.11

Tableau	c_b	c_j	20	10	0	M	0	
	c_b	Basis	x_1	x_2	x_3	A_1	x_4	Solution
Initial	0	x_3	1	2	1	0	0	40
	M	A_1	③	1	0	1	0	30 ←
	0	x_4	−4	−3	0	0	1	−60
		z_j	$3M$	M	0	M	0	$30M$
		$c_j - z_j$	$20 - 3M$	$10 - M$	0	0	0	
			↑					
Second	0	x_3	0	$\frac{5}{3}$	1	$-\frac{1}{3}$	0	30
	20	x_1	1	$\frac{1}{3}$	0	$\frac{1}{3}$	0	10
	0	x_4	0	$\left(-\frac{5}{3}\right)$	0	$\frac{4}{3}$	1	−20 ←
		z_j	20	$\frac{20}{3}$	0	$\frac{20}{3}$	0	200
		$c_j - z_j$	0	$\frac{10}{3}$	0	$M - \frac{20}{3}$	0	
				↑				
Third	0	x_3	0	0	1	1	1	10
	20	x_1	1	0	0	$\frac{3}{5}$	$\frac{1}{5}$	6
	10	x_2	0	1	0	$-\frac{4}{5}$	$-\frac{3}{5}$	12
		z_j	20	10	0	4	−2	240
		$c_j - z_j$	0	0	0	$M - 4$	2	

7.4 Sensitivity Analysis

One of the assumptions in the linear programming problems disussed in the preceding sections is that the parameters of the problem are known. We have assumed, for instance, that the contribution margin for a particular product was given for the problem. Similarly, we have taken for granted that the number of units of a resource available to use in the production of a product was known.

The business practitioner would be quick to point out that these parameters are not always known with certainty. For instance, the contribution margin in a product mix problem includes fixed costs that must be allocated among the products. A change in the method of allocating these costs affects the contribution margin. Similarly, changes in the cost of materials, cost of labor, or the price of a product would cause changes in the contribution margin.

The sensitivity of the optimal solution to changes in the contribution margins of the products is important to the decision maker. The sensitivity is measured by the range over which the contribution margin can vary without causing a change in the optimal solution. To illustrate, assume that the contribution margin for a product is $5.00 and that the product is included as a basic variable in the optimal solution. Furthermore, assume that sensitivity analysis shows that the current solution remains optimal as long as the contribution margin for the product is between $4.00 and $5.50. On the basis of this information, the decision maker knows that a decrease in the contribution margin of more than $1.00 or an increase of more than $0.50 would lead to a revised product mix.

Sensitivity analysis is also used to determine the effect of changes in the resource availabilities on the optimal production schedule. Delayed shipments from suppliers, strikes, spoilage, and other factors all lead to changes in the supply of resources. These changes could, in turn, affect the optimal solution to the linear programming problem.

In addition to investigating the sensitivity of the optimal solution to changes in the objective function coefficients and right-hand-side values, it is often necessary to determine the sensitivity of the optimal solution to changes in products or constraints. For instance, the analyst may want to determine the effect of adding a new product, a new constraint, or of changing the coefficients in the constraining equations. The effect of these types of changes on the optimal solution can also be determined by the techniques presented in this section.

7.4.1 THE OBJECTIVE FUNCTION

The sensitivity of the optimal solution to changes in the coefficients of the objective function is determined by adding a variable δ_j to the objective function coefficient c_j. The new objective function coefficient is $c_j' = c_j + \delta_j$. The domain of δ_j is determined from the requirement that $c_j - z_j$ be nonpositive for a maximization problem and nonnegative for a minimization problem. For the maximization problem, the current solution remains optimal as long as all $c_j - z_j$ entries are nonpositive. Conversely, the current solution is optimal in a minimization problem as long as all $c_j - z_j$ entries are nonnegative. The sensitivity is determined by calculating the values of δ_j for which these criteria are met.

To illustrate the technique, the optimal tableau for the linear programming problem

$$\text{Maximize:} \quad P = 3x_1 + 4x_2 + 5x_3 + 4x_4$$
$$\text{Subject to:} \quad 2x_1 + 5x_2 + 4x_3 + 3x_4 \le 224$$
$$5x_1 + 4x_2 - 5x_3 + 10x_4 \le 280$$
$$2x_1 + 4x_2 + 4x_3 - 2x_4 \le 184$$
$$x_j \ge 0 \quad \text{for } j = 1, 2, 3, 4$$

is given in Table 7.12. The optimal tableau shows that x_1, x_3, and x_4 are basic variables and x_2, x_5, x_6, and x_7 are nonbasic variables.

Table 7.12

c_b	c_j Basis	3 x_1	4 x_2	5 x_3	4 x_4	0 x_5	0 x_6	0 x_7	Solution
4	x_4	0	$\frac{1}{5}$	0	1	$\frac{1}{5}$	0	$-\frac{1}{5}$	8
3	x_1	1	1	0	0	$-\frac{1}{5}$	$\frac{2}{15}$	$\frac{11}{30}$	60
5	x_3	0	$\frac{3}{5}$	1	0	$\frac{1}{5}$	$-\frac{1}{15}$	$-\frac{1}{30}$	20
	z_j	3	$\frac{34}{5}$	5	4	$\frac{6}{5}$	$\frac{1}{15}$	$\frac{2}{15}$	312
	$c_j - z_j$	0	$-\frac{14}{5}$	0	0	$-\frac{6}{5}$	$-\frac{1}{15}$	$-\frac{2}{15}$	

We shall first determine the sensitivity of the optimal solution to changes in the objective function coefficients of the nonbasic variables. To illustrate, the sensitivity of the optimal solution to changes in the objective function coefficient of the nonbasic variable x_2 is determined by adding δ_2 to the objective function coefficient c_2. The objective function coefficient becomes $c_2' = 4 + \delta_2$. The current solution remains optimal as long as the $c_j' - z_j'$ entries are nonpositive. Table 7.13 shows that this criterion is met for $\delta_2 \le \frac{14}{5}$. This means that the objective function coefficient of x_2 can be increased by as much as $\frac{14}{5}$ without causing a change in the optimal solution. If the objective function coefficient of x_2 is increased by more than $\frac{14}{5}$, however, x_2 will enter the basis.

This argument is intuitively plausible, since increasing the objective function coefficient is equivalent to increasing the profit from a product. If profit is sufficiently increased, we would expect to improve the current optimal solution by producing the product.

The sensitivity of the optimal solution to changes in the objective function

Table 7.13

c_j		3	$4+\delta_2$	5	4	0	0	0	
c_b	Basis	x_1	x_2	x_3	x_4	x_5	x_6	x_7	Solution
4	x_4	0	$\frac{1}{5}$	0	1	$\frac{1}{5}$	0	$-\frac{1}{5}$	8
3	x_1	1	1	0	0	$-\frac{1}{5}$	$\frac{2}{15}$	$\frac{11}{30}$	60
5	x_3	0	$\frac{3}{5}$	1	0	$\frac{1}{5}$	$-\frac{1}{15}$	$-\frac{1}{30}$	20
	z_j'	3	$\frac{34}{5}$	5	4	$\frac{6}{5}$	$\frac{1}{15}$	$\frac{2}{15}$	312
	$c_j' - z_j'$	0	$\delta_2 - \frac{14}{5}$	0	0	$-\frac{6}{5}$	$-\frac{1}{15}$	$-\frac{2}{15}$	

coefficients of the remaining nonbasic variables is determined in the same manner. These values are $\delta_5 \le \frac{6}{5}$, $\delta_6 \le \frac{1}{15}$, and $\delta_7 \le \frac{2}{15}$.

The procedure for determining the sensitivity of the optimal solution to changes in the objective function coefficients of basic variables is illustrated in Table 7.14. The objective function coefficient for the basic variable x_1 has been replaced by $c_1' = 3 + \delta_1$ in this table. For the current solution to remain optimal, all $c_j' - z_j'$ entries in Table 7.14 must remain nonpositive, i.e.,

Table 7.14

c_j'		$3+\delta_1$	4	5	4	0	0	0	
c_b	Basis	x_1	x_2	x_3	x_4	x_5	x_6	x_7	Solution
4	x_4	0	$\frac{1}{5}$	0	1	$\frac{1}{5}$	0	$-\frac{1}{5}$	8
$3+\delta_1$	x_1	1	1	0	0	$-\frac{1}{5}$	$\frac{2}{15}$	$\frac{11}{30}$	60
5	x_3	0	$\frac{3}{5}$	1	0	$\frac{1}{5}$	$-\frac{1}{15}$	$-\frac{1}{30}$	20
	z_j'	$3+\delta_1$	$\frac{34}{5}+\delta_1$	5	4	$\frac{6}{5}-\frac{1}{5}\delta_1$	$\frac{1}{15}+\frac{2}{15}\delta_1$	$\frac{2}{15}+\frac{11}{30}\delta_1$	$312+60\delta_1$
	$c_j' - z_j'$	0	$-\frac{14}{5}-\delta_1$	0	0	$-\frac{6}{5}+\frac{1}{5}\delta_1$	$-\frac{1}{15}-\frac{2}{15}\delta_1$	$-\frac{2}{15}-\frac{11}{30}\delta_1$	

$c_j' - z_j' \le 0$. The values of δ_1 that satisfy this criterion are determined by solving the following system of linear inequalities:

$$-\tfrac{14}{5} - \delta_1 \le 0$$
$$-\tfrac{6}{5} + \tfrac{1}{5}\delta_1 \le 0$$
$$-\tfrac{1}{15} - \tfrac{2}{15}\delta_1 \le 0$$
$$-\tfrac{2}{15} - \tfrac{11}{30}\delta_1 \le 0$$

By solving each inequality for δ_1 we find that the smallest value of δ_1 that satisfies all inequalities is $\delta_1 = -\frac{4}{11}$ and that the largest value of δ_1 that satisfies all inequalities is $\delta_1 = 6$. The solution, therefore, is $-\frac{4}{11} \leq \delta_1 \leq 6$. This means that the current variables will remain in the basis for $(3 - \frac{4}{11}) \leq c_1 \leq (3 + 6)$, or equivalently, $\frac{29}{11} \leq c_1 \leq 9$.

It is interesting to notice that if the objective function coefficient of a basic variable is either increased or decreased significantly the current basic variables change. Although this result might at first seem odd, it is what one should expect. An increase in the objective function coefficient implies that the product has become more profitable. Resources should then be diverted from the manufacture of other products to the manufacture of the more profitable product. Conversely, if the product becomes less profitable, resources should be diverted from that product to the manufacture of other products. Either case requires a change of basis. In the example, as long as the contribution margin of product 1 is between $\frac{29}{11}$ and 9 the current solution is optimal. If $c_1 < \frac{29}{11}$ or $c_1 > 9$, however, the current solution is no longer optimal and a change of basis is required. The change of basis is made by applying Simplex Rules 1 and 2.

The sensitivity of the optimal solution to changes in the objective function coefficients of each variable is given in Table 7.15. The reader may want to

Table 7.15

Variable	δ_j	c_j
x_1	$-\frac{4}{11} \leq \delta_1 \leq 6$	$\frac{29}{11} \leq c_1 \leq 9$
x_2	$\delta_2 \leq \frac{14}{5}$	$c_2 \leq \frac{34}{5}$
x_3	$-\frac{14}{3} \leq \delta_3 \leq 1$	$\frac{1}{3} \leq c_3 \leq 6$
x_4	$-6 \leq \delta_4 \leq \frac{2}{3}$	$-2 \leq c_4 \leq \frac{14}{3}$
x_5	$\delta_5 \leq \frac{6}{5}$	$c_5 \leq \frac{6}{5}$
x_6	$\delta_6 \leq \frac{1}{15}$	$c_6 \leq \frac{1}{15}$
x_7	$\delta_7 \leq \frac{2}{15}$	$c_7 \leq \frac{2}{15}$

test his understanding of the technique by calculating the values shown in the table.

Example: Management of the McGraw Chemical Company has expressed concern over the accuracy of the contribution margins for the three chemical fertilizers manufactured by the company. The contribution margins are $8.00 per ton of 5–10–5, $11.40 per ton of 5–8–8, and $6.20 per ton of 8–12–12 (see Chapter 5, p. 161). Determine the sensitivity of the optimal solution to changes in each of the contribution margins.

The linear programming problem was

$$\text{Maximize: } P = 8.00x_1 + 11.40x_2 + 6.20x_3$$
$$\text{Subject to: } 0.05x_1 + 0.05x_2 + 0.08x_3 \leq 1200$$
$$0.10x_1 + 0.08x_2 + 0.12x_3 \leq 2000$$
$$0.05x_1 + 0.08x_2 + 0.12x_3 \leq 1500$$
$$x_2 \geq 8000$$
$$x_1, x_2, x_3 \geq 0$$

The optimal solution for the problem was given in Table 7.8, p. 249, and is repeated below. The optimal solution is $x_1 = 10{,}000$ tons of 5–10–5,

Table 7.16

c_b	Basis	c_j 8.00 x_1	11.40 x_2	6.20 x_3	0 x_4	0 x_5	0 x_6	0 x_i	$-M$ A_1	Solution
0	x_4	0	0	0.005	1	-0.375	-0.25	0	0	75
8.00	x_1	1	0	0	0	20	-20	0	0	10,000
0	x_7	0	0	1.5	0	-12.5	25	1	-1	4,500
11.40	x_2	0	1	1.5	0	-12.5	25	0	0	12,500
	z_j	8.00	11.40	17.10	0	17.5	125	0	0	222,500
	$c_j - z_j$	0	0	-10.90	0	-17.5	-125	0	$-M$	

$x_2 = 12{,}500$ tons of 5–8–8, and $x_3 = 0$ tons of 8–12–12 fertilizer. The sensitivity of the nonbasic variable, x_3, can be determined directly from the table. The table shows that the contribution margin of x_3 must increase from $6.20 to $17.10 per ton before x_3 could be profitably manufactured.

The sensitivity of the solution to changes in the contribution margin of x_1 is determined by substituting $c_1' = 8.00 + \delta_1$ for c_1. The $c_j - z_j$ entries remain nonpositive for

$$-17.5 - 20\delta_1 \leq 0$$

and

$$-125 + 20\delta_1 \leq 0$$

The solution to the two inequalities gives

$$-0.88 \leq \delta_1 \leq 6.25$$

Provided that the contribution margins of the other products are not changed, the current solution remains optimal for

$$\$7.12 \leq c_1 \leq \$14.25$$

The sensitivity of the optimal solution to changes in the contribution margin of x_2 is determined in the same manner. The current solution remains optimal for

$$\$6.40 \leq c_2 \leq \$12.80$$

7.4.2 RIGHT-HAND SIDE

The sensitivity of the basic variables to changes in the right-hand-side values of the constraining equations is often important to the analyst. The sensitivity is measured by an upper and a lower bound on a right-hand-side value. To illustrate, the sensitivity of the basic variables to changes in the right-hand-side value or resource r_1 is described by a constraint of the form

$$a \leq r_1 \leq b$$

As long as this inequality is satisfied, the current basic variables remain in the optimal tableau. If, however, the values of r_1 change so that $r_1 < a$ or $r_1 > b$, one or more basic variables are replaced.

To introduce the method for determining the bounds on the resource, consider the problem introduced in Chapter 6, p. 196. The problem was

$$\text{Maximize: } P = 3x_1 + 4x_2 + 5x_3 + 4x_4$$
$$\text{Subject to: } \quad 2x_1 + 5x_2 + 4x_3 + 3x_4 \leq 224$$
$$5x_1 + 4x_2 - 5x_3 + 10x_4 \leq 280$$
$$2x_1 + 4x_2 + 4x_3 - 2x_4 \leq 184$$
$$x_j \geq 0 \quad \text{for } j = 1, 2, 3, 4$$

The solution to this problem (Chapter 6, p. 203) is $x_1 = 60$, $x_2 = 0$, $x_3 = 20$, and $x_4 = 8$.

The sensitivity of the basic variables to changes in a right-hand-side value is determined by adding a variable δ_i to the ith resource. The ith resource r_i is replaced by $r_i + \delta_i$. Replacing r_1 by $r_1 + \delta_1$ in the example gives the right-hand-side vector

$$R_1^* = \begin{pmatrix} 224 + \delta_1 \\ 280 \\ 184 \end{pmatrix}$$

The bounds on the resource are determined by solving for the values of δ_1 for which the basic variables x_1, x_3, and x_4 remain in the optimal tableau.

The values of the basic variables are given by $X_b = B^{-1}R$. The upper and lower bounds for r_i are determined from the requirement that the solution vector X_b be nonnegative, i.e., $X_b = B^{-1}R_1^* \geq 0$. The matrix B^{-1} was given in Sec. 6.5 as

$$B^{-1} = \begin{pmatrix} \frac{1}{5} & 0 & -\frac{1}{5} \\ -\frac{1}{5} & \frac{2}{15} & \frac{11}{30} \\ \frac{1}{5} & -\frac{1}{15} & -\frac{1}{30} \end{pmatrix}$$

The values of δ_1 are determined from the system of linear inequalities, $B^{-1}R_1^* \geq 0$. This system is

$$\begin{pmatrix} \frac{1}{5} & 0 & -\frac{1}{5} \\ -\frac{1}{5} & \frac{2}{15} & \frac{11}{30} \\ \frac{1}{5} & -\frac{1}{15} & -\frac{1}{30} \end{pmatrix} \begin{pmatrix} 224 + \delta_1 \\ 280 \\ 184 \end{pmatrix} \geq \begin{pmatrix} 0 \\ 0 \\ 0 \end{pmatrix}$$

or alternatively,

$$8 + \tfrac{1}{5}\delta_1 \geq 0$$
$$60 - \tfrac{1}{5}\delta_1 \geq 0$$
$$20 + \tfrac{1}{5}\delta_1 \geq 0$$

The values of δ_1 that satisfy the three inequalities are $-40 \leq \delta_1 \leq 300$. This means that the basic variables x_1, x_3, and x_4 remain in solution for $184 \leq r_1 \leq 524$. If, however, r_1 is less than 184 or more than 524, a change of basic variables will occur.

The sensitivity of the basic variables to changes in the remaining two resources is determined in the same manner. Replacing the second resource by $280 + \delta_2$ gives the right-hand-side vector

$$R_2^* = \begin{pmatrix} 224 \\ 280 + \delta_2 \\ 184 \end{pmatrix}$$

The values of δ_2 are determined from the requirement that $B^{-1}R_2^* \geq 0$, i.e.,

$$\begin{pmatrix} \frac{1}{5} & 0 & -\frac{1}{5} \\ -\frac{1}{5} & \frac{2}{15} & \frac{11}{30} \\ \frac{1}{5} & -\frac{1}{15} & -\frac{1}{30} \end{pmatrix} \begin{pmatrix} 224 \\ 280 + \delta_2 \\ 184 \end{pmatrix} \geq \begin{pmatrix} 0 \\ 0 \\ 0 \end{pmatrix}$$

The system of inequalities is

$$8 + 0\delta_2 \geq 0$$
$$60 + \tfrac{2}{15}\delta_2 \geq 0$$
$$20 - \tfrac{1}{15}\delta_2 \geq 0$$

The solution of the system of inequalities gives $-450 \leq \delta_2 \leq 300$. This means that the basic variables x_1, x_3, and x_4 remain in the optimal tableau for $r_1 = 224$, $-170 \leq r_2 \leq 580$, and $r_3 = 184$. For the third resource, the basic variables x_1, x_3, and x_4 remain in the optimal solution for $r_1 = 224$, $r_2 = 280$, and $20 \leq r_3 \leq 224$.

Example: Determine the optimal solution in the preceding problem for $r_1 = 224$, $r_2 = 126$, and $r_3 = 184$.

The sensitivity analysis showed that the basic variables x_1, x_3, and x_4 remain in the optimal tableau for $r_1 = 224$, $-170 \leq r_2 \leq 580$, and $r_3 = 184$. The values of the basic variables are given by $B^{-1}R$. Thus,

$$X_b = B^{-1}R = \begin{pmatrix} \frac{1}{5} & 0 & -\frac{1}{5} \\ -\frac{1}{5} & \frac{2}{15} & \frac{11}{30} \\ \frac{1}{5} & -\frac{1}{15} & -\frac{1}{30} \end{pmatrix} \begin{pmatrix} 224 \\ 126 \\ 184 \end{pmatrix}$$

$$X_b = \begin{pmatrix} 8 \\ 39.5 \\ 30.3 \end{pmatrix}$$

The values of the basic variables are $x_1 = 39.5$, $x_3 = 30.3$, and $x_4 = 8$.

Example: Management of McGraw Chemical Company has received notice from their principal supplier of potash of a pending strike by their employees. McGraw is scheduled to receive 1500 tons of potash from this supplier. If the strike occurs, however, a portion of this shipment could be delayed. In order to plan production during the coming month, McGraw must determine the effect of possible changes in the supply of potash on their products.

The effect of a shortage of potash on the production of the three fertilizers can be determined from the sensitivity of the basic variables to changes in the supply of potash. In the original statement of the problem (Chapter 5, p. 191), McGraw's raw material inventory included 1200 tons of nitrates, 2000 tons of phosphates, and 1500 tons of potash. The optimal solution based on this inventory (Chapter 7, p. 249) was 10,000 tons of 5–10–5 fertilizer and 12,500 tons of 5–8–8 fertilizer. Production of 8–12–12 fertilizer was not scheduled.

The sensitivity of the basic variables to changes in a resource is determined from the requirement that $B^{-1}R \geq 0$. The inverse matrix (Table 7.8, p. 249) is

$$B^{-1} = \begin{pmatrix} 1 & -0.375 & -0.25 & 0 \\ 0 & 20 & -20 & 0 \\ 0 & -12.5 & 25 & -1 \\ 0 & -12.5 & 25 & 0 \end{pmatrix}$$

The right-hand-side vector is

$$R^* = \begin{pmatrix} 1200 \\ 2000 \\ 1500 + \delta_3 \\ 8000 \end{pmatrix}$$

The requirement that the solution values of the basic variables remain nonnegative gives

$$\begin{pmatrix} 1 & -0.375 & -0.25 & 0 \\ 0 & 20 & -20 & 0 \\ 0 & -12.5 & 25 & -1 \\ 0 & -12.5 & 25 & 0 \end{pmatrix} \begin{pmatrix} 1200 \\ 2000 \\ 1500 + \delta_3 \\ 8000 \end{pmatrix} \geq \begin{pmatrix} 0 \\ 0 \\ 0 \\ 0 \end{pmatrix}$$

The system of inequalities is

$$75 - 0.25\delta_3 \geq 0$$
$$10{,}000 - 20\delta_3 \geq 0$$
$$4500 + 25\delta_3 \geq 0$$
$$12{,}500 + 25\delta_3 \geq 0$$

and the solution of this system of inequalities is $-180 \leq \delta_3 \leq 300$.

The system of linear inequalities shows that the basic variables remain in solution for $r_1 = 1200$ tons of nitrates, $r_2 = 2000$ tons of phosphates, and $1320 \leq r_3 \leq 1800$ tons of potash. If the strike limits the supply of potash to less than 1320 tons, one or more of the current basic variables will be replaced in the optimal tableau.

Example: McGraw Chemical has received a shipment of 1000 tons of potash. If no additional chemicals are received during the month, determine that optimal product mix for the company.

From the preceding example, management knows that a change of basic variables is required in the simplex tableau. The values of the basic variables assuming 1000 tons of potash are given by $B^{-1}R$.

$$X_b = B^{-1}R = \begin{pmatrix} 1 & -0.375 & -0.25 & 0 \\ 0 & 20 & -20 & 0 \\ 0 & -12.5 & 25 & -1 \\ 0 & -12.5 & 25 & 0 \end{pmatrix} \begin{pmatrix} 1200 \\ 2000 \\ 1000 \\ 8000 \end{pmatrix} = \begin{pmatrix} 200 \\ 20{,}000 \\ -8000 \\ 0 \end{pmatrix}$$

Rather than resolving the entire problem, the values of the basic variables can be inserted directly into the optimal tableau of Table 7.8. This is shown in Table 7.17.

Table 7.17

c_b	Basis	c_j 8.00	11.40	6.20	0	0	0	0	$-M$	
		x_1	x_2	x_3	x_4	x_5	x_6	x_7	A_1	Solution
0	x_4	0	0	0.005	1	-0.375	-0.25	0	0	200
8.00	x_1	1	0	0	0	20	-20	0	0	20,000
0	x_7	0	0	1.5	0	-12.5	25	1	-1	-8000
11.40	x_2	0	1	1.5	0	-12.5	25	0	0	0
	z_j	8.00	11.40	17.10	0	17.5	125	0	0	
	$c_j - z_j$	0	0	-10.90	0	-17.5	-125	0	$-M$	

The solution given in Table 7.17 is dual feasible but not primal feasible. By applying the dual simplex algorithm, primal feasibility can be restored. The optimal solution for 1200 tons of nitrates, 2000 tons of phosphates, and 1000 tons of potash is given in Table 7.18 as 7200 tons of 5–10–5 and 8000 tons of 5–8–8.

Table 7.18

		8.00	11.40	6.20	0	0	0	0	$-M$	
c_b	Basis	x_1	x_2	x_3	x_4	x_5	x_6	x_7	A_1	Solution
0	x_4	0	0	-0.04	1	0	-1	-0.03	0.03	440
8.00	x_1	1	0	2.40	0	0	20	1.60	-1.60	7,200
0	x_5	0	0	-0.12	0	1	-2	-0.08	0.08	640
11.40	x_2	0	1	0	0	0	0	-1	1	8,000
	z_j	8.00	11.40	19.20	0	0	160	1.40	-1.40	148,800
	$c_j - z_j$	0	0	-13.00	0	0	-160	-1.40	$-M+1.40$	

The tableau also shows that 440 tons of nitrates and 640 tons of phosphates are not being used. Profit from this schedule is $148,800.

7.4.3 CHANGES TO A COLUMN

Changes can be made to one or more of the components of a column in a simplex tableau. The column entries, it will be remembered, usually represent the technological relationship between a product and the resources. Should these technological relationships change, the column of technological coefficients must reflect this change. Rather than resolving the entire linear programming problem, the change can be made beginning with the optimal tableau of the original problem.

The procedure for calculating the effect of changing a column vector or from adding an entirely new vector is determined from the matrix relationships introduced in Chapter 6. In Sec. 6.5, A_j represents the jth column in the initial tableau. The corresponding column in any succeeding tableau is given by $B^{-1}A_j$.

The effect of changing a column can be determined by replacing the original column vector A_j by the new column vector A_j^*. The jth column in the original optimal tableau now becomes $B^{-1}A_j^*$. The new optimal solution is found by applying the simplex rules to this modified tableau.

To illustrate this procedure, consider the problem

$$
\begin{aligned}
\text{Maximize:} \quad & P = 3x_1 + 4x_2 + 5x_3 + 4x_4 \\
\text{Subject to:} \quad & 2x_1 + 5x_2 + 4x_3 + 3x_4 \le 224 \\
& 5x_1 + 4x_2 - 5x_3 + 10x_4 \le 280 \\
& 2x_1 + 4x_2 + 4x_3 - 2x_4 \le 184 \\
& x_j \ge 0 \qquad \text{for } j = 1, 2, 3, 4
\end{aligned}
$$

The optimal solution to this problem is given in Chapter 6 in Table 6.10. The effect of changing a column is demonstrated by changing the components of the x_4 column to

$$
A_4^* = \begin{pmatrix} 5 \\ 8 \\ -1 \end{pmatrix}
$$

The x_4 column in Table 6.10 is replaced by $B^{-1}A_4^*$. The new x_4 column is

$$
B^{-1}A_4^* = \begin{pmatrix} \frac{1}{5} & 0 & -\frac{1}{5} \\ -\frac{1}{5} & \frac{2}{15} & \frac{11}{30} \\ \frac{1}{5} & -\frac{1}{15} & -\frac{1}{30} \end{pmatrix} \begin{pmatrix} 5 \\ 8 \\ -1 \end{pmatrix} = \begin{pmatrix} \frac{6}{5} \\ -\frac{3}{10} \\ \frac{1}{2} \end{pmatrix}
$$

This column, along with the new z_4 and $c_4 - z_4$ values, is shown by Table 7.18. Since the $c_j - z_j$ value for x_4 is nonpositive in Table 7.19, x_4 remains

Table 7.19

c_b	c_j Basis	3 x_1	4 x_2	5 x_3	4 x_4	0 x_5	0 x_6	0 x_7	Solution
4	x_4	0	$\frac{1}{5}$	0	$\frac{6}{5}$	$\frac{1}{5}$	0	$-\frac{1}{5}$	8
3	x_1	1	1	0	$-\frac{3}{10}$	$-\frac{1}{5}$	$\frac{2}{15}$	$\frac{11}{30}$	60
5	x_3	0	$\frac{3}{5}$	1	$\frac{1}{2}$	$\frac{1}{5}$	$-\frac{1}{15}$	$-\frac{1}{30}$	20
	z_j	3	$\frac{34}{5}$	5	$\frac{32}{5}$	$\frac{6}{5}$	$\frac{1}{15}$	$\frac{2}{15}$	312
	$c_j - z_j$	0	$-\frac{14}{5}$	0	$-\frac{12}{5}$	$-\frac{6}{5}$	$-\frac{1}{15}$	$-\frac{2}{15}$	

as a basic variable in the optimal tableau. The solution values of the basic variables are obtained by pivoting on the first row, fourth column. The optimal solution based on the new column is shown in Table 7.20.

Table 7.20

		3	4	5	4	0	0	0	
c_b	Basis	x_1	x_2	x_3	x_4	x_5	x_6	x_7	Solution
4	x_4	0	$\frac{1}{6}$	0	1	$\frac{1}{6}$	0	$-\frac{1}{6}$	$\frac{20}{3}$
3	x_1	1	$\frac{21}{20}$	0	0	$-\frac{3}{20}$	$\frac{2}{15}$	$\frac{19}{60}$	62
5	x_3	0	$\frac{31}{60}$	1	0	$\frac{7}{60}$	$-\frac{1}{15}$	$\frac{1}{20}$	$\frac{50}{3}$
	z_j	3	$\frac{32}{5}$	5	4	$\frac{4}{5}$	$\frac{1}{15}$	$\frac{8}{15}$	296
	$c_j - z_j$	0	$-\frac{12}{5}$	0	0	$-\frac{4}{5}$	$-\frac{1}{15}$	$-\frac{8}{15}$	

7.4.4 ADDING A NEW CONSTRAINT

It is sometimes necessary to modify a linear programming problem by adding one or more constraints. Rather than resolving the entire problem, the new constraints can be added directly to the final tableau of the original problem. The dual simplex algorithm is then used to find the solution to the revised problem.

To illustrate this procedure, consider the problem

$$\text{Maximize:} \quad P = 3x_1 + 4x_2 + 5x_3 + 4x_4$$
$$\text{Subject to:} \quad 2x_1 + 5x_2 + 4x_3 + 3x_4 \leq 224$$
$$5x_1 + 4x_2 - 5x_3 + 10x_4 \leq 280$$
$$2x_1 + 4x_2 + 4x_3 - 2x_4 \leq 184$$
$$x_j \geq 0 \quad \text{for } j = 1, 2, 3, 4$$

Suppose that the constraint

$$3x_1 + 6x_2 + 2x_3 + 4x_4 \leq 220$$

is added to this problem. The effect of the constraint on the optimal solution is determined by adding the constraint to the optimal tableau of the original problem. Adding this constraint to the optimal tableau (Table 6.10) gives the system of equations shown in Table 7.21. The solution to this system of equations is found by completing the pivoting process. This involves completing the identity matrix for the basic variables. The revised tableau is shown in Table 7.22.

Adding the new constraint to the optimal tableau of the original problem does not, by itself, change the values of the basic variables. The values of the basic variables continue to be $x_1 = 60$, $x_3 = 20$, and $x_4 = 8$. The additional variable included in the basis in Table 7.22 is x_8. Since x_8 is the slack variable

Table 7.21

c_b	c_j Basis	3 x_1	4 x_2	5 x_3	4 x_4	0 x_5	0 x_6	0 x_7	0 x_8	Solution
4	x_4	0	$\frac{1}{5}$	0	1	$\frac{1}{5}$	0	$-\frac{1}{5}$	0	8
3	x_1	1	1	0	0	$-\frac{1}{5}$	$\frac{2}{15}$	$\frac{11}{30}$	0	60
5	x_3	0	$\frac{3}{5}$	1	0	$\frac{1}{5}$	$-\frac{1}{15}$	$-\frac{1}{30}$	0	20
0	x_8	3	6	2	4	0	0	0	1	220
	z_j	3	$\frac{34}{5}$	5	4	$\frac{6}{5}$	$\frac{1}{15}$	$\frac{2}{15}$	0	312
	$c_j - z_j$	0	$-\frac{14}{5}$	0	0	$-\frac{6}{5}$	$-\frac{1}{15}$	$-\frac{2}{15}$	0	

Table 7.22

c_b	c_j Basis	3 x_1	4 x_2	5 x_3	4 x_4	0 x_5	0 x_6	0 x_7	0 x_8	Solution
4	x_4	0	$\frac{1}{5}$	0	1	$\frac{1}{5}$	0	$-\frac{1}{5}$	0	8
3	x_1	1	1	0	0	$-\frac{1}{5}$	$\frac{2}{15}$	$\frac{11}{30}$	0	60
5	x_3	0	$\frac{3}{5}$	1	0	$\frac{1}{5}$	$-\frac{1}{15}$	$-\frac{1}{30}$	0	20
0	x_8	0	1	0	0	-1	$-\frac{4}{15}$	$-\frac{7}{30}$	1	-32
	z_j	3	$\frac{34}{5}$	5	4	$\frac{6}{5}$	$\frac{1}{15}$	$\frac{2}{15}$	0	312
	$c_j - z_j$	0	$-\frac{14}{5}$	0	0	$-\frac{6}{5}$	$-\frac{1}{15}$	$-\frac{2}{15}$	0	

in the new constraint, the value of x_8 should be given by

$$x_8 = 220 - 3x_1 - 6x_2 - 2x_3 - 4x_4$$

Substituting the values of the basic variables x_1, x_3, and x_4 along with the value of the nonbasic variable x_2 in this equation gives

$$x_8 = 220 - 3(60) - 6(0) - 2(20) - 4(8) = -32$$

The fact that $x_8 = -32$ is also shown by the solution column in Table 7.22.

The revised tableau shown in Table 7.22 is dual feasible but not primal feasible. Primal feasibility can be regained, however, by applying the dual simplex algorithm. The solution is found after one iteration and is shown in Table 7.23.

Table 7.23

c_b	Basis	c_j 3 x_1	4 x_2	5 x_3	4 x_4	0 x_5	0 x_6	0 x_7	0 x_8	Solution
4	x_4	0	$\frac{1}{5}$	0	1	$\frac{1}{5}$	0	$-\frac{1}{5}$	0	8
3	x_1	1	$\frac{3}{2}$	0	0	$-\frac{7}{10}$	0	$\frac{1}{4}$	$\frac{1}{2}$	44
5	x_3	0	$\frac{7}{20}$	1	0	$\frac{9}{20}$	0	$\frac{1}{40}$	$-\frac{1}{4}$	28
0	x_6	0	$-\frac{15}{4}$	0	0	$\frac{15}{4}$	1	$\frac{7}{8}$	$-\frac{15}{4}$	120
	z_j	3	$\frac{141}{20}$	5	4	$\frac{19}{20}$	0	$\frac{3}{40}$	$\frac{1}{4}$	304
	$c_j - z_j$	0	$-\frac{61}{20}$	0	0	$-\frac{19}{20}$	0	$-\frac{3}{40}$	$-\frac{1}{4}$	

Example: The Hall Manufacturing Company produces four types of electronic subassemblies for use in aircraft avionic equipment. The parts and labor required for each product along with the resource availabilities and profit are given in Chapter 5, p. 159.

Hall recently received a directive from the Department of Transportation that establishes minimum requirements for the operational life of each subassembly. Although management is confident that their products equal or exceed these requirements, it will nevertheless be necessary to conduct an additional test on each subassembly. Unfortunately, test facilities for the test have a capacity of only 90 units. Determine the revised production schedule based on this new constraint. To simplify the calculations, assume that the original requirement for at least forty units of product 1 no longer applies.

The original linear programming problem was

$$\text{Maximize:} \quad P = 4.25x_1 + 6.25x_2 + 5.00x_3 + 4.50x_4$$
$$\text{Subject to:} \quad 6.0x_1 + 5.0x_2 + 3.5x_3 + 4.0x_4 \leq 600$$
$$1.0x_1 + 1.5x_2 + 1.2x_3 + 1.2x_4 \leq 120$$
$$4x_1 + 3x_2 + 3x_3 + 3x_4 \leq 400$$
$$2x_1 + 2x_2 + 2x_3 + 3x_4 \leq 300$$
$$x_j \geq 0 \quad \text{for } j = 1, 2, 3, 4$$

The optimal tableau for this problem is given in Table 7.24.

The profit from producing the subassemblies is maximum for $x_1 = 76.9$, $x_2 = 20.5$, $x_3 = 10.3$, and $x_4 = 0$ units. The total of 107.7 units manufactured, however, clearly exceeds the capacity of the test facilities. The requirement for testing each subassembly for minimum operational life can be included in the problem by adding the constraint

$$x_1 + x_2 + x_3 + x_4 \leq 90$$

Table 7.24

c_b	c_j Basis	4.25 x_1	6.25 x_2	5.00 x_3	4.50 x_4	0 x_5	0 x_6	0 x_7	0 x_8	Solution
4.25	x_1	1	0	0	0.115	0.231	−1.154	0.192	0	76.9
6.25	x_2	0	1	0	0.231	0.462	1.026	−0.949	0	20.5
5.00	x_3	0	0	1	0.615	−0.769	0.513	1.026	0	10.3
0	x_8	0	0	0	1.077	0.154	−0.769	−0.538	1	84.6
	z_j	4.25	6.25	5.00	5.01	0.024	4.07	0.015	0	506.46
	$c_j - z_j$	0	0	0	−0.51	−0.024	−4.07	−0.015	0	

to the tableau of Table 7.24. This gives the tableau shown in Table 7.25. The value of the slack variable for the new constraint was found by solving the system of constraining equations for the basic variables, i.e., by completing the identity matrix for the basic variables.

Table 7.25

c_b	c_j Basis	4.25 x_1	6.25 x_2	5.00 x_3	4.50 x_4	0 x_5	0 x_6	0 x_7	0 x_8	0 x_9	Solution
4.25	x_1	1	0	0	0.115	0.231	−1.154	0.192	0	0	76.9
6.25	x_2	0	1	0	0.231	0.462	1.026	−0.949	0	0	20.5
5.00	x_3	0	0	1	0.615	−0.769	0.513	1.026	0	0	10.3
0	x_8	0	0	0	1.077	0.154	−0.769	−0.538	1	0	84.6
0	x_9	0	0	0	0.039	0.076	−0.385	−0.269	0	1	−17.7
	z_j	4.25	6.25	5.00	5.008	0.024	4.073	0.015	0	0	506.46
	$c_j - z_j$	0	0	0	−0.508	−0.024	−4.073	−0.015	0	0	

The tableau shown in Table 7.25 is dual feasible but not primal feasible. Primal feasibility can be restored by the dual simplex algorithm. The solution, obtained after two iterations, is $x_1 = 30$, $x_2 = 60$, $x_3 = 0$, and $x_4 = 0$. Interestingly, the profit for this schedule decreases by only \$3.96 from the previous optimum. Profit, after including the new constraint, is \$502.50.

PROBLEMS

1. Express the following linear programming problems in the dual format.

   (a)   Minimize: $C = 2x_1 + 3x_2$
                       Subject to: $13x_1 + 18x_2 \geq 234$
   $$16x_1 + 14x_2 \geq 214$$
   $$x_1, x_2 \geq 0$$

   (b)   Minimize: $C = 4x_1 + 3x_2$
                       Subject to: $x_1 \geq 4$
   $$x_2 \geq 2$$
   $$3x_1 + 2x_2 \geq 18$$
   $$x_1, x_2 \geq 0$$

   (c)   Minimize: $C = 2x_1 + x_2$
                       Subject to: $12x_1 + 7x_2 \geq 42$
   $$5x_1 + 12x_2 \geq 60$$
   $$3x_1 + 21x_2 \geq 63$$
   $$x_1, x_2 \geq 0$$

   (d)   Maximize: $Z = 2x_1 + 4x_2 + 3x_3$
                       Subject to: $x_1 + 2x_2 \leq 80$
   $$x_1 + 4x_2 + 2x_3 \leq 120$$
   $$x_1, x_2, x_3 \geq 0$$

2. Determine the optimal values of the primal and dual variables in Problem 1.

3. Express the following problems in the dual format. Determine the optimal values of the primal and dual variables.

   (a)   Minimize: $C = 40x_1 + 120x_2$
                       Subject to: $x_1 - 2x_2 \leq 8$
   $$15x_1 + 44x_2 \leq 660$$
   $$3x_1 + 5x_2 = 90$$
   $$x_1, x_2 \geq 0$$

   (b)   Maximize: $Z = 8x_1 + 12x_2$
                       Subject to: $-4x_1 + 5x_2 \geq 20$
   $$11x_1 - 4x_2 = 0$$
   $$5x_1 + 4x_2 \leq 40$$
   $$x_1, x_2 \geq 0$$

4. Use the dual simplex algorithm to determine the solution to the following linear programming problems.

(a)
$$\text{Minimize:} \quad C = \quad x_1 + \quad x_2$$
$$\text{Subject to:} \quad 5x_1 + \quad 3x_2 \geq \quad 30$$
$$6x_1 + 16x_2 \geq 110$$
$$x_1, x_2 \geq 0$$

(b)
$$\text{Minimize:} \quad C = \quad 2x_1 + \quad 3x_2$$
$$\text{Subject to:} \quad 12x_1 + \quad 8x_2 \geq 48$$
$$5x_1 + 12x_2 \geq 64$$
$$3x_1 + 18x_2 \geq 60$$
$$x_1, x_2 \geq 0$$

(c)
$$\text{Minimize:} \quad C = 3x_1 + 5x_2 + 2x_3$$
$$\text{Subject to:} \quad x_1 + 2x_2 + 3x_3 \geq 64$$
$$3x_1 + 2x_2 + 4x_3 \geq 86$$
$$x_1, x_2, x_3 \geq 0$$

(d)
$$\text{Minimize:} \quad C = 3x_1 + \quad 5x_2$$
$$\text{Subject to:} \quad 5x_1 + 10x_2 \geq 60$$
$$8x_1 + \quad 6x_2 \geq 54$$
$$x_1, x_2 \geq 0$$

5. Solve the following linear programming problems by (1) using the primal simplex algorithm to restore dual feasibility and (2) using the dual simplex algorithm to restore primal feasibility. (*Hint:* Refer to Table 7.11 for an illustration of this approach).

(a)
$$\text{Minimize:} \quad C = 5x_1 + 6x_2$$
$$\text{Subject to:} \quad x_1 + 3x_2 \geq \quad 76$$
$$6x_1 + 7x_2 \leq 198$$
$$x_1, x_2 \geq 0$$

(b)
$$\text{Maximize:} \quad Z = 5x_1 + 4x_2 + 6x_3$$
$$\text{Subject to:} \quad 2x_1 + 3x_2 + 5x_3 \leq 125$$
$$x_1 + \quad x_2 + \quad x_3 \geq \quad 15$$
$$6x_1 + 2x_2 + 2x_3 \leq 150$$
$$x_1, x_2, x_3 \geq 0$$

6. A firm uses three machines in the manufacture of three products. Each unit of product 1 requires 3 hours on machine 1, 2 hours on machine 2, and 1 hour on machine 3. Each unit of product 2 requires 4 hours on machine 1, 1 hour on machine 2, and 3 hours on machine 3. Each unit of product 3 requires 2 hours on machine 1, 2 hours on machine 2, and 2 hours on machine 3. The contribution margin of the three products is $25, $42, and $30 per unit, respectively. Sixty hours of machine 1 time, 36 hours of machine 2 time, and 62 hours of machine 3 time are available for scheduling. The linear programming formulation of this problem is

$$\text{Maximize:} \quad Z = 25x_1 + 42x_2 + 30x_3$$
$$\text{Subject to:} \quad 3x_1 + 4x_2 + 2x_3 \leq 60$$
$$2x_1 + x_2 + 2x_3 \leq 36$$
$$x_1 + 3x_2 + 2x_3 \leq 62$$
$$x_1, x_2, x_3 \geq 0$$

and the optimal tableau for the problem is

	c_j	25	42	30	0	0	0	
c_b	Basis	x_1	x_2	x_3	x_4	x_5	x_6	Solution
42	x_2	$\frac{1}{3}$	1	0	$\frac{1}{3}$	$-\frac{1}{3}$	0	8
30	x_3	$\frac{5}{6}$	0	1	$-\frac{1}{6}$	$\frac{2}{3}$	0	14
0	x_6	$-\frac{5}{3}$	0	0	$-\frac{2}{3}$	$-\frac{1}{3}$	1	10
	z_j	39	42	30	9	6	0	756
	$c_j - z_j$	-14	0	0	-9	-6	0	

(a) Determine the sensitivity of the basic variables to changes in the contribution margins of each of the three products.

(b) Determine the sensitivity of the basic variables to changes in the right-hand-side values of the constraining equations.

7. A fourth product can be added to the product mix described in Problem 6. The contribution margin of this product is $36 and the product requires 2 hours on machine 1, 2 hours on machine 2, and 4 hours on machine 3. Add this product to the tableau in Problem 6 and determine the new solution.

8. A new process is required for the manufacture of the three products in Problem 6. Each unit of product 1 requires 3 hours of process time, each unit of product 2 requires 4 hours of process time, and each unit of product 3 requires 5 hours of process time. Ninety-three hours of process time are available. Add this constraint to the tableau in Problem 6 and determine the new optimal solution.

9. The manager of a large cattle ranch must select a mix of feed for beef cattle. The minimum daily requirements of the cattle of three important nutritional elements and the number of units of each of these elements in two feeds is given in the following table.

| Nutritional Element | Units of Nutritional Elements | | Minimum Requirement |
| | Contained in One Pound of | | |
	Feed 1	Feed 2	
A	6	2	54
B	2	2	30
C	2	4	60
Cost per pound	$0.04	$0.02	

Use the dual simplex algorithm to determine the minimum cost feed mix.

10. The manager of the cattle ranch referred to in Problem 9 has posed the following questions:

   (a) By how much could the price of either feed change before the optimal mix of feeds would change?

   (b) Would a 10 percent increase in the minimum daily requirements of each of the nutritional elements result in a 10 percent increase in cost? Why?

   (c) Suppose that a third feed, costing $0.03 per pound and having 4 units of nutritional element A, 3 units of nutritional element B, and 2 units of nutritional element C, become available. Would this change the optimal mix of feeds?

   (d) Assume that a fourth nutritional element became important. Each pound of feed 1 was found to contain 1.5 units of this element and each pound of feed 2 was found to contain 0.75 unit of the element. The feed mix must contain a minimum of 30 units of the nutritional element. Does this new requirement change the optimal solution in Problem 9? If so, determine the new solution.

SUGGESTED REFERENCES

The references for this chapter are listed in Chapter 5.

Chapter 8

The Transportation
and
Assignment Problems

By this point it should be obvious that linear programming is an important tool for the quantitative analysis of business decisions. The allocation of resources among competing products or activities so that profits are maximized or, alternatively, costs are minimized is a necessary function in a firm. Linear programming provides the manager the information necessary to make an optimal decision in these complex allocation problems.

This chapter considers problems that might be termed derivatives of the general linear programming problem. By this we mean that while certain problems can be formulated and solved as a linear programming problem by using the simplex algorithm, their special mathematical structure allows solution by much more efficient algorithms. Two types of problems for which such special algorithms exist are the transportation and the assignment problems. Examples of these important types of problems, along with the solution algorithms for the problems, are introduced in this chapter.

8.1   The Transportation Problem

The *transportation problem* derives its name from the problem of transporting homogeneous products from various sources of supply to several points of demand. The allocation of the products from the sources of supply to the points or demand is made with the objective of minimizing the total transportation costs or, alternatively, maximizing the profit from the sale of the products. To illustrate, consider a firm that has three factories (i.e., sources of supply) and four warehouses (i.e., points of demand). The cost of shipping from each factory to each warehouse depends on the distance the product must be shipped, freight rates, etc. These costs are shown in Table 8.1.

Table 8.1

Factory (F_i)	Warehouse (W_j)				Factory Capacity
	W_1	W_2	W_3	W_4	
F_1	0.30	0.25	0.40	0.20	100
F_2	0.29	0.26	0.35	0.40	250
F_3	0.31	0.33	0.37	0.30	150
Warehouse Requirement	100	150	200	50	500

The cost of shipping from factory i to warehouse j is represented by c_{ij}. The alternative values of c_{ij} are shown in the table. For instance, the cost of transporting one unit of product from factory 1 to warehouse 2 is $c_{12} = \$0.25$. The factory capacities and warehouse requirements are also given in the table. In this simplified example, the factory capacities and warehouse requirements both sum to 500 units. This represents an example of a *balanced* transportation problem. The more realistic case of the *unbalanced* transportation problem in which factory capacities and warehouse requirements are not equal is introduced later in this section. At this point it is necessary only to observe that an unbalanced transportation problem can be converted to a balanced transportation problem through the addition of an appropriate slack variable.

The objective in this problem is to develop a shipping schedule that minimizes the total transportation cost. The problem can be formulated by letting x_{ij} represent the number of units shipped from factory i to warehouse j. The objective is to minimize the total transportation costs subject to the constraints on factory capacities and warehouse requirements. For the example problem, this can be written as

$$
\begin{aligned}
\text{Minimize } C = \quad & 0.30x_{11} + 0.25x_{12} + 0.40x_{13} + 0.20x_{14} + 0.29x_{21} \\
& + 0.26x_{22} + 0.35x_{23} + 0.40x_{24} + 0.31x_{31} + 0.33x_{32} \\
& + 0.37x_{33} + 0.30x_{34}
\end{aligned}
$$

The warehouse constraints are

$$x_{11} + x_{21} + x_{31} = 100$$
$$x_{12} + x_{22} + x_{32} = 150$$
$$x_{13} + x_{23} + x_{33} = 200$$
$$x_{14} + x_{24} + x_{34} = 50$$

The factory constraints are

$$x_{11} + x_{12} + x_{13} + x_{14} = 100$$
$$x_{21} + x_{22} + x_{23} + x_{24} = 250$$
$$x_{31} + x_{32} + x_{33} + x_{34} = 150$$

and the nonnegativity constraints are

$$x_{ij} \geq 0, \quad \text{for } i = 1, 2, 3 \quad \text{and} \quad j = 1, 2, 3, 4$$

This transportation problem can be solved by using the simplex algorithm. Because of the special structure of the constraints, however, alternative algorithms that reduce the computational burden are available. Before introducing these algorithms, it will be worthwhile to formally state the problem mathematically and to provide additional examples of the transportation problem.

The standard transportation problem is to optimize an objective function

$$Z = \sum_{i=1}^{m} \sum_{j=1}^{n} c_{ij} x_{ij}, \quad \text{for } i = 1, 2, \ldots, m; j = 1, 2, \ldots, n \quad (8.1)$$

subject to the constraints that

$$\sum_{i=1}^{m} x_{ij} = D_j, \quad \text{for } j = 1, 2, \ldots, n \quad (8.2)$$

$$\sum_{j=1}^{n} x_{ij} = S_i, \quad \text{for } i = 1, 2, \ldots, m \quad (8.3)$$

and

$$x_{ij} \geq 0 \quad \text{for all } i \text{ and } j$$

D_j and S_i are nonnegative integers that represent, respectively, the demand at the jth facility (warehouse, retail store, etc.) and the supply at the ith source (factory, warehouse, supplier, etc.). An additional requirement is that the sum of the demands equal the sum of the supplies, i.e.,

$$\sum_{i=1}^{m} S_i = \sum_{j=1}^{n} D_j \quad (8.4)$$

For the unbalanced transportation problem, the requirement that the sum of the demands equal the sum of the supplies can be satisfied by creating a fictitious demand facility or supply facility. For instance, if supply exceeds demand, a fictitious demand column is established with zero transportation cost to absorb the excess supply. Similarly, if demand exceeds supply, a fictitious supply row is established with zero transportation cost to absorb the unsatisfied demand. The addition of a fictitious supply or demand to convert an unbalanced transportation problem to the required balanced transportation format is illustrated by the following three examples.

Example: A national firm has three factories and four warehouses. The cost of manufacturing a given product is the same at each of the factories. Because of the locations of the warehouses and factories, the cost of shipping the product from the different factories to the warehouses varies. The capacities of factories 1, 2, and 3 are 5000 units per month, 4000 units per month, and 7000 units per month. The requirements of warehouses 1, 2, 3, and 4 are, respectively, 3000 units per month, 2500 units per month, 3500 units per month, and 4000 units per month. The transportation costs, factory capacities, and warehouse requirements are given in the following table.

Factories	Warehouses W_1	W_2	W_3	W_4	Factory Capacity
F_1	15	24	11	12	5000
F_2	25	20	14	16	4000
F_3	12	16	22	13	7000
Warehouse Requirements	3000	2500	3500	4000	13,000 16,000

Since the total factory capacity exceeds the total warehouse requirement, the transportation problem is unbalanced. It can be converted to a balanced transportation problem by adding a fictitious warehouse W_5 with requirements of 3000 units. The cost of shipping from each factory to this fictitious warehouse is $0. The balanced tableau is shown below.

Factory	W_1	W_2	W_3	W_4	W_5	Factory Capacity
F_1	15	24	11	12	0	5000
F_2	25	20	14	16	0	4000
F_3	12	16	22	13	0	7000
Warehouse Requirements	3000	2500	3500	4000	3000	16,000

The procedure for converting an unbalanced problem in which the demand exceeds the supply to a balanced problem is similar to that shown in the preceding table. The only difference is that a fictitious row rather than column must be added to the table. The costs associated with this row are again $0. The supply available from this fictitious source equals the number of units required to balance the table.

Example: A firm has decided to expand its product line by producing one or more of five possible products. Three of the firm's current plants have the excess capacity to produce these products. The profit margin on each of the proposed products is the same; consequently, the firm wants to minimize the cost of production. The excess capacity of each plant, the potential sales of each product, and the estimated cost of producing the proposed products at the three plants are given in the following table.

Plant	Product 1	2	3	4	5	Excess Capacity (standard units)
1	22	20	16	23	17	70
2	16	19	14	19	18	80
3	19	16	X	20	21	100
Potential Sales (standard units)	60	70	90	65	85	250 / 370

This problem could be treated as a standard linear programming problem and solved by using the simplex algorithm. If, however, the plants are considered as supply facilities and potential sales of the five products are treated as demands, the problem can also be viewed as a transportation problem.

To convert the table into a balanced transportation table, a fictitious plant capable of supplying 120 units of capacity must be added. This involves adding a fourth row with a capacity of 120 units and production costs of $0. In addition to this modification, the table shows that product 3 cannot be manufactured at plant 3. To insure that the optimal solution does not include the use of plant 3 to produce product 3, we can assign an arbitrarily large cost of production for this plant-product combination. Although any large cost would insure that production of product 3 is not scheduled for plant 3, we shall follow the "Big M" convention introduced in Chapter 6 and assign the letter M as the cost of production. The balanced transportation table is given below.

Plant	Product 1	2	3	4	5	Excess Capacity (Standard Units)
1	22	20	16	23	17	70
2	16	19	14	19	18	80
3	19	16	M	20	21	100
4	0	0	0	0	0	120
Potential Sales (standard units)	60	70	90	65	85	370

Example: Bevitz Furniture Company has requested bids from four furniture manufacturers on five different styles of furniture. The quantities of the five styles of furniture required by Bevitz are shown below.

Style	A	B	C	D	E
Quantity	125	75	50	200	175

The four manufacturers have limited production capacities. The total quantities that can be supplied in the time span available are shown below.

Manufacturer	1	2	3	4
Quantity	275	225	175	200

On the basis of the quotes from each manufacturer, Bevitz estimates that the profit per unit for each item sold will vary as shown in the following table.

		Style A	B	C	D	E
	1	28	35	42	23	15
Manufacturer	2	30	33	45	18	10
	3	25	35	48	20	13
	4	33	28	40	26	18

Determine the allocation that maximizes profit.

This problem can be formulated as a transportation problem. Instead of minimizing transportation cost, however, the objective is to maximize profit. To apply the transportation algorithm, the problem must be converted to a balanced transportation problem. Since the capacity of the manufacturers exceeds the quantity demanded by Bevitz Furniture, the balance is achieved by adding a fictitious style. The demand for the fictitious style is equal to the unused capacity of the manufacturers, i.e., 250 units. The profit from the fictitious demand is $0. The balanced transportation table is shown below.

Manufacturer	Style						Quantity Supplied
	A	B	C	D	E	F	
1	28	35	42	23	15	0	275
2	30	33	45	18	10	0	225
3	25	35	48	20	13	0	175
4	33	28	40	26	18	0	200
Demand	125	75	50	200	175	250	875

8.2 The Initial Basic Feasible Solution

The first step in solving the transportation problem is to develop an initial basic feasible solution. Three methods for obtaining this solution are illustrated in this section. The first method, termed the *northwest corner rule*, provides a straightforward technique for obtaining the initial solution. It suffers, however, when compared to the *penalty method* and the *minimum (maximum) cell method* in that more iterations are normally required to obtain the optimal solution. The penalty method and the minimum cell method are introduced later in this section.

8.2.1 THE NORTHWEST CORNER RULE

To illustrate the northwest corner rule, consider the transportation problem summarized in Table 8.1, p. 275. The factory capacities and warehouse requirements for the problem are shown in Table 8.2. The transportation costs (i.e., the c_{ij}'s) are not included in this table. These costs are not relevant when one is using the northwest corner rule to determine the initial solution.

The initial solution is found by beginning in the upper left-hand (i.e., northwest) corner of the tableau and allocating the resource of the first row to the cells in the first row until the resource is exhausted. If a resource is exhausted and a requirement is satisfied by a single allocation, a zero is placed in a neighboring cell. We then move to the second row and continue the allocation until the resources of that row are fully allocated. Again, if a single allocation exhausts both the supply of resource and satisfies the requirement

Table 8.2

Factory	Warehouse W_1	W_2	W_3	W_4	Factory Capacity
F_1					100
F_2					250
F_3					150
Requirements	100	150	200	50	500

of a column, a zero must be placed in a neighboring cell. This process is continued until all resources are exhausted and all requirements are satisfied.

In this example, the 100 units of capacity of factory 1 is assigned to warehouse 1. This allocation exhausts the capacity of the first row and satisfies the requirements of the first column; consequently, we place a zero in the row 1, column 2 cell and move to the second row. Since the requirements of warehouse 1 have been met, the next allocation is 150 units of factory 2 capacity to warehouse 2. The remaining 100 units capacity of factory 2 is allocated to warehouse 3. One hundred units of factory 3 capacity completes the requirements of warehouse 3. The final 50 units of factory 3 capacity satisfies the requirements of warehouse 4. The initial solution is shown in Table 8.3.

Table 8.3

Factory	Warehouse W_1	W_2	W_3	W_4	Factory Capacity
F_1	100	0			100
F_2		150	100		250
F_3			100	50	150
Requirements	100	150	200	50	500

By comparing the solution in Table 8.3 with the constraints for the problem given on p. 275, it can be seen that the solution is feasible. Both the warehouse and the factory constraints are satisfied. The cost of this solution is found by multiplying the cell allocations in Table 8.3 by the transportation costs from Table 8.1 and summing these products. The cost of this initial solution is

$0.30(100) + $0.26(150) + $0.35(100) + $0.37(100) + $0.30(50) = $156.00

Example: Use the northwest corner rule to determine an initial solution for the transportation table shown on p. 277.

The factory capacities and the warehouse requirements, including the fictitious warehouse W_5, are given in the transportation table. The allocation is again made by beginning in the upper left-hand corner of the table and assigning capacities to requirements so as to exhaust all capacities and satisfy all requirements. The resulting feasible solution is shown below.

Factories	Warehouses W_1	W_2	W_3	W_4	W_5	Factory Capacity
F_1	3000	2000				5000
F_2		500	3500	0		4000
F_3				4000	3000	7000
Requirements	3000	2500	3500	4000	3000	16,000

The cost of this solution, again found by summing the products of the cell entries and the transportation costs, is $204,000.

The northwest corner rule provides a systematic, easily understandable method of obtaining an initial basic feasible solution. As mentioned earlier, however, it is quite inefficient in comparison with alternative methods for obtaining the initial solution. This inefficiency occurs because the costs (profits) are not considered in determining the initial solution. Although the algorithm used to determine the optimal solution is not introduced until the following section, it is reasonable to conclude that the number of iterations required to obtain the optimal solution is dependent upon how near the initial solution is to being optimal. This implies that the number of iterations can be reduced by beginning the iterative process from a "near optimal" initial solution. Costs (profits) are used to determine the initial solution in both the minimum (maximum) cell method and the penalty method. Consequently, the number of iterations required to obtain the optimal solution is normally reduced by beginning with an initial solution found by using one of these methods.

8.2.2   THE MINIMUM (MAXIMUM) CELL METHOD

To illustrate the minimum (maximum) cell method of obtaining an initial solution, we again consider the transportation problem summarized in Table 8.1. This table, complete with transportation costs, capacities, and requirements, is reproduced as Table 8.4. Notice that the transportation costs from Table 8.1 are placed in the upper left-hand corner of each cell in Table 8.4.

Table 8.4

| Factory | Warehouse | | | | Factory Capacity |
	W_1	W_2	W_3	W_4	
F_1	0.30	0.25	0.40	0.20	100
F_2	0.29	0.26	0.35	0.40	250
F_3	0.31	0.33	0.37	0.30	150
Requirements	100	150	200	50	500

The initial solution is obtained by sequentially allocating the resources to the cells with the minimum cost (or alternatively, with the maximum profit). As in the northwest corner rule, if a single allocation exhausts both the capacity of a row and satisfies the requirements of a column, a zero is placed in one of the cells that borders the allocation. Referring to Table 8.4, we see that the minimum transportation cost of $0.20 per unit is between factory 1 and warehouse 4. Since the requirements of warehouse 4 are 50 units, this allocation is entered in cell (1, 4). The allocation is shown in Table 8.5 in the lower right-hand corner of cell (1, 4).

Referring again to Table 8.4, we see that the next least costly allocation is the $0.25 per unit transportation cost from factory 1 to warehouse 2. The 50 units of unallocated capacity from factory 1 are, therefore, allocated to warehouse 2. This allocation is shown in cell (1, 2) in Table 8.5.

Again with reference to the table, the next least costly allocation is the $0.26 per unit transportation cost from factory 2 to warehouse 2. The remaining 100 units required by warehouse 2 thus come from factory 2. The allocation is entered in cell (2, 2) in Table 8.5.

The procedure continues by allocating 100 units from factory 2 to warehouse 1, thereby satisfying the demand of warehouse 1. The remaining 50 units capacity of factory 2 are next allocated to warehouse 3. The initial solution is completed by allocating the 150 units of factory 3 capacity to warehouse 3. The total cost of this initial solution is $150.50. This cost is $5.50 less than the cost of the initial solution obtained by using the northwest corner method.

Table 8.5

Factory	Warehouse W_1	W_2	W_3	W_4	Factory Capacity
F_1	0.30	0.25 50	0.40	0.20 50	100
F_2	0.29 100	0.26 100	0.35 50	0.40	250
F_3	0.31	0.33	0.37 150	0.30	150
Requirements	100	150	200	50	500

Example: Use the minimum (maximum) cell method to obtain an initial solution for the transportation problem described on p. 275. Compare this solution with that found by using the northwest corner method.

The transportation table for this problem is reproduced below. The transportation costs are entered in the upper left-hand corner of each cell. The initial allocation of factory capacity to warehouse requirements will be entered in the lower right-hand corner of the cell.

Factory	Warehouse W_1	W_2	W_3	W_4	W_5	Factory Capacity
F_1	15	24	11	12	0	5000
F_2	25	20	14	16	0	4000
F_3	12	16	22	13	0	7000
Requirements	3000	2500	3500	4000	3000	16,000

The transportation costs from each source to the fictitious warehouse W_5 are $0. Using the minimum cell method, we can allocate units from any of the three factories to W_5. We arbitrarily allocate 3000 units from F_3 to W_5.

The next least costly transportation charge is from factory 1 to warehouse 3. Allocating 3500 units of F_1 capacity to W_3 satisfies the requirements of W_3. The process of allocating the factory capacities to the warehouses with the smallest transportation cost is continued until all factory capacities are exhausted and warehouse requirements are satisfied. The initial solution is shown below.

Factory	W_1	W_2	Warehouse W_3	W_4	W_5	Factory Capacity
F_1	15	24	11 3500	12 1500	0	5000
F_2	25	20 2500	14	16 1500	0	4000
F_3	12 3000	16	22	13 1000	0 3000	7000
Requirements	3000	2500	3500	4000	3000	16,000

The total transportation cost of this initial solution is $179,500. This represents a reduction of $24,500 from the initial solution of $204,000 obtained by using the northwest corner rule.

Example: Use the minimum (maximum) cell method to determine the initial solution for the Bevitz Furniture Company problem.

The balanced transportation table for the Bevitz Furniture Company was given on p. 280. Since the Bevitz problem involves maximizing profit rather than minimizing transportation cost, the allocations are made to the cells with the largest c_{ij} values rather than to those with the smallest. The initial solution using the maximum cell method is given below. The total profit of this initial solution is $16,100.

Manufacturer	A	B	Style C	D	E	F	Quantity Supplied
1	28	35 75	42	23 125	15 75	0	275
2	30	33	45	18	10 100	0 125	225
3	25	35	48 50	20	13	0 125	175
4	33 125	28	40	26 75	18	0	200
Demand	125	75	50	200	175	250	875

8.2.3 THE PENALTY METHOD

The penalty method, also known as *Vogel's approximation method*, involves determining the penalty for not assigning a resource to a requirement. For a minimization problem, the penalties for each resource (row) are determined by subtracting the smallest c_{ij} value in the row from the next smallest. Similarly, the penalty values for each requirement (column) are determined by subtracting the smallest c_{ij} value in each column from the next smallest. The objective is to make the initial allocation such that the penalties from not using the cells are minimized. This can be done by allocating the resources to the requirements so as to avoid large penalties. The rules for determining the initial allocation in a minimization problem are

1. Subtract the smallest c_{ij} value in each row from the next smallest. Place this number at the end of the row.
2. Subtract the smallest c_{ij} value in each column from the next smallest. Place this number at the foot of the column.
3. Determine the largest penalty for either row or column. Allocate as many units as possible to the cell with the largest penalty cost so as to avoid this penalty. If a row resource is exhausted and a column requirement satisfied by a single allocation, place a zero in a neighboring cell.
4. Recalculate the penalty cost, ignoring any row whose resources are exhausted or column whose requirements are satisfied.
5. Repeat Steps 3 and 4 until all allocations have been made.

The procedure for determining the initial allocation in a maximization problem is quite similar to that described above for the minimization problem. The only differences are (1) the penalties are calculated from the largest and next largest c_{ij} values in each row and each column, and (2) the allocations are made so as to maximize, rather than minimize, these penalties.

To illustrate the penalty method, consider again the transportation problem given by Table 8.1. The transportation table together with the initial penalties are shown in Table 8.6.

The largest penalty is that associated with the W_4 column. In order to avoid this \$0.10 penalty, units from factory 1 must be allocated to warehouse 4. If this allocation is not made, an additional cost of at least \$0.10 per unit will be incurred in transporting units to warehouse 4. This allocation, together with penalties for rows with unexhausted capacity and the columns with unsatisfied demand, is shown in Table 8.7.

The largest penalty in Table 8.7 is the \$0.05 associated with factory 1. This penalty can be avoided by allocating factory 1 capacity to warehouse 2. No-

Table 8.6

Factory	Warehouse				Factory Capacity	Penalty
	W_1	W_2	W_3	W_4		
F_1	0.30	0.25	0.40	0.20	100	0.05
F_2	0.29	0.26	0.35	0.40	250	0.03
F_3	0.31	0.33	0.37	0.30	150	0.01
Requirements	100	150	200	50	500	
Penalty	0.01	0.01	0.02	0.10		

Table 8.7

Factory	Warehouse				Factory Capacity	Penalty
	W_1	W_2	W_3	W_4		
F_1	0.30	0.25	0.40	0.20 50	100	0.05
F_2	0.29	0.26	0.35	0.40	250	0.03
F_3	0.31	0.33	0.37	0.30	150	0.02
Requirements	100	150	200	50	500	
Penalty	0.01	0.01	0.02	—		

tice that since warehouse 4 requirements are satisfied, no penalty is given for warehouse 4. Furthermore, the transportation cost from factory 1 to warehouse 4 cannot be used in determining the penalty for factory 1. The allocation of factory 1 capacity to warehouse 2 is shown in Table 8.8.

The largest penalty in Table 8.8 is the $0.07 associated with warehouse 2. This penalty is avoided by allocating factory 2 capacity to warehouse 2. The results are shown in Table 8.9.

The largest remaining penalties are the $0.06 associated with factories 2 and 3. The penalty associated with factory 2 can be avoided by allocating

Table 8.8

Factory	Warehouse W_1	W_2	W_3	W_4	Factory Capacity	Penalty
F_1	0.30	0.25 50	0.40	0.20 50	100	—
F_2	0.29	0.26	0.35	0.40	250	0.03
F_3	0.31	0.33	0.37	0.30	150	0.02
Requirements	100	150	200	50	500	
Penalty	0.01	0.07	0.02	—		

Table 8.9

Factory	Warehouse W_1	W_2	W_3	W_4	Factory Capacity	Penalty
F_1	0.30	0.25 50	0.40	0.20 50	100	—
F_2	0.29	0.26 100	0.35	0.40	250	0.06
F_3	0.31	0.33	0.37	0.30	150	0.06
Requirements	100	150	200	50	500	
Penalty	0.02	—	0.02	—		

factory 2 capacity to warehouse 1. The initial assignment is completed by allocating the unused capacity of factories 2 and 3 to warehouse 3. The initial solution is shown in Table 8.10.

The solutions found by using the penalty method and the minimum cell method are identical in this example. Although this often happens in relatively small problems, this is not a general rule.

The penalty method and the minimum cell method are both used in practice to determine the initial solution. Seeing the number of tableaus required

Table 8.10

Factory	W_1	W_2	W_3	W_4	Factory Capacity	Penalty
			Warehouse		*Factory*	
F_1	0.30	0.25 ⟨50⟩	0.40	0.20 ⟨50⟩	100	—
F_2	0.29 ⟨100⟩	0.26 ⟨100⟩	0.35 ⟨50⟩	0.40	250	—
F_3	0.31	0.33	0.37	0.30 ⟨150⟩	150	—
Requirements	100	150	200	50	500	
Penalty	—	—	—	—		

to obtain the initial solution, the reader may question the need for the penalty method. It should be pointed out that much of the repetition required to explain the technique becomes unnecessary after the technique is understood. To verify this statement, the reader should return to the examples in Sec. 8.1 and determine an initial solution, using this method.

8.3 The Stepping Stone Algorithm

After an initial solution to the transportation problem is obtained, alternative solutions must be evaluated. A straightforward method of calculating the effect of alternative allocations is provided by the *stepping stone* algorithm.

To apply the stepping stone algorithm, we must first verify that the initial solution is a basic solution. The reader will remember that in the linear programming problem, a basic solution is found for a problem of n variables and m constraints by equating $n - m$ of the variables to zero and solving the resulting system of m equations and m variables simultaneously. The procedure differs slightly in a transportation problem.

The transportation problem is constructed so that there are $m + n$ equations, where n represents the number of requirements and m represents the number of resources. It can be shown that one of these $m + n$ equations is redundant and that there are only $m + n - 1$ independent equations. For a solution to be basic, therefore, $m + n - 1$ cells must be occupied. This means that $m + n - 1$ cells in the initial solution must contain either an allocation or a zero.

Example: Show that one of the equations is redundant in the following transportation problem.

Factory	Warehouse W_1	W_2	Factory Capacity
F_1	0.30	0.25	100
F_2	0.29	0.26	150
Requirements	150	100	250

The problem is

Minimize: $C = 0.30x_{11} + 0.25x_{12} + 0.29x_{21} + 0.26x_{22}$

Subject to:
$$x_{11} + x_{12} = 100$$
$$x_{21} + x_{22} = 150$$
$$x_{11} + x_{21} = 150$$
$$x_{12} + x_{22} = 100$$
$$x_{ij} \geq 0$$

Adding the first two equations gives

$$x_{11} + x_{12} + x_{21} + x_{22} = 250$$

Subtracting the third equation from this sum gives

$$x_{12} + x_{22} = 100$$

Since the first three equations can be combined to produce the fourth, one of the equations is redundant and there are only $m + n - 1$ independent equations, i.e., three independent equations. The initial solution to this problem should, therefore, contain $n + m - 1$, or three, occupied cells.

If the initial solution contains less than $n + m - 1$ occupied cells, the solution is termed *degenerate*. Degeneracy occurs only when the resources of a row are exhausted and the requirements of a column are satisfied by a single allocation. It is eliminated by placing a zero in a cell that borders the allocation. The zero-valued cell is then considered occupied when applying the stepping stone algorithm.

The stepping stone algorithm involves transferring one unit from an occupied to an unoccupied cell and calculating the change in the objective function. The transfer must be made so as to retain the column and row equalities of the problem. After all unoccupied cells have been evaluated, a reallocation is made to the cell that provides the greatest per unit change in the objective function. Any degeneracies caused by the transfer of units must be removed by placement of zeros in the appropriate cells. The process of evaluating the empty cells and reallocating the units is continued until no further improvement in the objective function is possible. This final allocation is the optimal solution.

To illustrate the stepping stone algorithm, consider the transportation problem summarized by Table 8.1. The initial solution to this problem, found

Table 8.11

Factory	Warehouse				Factory Capacity
	W_1	W_2	W_3	W_4	
F_1	0.30 100	0.25 0	0.40	0.20	100
F_2	0.29	0.26 150	0.35 100	0.40	250
F_3	0.31	0.33	0.37 100	0.30 50	150
Requirements	100	150	200	50	500

by using the northwest corner rule, was given in Table 8.3. This solution is repeated in Table 8.11. The transportation costs, i.e., the c_{ij}'s, are included in the table.

To determine the effect on the objective function of transferring one unit to an unoccupied cell, we must find a *closed path* between the unoccupied cell and occupied cells. The path consists of a series of steps leading from the unoccupied cell to occupied cells and back to the unoccupied cell. In the case of cell (2, 1), for instance, a closed path consists of the series of steps from this cell to cell (1, 1), from cell (1, 1) to cell (1, 2), from cell (1, 2) to cell (2, 2), and from cell (2, 2) back to cell (2, 1). By following this path, we can determine the effect on the objective function of allocating a unit to cell (2, 1).

To illustrate, assume that one unit is allocated to cell (2, 1). In order to maintain the column and row equalities in the problem, a unit must be subtracted from cell (1, 1), added to cell (2, 1), and subtracted from cell (2, 2). Notice that this reallocation of units follows the closed path for cell (2, 1).

The net change in the objective function from the reallocation of one unit to cell (2, 1) can be found by adding and subtracting the appropriate transportation costs. Again, the closed path is followed; adding one unit to cell (2, 1) increases the objective function by $0.29, subtracting the unit from cell (1, 1) reduces the objective function by $0.30, adding the unit to cell (1, 2) increases the objective function by $0.25, and subtracting the unit from cell

(2, 2) reduces the objective function by \$0.26. The net decrease in the objective function is, therefore, \$0.02. This decrease can be represented in equation form by

$$F_2W_1 = +F_2W_1 - F_1W_1 + F_1W_2 - F_2W_2$$

or

$$F_2W_1 = +0.29 - 0.30 + 0.25 - 0.26 = -\$0.02$$

The net decrease of $-\$0.02$ is entered in the lower right-corner of cell (2, 1) in Table 8.12.

The effect of reallocating one unit to each of the other unoccupied cells is determined in the same manner. The computations are shown below. It is important to remember that the closed paths are established so as to maintain both column and row equalities.

$$F_3W_2 = +F_3W_2 - F_3W_3 + F_2W_3 - F_2W_2$$
$$F_3W_2 = +0.33 - 0.37 + 0.35 - 0.26 = +\$0.05$$
$$F_1W_3 = +F_1W_3 - F_2W_3 + F_2W_2 - F_1W_2$$
$$F_1W_3 = +0.40 - 0.35 + 0.26 - 0.25 = +\$0.06$$
$$F_2W_4 = +F_2W_4 - F_3W_4 + F_3W_3 - F_2W_3$$
$$F_2W_4 = +0.40 - 0.30 + 0.37 - 0.35 = +\$0.12$$
$$F_3W_1 = +F_3W_1 - F_3W_3 + F_2W_3 - F_2W_2 + F_1W_2 - F_1W_1$$
$$F_3W_1 = +0.31 - 0.37 + 0.35 - 0.26 + 0.25 - 0.30 = -\$0.02$$
$$F_1W_4 = +F_1W_4 - F_3W_4 + F_3W_3 - F_2W_3 + F_2W_2 - F_1W_2$$
$$F_1W_4 = +0.20 - 0.30 + 0.37 - 0.35 + 0.26 - 0.25 = -\$0.07$$

The net change in the objective function caused by reallocating one unit to each unoccupied cell is entered in Table 8.12.

Table 8.12

Factory	W_1		W_2		W_3		W_4		Factory Capacity
	Warehouse								
F_1	0.30	100	0.25	0	0.40	+\$0.06	0.20	−\$0.07	100
F_2	0.29	−\$0.02	0.26	150	0.35	100	0.40	+\$0.12	250
F_3	0.31	−\$0.02	0.33	+\$0.05	0.37	100	0.30	50	150
Requirements	100		150		200		50		500

The dollar entries in the lower right-hand corner of each unoccupied cell in Table 8.12 represent the net change in the objective function from reallocating one unit to the cell. These are, in effect, equivalent to the $c_j - z_j$ values in the simplex tableau. The reader will remember that in the simplex algorithm for a minimization problem, the variable with the most negative $c_j - z_j$ entry is introduced into the basis. The same procedure is followed in the stepping stone algorithm. Namely, a reallocation is made to the most favorable evaluation, i.e., the cell that provides the largest per unit decrease in the objective function for a minimization problem and the largest per unit increase in the objective function for a maximization problem. The reallocation follows the closed path used to calculate the change in the objective function. As in the simplex algorithm, as many units as possible are reallocated to the cell.

Referring to Table 8.12, note that the largest per unit decrease in the objective function comes from reallocating units to cell $(1, 4)$. The closed path used to evaluate cell $(1, 4)$ was

$$F_1 W_4 = +F_1 W_4 - F_3 W_4 + F_3 W_3 - F_2 W_3 + F_2 W_2 - F_1 W_2$$

The limit on the number of units that can be reallocated to cell $(1, 4)$ is equal to the minimum of the current allocations to cells $(3, 4)$, $(2, 3)$, and $(1, 2)$. The table shows that 50 units can be subtracted from cell $(3, 4)$, 100 units can can be subtracted from cell $(2, 3)$, and 0 units can be subtracted from cell $(1, 2)$. Unfortunately, the closed path used to evaluate cell $(1, 4)$ involved subtracting units from cell $(1, 2)$. Since cell $(1, 2)$ has a zero allocation in the initial solution, allocating units to cell $(1, 4)$ would be equivalent to transferring the 0 entry in cell $(1, 2)$ to cell $(1, 4)$. This, of course, would not decrease the value of the objective function.

Rather than merely transferring the 0 entry from cell $(1, 2)$ to cell $(1, 4)$, units can be reallocated to a cell that decreases the value of the objective function. An allocation to cell $(2, 1)$ or cell $(3, 1)$ would decrease the objective function by \$0.02 per unit. We arbitrarily select cell $(2, 1)$ for reallocation.

The closed path used to evaluate cell $(2, 1)$ was

$$F_2 W_1 = +F_2 W_1 - F_1 W_1 + F_1 W_2 - F_2 W_2$$

The limit on the number of units that can be added to cell $(2, 1)$ is the 100 units initially assigned to cell $(1, 1)$. This is due to the fact that units must be subtracted from cell $(1, 1)$, and cell $(1, 1)$ contains only 100 units. These units are reallocated to cell $(2, 1)$. In order to maintain the column and row equalities, 100 of the 150 units in cell $(2, 2)$ are reallocated to cell $(1, 2)$. This leaves 50 units in cell $(2, 2)$. The new transportation table is shown by Table 8.13.

The solution shown in Table 8.13 contains six occupied cells. Since there are three rows and four columns in the problem and the number of occupied

Table 8.13

Factory	Warehouse				Factory Capacity
	W_1	W_2	W_3	W_4	
F_1	0.30	0.25 100	0.40	0.20	100
F_2	0.29 100	0.26 50	0.35 100	0.40	250
F_3	0.31	0.33	0.37 100	0.30 50	150
Requirements	100	150	200	50	500

cells is equal to $m + n - 1$, the solution is not degenerate. Therefore, we need not consider any of the blank cells as occupied at a zero level.

The stepping stone algorithm is used to evaluate the unoccupied cells in Table 8.14. The calculations are shown below.

$$F_1W_1 = +F_1W_1 - F_1W_2 + F_2W_2 - F_2W_1 = +\$0.02$$
$$F_3W_1 = +F_3W_1 - F_3W_3 + F_2W_3 - F_2W_1 = \quad \$0.00$$
$$F_3W_2 = +F_3W_2 - F_3W_3 + F_2W_3 - F_2W_2 = +\$0.05$$
$$F_1W_3 = +F_1W_3 - F_2W_3 + F_2W_2 - F_1W_2 = +\$0.06$$
$$F_1W_4 = +F_1W_4 - F_3W_4 + F_3W_3 - F_2W_3 + F_2W_2 - F_1W_2 = -\$0.07$$
$$F_2W_4 = +F_2W_4 - F_3W_4 + F_3W_3 - F_2W_3 = +\$0.12$$

The evaluations are entered in Table 8.14.

Table 8.14

Factory	Warehouse				Factory Capacity
	W_1	W_2	W_3	W_4	
F_1	0.30 +\$0.02	0.25 100	0.40 +\$0.06	0.20 -\$0.07	100
F_2	0.29 100	0.26 50	0.35 100	0.40 +\$0.12	250
F_3	0.31 \$0.00	0.33 +\$0.05	0.37 100	0.30 50	150
Requirements	100	150	200	50	500

The table shows that units should be reallocated to cell (1, 4). The reallocation is made by following the closed path that gave the $0.07 per unit decrease in the objective function for the cell. The reallocation is shown in Table 8.15.

Table 8.15

Factory	Warehouse				Factory Capacity
	W_1	W_2	W_3	W_4	
F_1	0.30	0.25 50	0.40	0.20 50	100
F_2	0.29 100	0.26 100	0.35 50	0.40	250
F_3	0.31	0.33	0.37 150	0.30	150
Requirements	100	150	200	50	500

The stepping stone is again applied to evaluate the unoccupied cells in Table 8.15. The calculations are shown below.

$$F_1W_1 = +F_1W_1 - F_2W_1 + F_2W_2 - F_1W_2 = +\$0.02$$
$$F_3W_1 = +F_3W_1 - F_3W_3 + F_2W_3 - F_2W_1 = \$0.00$$
$$F_3W_2 = +F_3W_2 - F_3W_3 + F_2W_3 - F_2W_2 = +\$0.05$$
$$F_1W_3 = +F_1W_3 - F_2W_3 + F_2W_2 - F_1W_2 = +\$0.06$$
$$F_2W_4 = +F_2W_4 - F_2W_2 + F_1W_2 - F_1W_4 = +\$0.19$$
$$F_3W_4 = +F_3W_4 - F_3W_3 + F_2W_3 - F_2W_2 + F_1W_2 - F_1W_4 = +\$0.07$$

Since all the evaluations are nonnegative, the solution is optimal. This is not, however, a unique solution. The calculations show that the objective function does not change as units are allocated to cell (3, 1). Thus, multiple optimal solutions exist for this problem. The value of the objective function, found by calculating the total transportation cost, is $150.50.

Example: Use the stepping stone algorithm to determine the optimal allocation for the transportation problem introduced on p. 277.

The initial solution for this problem, found by using the northwest corner rule, was determined on p. 287 and is reproduced on p. 296.

Factory	Warehouse					Factory Capacity
	W_1	W_2	W_3	W_4	W_5	
F_1	15 3000	24 2000	11	12	0	5000
F_2	25	20 500	14 3500	16 0	0	4000
F_3	12	16	22	13 4000	0 3000	7000
Requirements	3000	2500	3500	4000	3000	16,000

The unoccupied cells are evaluated by establishing closed paths between each unoccupied cell and the occupied cells. The evaluations are

$$F_1W_3 = +F_1W_3 - F_2W_3 + F_2W_2 - F_1W_2 = -\$7$$
$$F_1W_4 = +F_1W_4 - F_2W_4 + F_2W_2 - F_1W_2 = -\$8$$
$$F_1W_5 = +F_1W_5 - F_3W_5 + F_3W_4 - F_2W_4 + F_2W_2 - F_1W_2 = -\$7$$
$$F_2W_1 = +F_2W_1 - F_1W_1 + F_1W_2 - F_2W_2 = +\$14$$
$$F_2W_5 = +F_2W_5 - F_3W_5 + F_3W_4 - F_2W_4 = -\$3$$
$$F_3W_1 = +F_3W_1 - F_3W_4 + F_2W_4 - F_2W_2 + F_1W_2 - F_1W_1 = +\$4$$
$$F_3W_2 = +F_3W_2 - F_3W_4 + F_2W_4 - F_2W_2 = -\$1$$
$$F_3W_3 = +F_3W_3 - F_3W_4 + F_2W_4 - F_2W_3 = +\$11$$

The most favorable evaluation is the $-\$8$ for cell $(1, 4)$. It is impossible to reallocate units to this cell, however, because of the zero allocation in cell $(2, 4)$.

The next most favorable evaluations are for cells $(1, 3)$ and $(1, 5)$. Reallocating units to cell $(1, 5)$ also requires subtracting units from cell $(2, 4)$. Since this would only mean transferring the zero from cell $(2, 4)$ to cell $(1, 5)$, cell $(1, 3)$ rather than cell $(1, 5)$ is selected to receive the allocation. The result is shown below. The evaluations of the unoccupied cells in the new solution are also given. The reader should verify these evaluations by determining the closed paths for each unoccupied cell and calculating the effect on the objective function of reallocating one unit to the unoccupied cell.

Total transportation cost can be reduced by reallocating units to cell $(3, 1)$. The reallocation, together with the evaluation of the empty cells, is given in the following table.

Factory	Warehouse W_1	W_2	W_3	W_4	W_5	Factory Capacity
F_1	15 — 3000	24 — +$7	11 — 2000	12 — −$1	0 — $0	5000
F_2	25 — +$7	20 — 2500	14 — 1500	16 — 0	0 — −$3	4000
F_3	12 — −$3	16 — −$1	22 — +$11	13 — 4000	0 — 3000	7000
Requirements	3000	2500	3500	4000	3000	16,000

Factory	Warehouse W_1	W_2	W_3	W_4	W_5	Factory Capacity
F_1	15 — 1500	24 — +$4	11 — 3500	12 — −$4	0 — −$3	5000
F_2	25 — +$10	20 — 2500	14 — +$3	16 — 1500	0 — −$3	4000
F_3	12 — 1500	16 — −$1	22 — +$14	13 — 2500	0 — 3000	7000
Requirement	3000	2500	3500	4000	3000	16,000

Transportation cost can be further reduced by reallocating units to cell (1, 4). This allocation is shown below.

Factory	Warehouse W_1	W_2	W_3	W_4	W_5	Factory Capacity
F_1	15 — $4	24 — +$8	11 — 3500	12 — 1500	0 — +$1	5000
F_2	25 — +$10	20 — 2500	14 — −$1	16 — 1500	0 — −$3	4000
F_3	12 — 3000	16 — −$1	22 — +$10	13 — 1000	0 — 3000	7000
Requirement	3000	2500	3500	4000	3000	16,000

An additional reduction in transportation cost is possible by allocating units to cell (2, 5).

Factory	W₁	W₂	Warehouse W₃	W₄	W₅	Factory Capacity
F_1	15 +$4	24 +$5	11 3500	12 1500	0 +$1	5000
F_2	25 +$13	20 2500	14 +$2	16 +$3	0 1500	4000
F_3	12 3000	16 −$4	22 +$10	13 2500	0 1500	7000
Requirement	3000	2500	3500	4000	3000	16,000

Allocating units to cell (3, 2) gives the following table.

Factory	W₁	W₂	Warehouse W₃	W₄	W₅	Factory Capacity
F_1	15 +$4	24 +$9	11 3500	12 1500	0 +$5	5000
F_2	25 +$9	20 1000	14 −$2	16 −$1	0 3000	4000
F_3	12 3000	16 1500	22 +$10	13 2500	0 +$4	7000
Requirement	3000	2500	3500	4000	3000	16,000

The transportation costs is again reduced by reallocating units to cell (2, 3). The final reallocation, shown on the following page, gives the optimal solution. The total transportation cost is $167,000.

We stated earlier that the number of iterations required to obtain the optimal solution is dependent on how near the initial solution is to being optimal. This is demonstrated by the preceding example. Notice in this example that the fourth tableau (p. 297) contains the same allocations as the initial tableau found using the minimum cell method (p. 285). The iterations in the example leading to the fourth tableau would not have been necessary had we begun with the initial solution obtained by using the minimum cell method instead of the initial solution from the Northwest corner rule.

| Factory | Warehouse | | | | | Factory Capacity |
	W_1	W_2	W_3	W_4	W_5	
F_1	15 +$4	24 +$9	11 2500	12 2500	0 +$3	5000
F_2	25 +$11	20 +$2	14 1000	16 +$1	0 3000	4000
F_3	12 3000	16 2500	22 +$10	13 1500	0 +$2	7000
Requirement	3000	2500	3500	4000	3000	16,000

8.4 The Modified Distribution Algorithm†

The *modified distribution* algorithm, also referred to as the MODI method, offers an alternative approach for evaluating the unoccupied cells in a transportation tableau. The algorithm is introduced for two reasons. First, it provides a shortcut for evaluating the unoccupied cells. Second, and of equal importance, it can be used to demonstrate the relationship between the dual theorem of linear programming and the transportation problem.

To illustrate the modified distribution algorithm, consider the transportation problem first introduced in Table 8.1. The initial solution to this problem, found by using the northwest corner rule, was given in Table 8.3. This solution is repeated in Table 8.16.

In addition to the initial solution, Table 8.16 includes a column labeled u_i and a row labeled v_j. The numerical values of u_i and v_j are calculated from the equation

$$\Delta_{ij} = c_{ij} - u_i - v_j \tag{8.5}$$

where Δ_{ij} represents the net change in the objective function of the transportation problem from reallocating one unit to cell (i, j) and c_{ij} represents the transportation cost from factory i to warehouse j.

The modified distribution algorithm consists of determining the values of u_i for the $i = 1, 2, \ldots, m$ rows and the values of v_j for the $j = 1, 2, \ldots, n$ columns of the transportation table. After the values of u_i and v_j have been determined, the change in the objective function caused by reallocating units can be determined by calculating the value of Δ_{ij} for each unoccupied cell. As in the stepping stone algorithm, the values of Δ_{ij} are entered in the unoccupied

---

† This section is optional at the discretion of the instructor.

Table 8.16

Factory	Warehouse W_1	W_2	W_3	W_4	Factory Capacity	u_i
F_1	0.30 100	0.25 0	0.40	0.20	100	
F_2	0.29	0.26 150	0.35 100	0.40	250	
F_3	0.31	0.33	0.37 100	0.30 50	150	
Requirement	100	150	200	50	500	
v_j						

cells. Units are then reallocated to the cell with the most favorable Δ_{ij}, i.e., the most favorable evaluation. After the reallocation, the algorithm is re-applied to the subsequent tableau. The procedure continues until no improvement in the objective function through reallocation of units is possible.

The method of determining the values of u_i and v_j is derived from the linear programming formulation of the transportation problem. The reader will remember that the change in the objective function from introducing a variable into the basis is given by the $c_j - z_j$ row of the simplex tableau. The change in the objective function (i.e., the $c_j - z_j$ value) associated with a variable currently in the basis is always zero. Since an occupied cell represents a variable in the current basis, the change in the objective function (i.e., the Δ_{ij} value) associated with that cell is zero. Consequently, the Δ_{ij} values for all occupied cells are zero.

In order to solve for u_i and v_j, we merely need to make Δ_{ij} equal zero for each occupied cell and solve the resulting system of equations simultaneously. The system of equations for the solution shown in Table 8.16 is

$$u_1 + v_1 = 0.30$$
$$u_1 + v_2 = 0.25$$
$$u_2 + v_2 = 0.26$$
$$u_2 + v_3 = 0.35$$
$$u_3 + v_3 = 0.37$$
$$u_3 + v_4 = 0.30$$

The system has six equations and seven variables. In order to obtain a solution, one of the variables must be specified. In the modified distribution algorithm, u_1 is customarily specified as equaling zero. In this example, equating u_1 with zero gives the following solution: $u_1 = \$0.00$; $v_1 = \$0.30$; $v_2 = \$0.25$; $u_2 = \$0.01$; $v_3 = \$0.34$; $u_3 = \$0.03$; and $v_4 = \$0.27$. These values are shown in Table 8.17.

Table 8.17

Factory	Warehouse W_1	W_2	W_3	W_4	Factory Capacity	u_i
F_1	0.30 100	0.25 0	0.40 +$0.06	0.20 −$0.07	100	$0.00
F_2	0.29 −$0.02	0.26 150	0.35 100	0.40 +$0.12	250	0.01
F_3	0.31 −$0.02	0.33 +$0.05	0.37 100	0.30 50	150	0.03
Requirement	100	150	200	50	500	
v_j	$0.30	0.25	0.34	0.27		

The change in the objective function from reallocating one unit to the unoccupied cells is also shown in Table 8.17. These values were calculated by formula (8.5), i.e., $\Delta_{ij} = c_{ij} - u_i - v_j$. The reader should note that the evaluations of unoccupied cells are the same for both the stepping stone algorithm, Table 8.12, and the modified distribution algorithm, Table 8.17.

The most favorable evaluation in Table 8.17 is $\Delta_{14} = -\$0.07$. As explained on p. 293, a reallocation to this cell would merely involve reassigning the zero allocation from cell (1, 2) to cell (1, 4). Since cells (2, 1) and (3, 1) were also favorably evaluated, a reallocation can instead be made to one of these cells. We again arbitrarily select cell (2, 1). The resulting tableau is given in Table 8.18.

The values of u_i and v_j in Table 8.19 are again calculated by equating Δ_{ij} to zero for the occupied cells and solving the resulting system of equations. Although the reader may wish to write out the system of equations, the structure of the transportation problem makes this step unnecessary. In Table 8.18, u_1 is specified as $0.00. Since $c_{12} = u_1 + v_2$, it follows that $v_2 = \$0.25$. Similarly, $c_{22} = u_2 + v_2$, and u_2 is, therefore, $0.01. The remaining values of u_i and v_j are determined in the same manner. The values of Δ_{ij} for the unoccupied cells are determined to complete the table. The evaluations found by using the modified distribution algorithm, Table 8.18, are again identical to those given by the stepping stone algorithm, Table 8.14.

Table 8.18

Factory	W_1	Warehouse W_2	W_3	W_4	Factory Capacity	u_i
F_1	0.30 +$0.02	0.25 100	0.40 +$0.06	0.20 −$0.07	100	$0.00
F_2	0.29 100	0.26 50	0.35 100	0.40 +$0.12	250	0.01
F_3	0.31 $0.00	0.33 +$0.05	0.37 100	0.30 50	150	0.03
Requirement	100	150	200	50	500	
v_j	$0.28	0.25	0.34	0.27		

The final solution is obtained by reapplying the modified distribution algorithm. One additional tableau is required. This tableau is shown in Table 8.19. The solution is, of course, identical to that given by the stepping stone algorithm in Table 8.15.

Table 8.19

Factory	W_1	Warehouse W_2	W_3	W_4	Factory Capacity	u_i
F_1	0.30 +$0.02	0.25 50	0.40 +$0.06	0.20 50	100	$0.00
F_2	0.29 100	0.26 100	0.35 50	0.40 +$0.19	250	0.01
F_3	0.31 $0.00	0.33 +$0.05	0.37 150	0.30 +$0.07	150	0.03
Requirements	100	150	200	50	500	
v_j	$0.28	0.25	0.34	0.20		

8.5 The Dual Theorem and the Transportation Problem†

In the discussion of the dual theorem in Chapter 7, we pointed out that this theorem provides the basis for many important extensions of linear pro-

---

† This section is optional at the discretion of the instructor.

gramming. Although many of these extensions must be reserved for advanced studies in operations research, we can offer the reader an indication of the importance of this theorem by illustrating the relationship between the dual theorem and the transportation problem.

To introduce the importance of the dual theorem to the transportation problem, consider the transportation problem summarized in Table 8.20.

Table 8.20

Factory	W_1	Warehouse W_2	W_3	Resource
F_1	5	4	4	200
F_2	5	3	6	300
Requirements	100	200	200	500

The linear programming formulation of this problem is

Minimize: $Z = 5x_{11} + 4x_{12} + 4x_{13} + 5x_{21} + 3x_{22} + 6x_{23}$
Subject to:
$$x_{11} + x_{12} + x_{13} = 200$$
$$x_{21} + x_{22} + x_{23} = 300$$
$$x_{11} + x_{21} = 100$$
$$x_{12} + x_{22} = 200$$
$$x_{13} + x_{23} = 200$$
$$x_{ij} \geq 0$$

Five variables are required for the dual formulation of this problem, two for the resource constraints and three for the requirement constraints. In Chapter 7 the dual variables were symbolized by w_i. We depart from this convention in this example by letting u_i represent the dual variables for the resource equalities (i.e., the first two equations) and v_j represent the dual variables for the requirement equalities (i.e., the final three equations). Based on this convention, the dual formulation is

Maximize: $P = 200u_1 + 300u_2 + 100v_1 + 200v_2 + 200v_3$
Subject to:
$$u_1 + v_1 \leq 5$$
$$u_1 + v_2 \leq 4$$
$$u_1 + v_3 \leq 4$$
$$u_2 + v_1 \leq 5$$
$$u_2 + v_2 \leq 3$$
$$u_2 + v_3 \leq 6$$
u_i, v_j unrestricted in sign.

The optimal solution could be determined by applying the simplex algorithm to either the dual or the primal formulation of the problem. Both formulations are shown above. Instead, an important relationship between the primal and dual problems is used to develop the modified distribution algorithm.

This relationship can be explained as follows. A basic solution to the primal problem occurs when $n + m - 1$ of the cells are occupied, i.e., when $n + m - 1$ of the primal variables are in the basis. By referring to the discussion of the dual formulation of the primal problem in Chapter 7, the reader will remember that the number of dual inequalities that are strictly satisfied is equal to the number of variables in the primal basis. Therefore, $n + m - 1$ of the dual inequalities are strictly satisfied. This means that the values of the dual variables corresponding to any primal basic solution can be determined by selecting the appropriate subset of $m + n - 1$ dual equations and solving for the $m + n$ dual variables. Since the subset of $m + n - 1$ dual equations has $m + n$ dual variables, a solution can be obtained only if the value of one of the dual variables is specified. As stated earlier, it is customary to specify $u_1 = \$0.00$.

The dual formulation of the transportation problem given in Table 8.20 has six variables and five inequalities. Since there are four primal variables in the basis, four of the dual inequalities are strictly satisfied. The four equations corresponding to the northwest corner solution to the problem are

$$u_1 + v_1 = 5$$
$$u_1 + v_2 = 4$$
$$u_2 + v_2 = 3$$
$$u_2 + v_3 = 6$$

The solution to the system of equations, obtained by specifying $u_1 = 0$, is $u_1 = 0$, $v_1 = 5$, $v_2 = 4$, $u_2 = -1$, and $v_3 = 7$.

The right-hand-side constants in the system of dual inequalities are the c_{ij} values in the transportation table. Suppose that we again define Δ_{ij} by the equation

$$\Delta_{ij} = c_{ij} - u_i - v_j \tag{8.5}$$

Δ_{ij} represents the difference between the right-hand-side constant c_{ij} and the dual variables u_i and v_j. Δ_{ij} is, of course, equal to zero if cell (i, j) is occupied or, alternatively, if variable x_{ij} is in the basis of the primal problem. If cell (i, j) is not occupied, however, Δ_{ij} gives the change in the objective function in the primal problem from reallocating one unit to cell (i, j).

The values of Δ_{ij} in the example are $\Delta_{11} = 0$, $\Delta_{12} = 0$, $\Delta_{13} = -3$, $\Delta_{21} = 1$, $\Delta_{22} = 0$, and $\Delta_{23} = 0$. Since Δ_{13} is negative, the initial northwest corner solution can be improved by reallocating units to cell $(1, 3)$. Following the reallocation, a different subset of four equations is solved for the dual variables. The iterations continue until all Δ_{ij} values are positive.

The reader should by now be aware that the modified distribution algorithm consists of nothing more than solving selected subsets of dual equations to determine the values of the dual variables. At each iteration the subset of dual equations are those that are associated with the occupied cells. Rather than writing out the entire set of dual inequalities and selecting the appropriate subset of dual equations, the modified distribution algorithm can be used. This algorithm permits one to solve for the dual variables u_i and v_j directly from the transportation table.

Example: Show that the Δ_{ij} values in the modified distribution algorithm are equal to the $c_{ij} - z_{ij}$ values of the primal simplex solution to the transportation problem.

The fact that Δ_{ij} and $c_{ij} - z_{ij}$ are equivalent measures of the change in the objective function from reallocating one unit to cell (i, j) can be shown by referring to the problem described in Table 8.20. The primal formulation of this problem was given on p. 303. One of the five equations in the primal formulation is redundant and is, therefore, omitted from the simplex tableau. In this example we omit the final equation.

The simplex tableau that corresponds to the northwest corner solution is given below. This tableau was obtained by solving the first four equations for the basic variables shown in the tableau. These variables correspond, of course, to the northwest corner solution to the problem.

	c_{ij}	5	4	4	5	3	6	
c_b	Basis	x_{11}	x_{12}	x_{13}	x_{21}	x_{22}	x_{23}	Solution
5	x_{11}	1	0	0	1	0	0	100
4	x_{12}	0	1	1	−1	0	0	100
3	x_{22}	0	0	−1	1	1	0	100
6	x_{23}	0	0	1	0	0	1	200
	z_{ij}	5	4	7	4	3	6	
	$c_{ij} - z_{ij}$	0	0	−3	1	0	0	

The Δ_{ij} values were found earlier to be $\Delta_{11} = 0$, $\Delta_{12} = 0$, $\Delta_{13} = -3$, $\Delta_{21} = 1$, $\Delta_{22} = 0$, and $\Delta_{23} = 0$. These are equivalent to the $c_{ij} - z_{ij}$ values given in the simplex tableau.

8.6 The Assignment Problem

The *assignment problem*, like the transportation problem, is a special case of the linear programming problem. The assignment problem occurs when n jobs must be assigned to n facilities on a one-to-one basis. The assignment is made with the objective of minimizing the overall cost of completing the jobs, or, alternatively, of maximizing the overall profit from the jobs.

To illustrate the assignment problem, consider the following example. A firm has five jobs that must be assigned to five work crews. Because of varying experience of the work crews, each work crew is not able to complete each job with the same effectiveness. The cost of each work crew to do each job is given by the cost matrix shown in Table 8.21. The objective is to assign the jobs to the work crews so as to minimize the total cost of completing all jobs.

Table 8.21

Job

	i \\ j	1	2	3	4	5
	1	41	72	39	52	25
	2	22	29	49	65	81
Work	3	27	39	60	51	40
Crew	4	45	50	48	52	37
	5	29	40	45	26	30

8.6.1 ENUMERATING ALL POSSIBLE ASSIGNMENTS

One way to approach this problem would be to enumerate all possible assignments of work crew i to job j. Although this approach is possible for small problems, it rapidly becomes unmanageable as the size of the problem increases.

The number of possible assignments of n facilities to n jobs on a one-to-one basis is equal to $n(n - 1)(n - 2) \ldots 1$, or equivalently, by $n!$. Thus, the number of possible assignments in Table 8.21 is $5! = 120$. If only one additional work crew and job were added to the table, the possible number of assignments would increase to $6! = 720$. Quite obviously, enumerating all possible assignments is feasible for only very small problems. Consequently, it is necessary to develop an alternative solution technique.

8.6.2 LINEAR PROGRAMMING FORMULATION

One alternative to complete enumeration is to formulate the assignment problem as a linear programming problem. If c_{ij} is defined as the cost of assigning facility i to job j and x_{ij} is defined as the proportion of time that facility i is assigned to job j, the linear programming problem is

$$\text{Minimize:} \quad Z = \sum_{i=1}^{n} \sum_{j=1}^{n} c_{ij} x_{ij} \tag{8.6}$$

$$\text{Subject to:} \quad \sum_{i=1}^{n} x_{ij} = 1 \quad \text{for } j = 1, 2, \ldots, n \tag{8.7}$$

$$\sum_{j=1}^{n} x_{ij} = 1 \quad \text{for } i = 1, 2, \ldots, n \tag{8.8}$$

$$x_{ij} \geq 0$$

The first set of constraints is necessary to assure that each of the j jobs is assigned. The second set of constraints is required to make certain that exactly 100 percent of the time available to a facility is accounted for. For instance, the second set of constraints eliminates the possibility of assigning all jobs to the most efficient facility and no jobs to the other facilities. The final constraint eliminates the possibility of negative-valued variables.

An interesting result of the linear programming solution of the assignment problem is that the variables x_{ij} will have values of either zero or one. This occurs because of the fact that both the coefficients in the constraining equations and the right-hand-side values are equal to one. The implication of this result is that the optimal assignment always involves a one-to-one matching of facility to job. For instance, if $x_{12} = 1$, then facility 1 is assigned to job 2. On the other hand, if $x_{12} = 0$, then facility 1 is not assigned to job 2. Since the optimal solution can never include variables that have a fractional value, one facility will always be assigned to one and only one job. Conversely, one job will always be assigned to one and only one facility.

8.7 The Assignment Algorithm

The assignment problem can be solved by writing the problem as a linear programming problem and using the simplex algorithm. Fortunately, however, an algorithm that eliminates much of the computational burden required by the simplex algorithm has been developed for the assignment problem.

The algorithm is based on two facts. First, each facility must be assigned to one of the jobs. Second, the relative cost of assigning facility i to job j is not changed by the subtraction of a constant from either a column or a row of the cost matrix. To illustrate, consider the first row in Table 8.21. Since work crew 1 must be assigned to one of the five jobs, and since the relative costs of the assignment are not changed by the subtraction of a constant from each element in the row, we can subtract the minimum element in the row from all other elements in the row. Similarly, the relative costs of assigning work crew 2 to each of the five jobs is not changed by subtracting the minimum element in the second row from all other elements in the second row. The same principle is true for all rows of the assignment table. Table 8.22 gives the *reduced*

Table 8.22

Job

i \\ j	1	2	3	4	5	Number Subtracted
1	16	47	14	27	0	25
2	0	7	27	43	59	22
3	0	12	33	24	13	27
4	8	13	11	15	0	37
5	3	14	19	0	4	26

Work Crew (rows 1–5)

cost matrix obtained by subtracting the minimum element in each row from all other elements in the row.

In some instances an optimal assignment is possible from the first reduced matrix. This occurs when an assignment can be made such that the total reduced cost of the assignment is zero. Since this is not the case in Table 8.22. we must further reduce the cost matrix.

A second reduced cost matrix can be obtained by subtracting the minimum element in each column from all other elements in the column. The rationale for this step is analogous to that used to obtain the first reduced matrix. Since each job must be assigned to one of the work crews, the relative cost of the assignment is not changed by the subtraction of a constant from all elements in the job column. The second reduced cost matrix, obtained by subtracting the minimum element in each column from all other elements in the column, is shown in Table 8.23.

An optimal assignment is possible from Table 8.23. The assignment is work crew 1 to job 5, 2 to 2, 3 to 1, 4 to 3, and 5 to 4. The cost of this assignment is found by summing the cost of the assignments in the original cost matrix, Table 8.21, and is $155.

It is not always possible to obtain an optimal assignment from the second

Table 8.23

Job

j i	1	2	3	4	5
1	16	40	3	27	0
2	0	0	16	43	59
3	0	5	22	24	13
4	8	6	0	15	0
5	3	7	8	0	4
Number Subtracted	0	7	11	0	0

Work Crew labels rows 1–5.

reduced cost matrix. To illustrate this case, consider the assignment problem given by Table 8.24. This problem is identical to the original problem with

Table 8.24

Job

j i	1	2	3	4	5
1	41	72	39	52	25
2	22	29	49	65	81
3	27	39	60	51	40
4	45	50	48	26	37
5	29	40	45	52	30

Work Crew labels rows 1–5.

the exception that the final two elements in the fourth column have been interchanged. The reduced cost matrices are obtained in the manner described and are shown in Tables 8.25 and 8.26.

Table 8.25

Job

j i	1	2	3	4	5	Number Subtracted
1	16	47	14	27	0	25
2	0	7	27	43	59	22
3	0	12	33	24	13	27
4	19	24	22	0	11	26
5	0	11	16	23	1	29

Work Crew labels rows 1–5.

Table 8.26

Job

i \ j	1	2	3	4	5
1	16	40	0	27	0
2	0	0	13	43	59
3	0	5	19	24	13
4	19	17	8	0	11
5	0	4	2	23	1
Number Subtracted	0	7	14	0	0

(Work Crew labels rows 1–5)

An optimal assignment exists if the total reduced cost of the assignment is zero. Since the total reduced cost is not zero, an optimal assignment is not present. Consequently, the cost matrix must be reduced even further.

Table 8.26 can be reduced by use of a very simple technique. The technique involves drawing straight lines, either horizontally or vertically, that cover all zeroes in the reduced cost matrix. The zeros must be covered with as few lines as possible. If the minimum number of lines necessary to cover all zeros equals the number of assignments that must be made, an optimal solution has already been found. If the number of lines is less than the number of assignments, an additional reduction is necessary.

Four lines are required to cover the zeros in the reduced cost matrix of Table 8.26. Since five assignments must be made (i.e., the matrix has dimensions of 5 by 5), an additional reduction is necessary. The lines used to cover the zeros are shown in Table 8.27.

Table 8.27

Job

i \ j	1	2	3	4	5
1	16	40	0	27	0
2	0	0	13	43	59
3	0	5	19	24	13
4	19	17	8	0	11
5	0	4	2	23	1

(Work Crew labels rows 1–5)

The reduction of Table 8.27 is made by first determining the smallest element not covered by a line. This number is subtracted from each uncovered element in the matrix and is added to those elements covered by two lines. Since the smallest uncovered number in Table 8.27 is 1, each uncovered number is reduced by 1 and each number covered by two lines is increased by 1. The resulting reduced cost matrix is given in Table 8.28.

Table 8.28

Job

	j *i*	*1*	*2*	*3*	*4*	*5*
	1	17	40	0	27	0
	2	1	0	13	43	59
Work	3	0	4	18	23	12
Crew	4	20	17	8	0	11
	5	0	3	1	22	0

An assignment with a total reduced cost of zero is possible in Table 8.28. The assignment is work crew 1 to job 3, 2 to 2, 3 to 1, 4 to 4, and 5 to 5. The cost of this assignment is found from the original cost matrix, Table 8.24, and is $151.

The logic underlying the reduction of Table 8.27 to obtain Table 8.28 can be explained as follows. An optimal assignment is not possible in Table 8.27. Therefore, an additional reduction is necessary. This reduction is made by subtracting the smallest nonzero element from all elements in the matrix. Subtracting 1 from each number in the reduced cost matrix, Table 8.27, gives the cost matrix shown by Table 8.29.

Table 8.29

Job

	j *i*	*1*	*2*	*3*	*4*	*5*
	1	15	39	−1	26	−1
	2	−1	−1	12	42	58
Work	3	−1	4	18	23	12
Crew	4	18	16	7	−1	10
	5	−1	3	1	22	0

Table 8.29 contains negative values. Since the objective is to obtain an assignment with reduced cost of zero, the negative numbers must be elimi-

nated. This can be done by adding a constant to the appropriate rows and/or columns. Adding 1 to each element in the first column gives the matrix shown in Table 8.30.

Table 8.30

Job

	j	*1*	*2*	*3*	*4*	*5*
	i					
Work Crew	1	16	39	−1	26	−1
	2	0	−1	12	42	58
	3	0	4	18	23	12
	4	19	16	7	−1	10
	5	0	3	1	22	0

The negative numbers in the first, second, and fourth rows of Table 8.30 can be eliminated by adding 1 to each element in those rows. The resulting reduced cost matrix is identical to the matrix shown by Table 8.28.

The assignment algorithm can be summarized by a series of steps. These steps are:

1. Subtract the minimum element of each row in the assignment from all elements of the row. This gives the first reduced cost matrix.
2. Subtract the minimum element in each column of the first reduced matrix from all elements in the column. This gives the second reduced cost matrix.
3. Determine if an assignment with a total reduced cost of zero is possible from the second reduced matrix. If so, this assignment is optimal. If not, proceed to step 4.
4. Draw horizontal and vertical lines to cover all zeros. Use as small a number of lines as possible. Subtract the smallest element not covered by a line from all the elements not covered and add this element to all elements lying at the intersection of two lines.
5. Determine if an optimal assignment is possible. If not, repeat steps 4 and 5 until an optimal assignment is found.

The assignment algorithm, as described, applies only to minimization problems. In order to solve a maximization problem, the assignment matrix must be converted to a new matrix whose elements have reversed magnitudes.

The easiest way of reversing the magnitudes of the elements in the matrix is to subtract each element in the matrix from the largest element of the matrix. This has the effect of converting the maximization problem to a minimization problem. The algorithm can then be applied to the matrix with reversed magnitudes.

We have indicated that the assignment matrix must be square; i.e., the number of facilities must equal the number of jobs. There are many "real world" problems, however, in which the number of facilities is greater than the number of jobs or vice versa. These problems can be solved by adding a "dummy" row or column and applying the assignment algorithm to the modified square matrix. For instance, if there had been five jobs and six work crews in the problem discussed earlier in this section, we could have modified the problem to make the matrix square by adding a sixth column representing "idleness" (i.e., job 6). The c_{ij} entries in the column would be either zeros or some measure of the cost of the idle work crew. Conversely, had there been six jobs and five work crews, an additional row representing an imaginary work crew would have been necessary. In these cases, the optimal solution would include either an imaginary job (idle work) or an imaginary work crew.

Example: A management consulting firm has a backlog of four contracts. Work on these contracts must be started immediately. Three project leaders are available for assignment to the contracts. Because of varying work experience of the project leaders, the profit to the consulting firm will vary based on the assignment as shown below. The unassigned contract can be completed by subcontracting the work to an outside consultant. The profit on the subcontract is zero.

Contract

		1	2	3	4
	A	13	10	9	11
Project	B	15	17	13	20
Leader	C	6	8	11	7
	sub.	0	0	0	0

In order to determine the optimal assignment for this maximization problem, the magnitude of the profit matrix must be reversed. This is done by subtracting each element in the profit matrix from the largest element of the matrix. The resulting matrix is shown below.

Contract

		1	2	3	4
	A	7	10	11	9
Project	B	5	3	7	0
Leader	C	14	12	9	13
	sub.	20	20	20	20

The optimal solution is found by applying the assignment algorithm to the reversed magnitude matrix. The reader should verify that the optimal solution is project leader A to contract 1, B to 4, C to 3, and contract 2 to the outside consultant.

Example: In designing a production facility, it is important to locate the work centers so as to minimize the materials handling cost. In a specific example, three work centers are required to manufacture, assemble, and package a product. Four locations are available within the plant. The materials handling cost at each location for the work centers is given by the following cost matrix. Determine the location of work centers that minimizes total materials handling cost.

Location

		1	2	3	4
	Man.	18	15	16	13
Job	Ass'bly	16	11	X	15
	Pkg.	9	10	12	8

To apply the assignment algorithm, two modifications to the cost matrix are required. First, the matrix shows that assembly cannot be performed in location 3. Therefore, the cost of assembly at location 3 is represented by M. Second, a dummy job must be added to the matrix. The cost of this job at each location is zero. The modified cost matrix is given by the following table.

Location

		1	2	3	4
	Man.	18	15	16	13
Job	Ass'bly	16	11	M	15
	Pkg.	9	10	12	8
	Dummy	0	0	0	0

The optimal assignment is manufacturing at location 4, assembly at location 2, and packaging at location 1. Location 3 is not assigned to any of the three jobs; i.e., location 3 is assigned to the dummy row.

PROBLEMS

1. A manufacturer of inboard motor boats has three assembly plants where different models of the boats are made. The engines for the boats are purchased from a vendor and shipped to the assembly plants from the vendor's two manufacturing facilities. The cost of the engines, with the exception of shipping charges, is the same at each of the vendor's two manufacturing facilities. The supply of engines along with the number of boats scheduled for assembly during the scheduling period is shown below. The shipping costs from each engine plant to the three assembly plants are also shown. The manufacturer wishes to develop a shipping schedule that minimizes the total shipping cost while meeting the assembly requirements.

Engine Plant	Cost of Shipping			Engines Available
	1	2	3	
1	$40	$30	$20	500
2	$15	$25	$35	500
Boats scheduled for assembly	300	300	400	

(a) Establish the northwest corner solution.

(b) Determine the optimal solution, using the stepping stone algorithm.

2. A farm implement company has manufacturing facilities in Tulsa, Phoenix, and Portland, Oregon. Each of these facilities has the capability to manufacture the four major products made by the company. However, because of differences in equipment and plant design, the cost of manufacturing the four products differs at the three plants. These costs are shown in the table below.

The capacity of the plants also differs. The capacity of each plant and the requirements for each product are shown below. Both capacity and requirements have been expressed in terms of standard units. (This means that capacity and requirements are directly comparable.)

Plant	Product				Capacity (std. units)
	A	B	C	D	
Tulsa	$300	$200	$500	$200	75
Phoenix	200	100	200	400	120
Portland	200	300	400	300	105
Requirements (std. units)	65	60	80	95	300

(a) Establish the Northwest Corner solution.

(b) Determine the optimal solution using the stepping stone algorithm.

3. The Baxter Glass Company produces disposable glass containers that are purchased by five soft-drink bottlers. The bottles are sold at a fixed delivered price of $0.25 per case. Orders for the current scheduling period have been received from the five bottlers and are as follows: 30,000 cases from bottler 1; 30,000 cases from bottler 2; 100,000 cases from bottler 3; 50,000 cases from bottler 4; and 40,000 cases from bottler 5.

The bottles are made at three plants. The monthly production capacities of the plants are 75,000 cases at plant 1, 100,000 cases at plant 2, and 125,000 cases at plant 3. The direct costs of production are $0.10 at plant 1, $0.09 at plant 2, and $0.08 at plant 3. The transportation costs of shipping a case from a plant to a bottler are shown below.

Plant	Bottler				
	1	2	3	4	5
1	$0.04	$0.06	$0.10	$0.13	$0.13
2	0.07	0.04	0.09	0.10	0.11
3	0.09	0.08	0.09	0.08	0.12

(a) Formulate the problem as a balanced transportation problem.

(b) Use the minimum cell method to determine the initial solution.

(c) Determine the optimal solution, using the stepping stone algorithm.

4. The Econo Car Rental Company has a shortage of rental cars in certain cities and an oversupply of cars in other cities. The imbalances are shown in the following table.

City	Cars Short	Cars Excess
Albany	—	15
Boston	—	20
Chicago	28	—
Cleveland	—	15
Dallas	—	20
Detroit	—	30
Kansas City	13	—
Miami	15	—
Philadelphia	24	—

The costs of transporting a car from the cities with an excess to those cities with a shortage are shown below.

From \ To:	Chicago	Transportation Cost ($) per Car Kansas City	Miami	Philadelphia
Albany	$140	$210	$240	$ 50
Boston	150	220	230	60
Cleveland	70	190	200	80
Dallas	160	90	180	130
Detroit	60	110	190	90

The company wants to correct the imbalances at the minimum cost.
(a) Formulate the problem as a balanced transportation problem.
(b) Use the minimum cell method to determine the initial solution.
(c) Determine the optimal solution, using the stepping stone algorithm.

5. The Superior Oil Company has three oil refineries. These refineries are currently operating at less than maximum capacity. Superior has the opportunity to sell certain oil products to a competing oil company. Since the competing firm has alternative sources of supply, the sale will not change Superior's current market position. The potential profit from from the sale of the products is shown in the following table.

Plant	Gasoline	Kerosene	Product Diesel	Jet Fuel	Asphalt
1	$0.165	—	$0.140	$0.125	$0.128
2	0.140	$0.146	0.126	—	0.133
3	—	0.139	0.134	0.125	0.130

As indicated in the table, the profit on a product differs at the several plants. In addition, certain products are not available at all plants. This is indicated by the dashes. Superior has excess capacity of 70,000 barrels at plant 1, 90,000 barrels at plant 2, and 40,000 barrels at plant 3. They have been requested to supply part or all of the following products: 60,000 barrels of gasoline, 15,000 barrels of kerosene, 40,000 barrels of diesel fuel, 25,000 barrels of jet fuel, and 20,000 barrels of asphalt. Management wants to schedule production of these products at the three plants so as to maximize profits.

(a) Formulate the problem as a balanced transportation problem.

(b) Use the penalty method to obtain the initial solution and verify that the initial solution is optimal.

6. The Quapaw Company operates three rock quarries that are located in central Oklahoma. Because of the lack of suitable concrete rock in that portion of the state, this company will be the principal supplier of concrete rock for a new turnpike currently under construction.

The concrete rock must be quarried and delivered to six concrete ready-mix plants that are situated along the road site. The quantity of rock that must be delivered to each of these plants is shown below.

	Concrete Plant Site					
	1	*2*	*3*	*4*	*5*	*6*
Rock required (cu. yd.)	120,000	80,000	160,000	100,000	80,000	100,000

The cost of the delivered rock is dependent on both the cost of quarrying the rock and the cost of trucking the rock to the ready-mix plant sites. The cost of quarrying the rock is related to factors such as the size of the quarry, the amount of overburden that must be removed, the amount of washing required, etc. These costs, per cubic yard of rock, are $2.40 at quarry 1, $3.20 at quarry 2, and $2.85 at quarry 3. The trucking cost per cubic yard is given in the following table.

From	*To:*	*1*	*2*	*3*	*4*	*5*	*6*
1		$1.80	$1.60	$0.80	$1.30	$1.40	$2.40
2		1.20	0.90	0.70	1.20	1.30	1.50
3		1.70	1.60	1.30	1.00	1.00	1.50

Because of differences in both equipment and rock, the capacities of the three quarries differ. The capacity of quarry 1 is 300,000 cubic yards, the capacity of quarry 2 is 180,000 cubic yards, and the capacity of quarry 3 is 240,000 cubic yards. The objective is to minimize the cost of the delivered rock.

(a) Formulate the problem as a balanced transportation problem.

(b) Use the penalty method to obtain the initial solution.

(c) Determine the optimal solution, using the modified distribution algorithm.

7. The Graham Electron Company has received a contract from Transcontinental Airlines, Inc., to provide metal detection devices. These devices are used by Transcontinental in the screening of passengers for both domestic and international flights. The contract specifies delivery of the metal detection devices according to the following schedule.

		Quarter		
	First	Second	Third	Fourth
Number	20	30	40	40

Graham has sufficient production capacity to meet the above schedule. However, because of contracts with other customers, the number of metal detection devices that can be manufactured in the next four quarters varies. In addition, Graham expects the cost of manufacture to increase. Maximum production, the unit cost of production, and the unit storage cost are given in the following table.

Quarter	Production	Unit Cost of Production	Unit Storage Cost per Quarter
1	30	$14,000	$1000
2	45	16,000	1000
3	40	15,000	1000
4	25	17,000	1000

No storage costs are incurred for devices that are delivered in the same quarter as produced. The objective is to schedule production and delivery so as to minimize cost.

(a) Formulate the problem as a balanced transportation problem.

(b) Verify that the initial solution found by using the minimum cell method is optimal.

8. Lacy's Department Store must place orders for five styles of their very popular hand-carved wooden statues. Lacy's has been able to find only three manufacturers of these particular statues. The number of statues available from each manufacturer is shown below.

Manufacturer	No. of Statues
1	200
2	300
3	450

All statues, regardless of style, are sold at a retail price of $225. The cost of the statues varies, however, from one manufacturer to another. These costs are shown in the following table.

Manufacturer	1	2	Style 3	4	5
1	$200	$190	$140	$210	$160
2	150	200	130	190	160
3	180	150	180	200	250

The potential sales of the statues exceeds the total supply. The potential sales are: style 1, 150; style 2, 200; style 3, 350; style 4, 200; style 5, 300. Since the price of all five styles of statues is the same, Lacy's objective is to minimize cost.

(a) Formulate the problem as a balanced transportation problem.

(b) Determine the initial solution, using the penalty method.

(c) Find the optimal solution, using the modified distribution algorithm.

9. A large oil company has oil reserves in four different locations. This oil must be transported from these locations to five oil refineries. The transportation costs, supplies, and requirements are shown in the following balanced transportation table.

		Transportation Cost (per barrel) Refinery				Maximum Yearly Supply (thousands of barrels)
Reserve	1	2	3	4	5	
1	$6.50	$2.50	$5.00	$3.00	$5.00	1100
2	4.00	0.50	3.00	3.00	3.50	920
3	0.50	6.00	2.00	3.50	3.50	620
4	5.50	8.00	3.50	5.00	1.00	1020
Yearly demand (thousands of barrels)	840	400	1000	600	820	3660

Use the modified distribution algorithm to determine the shipments from the reserves to the refineries that minimize total shipping cost.

10. Five girls in a secretarial pool must be assigned to five different jobs. From past records, the time that each girl takes to do each job is known. These times, in hours, are shown in the following table.

Girl	Job 1	2	3	4	5
1	3	7	8	4	6
2	4	8	7	5	6
3	5	7	9	5	7
4	4	6	8	6	8
5	6	9	10	7	9

Assuming that each girl can be assigned to only one job, determine the optimal assignment.

11. A television repair shop employs six technicians. These technicians must be assigned to five different repair jobs. Because of their different specialties and levels of skill, the time required for each technician to complete each job varies. The times required are shown in the following table. All entries are in minutes.

Technician	Job 1	2	3	4	5
1	180	160	240	300	150
2	190	150	250	320	170
3	150	170	230	340	140
4	170	200	210	310	190
5	220	190	260	330	160
6	190	180	300	310	210

The objective is to minimize the total repair time. The unassigned technician will remain idle. Formulate the problem as an assignment problem and determine the optimal solution.

12. A major aerospace company has received a government contract for the production of an advanced fighter/bomber. As is usually the case, a large number of the systems required for the aircraft must be obtained from subcontractors. To illustrate, the avionics have been divided into six different packages and bids have been requested for each package. Eight major avionics manufacturers have submitted bids on the different packages. These bids are shown in the following table. A dash indicates that the avionics manufacturer declined to bid on an individual package. All dollar figures are in millions of dollars.

| Manufacturer | Avionics Package | | | | | |
	1	2	3	4	5	6
1	$4.2	$6.8	$1.4	$3.6	$9.4	$6.2
2	4.4	—	1.6	3.5	9.5	6.4
3	4.3	7.0	1.5	3.5	9.6	6.1
4	5.0	7.0	1.4	3.7	10.0	6.0
5	5.0	6.9	1.6	3.6	9.5	5.9
6	3.9	6.7	1.7	—	9.6	6.1
7	4.1	6.8	1.5	3.5	9.4	6.1
8	4.2	6.9	1.7	3.5	—	6.0

The scheduling is such that only one avionics package can be subcontracted to any single manufacturer. The aerospace company's objective is, of course, to minimize the cost of procuring the entire avionics system.
(a) Formulate the problem as an assignment problem.
(b) Determine the optimal solution.

13. KECT, the local educational television station, must schedule their weekend television programming. In addition to KECT, there are three network affiliate stations in the local viewing area. The programming for these three network affiliates has already been announced.

   On the basis of past experience, KECT realizes that they can attract only a small portion of the total viewing public. Nevertheless, their objective is to maximize their total program exposure.

   The station has eight, one-hour programs that have been selected for the weekend prime-time viewing hours (i.e., 7:00-11:00 p.m. on Saturdays and Sundays). The number of viewers they can expect depends on the popularity of the network shows scheduled during these hours as well as the popularity of KECT's shows. Management's estimates of the number of viewers each show can attract during the prime-time viewing hours are given in the following table. All entries in the table are in thousands of viewers.

| Day & Time | Show | | | | | | | |
	1	2	3	4	5	6	7	8
Sat. 7:00– 8:00	18	12	25	16	30	5	15	8
8:00– 9:00	14	14	30	18	32	6	18	6
9:00–10:00	12	15	22	18	36	19	17	4
10:00–11:00	10	20	16	14	40	13	16	4
Sun. 7:00– 8:00	20	11	20	15	28	7	14	7
8:00– 9:00	16	12	26	17	34	11	17	5
9:00–10:00	14	14	20	25	40	10	16	3
10:00–11:00	12	16	14	14	30	6	13	2

As an illustration of the table, management expects show 1 to have 18,000 viewers if scheduled on Saturdays from 7:00–8:00 p.m. The number of viewers would drop to 14,000, however, if the show were scheduled from 8:00 to 9:00 p.m. on the same day.

(a) Formulate the problem as an assignment problem.

(b) Determine the optimal solution.

14. Nash, Smith, and Co., Certified Public Accountants, provides auditing and tax services for a large number of firms in Northern California. Mr. Nash, the managing senior partner of the firm, must assign jobs to the staff (junior) and senior members of the firm.

The accounting firm currently has six staff accountants and three senior accountants available for assignments. The major differences in the two classifications is in experience, the staff accountants normally having less than three years' experience and the senior accountants three to six years' experience.

The firm currently has a backlog of five auditing jobs and three tax jobs that must be assigned. Mr. Nash has estimated the time required by each accountant to complete each job. These entries are shown in the following table in units of days. A dash indicates that the accountant is not qualified for the job.

Staff Acc't†	Audit					Tax		
	1	*2*	*3*	*4*	*5*	*1*	*2*	*3*
Mr. Hill	10	—	5	21	13	—	—	—
Mr. Nutter	12	20	7	18	15	—	15	—
Mr. Hamilton	10	18	6	17	16	18	13	18
Mr. Pryor	8	—	8	19	15	—	19	22
Mr. Pratt	12	22	6	19	13	16	14	20
Mr. Wagner	14	21	7	20	16	—	16	19
Senior Acc't								
Mr. Redman	7	15	5	16	12	10	12	15
Mr. Lochen	8	17	5	15	11	12	14	17
Mr. Savich	8	14	5	15	11	13	14	17

† Entries are in 8-hour days.

The billing rate is $15 per hour for staff accountants and $20 per hour for senior accountants. The objective is to minimize the total billings to the customers.

(a) Formulate the problem as an assignment problem.
(b) Determine the optimal solution.

SUGGESTED REFERENCES

The references for this chapter are listed in Chapter 5.

Chapter 9

Differential
Calculus

Many of the important modern quantitative models are based on a branch of mathematics termed *differential* and *integral calculus*. The roots of differential and integral calculus can be traced to problems posed by the early Greek geometers and philosophers. It was not, however, until the development of fundamental and unifying principles by Sir Isaac Newton (1642–1727) and Gottfried Wilhelm von Leibnitz (1646–1716), that solution techniques for these and related problems became available. These solution techniques and principles were refined until, by the middle of the nineteenth century, the foundations of calculus were firmly established.

The next six chapters of this text provide an introduction to calculus and some of the more important quantitative managerial decision models that are based on calculus. Differential calculus is introduced in this chapter and integral calculus in Chapter 14. The importance of differential and integral calculus in modern quantitative business models is illustrated in Chapters 10 through 13 and in Chapter 16.

Before we can define the meaning of differential calculus, we must define three important concepts. These are rate of change, limit, and continuity. These concepts, along with the concept of a function, are fundamental to an understanding of differential calculus.

9.1   Rate of Change

The *rate of change* of a function is defined as the change in the value of the dependent variable divided by the change in the value of the independent variable. The formula for the rate of change of a linear function can be developed from the definition of rate of change. The functional form of the linear function is given by Formula (9.1) as

$$f(x) = a + bx \tag{9.1}$$

If the independent variable x is increased by the arbitrary amount Δx, the value of the independent variable becomes $(x + \Delta x)$. Replacing x by $(x + \Delta x)$, we obtain

$$f(x + \Delta x) = a + b(x + \Delta x) \tag{9.2}$$

To determine the increase in the dependent variable due to the increase Δx in the independent variable, we subtract (9.1) from (9.2).

$$f(x + \Delta x) - f(x) = (a + b(x + \Delta x)) - (a + bx)$$
$$f(x + \Delta x) - f(x) = a + bx + b\,\Delta x - a - bx \tag{9.3}$$
$$f(x + \Delta x) - f(x) = b\,\Delta x$$

Remembering that the rate of change of a linear function is defined as the change in the value of the function divided by the change in the value of the independent variable, we see that

$$\frac{f(x + \Delta x) - f(x)}{\Delta x} = \frac{b\,\Delta x}{\Delta x} = b$$

which is the slope of the linear function. Using the symbol Δ to represent change, an equivalent formula for the rate of change of a linear function is

$$\text{Rate of change} = \frac{\Delta f(x)}{\Delta x} = \frac{f(x + \Delta x) - f(x)}{\Delta x} \tag{9.4}$$

In deriving the formula for the slope or rate of change of the linear function, the symbol Δ was used to represent change. The symbol Δx refers to an incremental quantity of the analyst's choosing. For example, if $x = 2.0$ and $\Delta x = 0.1$, then $x + \Delta x = 2.1$. The symbol $\Delta f(x)$ represents the change in $f(x)$ which is associated with the change in x; that is,

$$\Delta f(x) = f(x + \Delta x) - f(x)$$

The formula for the slope or rate of change of a linear function is illustrated by the following examples.

Example: $f(x) = -2x + 1$.

$$\text{Rate of change} = \frac{\Delta f(x)}{\Delta x} = \frac{f(x + \Delta x) - f(x)}{\Delta x}$$

$$= \frac{-2(x + \Delta x) + 1 - (-2x + 1)}{\Delta x}$$

$$= \frac{-2x - 2\,\Delta x + 1 + 2x - 1}{\Delta x} = \frac{-2\,\Delta x}{\Delta x} = -2$$

Example: $f(x) = \dfrac{x - 2}{3}$

$$\text{Rate of change} = \frac{\Delta f(x)}{\Delta x} = \frac{f(x + \Delta x) - f(x)}{\Delta x}$$

$$= \frac{\dfrac{(x + \Delta x) - 2}{3} - \dfrac{(x - 2)}{3}}{\Delta x}$$

$$= \frac{x + \Delta x - 2 - x + 2}{3\,\Delta x} = \frac{\Delta x}{3\,\Delta x} = \frac{1}{3}$$

9.2 Limit of a Function

The second important concept that must be introduced before we can define the meaning of differential calculus is the concept of the *limiting value* of a function. The limiting value, or simply the *limit* of a function, can be defined as follows. Let $y = f(x)$, where y is the dependent and x the independent variable. Furthermore, let a be some fixed number. If y approaches some fixed number L as x approaches a, then L is the limiting value of the dependent variable as x approaches a. L is also said to be the limit of the function $f(x)$ as x approaches a. This is expressed as

$$\lim_{x \to a} y = \lim_{x \to a} f(x) = L$$

An alternative formulation is simply

$$\lim_{x \to a} f(x) = L \tag{9.5}$$

The existence of a limiting value of a function for certain types of functions can easily be demonstrated. As an example, consider the function

$$y = 1 + (\tfrac{1}{2})^x \qquad \text{for } x = 0, 1, 2, 3, 4, \ldots, \infty$$

As x approaches infinity, written $x \to \infty$, y assumes the values

$$2, 1\tfrac{1}{2}, 1\tfrac{1}{4}, 1\tfrac{1}{8}, 1\tfrac{1}{16}, 1\tfrac{1}{32}, 1\tfrac{1}{64}, \ldots$$

It is seen that as $x \to \infty$, $y \to 1$; i.e., the limiting value of y as x approaches infinity is 1. Mathematically, this is expressed as

$$\operatorname*{limit}_{x \to \infty} y = \operatorname*{limit}_{x \to \infty} [1 + (\tfrac{1}{2})^x] = 1$$

The selection of a determines the value of the limit. Given the linear function,

$$y = x + 2$$

we see that the limit of y as x approaches 3 is 5.

$$\operatorname*{limit}_{x \to 3} y = \operatorname*{limit}_{x \to 3} (x + 2) = 5$$

The limit of 5 as x approaches 3 means that as x comes closer and closer to 3, y comes closer and closer to 5. However, for the same function, the limit as x approaches infinity does not exist. That is, there is no fixed number L that y approaches as x approaches infinity. This function, therefore, has no limiting value as x approaches infinity. This illustrates the fact that a function which has a limiting value for certain values of the independent variable need not have a limiting value for all possible values of the independent variable.

The following examples illustrate limiting values of selected functions.

Example: $\operatorname*{limit}_{x \to \infty} \left(\dfrac{1}{x + 1} \right) = 0$

Example: $\operatorname*{limit}_{x \to 6} (x^2) = 36$

Example: $\operatorname*{limit}_{x \to 2} \left(\dfrac{x^2 - 4}{x - 2} \right) = \operatorname*{limit}_{x \to 2} \left(\dfrac{(x + 2)(x - 2)}{(x - 2)} \right) = \operatorname*{limit}_{x \to 2} (x + 2) = 4$

It is tempting in determining limits simply to set $x = a$ and determine $f(a)$. The preceding example illustrates, however, that this is not the proper procedure. If we set $x = 2$, we are attempting to divide $(x^2 - 4)$ by 0. *Division by 0 is not permitted.* The correct procedure is to eliminate the term $(x - 2)$ from the denominator by division of the numerator by the denominator. This results in $L = 4$ rather than the mathematically undefined term $(\tfrac{0}{0})$.

Figure 9.1 illustrates several possible situations regarding the limit of $f(x)$ as x approaches a.

Diagrams (a) and (b) both show functions that have the limit L as x approaches a. Diagrams (c) and (d) illustrate functions for which limits do not exist as x approaches a. In diagram (c), the limit as x approaches a from the left is L_1, whereas the limit as x approaches a from the right is L_2. *An additional requirement for a limit is that the left side and right side limits must be*

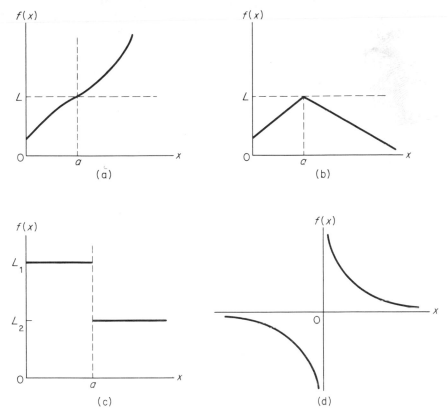

Figure 9.1

the same. This is not the case in diagram (c). Consequently, the function does not have a limit as x approaches a.

Diagram (d) shows a function that has a limit for all values of x except as x approaches 0. As x approaches 0, the function $f(x)$ approaches $+\infty$ or $-\infty$. Infinity is not a limiting value of the function; i.e., ∞ is not considered as a limit. Consequently, the function has no limit as x approaches 0. The function in diagram (d) does, however, have a limit as x approaches any value other than 0.

FORMAL DEFINITION OF LIMIT. A function $f(x)$ has a limit L as x approaches a when, as x is given in a sequence of values approaching a from either the left or the right, the corresponding value of $f(x)$ can be made to approach the constant L as closely as desired. This definition is illustrated in Fig. 9.2. Let $[L - f(x - \Delta x)] = f_1$ and $[f(x + \Delta x) - L] = f_2$. Note that f_1 or f_2 can be made as close to 0 as desired by permitting Δx to approach 0. As

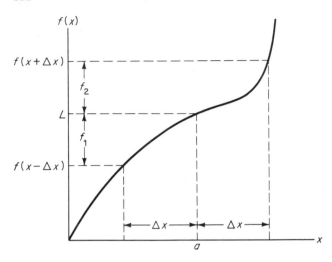

Figure 9.2

f_1 and f_2 approach 0, $f(x)$ approaches the constant L. This illustrates the requirement in the definition that $f(x)$ can be made to approach the constant L as closely as desired.

9.3 Continuity

The concept of *continuity* of a function is the third important concept that must be introduced prior to defining differential calculus. Simply stated, a function is continuous in an interval from $x = b$ to $x = c$ if the function has no breaks or jumps in the interval. Figure 9.3 shows examples of both continuous and discontinuous functions.

The function in diagram (a) is continuous. The function in diagram (b) is also continuous, even though it reaches a point at $x = a$. There is a break or step in the function depicted in diagram (c), which causes a discontinuity at $x = a$. Similarly, the function in diagram (d) has a break at $x = a$ and is therefore discontinuous at $x = a$.

The requirements for continuity of a function at a point $x = a$ are

1. $f(a)$ is defined, i.e., the domain of x includes $x = a$.
2. $\lim\limits_{x \to a} f(x)$ exists.
3. $\lim\limits_{x \to a} f(x) = f(a)$, whether x approaches a from the left or the right.

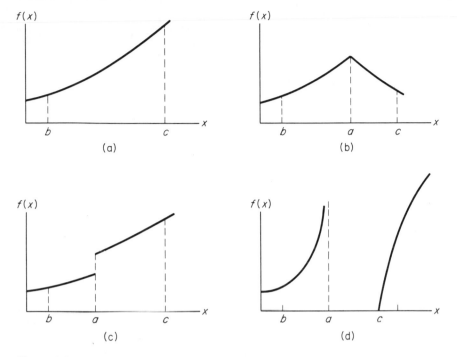

Figure 9.3

The following examples illustrate the requirements for continuity.

Example: $f(x) = x + 4$, for $0 \le x \le 4$. Specify the continuity at $x = 6$. The function is not defined at $x = 6$. Consequently, we cannot state that the function is continuous at that point.

Example: $f(x) = \dfrac{10}{x - 2}$ for $0 \le x \le \infty$. Specify the continuity at $x = 2$. The limit of $f(x)$ as x approaches 2 does not exist. Therefore, the function is discontinuous at $x = 2$.

Example: $f(x) = x^2$, for $0 \le x \le \infty$. Specify the continuity at $x = 6$. $f(6)$ is defined and the limit as x approaches 6 exists. Furthermore, the limit as x approaches 6 from the left equals the limit as x approaches 6 from the right. Therefore, the function is continuous at $x = 6$.

9.4 The Derivative

We can now introduce the concept of the derivative. The *derivative* of the function $f(x)$ is a function $f'(x)$ that is defined as

$$f'(x) = \lim_{\Delta x \to 0} \frac{\Delta f(x)}{\Delta x}$$

or, alternatively,

$$f'(x) = \lim_{\Delta x \to 0} \frac{f(x + \Delta x) - f(x)}{\Delta x} \tag{9.6}$$

With one exception, the formula for the derivative is identical to the formula for the rate of change of a function. The exception, of course, is that the formula for the derivative includes the term

$$\lim_{\Delta x \to 0}$$

This exception is very important. The formula for the rate of change is merely the change in the dependent variable divided by the change in the independent variable. The formula for the derivative of a function, however, states that the derivative is defined as the limiting value of the change in the dependent variable divided by the change in the independent variable. The magnitude of Δx is not specified in the formula for the rate of change, whereas the magnitude of Δx approaches zero in the formula for the derivative. Comparing the formula for the rate of change (9.4) and that for the derivative (9.6), it can be seen that the formula for the derivative gives the rate of change as the interval over which the change is measured, Δx, approaches zero. For this reason, the derivative is also referred to as the *instantaneous rate of change*.

The formula for the derivative or, alternatively, the instantaneous rate of change can be used to determine the slope of a function. The reader will remember that the slope of a function is defined as the change in the value of the dependent variable divided by the magnitude of the change in the independent variable. The slope of the linear function in Fig. 9.4(a) is constant for any arbitrary Δx. The slope of the curvilinear function in Fig. 9.4(b) depends, however, upon the selection of Δx. If the objective is to determine the slope of the function at $x = a$, or, alternatively, the instantaneous rate of change of the function at $x = a$, selection of a large interval for Δx provides a poor approximation of the true slope for the curvilinear function at $x = a$. Consequently, Δx must be allowed to approach 0 in order accurately to determine the slope of the function at $x = a$.

The derivative provides a method for determining the slope or rate of change of a curvilinear function at a value of the independent variable $x = a$.

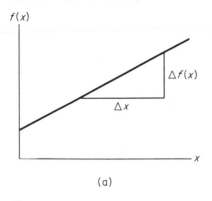
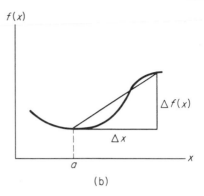

Figure 9.4

As an example, the function $f(x) = x^2/4$ is plotted in Fig. 9.5. Assume that we wish to determine the slope of this function for different values of the independent variable. Applying Formula (9.4) for the slope of a linear function, we can determine the average slope of the function between two values of the independent variable. Thus, between $x = 2$ and $x = 4$, the average slope is $\Delta f(x)/\Delta x = 3/2 = 1.5$ (see Fig. 9.5). Note that there is no single

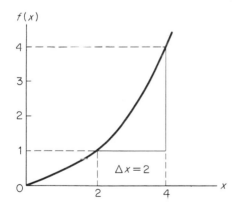

$$f(x) = x^2/4$$
$$f(x + \Delta x) - f(x) = 3$$

Figure 9.5

slope in this interval of Δx. For each value of x in the interval ($x = 2.0$, $x = 2.5$, $x = 3.0$, etc.) the slope of the function differs. However, if we let Δx approach 0, we can obtain an accurate approximation of the slope for a selected value of x. Applying the formula for the derivative, Formula (9.6), to the function $f(x) = x^2/4$, we obtain

$$f'(x) = \lim_{\Delta x \to 0} \frac{f(x + \Delta x) - f(x)}{\Delta x} = \lim_{\Delta x \to 0} \frac{\frac{(x + \Delta x)^2}{4} - \frac{x^2}{4}}{\Delta x}$$

$$f'(x) = \lim_{\Delta x \to 0} \frac{x^2 + 2x(\Delta x) + (\Delta x)^2 - x^2}{4(\Delta x)} = \lim_{\Delta x \to 0} \left(\frac{2x(\Delta x)}{4(\Delta x)} + \frac{(\Delta x)^2}{4(\Delta x)} \right)$$

$$f'(x) = \lim_{\Delta x \to 0} \left(\frac{2x}{4} + \frac{\Delta x}{4} \right)$$

The limiting value of $2x/4$ as Δx approaches 0 is $2x/4$, and the limiting value of $\Delta x/4$ as Δx approaches 0 is 0. Therefore, the derivative of $f(x) = x^2/4$ is

$$f'(x) = \frac{2x}{4}$$

The slope of $f(x)$ can be determined at any value $x = a$, simply by evaluating the derivative at the value of $x = a$. Remembering that $x = a$ is a value of the independent variable, we represent the value of the derivative of the function at $x = a$ by the symbol $f'(a)$. The slope of the function $f(x) = x^2/4$ at $x = 2$, $x = 3$, and $x = 4$ is $f'(2) = 2(2)/4 = 1$, and similarly $f'(3) = 1.5$ and $f'(4) = 2$.

We can check our formula by selecting an arbitrarily small value for Δx, and determining the *average slope* of the function between $x = a$ and $x = a + \Delta x$. If, for example, $\Delta x = 0.01$ and $x = 2$, the average slope is

$$\text{Slope} = \frac{f(x + \Delta x) - f(x)}{\Delta x} = \frac{(2.01)^2 - (2.0)^2}{4(0.01)}$$

$$\text{Slope} = \frac{4.0401 - 4.0000}{0.04} = 1.0025$$

This is a close approximation to the actual slope at $x = 2$ of $f'(2) = 1.0$.

The function graphed in Fig. 9.5 is of the form $f(x) = cx^2$. It is possible to apply the formula for the derivative directly to the general form of the function. Once the derivative of the general function $f(x) = cx^2$ has been determined, we can determine the derivative of all functions of this form without resulting to continual reapplication of Formula (9.6). The derivative of $f(x) = cx^2$ is

$$f'(x) = \lim_{\Delta x \to 0} \frac{c(x + \Delta x)^2 - cx^2}{\Delta x}$$

$$f'(x) = \lim_{\Delta x \to 0} \frac{c(x^2 + 2x \, \Delta x + \Delta x^2) - cx^2}{\Delta x}$$

$$f'(x) = \lim_{\Delta x \to 0} \frac{cx^2 + 2cx \, \Delta x + c \, \Delta x^2 - cx^2}{\Delta x}$$

$$f'(x) = \lim_{\Delta x \to 0} \frac{2cx \, \Delta x + c \, \Delta x^2}{\Delta x}$$

$$f'(x) = \lim_{\Delta x \to 0} (2cx + c\,\Delta x)$$

$$f'(x) = 2cx$$

Using this rule for the derivative of $f(x) = cx^2$, we find that the derivative of $f(x) = x^2/4$ is $f'(x) = 2x/4$. The use of rules for determining derivatives is discussed in the following section.

The concepts of continuity and limit have been necessary for the development of the concept of the derivative. *The derivative of a function gives the slope of the function at a value of the independent variable.* If a function, as shown in Fig. 9.3(c), is discontinuous, it has no slope for those values of the independent variable in the interval of the discontinuity. Consequently, a function must be continuous to have a derivative.

The slope is given by the change in $f(x)$ divided by the change in x. The limit allows us to determine the slope for progressively smaller values of a change in x. The limit of the ratio $\Delta f(x)/\Delta x$ as Δx approaches 0 gives the slope of the function at a particular value of the independent variable. The limit of this ratio is termed the derivative and is given by Formula (9.6).

That a function is continuous is a necessary but not sufficient condition for it to have a derivative. The function shown in Fig. 9.1(b) is continuous at the point $x = a$. It does not, however, have a derivative at this point. Remembering that the derivative gives the slope of a function at a point, observe that the function does not have a slope at $x = a$. Consequently, the derivative does not exist at the point $x = a$. The function does, however, have a derivative for other values of x in the domain.

The mathematical requirement that must be met for a derivative to exist is defined in terms of the derivative formula. The requirement is that

$$\lim_{\substack{\Delta x \to 0 \\ \text{from the left}}} \left[\frac{f(x + \Delta x) - f(x)}{\Delta x}\right] = \lim_{\substack{0 \leftarrow \Delta x \\ \text{from the right}}} \left[\frac{f(x + \Delta x) - f(x)}{\Delta x}\right]$$

This requirement is not satisfied by the function in Fig. 9.1(b). The slope for an incremental Δx to the right of a differs from that for an incremental Δx to the left of a.

It should be remembered that the derivative gives the rate of change or slope of the function. To determine the slope of the function at a particular value of the independent variable, the derivative must be evaluated for this value of the independent variable. The following symbols are used for the function, the derivative, and the slope of the function:

> $f(x)$ represents the function.
> $f'(x)$ represents the derivative of the function.
> $f'(a)$ represents the slope of the function
>   at the value $x = a$ of the independent variable.

Alternative symbols that are commonly used are $y = f(x)$ for the function, dy/dx for the derivative, $dy/dx\,(x = a)$ for the slope of the function at $x = a$. Note that $f'(x)$ and dy/dx are used interchangeably to indicate the derivative. Other commonly used symbols that indicate derivatives are $y', f_x,$ and df/dx.

9.4.1 RULES FOR DETERMINING DERIVATIVES

Rules for determining the derivatives of algebraic functions can now be given.† The formula for the derivative is used to derive the first two rules. All of the remaining rules are given without derivation.

Rule 1. The derivative of a constant is 0.

$$f(x) = c, \qquad f'(x) = 0$$

Derivation: Let $f(x) = c$; then

$$f'(x) = \lim_{\Delta x \to 0} \frac{f(x + \Delta x) - f(x)}{\Delta x} = \lim_{\Delta x \to 0} \frac{c - c}{\Delta x} = 0$$

Example: $f(x) = 6$; then $f'(x) = 0$.

Example: $f(x) = 3$; then $f'(x) = 0$.

The slope of each of these functions is 0. That is, $f'(a) = 0$ for all values of $x = a$ in the domain. That the slope of the function is 0 for all values of x

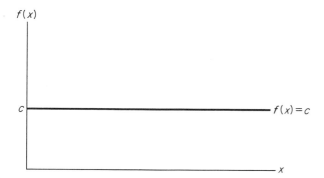

Figure 9.6

is easily seen from Fig. 9.6. The function $f(x) = c$ is plotted in this figure. Since the function plots as a horizontal line, the slope of the function is 0 at all values of x.

---

† Algebraic functions are an important class of functions. These are defined in Sec. 9.4.3.

Rule 2. The derivative of a variable that is raised to the first power is 1.

$$f(x) = x, \qquad f'(x) = 1$$

Derivation: Let $f(x) = x$; then

$$f'(x) = \lim_{\Delta x \to 0} \frac{f(x + \Delta x) - f(x)}{\Delta x} = \lim_{\Delta x \to 0} \frac{x + \Delta x - x}{\Delta x}$$

$$f'(x) = \frac{\Delta x}{\Delta x} = 1$$

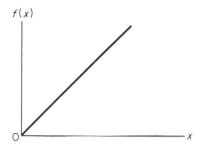

$f(x)$

x

O

Figure 9.7

The functional relationship $f(x) = x$ is plotted in Fig. 9.7. It can be seen from this figure that the function plots as a straight line passing through the origin with a slope of 1. For any value of x, the slope of the linear function is constant and equals 1. This can also be determined from the derivative of the function. The derivative of $f(x) = x$ from Rule 2 is $f'(x) = 1$. For any value of $x = a$, the derivative of the function has the constant value of 1.

Rule 3. The derivative of a variable raised to the constant power n is equal to the power n multiplied by the variable raised to the power $n - 1$.

$$f(x) = x^n, \qquad f'(x) = nx^{n-1}$$

Example: $f(x) = x^3$; determine the derivative.

$$f'(x) = 3x^2$$

Example: $y = x^{-3}$; determine the slope at $x = 3$.

$$\frac{dy}{dx} = -3x^{-4} = -\frac{3}{x^4}$$

and

$$\frac{dy}{dx}(x = 3) = -\frac{3}{(3)^4} = -\frac{3}{81}$$

Example: $f(x) = x^1$; determine the derivative.

$$f'(x) = 1x^0 = 1$$

which verifies Rule 2.

Example: $f(x) = \sqrt{x} = x^{1/2}$; determine the slope at $x = 4$.

$$f'(x) = \frac{1}{2}x^{-1/2} = \frac{1}{2\sqrt{x}}$$

and

$$f'(4) = \frac{1}{2\sqrt{4}} = \frac{1}{4}$$

Rule 4. The derivative of a constant times a function equals the constant times the derivative of the function.

$$f(x) = c \cdot g(x), \qquad f'(x) = c \cdot g'(x)$$

Example: $f(x) = 3x^2$; determine the derivative.

$$f'(x) = 3 \cdot 2 \cdot x^1 = 6x$$

Example: $y = -4x^{-3}$; determine the slope at $x = -2$.

$$\frac{dy}{dx} = -4(-3)x^{-4} = 12x^{-4} = \frac{12}{x^4}$$

and

$$\frac{dy}{dx}(x = -2) = \frac{12}{(-2)^4} = \frac{12}{16} = \frac{3}{4}$$

Example: $f(x) = c^2x^4$.

$$f'(x) = c^2(4)x^3 = 4c^2x^3$$

since c is a constant, c^2 is also a constant, and Rule 4 applies.

Rule 5. The derivative of the sum (or difference) of two or more functions equals the sum (or difference) of their respective derivatives.

$$h(x) = f(x) + g(x), \qquad h'(x) = f'(x) + g'(x)$$
$$h(x) = f(x) - g(x), \qquad h'(x) = f'(x) - g'(x)$$

Example: $f(x) = x^2 + 3x + 6$; determine the derivative.

$$f'(x) = 2x + 3$$

Example: $f(x) = 3x + 4$; determine the slope at $x = 5$.

$$f'(x) = 3$$

The slope is constant for all values of x, therefore $f'(5) = 3$.

Example: $f(x) = 2x^2 + 3x - 4$; determine the slope at $x = 4$.

$$f'(x) = 4x + 3$$

and

$$f'(4) = 4(4) + 3 = 19$$

Example: $y = 2x^2 - 3x + 10$; determine the value of x such that the slope is 0. For the function to have a slope of 0, the derivative evaluated at $x = a$ must be 0.

$$\frac{dy}{dx} = 4x - 3$$

Equating the derivative with 0 gives

$$4x - 3 = 0$$

$$x = \tfrac{3}{4}$$

The derivative is 0 at $x = \tfrac{3}{4}$. Consequently, the function has a slope of 0 at $x = \tfrac{3}{4}$.

Rule 6. The derivative of a product of two functions is equal to the derivative of the first function times the second function plus the first function times the derivative of the second function.

$$h(x) = f(x) \cdot g(x), \qquad h'(x) = f'(x) \cdot g(x) + f(x) \cdot g'(x)$$

Example: $h(x) = x^3(3x^2 + 4x)$; determine the derivative.

$$h'(x) = 3x^2(3x^2 + 4x) + x^3(6x + 4)$$

Example: $y = (x^2 + 3)(x + 2)$; determine the derivative.

$$\frac{dy}{dx} = (2x)(x + 2) + (x^2 + 3)(1)$$

Example: $z = u^2(u - 3)$; determine values of u such that the slope of the function is 0.

$$\frac{dz}{du} = 2u(u - 3) + u^2(1) = 3u^2 - 6u$$

The slope of z is 0 for values of u that make the derivative equal 0. The derivative is equated with 0 to determine the values of u.

$$3u^2 - 6u = 0$$

$$3u(u - 2) = 0$$

The function z has a slope of 0 for $u = 0$ and for $u = 2$.

Rule 7. The derivative of the quotient of two functions is equal to the denominator times the derivative of the numerator minus the numerator times the derivative of the denominator, all divided by the square of the denominator.

$$h(x) = \frac{f(x)}{g(x)}, \qquad h'(x) = \frac{g(x)f'(x) - f(x)g'(x)}{g(x)^2}$$

Example: $h(x) = \dfrac{2x^2 + x}{x^3 - 3}$; determine the derivative.

$$h'(x) = \frac{(x^3 - 3)(4x + 1) - (2x^2 + x)(3x^2)}{(x^3 - 3)^2}$$

Example: $h(x) = \dfrac{(2x + 3)}{(x + 2)}$; determine the slope at $x = 2$.

$$h'(x) = \frac{(x + 2)(2) - (2x + 3)(1)}{(x + 2)^2} = \frac{1}{(x + 2)^2}$$

$$h'(2) = \frac{1}{(2 + 2)^2} = \frac{1}{16}$$

Example: $y = \dfrac{x}{x^2}$; determine the derivative.

$$\frac{dy}{dx} = \frac{x^2(1) - x(2x)}{x^4} = \frac{-x^2}{x^4} = \frac{-1}{x^2}$$

Note that this checks with the result obtained by reducing the function to $y = x^{-1}$ and using Rule 3.

9.4.2   DERIVATIVES OF COMPOSITE FUNCTIONS

A composite function occurs when one function can be considered as a variable in another function. That is,

$$y = f(g(x)) \tag{9.7}$$

can be decomposed into

$$y = f(u) \quad \text{and} \quad u = g(x)$$

Example: $y = \sqrt{x^2 + 2x + 4}$ can be decomposed into

$$y = \sqrt{u} \quad \text{and} \quad u = x^2 + 2x + 4$$

Example: $y = (2x + 1)^3$ can be decomposed into

$$y = u^3 \quad \text{and} \quad u = 2x + 1$$

Example: $y = (x + 2)^3$ can be decomposed into

$$y = u^3 \quad \text{and} \quad u = x + 2$$

Example: $y = (4 - x^3)^6$ can be decomposed into

$$y = u^6 \quad \text{and} \quad u = 4 - x^3$$

Derivatives of composite functions of the form

$$y = f(g(x))$$

where

$$y = f(u) \quad \text{and} \quad u = g(x)$$

can be obtained by first decomposing the function and applying the rule for determining the derivatives of such functions. This rule, given below, is termed the *chain rule*.

Rule 8. If y is a differentiable function of u, $y = f(u)$, and u is a differentiable function of x, $u = g(x)$, then

$$\frac{dy}{dx} = \frac{dy}{du} \cdot \frac{du}{dx}$$

Example: $y = (3x)^2$; determine $\frac{dy}{dx}$. Let

$$y = u^2 \quad \text{and} \quad u = 3x$$

$$\frac{dy}{dx} = \frac{dy}{du} \cdot \frac{du}{dx}$$

where

$$\frac{dy}{du} = 2u; \qquad \frac{du}{dx} = 3$$

$$\frac{dy}{dx} = (2u)3 = 6u, \quad \text{or} \quad \frac{dy}{dx} = 6(3x) = 18x$$

Example: $y = (x^2 + 3x + 4)^4$. Let

$$y = u^4 \quad \text{and} \quad u = x^2 + 3x + 4$$

$$\frac{dy}{dx} = \frac{dy}{du} \cdot \frac{du}{dx}$$

where

$$\frac{dy}{du} = 4u^3; \qquad \frac{du}{dx} = 2x + 3$$

$$\frac{dy}{dx} = 4u^3(2x + 3) = 4(x^2 + 3x + 4)^3(2x + 3)$$

We shall use the chain rule to develop a rule for determining the derivative of composite functions. We have shown that a composite function of the type

$$y = (x^2 + 4x)^3$$

can be decomposed and written as $y = u^3$ and $u = x^2 + 4x$. If y is a differentiable function of u and u is a differentiable function of x, then $y = (x^2 + 4x)^3$ has a derivative of y with respect to x, which is given by the chain rule. Thus,

$$\frac{dy}{dx} = \frac{dy}{du} \cdot \frac{du}{dx} = 3u^2(2x + 4) = 3(x^2 + 4x)^2(2x + 4)$$

Examination of this derivative verifies the following general rule for differentiating composite functions.

Rule 9. The derivative of a function raised to the power n is equal to the power n times the function raised to the power $n - 1$, all multiplied by the derivative of the function.

$$h(x) = [f(x)]^n \qquad h'(x) = n[f(x)]^{n-1} \cdot f'(x)$$

Example: $h(x) = [f(x)]^n = (x^2 + 4x)^3$; determine the derivative.
$$h'(x) = n[f(x)]^{n-1}f'(x) = 3(x^2 + 4x)^2(2x + 4)$$

Example: $y = (x^3 + 2x^2 + 4)^5$; determine the derivative.

$$\frac{dy}{dx} = 5(x^3 + 2x^2 + 4)^4(3x^2 + 4x)$$

Example: $y = (x^2 - 4x + 4)^3$; determine the derivative.

$$\frac{dy}{dx} = 3(x^2 - 4x + 4)^2(2x - 4)$$

Example: $y = \sqrt{(2x^2 - x + 4)}$; determine the slope at $x = 2$.
$$y = (2x^2 - x + 4)^{1/2}$$
$$\frac{dy}{dx} = \frac{1}{2}(2x^2 - x + 4)^{-1/2}(4x - 1) = \frac{4x - 1}{2\sqrt{2x^2 - x + 4}}$$
$$\frac{dy}{dx}(x = 2) = \frac{8 - 1}{2\sqrt{8 - 2 + 4}} = \frac{7}{2\sqrt{10}} = 1.11$$

9.4.3 COMBINING THE RULES

The rules for determining derivatives discussed in this chapter apply to functions of the form

$$y = a_n x^n + a_{n-1} x^{n-1} + a_{n-2} x^{n-2} + \ldots + a_0 \qquad (9.8)$$

where $a_n, a_{n-1}, \ldots, a_0$ represent constants and x represents the independent variable. Functions of this form are termed *algebraic* functions. The derivative of the function is determined by combining the appropriate rules for derivatives given in this chapter and applying these rules to the function.

Example: $y = x^4 + 4x^3 - 2x^2 + 10x - 25$

$$\frac{dy}{dx} = 4x^3 + 12x^2 - 4x + 10$$

Example: $y = 3x^{3/2} - x^{1/2} + 2x^{-1/2} + 10$

$$\frac{dy}{dx} = \frac{9}{2}x^{1/2} - \frac{1}{2}x^{-1/2} - x^{-3/2}$$

Algebraic functions also are of the form

$$y = a_n u_n^n + a_{n-1} u_{n-1}^{n-1} + \ldots + a_0 \tag{9.9}$$

where $a_n, a_{n-1}, \ldots, a_0$ again represent constants. $u_n, u_{n-1}, \ldots$ are functions of x. Note that u_n and u_{n-1} need not have the same functional form. The derivative of y with respect to x is determined by applying the appropriate rules for derivatives.

Example: $y = 6(x^2 + 2x + 10)^4 + 100x^2 + 500$

$$\frac{dy}{dx} = 24(x^2 + 2x + 10)^3(2x + 2) + 200x$$

Example: $y = 4(2x + 2)^2 + 6(2x + 2) + 2$

$$\frac{dy}{dx} = 8(2x + 2)(2) + 6(2)$$

$$\frac{dy}{dx} = 16(2x + 2) + 12$$

Example: $y = 3(x^2 + 2x)^3 + 4(x + 1)^2$

$$\frac{dy}{dx} = 9(x^2 + 2x)^2(2x + 2) + 8(x + 1)$$

Still another type of algebraic function is of the form

$$y = a_n f_n(x)g_n(x) + a_{n-1}f_{n-1}(x)g_{n-1}(x) + \ldots + a_0 \tag{9.10}$$

where $a_n, a_{n-1}, \ldots, a_0$ again represent constants, and $f_n(x)$ and $g_n(x)$ represent functions of x. The derivative is determined by combining the appropriate rules as illustrated by the following examples.

Example: $y = 3x^2(x^2 + 2x)^3 + 2x(x + 5)^2$

$$\frac{dy}{dx} = 6x(x^2 + 2x)^3 + 3x^2[3(x^2 + 2x)^2(2x + 2)]$$
$$+ 2(x + 5)^2 + 2x[2(x + 5)(1)]$$

Example: $y = (x + 3)^2(x^2 + 2x)^3$

$$\frac{dy}{dx} = 2(x + 3)(x^2 + 2x)^3 + (x + 3)^2[3(x^2 + 2x)^2(2x + 2)]$$

These examples, it should be noted, required the use of the derivative rules for a constant times a function, the sum of two functions, the product of two functions, and the composite function. In fact, the only rule discussed in this chapter that was not applied in one of the preceding two examples was that for the quotient of two functions. We shall next consider algebraic functions whose derivative requires application of the quotient rule.

The function

$$y = \frac{f(x)}{g(x)}$$

where $f(x)$ and $g(x)$ are functions similar to those defined by (9.8), (9.9), or (9.10), is an algebraic function. Again, the derivative of this function is determined by application of the appropriate rules of derivatives. This type of function is illustrated by the following examples.

Example: $y = \frac{(x^2 + 3x)^3}{(x^3 + 4x^2)^2}$; determine the derivative. To determine the

derivative of this function requires application of the quotient rule, the composite function rule, the power rule, and the rule for a constant times a function. The derivative of the function is

$$\frac{dy}{dx} = \frac{(x^3 + 4x^2)^2[3(x^2 + 3x)^2(2x + 3)] - (x^2 + 3x)^3[2(x^3 + 4x^2)(3x^2 + 8x)]}{(x^3 + 4x^2)^4}$$

Example: $y = \frac{(2x^3 + 4x^2)^2}{x^2 + 3}$; determine the derivative. We again apply

the rules illustrated in this chapter. The derivative is

$$\frac{dy}{dx} = \frac{(x^2 + 3)[2(2x^3 + 4x^2)(6x^2 + 8x)] - (2x^3 + 4x^2)^2(2x)}{(x^2 + 3)^2}$$

These examples illustrate determining the derivatives of algebraic functions. The procedure is to apply the appropriate combination of rules as required by the function. This, of course, requires the ability to recognize the functional form of the function and to be able to select the appropriate rule. This skill normally comes with a reasonable amount of practice.

9.4.4 INVERSE FUNCTIONS

The algebraic functions discussed in this chapter have been of the form $y = f(x)$, where y is the dependent variable and x is the independent variable. If, given the function $y = f(x)$, we rewrite the function with x as the dependent variable and y as the independent variable, the resulting function $x = g(y)$ is termed the inverse of $y = f(x)$.

As an example of inverse functions, consider the function $y = 4x + 10$. The inverse of this function is $x = 0.25y - 2.5$. The inverse was found by solving the function for x in terms of y; that is, $x = g(y)$.

The functions considered in this chapter have been *explicit* functions. Functions such as $y = 4x - 3$, in which the dependent variable is clearly designated as y and the independent variable as x, are termed explicit functions. A function can also be defined *implicitly*. As an example, the equation $5x - 2y = 14$ implicitly defines x in terms of y. If y is allowed to assume a certain value, then x is implicitly defined by the equation. Similarly, y is an implicit function of x in that if x is permitted to take on certain values, the value of y is implicitly established by the equation. If we are given the equation $h(x, y) = 0$, then $y = f(x)$ and $x = g(y)$ are inverse functions. The relationship between implicit functions and inverse functions is illustrated by the following examples.

Example: Consider the implicit function $3x + 4y - 6 = 0$. Determine $y = f(x)$ and $x = g(y)$ for this function. The explicit function of y in terms of x is

$$y = -0.75x + 1.5$$

and the explicit function of x in terms of y is

$$x = -1.33y + 2$$

$y = f(x)$ and $x = g(y)$ are inverse functions.

Example: For the implicit function $x + 2y - 10 = 0$, determine the explicit functions $y = f(x)$ and $x = g(y)$. The explicit function $y = f(x)$ is

$$y = 5 - 0.5x$$

and the explicit function $x = g(y)$ is

$$x = 10 - 2y$$

Example: Determine the inverse of the explicit function $y = x^2$. This is an example of a function that has more than one inverse function. By expressing x in terms of y, we find that

$$x = +\sqrt{y} \quad \text{and} \quad x = -\sqrt{y}$$

are both inverse functions of $y = x^2$.

It is not necessary to determine the inverse of a function in order to determine the derivative of that inverse. The derivatives of inverse functions can be determined by application of the following rule.

Rule 10. If $x = g(y)$ is the inverse function of $y = f(x)$, the derivative of the inverse function, if it exists, is given by the reciprocal of the derivative of $y = f(x)$. Thus

$$\frac{dx}{dy} = \frac{1}{dy/dx}, \quad \text{provided that} \quad \frac{dy}{dx} \neq 0$$

The rule for determining derivatives of inverse functions is illustrated by the following examples.

Example: For the implicit function $y - 3x - 4 = 0$, determine $y = f(x)$, dy/dx, $x = g(y)$, and dx/dy. The explicit function $y = f(x)$ is

$$y = 3x + 4$$

and has the derivative

$$\frac{dy}{dx} = 3$$

The explicit function $x = g(y)$ is

$$x = \frac{y}{3} - \frac{4}{3}$$

and has the derivative

$$\frac{dx}{dy} = \frac{1}{3}$$

The derivative dx/dy can also be determined from the rule for derivatives of inverse functions. Thus

$$\frac{dx}{dy} = \frac{1}{dy/dx} = \frac{1}{3}$$

Example: For $y = 2x^2 + 3x + 6$, determine dy/dx and dx/dy.

$$\frac{dy}{dx} = 4x + 3$$

From the rule for derivatives of inverse functions

$$\frac{dx}{dy} = \frac{1}{dy/dx} = \frac{1}{4x + 3}, \quad \text{provided that} \quad x \neq -\frac{3}{4}$$

Example: For the quadratic function $y = x^2 + 2x + 6$, determine both dy/dx and dx/dy. The derivative of y with respect to the independent variable x is

$$\frac{dy}{dx} = 2x + 2$$

The derivative of x with respect to the independent variable y is

$$\frac{dx}{dy} = \frac{1}{dy/dx} = \frac{1}{2x + 2}, \quad \text{provided that} \quad x \neq -1$$

PROBLEMS

1. Determine the derivative of the following functions.
   (a) $f(x) = x^2$ (b) $f(x) = 2x^2$
   (c) $f(x) = x^5$ (d) $f(x) = x^{21}$
   (e) $f(x) = x^{-4}$ (f) $f(x) = x^{-1/2}$
   (g) $f(x) = x^{1/2}$ (h) $f(x) = x^{0.35}$
   (i) $f(x) = x^{1/4}$ (j) $f(x) = x^{-1/4}$
   (k) $f(x) = 1/x^2$ (l) $f(x) = 1/x^4$
   (m) $f(x) = x^n$ (n) $f(x) = x^a$
   (o) $f(x) = a^b$ (p) $f(x) = b^a$
   (q) $f(x) = \sqrt{x}$ (r) $f(x) = 3\sqrt{x}$

2. Determine the derivative of the following functions.
   (a) $f(x) = 4x$ (b) $f(x) = x/2$
   (c) $f(x) = -3x$ (d) $f(x) = 2x^{1/2}$
   (e) $f(x) = x^4/4$ (f) $f(x) = 3x^2/2$
   (g) $f(x) = 2x^{-3}/5$ (h) $f(x) = 6x^{-1}/3$
   (i) $f(x) = ax^b$ (j) $f(x) = -ax^{-b}$
   (k) $f(x) = 3/x^2$ (l) $f(x) = 4/x^7$
   (m) $f(x) = k/x^{1/2}$ (n) $f(x) = ab/x^4$

3. Determine the derivative of the following functions.
   (a) $f(x) = 3x^2 + 2x + 5$ (b) $f(x) = 4x^3 + 3x^2 - 10$
   (c) $f(x) = 25x^3 + 16x^2 + 6x + 7$ (d) $f(x) = -6x^{-2} - 5x^{-3}$
   (e) $f(x) = 3x^{1/4} + 2x^{1/2}$ (f) $f(x) = 1/x^2 + 3/x^3$
   (g) $f(x) = ax^4 + bx^3 + cx^2 + dx + k$
   (h) $f(x) = 5x^5 + 6x^4 - 10x^3 + 16x^2 + 25x + 100$

4. Determine the derivative of the following functions.
   (a) $f(x) = 2x^2(x + 3)$ (b) $f(x) = 3x(x^3 + 2x + 4)$
   (c) $f(x) = (x + 6)(2x + 3)$ (d) $f(x) = (x^2 + 2x)(x^3 + 3x + 5)$
   (e) $f(x) = (x^5 - 6)(x^4 - 5)$ (f) $f(x) = (x^{1/2} + x)(x^{1/4} + 2x)$
   (g) $f(x) = (x^3 - 2x^2 + 4x + 10)(x^2 + 4x + 17)$
   (h) $f(x) = (x^{-2} + 4x^{-3})(x^3 + 6x^2 + 12x + 24)$

5. Determine the derivative of the following functions.
   (a) $f(x) = \dfrac{x - 2}{x + 3}$ (b) $f(x) = \dfrac{2x + 4}{x + 6}$

   (c) $f(x) = \dfrac{2x^2 - 5x}{x^2 + 2}$ (d) $f(x) = \dfrac{2x^3 + 3x^2 + 2x}{x^2 + 2x + 1}$

   (e) $f(x) = \dfrac{x^{-3} + x^{-2}}{x^4 + 3x^2}$ (f) $f(x) = \dfrac{x^{-1/2} + x^{-1/2}}{x^{-1/4} + x^{-1/2}}$

(g) $f(x) = \dfrac{(x^2 + 4x)(x^3 + 3x^2)}{x^2 + 2x}$　　(h) $f(x) = \dfrac{2x^2(x^4 + 4x)}{x + 3}$

(i) $f(x) = \dfrac{(x^2 + 5x)(3x + 4)x^2}{(x + 3)(x^2 + 3x)}$

6. Determine the derivative of the following functions.

(a) $f(x) = (x^2 + 2x)^3$　　　　　　　　(b) $f(x) = (x^3 + 3x^2 + 2x + 5)^4$
(c) $f(x) = (2x + 3)^{1/2}$　　　　　　　　(d) $f(x) = (x^2 + 4x)^{-5}$
(e) $f(x) = (x^4 - 4x^2)^3$　　　　　　　(f) $f(x) = x^2(4x^3 + 3x)^2$
(g) $f(x) = (x^2 + 3)^3(x + 4)^2$　　　　(h) $f(x) = (x + 3)^2(x^2 - 3)^{-2}$
(i) $f(x) = (x + 3)^{1/2}(x + 3)^{-1/2}$
(j) $f(x) = (6x^3 + 3x)^{-3}(x^2 + 4x + 7)$

7. Find the slope of the function at the specified value of the independent variable.

(a) $f(x) = x^2 + 2x,$ 　　　　$f'(3) =$
(b) $f(x) = 3x^2 + 2x - 5,$ 　　$f'(4) =$
(c) $f(x) = 2x^2 - 3x + 6,$ 　　$f'(1) =$
(d) $f(x) = (x + 3)^3,$ 　　　　$f'(5) =$
(e) $f(x) = (2x^2 + 4x)^{1/2},$ 　$f'(2) =$
(f) $f(x) = x^3 + 2x^2 + 6x,$ 　$f'(-2) =$
(g) $f(x) = \sqrt{3x + 4},$ 　　　$f'(-1) =$

(h) $y = 2x^3 + 3x,$ 　　　　$\dfrac{dy}{dx}(x = 3) =$

(i) $y = (x^2 + 5)^2,$ 　　　　$\dfrac{dy}{dx}(x = 2) =$

(j) $y = (x^2 + 2x)^{-2},$ 　　　$\dfrac{dy}{dx}(x = 1) =$

8. Find the value of the independent variable for which the slope of the function is 0.

(a) $f(x) = x^2 - 3x + 6$　　　　　(b) $f(x) = x^2 + 4x + 8$
(c) $f(x) = x^2 - 5x + 10$　　　　(d) $f(x) = x^2 - 2x + 4$

(e) $f(x) = \dfrac{x^3}{3} + \dfrac{3x^2}{2}$　　　　　(f) $f(x) = x^3 - 2x^2$

(g) $f(x) = \dfrac{x^3}{3} + \dfrac{x^2}{2} - 12x$　　　(h) $f(x) = \dfrac{x^3}{3} - \dfrac{5x^2}{2} + 6x + 10$

SUGGESTED REFERENCES

AYRES, FRANK, JR., *Calculus*, 2nd ed., Schaum's Outline Series (New York, N.Y.: McGraw-Hill Book Company, Inc., 1964).

DE LEEUW, KAREL, *Calculus* (New York, N.Y.: Harcourt, Brace, & World, Inc., 1966).

Fisher, R. C. and A. D. Ziebur, *Calculus and Analytic Geometry*, 2nd ed. (Englewood Cliffs, N.J.: Prentice-Hall, Inc., 1965).

Frank, Peter, et al., *A Brief Course in Calculus with Applications* (New York, N.Y.: Harper and Row, Publishers, 1971).

Meyer, Herman and Robert V. Mendenhall, *Techniques of Differentiation and Integration: A Program for Self-instruction* (New York, N.Y.: McGraw-Hill Book Company, Inc., 1966).

Rodin, Burton, *Calculus with Analytic Geometry* (Englewood Cliffs, N.J.: Prentice-Hall, Inc., 1970).

Shockley, James E., *The Brief Calculus* (New York, N.Y.: Holt, Rinehart and Winston, Inc., 1971).

Whipkey, K. L. and M. N. Whipkey, *The Power of Calculus* (New York, N.Y.: John Wiley and Sons, Inc., 1972).

Youse, B. K. and A. W. Stalnaker, *Calculus for the Social and Natural Sciences* (Scranton, Pa.: International Textbook Company, 1969).

Chapter 10

Optimization
Using Calculus

10.1 Optimum Values of Functions

In this chapter we develop a technique based upon differential calculus for determining the optimum value of a function. For example, suppose that profit and quantity are related as shown in Fig. 10.1. Our objective is to determine the quantity so that profits are a maximum.

Similarly, we might want to minimize the cost of purchasing an object. When both storage costs and ordering costs are taken into account, the cost of purchasing and storing the inventory varies with the quantity ordered as shown in Fig. 10.2. Our objective is to determine the quantity so that cost is a minimum. It is customary to use the expressions value of the function and value of the dependent variable interchangeably.

The optimum value of a function can be either a maximum or a minimum. A function is said to reach a local maximum at the value $x = a$ of the independent variable if the value of the dependent variable at $x = a$ is greater than the value of the dependent variable at any adjacent point. Similarly, a function reaches a local minimum at the point $x = a$ if the value of the dependent variable at that point is less than the dependent variable at adjacent points.

As an example of local optimum, consider the function graphed in Fig. 10.3

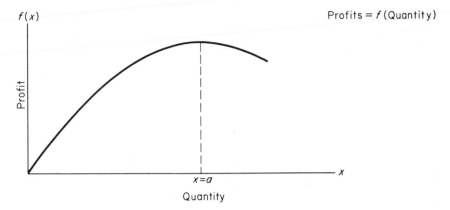

Figure 10.1

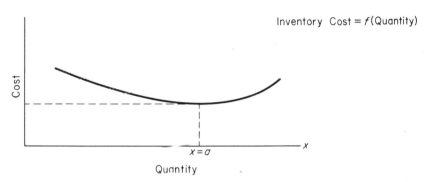

Figure 10.2

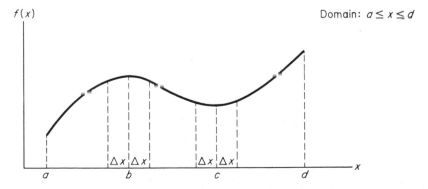

Figure 10.3

The function evaluated at $x = b$ is a local maximum, since the value of the function is greater at $x = b$ than at any adjacent points, shown as $b + \Delta x$ and $b - \Delta x$. Similarly, the function evaluated at $x = c$ is a local minimum. The quantity Δx is an arbitrarily small increment of the analyst's choosing.

What about the value of the dependent variable at $x = a$ and $x = d$? Clearly, the value of the dependent variable is less at $x = a$ than at $x = c$. Also, the value of the dependent variable is greater at $x = d$ than at $x = b$. The function evaluated at $x = a$ is thus termed the *absolute minimum*, and the function evaluated at $x = d$ is termed the *absolute maximum*. An absolute minimum of the function occurs for that value of $x = a$ for which the function evaluated at $x = a$ is smaller than for any other value of x in the domain. Similarly, an absolute maximum of the function occurs for that value of $x = d$ for which the function evaluated at $x = d$ is larger than for any other value of x in the domain.

Example: Specify all maximum and minimum values of $f(x)$ for the following function.

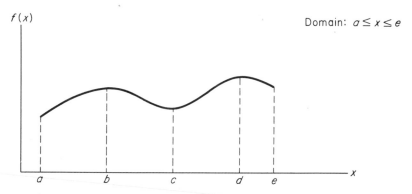

It can be seen that $f(a) =$ absolute minimum, $f(b) =$ local maximum, $f(c) =$ local minimum, and $f(d) =$ absolute maximum.

In the preceding example, the domain of the function is $a \leq x \leq e$. When the end points of the domain, such as a and e, are included as possible values of x, these values of x must be evaluated to determine if the dependent variable is an absolute maximum or minimum. When the end points are included in the domain, the interval between the end points is termed *closed*. If the end points are not included, the interval is said to be *open*. If one end point is included and the other is not, the interval is said to be *mixed*. A closed interval is written as $a \leq x \leq e$, the open interval is $a < x < e$, and a mixed interval is $a \leq x < e$ or $a < x \leq e$.

End points are not considered as candidates for *local* maxima or minima. The reason for excluding end points is that the function is not defined at both adjacent points. For instance, in the preceding example the function is not defined for values of x larger than $x = e$ or smaller than $x = a$. End points are, however, candidates for *absolute* maxima and minima and must be evaluated. The examples that follow illustrate the concept of optimum values of functions in both open and closed domains of the function.

Example: Find the maximum and minimum value of $f(x)$ with the domain $a \leq x \leq b$.

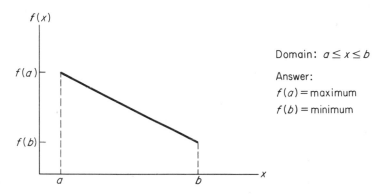

Domain: $a \leq x \leq b$

Answer:
$f(a) = $ maximum
$f(b) = $ minimum

Example: Determine the maximum and minimum values of $f(x)$ with the domain $a < x < d$.

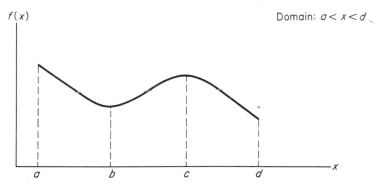

Domain: $a < x < d$

By convention, end points are not considered in determining absolute maxima and minima if the domain of the function is an open interval. Therefore, $f(b)$ is the local minimum and $f(c)$ is the local maximum. We do not specify absolute maxima or minima for functions for which the domain of the function is an open interval.

The preceding example illustrates the case in which the end points of the domain of the function are not considered in determining optimum values of the function. It is quite common in business situations for the analyst to be interested only in local maximum or minimum values of the function. In such cases the end points of the domain are not specified. In this text, we adopt the convention that a function cannot be said to achieve an *absolute* maximum or minimum unless the end points of the independent variable are given and the function is evaluated at these end points. If the end points of the domain are not specified, the function can reach local maxima or minima but not absolute maxima or minima.

If the end points are of no consequence in a problem, the domain of the function is often not specified. For functions for which the domain is not specified, we shall assume that the domain of the function is the open interval $-\infty < x < \infty$. Since the interval is open, we can determine values of the independent variable x for which the function reaches a local optimum. We follow the convention stated above, however, and do not specify an absolute maximum or minimum of a function for which the domain is not explicitly stated.

The domain of the independent variable is often determined from the structure of the problem. For instance, if the independent variable is manufacturing output, the domain of output would be limited on the upper side by the output capacity of the plant. The lower limit of the domain could possibly be that level of output such that the production costs are covered. An example of this type of problem follows.

Example: The Far Western Manufacturing Company is engaged in the business of manufacturing souvenirs that are sold in various shops throughout the western states. The company has found that the volume in standard units of sales is highly dependent upon price; the lower the price the higher the volume. They have also found that if the price is too low, profits tend to fall in spite of an increased volume of sales. Assume that we have been assigned the task of recommending a procedure for establishing the proper volume-profit relationship.

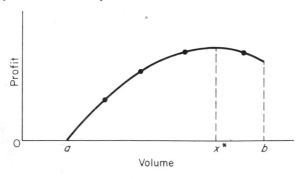

Profit $= f(\text{Volume})$
Domain: $a \leq x \leq b$

An explanation of the price-profit-volume relationship is presented in Chapter 11. At this point, however, our understanding of the relationship between these variables is sufficient to recommend a study of the historical relationship between changes in profit and volume. Assume that, based upon this study, the profit-volume data plotted on the accompanying graph are obtained. By drawing a curve through these data points, we obtain profit as a function of volume. The volume that leads to maximum profit is indicated by x^*. The price that must be charged to obtain this volume can be determined from company records. The curve is a quadratic function. If data points were given in this example, the parameters of the quadratic function could be determined by the method explained in Chapter 2.

10.2   Interpretation of Derivatives

The derivative of a function gives the slope of the function for alternative values of the independent variable. The slope of the function at a particular value of $x = a$ is determined by evaluating the derivative function at the particular value of $x = a$. Expressed in mathematical notation, the slope is given by $f'(a)$ or $\frac{dy}{dx}(x = a)$.

In understanding the derivative, it is helpful to examine graphs of several functions. First consider the quadratic function

$$y = x^2 - 6x + 8 \qquad \text{for } 0 \leq x \leq 6$$

The graph of this function is shown in Fig. 10.4(a). This quadratic function plots with the minimum value of y occurring at $x = 3$. The slope of the function is negative for values of x less than 3 and positive for values of x greater than 3. The slope of the function is 0 for $x = 3$.

The derivative of the function is $\frac{dy}{dx} = 2x - 6$. This derivative function can be used to determine the slope of the original function, $y = x^2 - 6x + 8$, at all values of x in the domain. For example, the slope of the original function for $x = 0$ is -6. Similarly, the slope for $x = 1$ is $\frac{dy}{dx}(x = 1) = -4$. The slope of the original function for any value of x in the domain can be determined simply by evaluating the derivative function for that value of x.

From Fig. 10.4(a) it is apparent that the function reaches a minimum at $x = 3$. This can also be seen from the plot of the derivative in Fig. 10.4(b). For values of x less than 3, the derivative is negative. This indicates that the slope of the original function is negative for x less than 3. For values of x greater than 3, the derivative is positive. This indicates that the slope of the original function is positive for x greater than 3. From Fig. 10.4(a), it can be

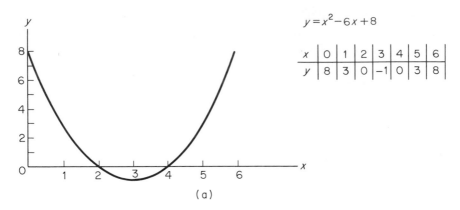

$$y = x^2 - 6x + 8$$

x	0	1	2	3	4	5	6
y	8	3	0	-1	0	3	8

(a)

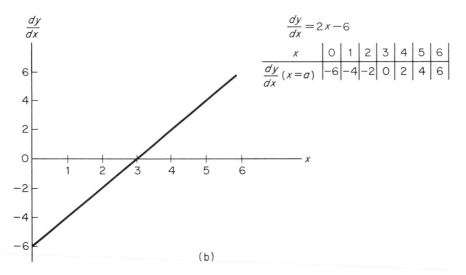

$$\frac{dy}{dx} = 2x - 6$$

x	0	1	2	3	4	5	6
$\frac{dy}{dx}(x=a)$	-6	-4	-2	0	2	4	6

(b)

Figure 10.4

seen that a function whose slope goes from negative to positive reaches a minimum value. This minimum occurs at that value of x for which the slope of the original function is 0. In our example, the slope of the function $y = x^2 - 6x + 8$ is 0 at $x = 3$.

Values of the independent variable for which the slope of the function equals 0 are termed *critical points*. In the last example, $x = 3$ is a critical point. These values of x can be determined from the derivative. To determine the critical point we equate the derivative to 0 and solve for x. For our example

$$\frac{dy}{dx} = 2x - 6 = 0$$

Therefore,

$$x = 3$$

Critical points are designated by an asterisk, i.e., $x^* = 3$.

A critical point is a necessary requirement for a local optimum. It does not follow, however, that all critical points result in local optima. Certain critical points result in *points of inflection* of a function. Since critical points can result in points of inflection rather than local optima, a critical point is a necessary but not sufficient condition for a local optimum. It is necessary in that a function cannot reach a local optimum for a value of x unless the slope of the function is 0 for that value of x. A critical point is not sufficient, however, in that critical points can result in points of inflection of a function rather than local optima. A critical point is thus a candidate for a local optimum. Each critical point must be investigated to determine if the critical point is a local optimum or a point of inflection. A critical point that leads to a point of inflection of a function is given by the following example.

Example: Investigate the function $y = x^3$ for local optima. The derivative of the function is $\dfrac{dy}{dx} = 3x^2$. Equating the derivative function with 0 and solving for the critical point (that is, the value of x such that the derivative function equals 0) gives the critical point $x^* = 0$. From a plot of the function,

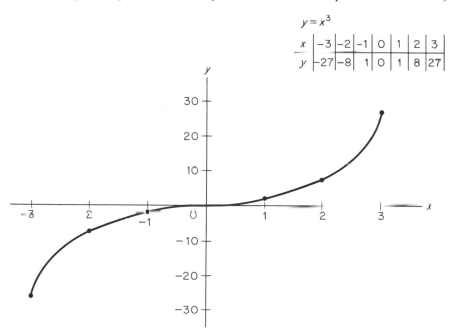

$$y = x^3$$

x	-3	-2	-1	0	1	2	3
y	-27	-8	1	0	1	8	27

it can be seen that the critical point $x^* = 0$ is not a local optimum. Rather, the function reaches a point of inflection at this critical point. Points of inflection are defined in Sec. 10.2.4 of this chapter.

We can summarize the requirements for a local optimum.† A function reaches a local optimum at the critical point $x^* = a$ when $f'(a) = 0$, provided that one of the following conditions is met: (1) The function is a local maximum if $f'(a - \Delta x) > 0$ and $f'(a + \Delta x) < 0$. (2) The function is a local minimum if $f'(a - \Delta x) < 0$ and $f'(a + \Delta x) > 0$. The function reaches an inflection point rather than a local optimum at the critical point $x = a$ if $f'(a - \Delta x)$ and $f'(a + \Delta x)$ are both greater than 0 or are both less than 0.

The relationship between a function and its derivative is further illustrated by the cubic function

$$y = \tfrac{1}{3}x^3 - 4x^2 + 12x + 5$$

The function is plotted in Fig. 10.5(a), and the derivative function is plotted in Fig. 10.5(b). The function reaches a local maximum at $x = 2$ and a local minimum at $x = 6$. For values of x less than 2, the derivative is positive. This indicates that the function has a positive slope for values of x less than 2. For x greater than 2 but less than 6, the derivative is negative, thus indicating that the slope of the function is negative. Since the slope of the function goes from positive to negative and passes through 0 at $x = 2$, we conclude from the derivative that the function achieves a local maximum at $x = 2$. This conclusion is, of course, verified by inspection of the plot of the function in Fig. 10.5(a).

For values of x between $x = 2$ and $x = 6$, the derivative is negative. For x greater than 6 the derivative is positive. A function whose slope changes from negative to positive necessarily reaches a local minimum. The minimum occurs when the derivative is 0. Thus, from inspection of the derivative at various values of x, we conclude that the function reaches a local minimum at $x = 6$. This conclusion is verified by inspection of the plot of the function.

The values of x for which the function has a slope of 0 can be determined by equating the derivative with 0 and solving the resulting equation for x. In our example,

$$\frac{dy}{dx} = x^2 - 8x + 12 = 0$$

which can be factored to give

$$(x - 2)(x - 6) = 0$$

---

† The use of the second derivative to establish maxima and minima is discussed in Sec. 10.2.3.

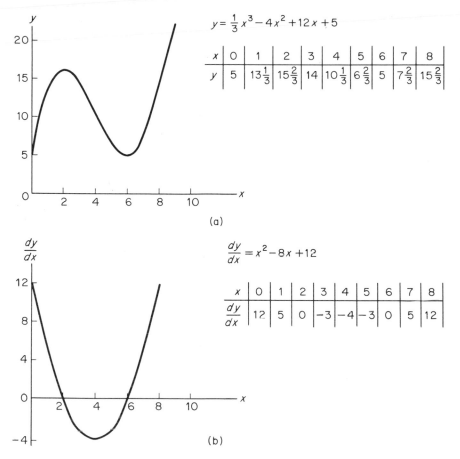

$$y = \frac{1}{3}x^3 - 4x^2 + 12x + 5$$

x	0	1	2	3	4	5	6	7	8
y	5	$13\frac{1}{3}$	$15\frac{2}{3}$	14	$10\frac{1}{3}$	$6\frac{2}{3}$	5	$7\frac{2}{3}$	$15\frac{2}{3}$

(a)

$$\frac{dy}{dx} = x^2 - 8x + 12$$

x	0	1	2	3	4	5	6	7	8
$\frac{dy}{dx}$	12	5	0	-3	-4	-3	0	5	12

(b)

Figure 10.5

For the left side to equal 0, either $x = 2$ or $x = 6$. The function has a slope of 0 at both $x^* = 2$ and $x^* = 6$.

To determine if the critical point $x^* = 2$ is a local optimum, the derivative can be evaluated at $2 - \Delta x$ and at $2 + \Delta x$. If the slope is positive for $2 - \Delta x$ and negative for $2 + \Delta x$, the function has reached a local maximum at the critical point $x^* = 2$. If, however, the derivative is negative at $2 - \Delta x$ and positive at $2 + \Delta x$, the function has reached a minimum at $x^* = 2$. The same procedure can be applied for $x = 6$ to determine if the function is a maximum, minimum, or point of inflection at the critical point $x^* = 6$.

Example: Find critical points for the following function and determine if the function when valued at the critical points is a maximum, minimum, or point of inflection.

$$y = \frac{x^3}{3} + 4x^2 - 20x + 10$$

$$\frac{dy}{dx} = x^2 + 8x - 20 = 0$$

$$(x - 2)(x + 10) = 0$$

Therefore, $x^* = 2$ and $x^* = -10$ are critical points. To determine if $x^* = 2$ is a local optimum, use $\Delta x = 0.01$ and evaluate the derivative function at $x - \Delta x$ and $x + \Delta x$.

$$\frac{dy}{dx} (x = 1.99) = (1.99)^2 + 8(1.99) - 20$$

$$= 3.9601 + 15.9200 - 20 = -0.1199$$

$$\frac{dy}{dx} (x = 2.01) = (2.01)^2 + 8(2.01) - 20$$

$$= 4.0401 + 16.08 - 20 = 0.1201$$

We conclude that for $x^* = 2$ the function is a local minimum since the slope of the function is less than 0 at $x - \Delta x$, equals 0 at the critical point, and is greater than 0 at $x + \Delta x$. By a similar process, we find that the function is a local maximum for $x^* = -10$. The value of the function for $x^* = -10$ and $x^* = 2$ is

$$y(-10) = 276.7 \quad \text{and} \quad y(2) = -11.3$$

Example: Determine the values of x such that the slope of y is 0.

$$y = x^3 - 6x^2 + 9x + 10$$

$$\frac{dy}{dx} = 3x^2 - 12x + 9 = 0$$

Dividing through by 3 gives

$$x^2 - 4x + 3 = 0$$

which can be factored to yield

$$(x - 3)(x - 1) = 0$$

Thus

$$x^* = 3, \qquad x^* = 1$$

Example: Determine the values of x such that the slope of y is 0.

$$y = -2x^3 + 4x^2 + 16x + 20$$

$$\frac{dy}{dx} = -6x^2 + 8x + 16 = 0$$

$$-3x^2 + 4x + 8 = 0$$

This equation cannot be factored. It is possible, however, to determine the values of x that satisfy the equation by using the quadratic formula.

By way of review, the reader will remember that the quadratic function was given in Chapter 2 as

$$f(x) = a + bx + cx^2 \qquad (2.7)$$

where a, b, and c are parameters and x is the independent variable. The quadratic formula is used to determine the values of x for which the function is equal to zero. Equating $f(x)$ to zero in (2.7) gives

$$a + bx + cx^2 = 0 \qquad (10.1)$$

Equation (10.1) is termed the *quadratic equation*. The values of x that satisfy this equation are found either by factoring the equation or by using the quadratic formula. The quadratic formula is

$$x = \frac{-(b) \pm \sqrt{b^2 - 4ac}}{2c} \qquad (10.2)$$

Notice that two values of x are given by the quadratic formula. These values, called the *roots* of the equation, are

$$x = \frac{-(b) + \sqrt{b^2 - 4ac}}{2c}$$

and

$$x = \frac{-(b) - \sqrt{b^2 - 4ac}}{2c}$$

Comparison of the example problem and the general form of the quadratic equation shows that $c = -3$, $b = 4$, and $a = 8$. Therefore,

$$x = \frac{-4 \pm \sqrt{(4)^2 - 4(-3)(8)}}{2(-3)} = \frac{-4 \pm \sqrt{16 + 96}}{-6}$$

$$x = \frac{-4 \pm \sqrt{112}}{-6} - \frac{-4 \pm 10.58}{-6}$$

$$x = \frac{-4 + 10.58}{-6} = \frac{6.58}{-6} = -1.10$$

is one root, and

$$x = \frac{-4 - 10.58}{-6} = \frac{-14.58}{-6} = 2.43$$

is the other root. The slope of the function is 0 at $x = -1.10$ and $x = 2.43$.

Example: Determine the critical points of the following function and specify if the critical points are local optima.

$$y = 4x^3 - 15x^2 + 10x + 100$$

$$\frac{dy}{dx} = 12x^2 - 30x + 10 = 0$$

$$\frac{dy}{dx} = 6x^2 - 15x + 5 = 0$$

$$x = \frac{-(-15) \pm \sqrt{(15)^2 - 4(5)(6)}}{2(6)} = \frac{15 \pm \sqrt{225 - 120}}{12}$$

$$x = \frac{15 \pm \sqrt{105}}{12} = \frac{15 \pm 10.25}{12}$$

$$x = \frac{25.25}{12} = 2.10 \quad \text{and} \quad x = \frac{4.75}{12} = 0.40$$

The critical points of the function are $x^* = 2.10$ and $x^* = 0.40$. To determine if the critical points are local optima, the derivative is evaluated at $x - \Delta x$ and $x + \Delta x$. If we specify Δx as 0.10, the calculations for determining the maximum and minimum are as follows.
For $x = 2.10$:

$$\frac{dy}{dx}(x = 2.00) = 12(2.00)^2 - 30(2.00) + 10 = -2.00$$

$$\frac{dy}{dx}(x = 2.20) = 12(2.20)^2 - 30(2.20) + 10 = +2.08$$

For $x = 0.40$:

$$\frac{dy}{dx}(x = 0.30) = 12(0.30)^2 - 30(0.30) + 10 = +2.08$$

$$\frac{dy}{dx}(x = 0.50) = 12(0.50)^2 - 30(0.50) + 10 = -2.00$$

The function reaches a local maximum at $x = 0.40$ and a local minimum at $x = 2.10$. The local optima are

$$y(x = 0.40) = 4(0.40)^3 - 15(0.40)^2 + 10(0.40) + 100 = 101.9$$
$$y(x = 2.10) = 4(2.10)^3 - 15(2.10)^2 + 10(2.10) + 100 = 91.9$$

We can also determine if a function is a local maximum or minimum from the definitions of local maximum and local minimum. A function is said to reach a local maximum at the value $x = a$ if the function evaluated at $x = a$ is greater than the function evaluated at any adjacent point. The function is a local maximum at $x^* = a$ if $f(a - \Delta x) < f(a)$ and $f(a + \Delta x) < f(a)$. The function is a local minimum if $f(a - \Delta x) > f(a)$ and $f(a + \Delta x) > f(a)$. This procedure for specifying whether a local optimum is a maximum or minimum is illustrated in the following examples.

Example: The profits of the Woodly Manufacturing Company vary with the quantity of product produced and sold according to the profit function

$$f(x) = -x^2 + 600x - 10,000$$

where x represents quantity and $f(x)$ represents profits. Determine the quantity that leads to maximum profits.

The critical points of the profit function are determined by equating the derivative of the profit function with 0. Thus

$$f'(x) = -2x + 600 = 0$$

and

$$x^* = 300$$

Profits are a maximum for Woodly Manufacturing Company when $x^* = 300$ units are produced. The profits for this output are $f(300) = \$80,000$. The fact that this quantity is a maximum rather than a minimum or point of inflection is verified by determining profit for $x = 299$ units and $x = 301$ units.

Example: The average costs per unit of manufacturing for Brown, Inc. varies with the percentage of capacity utilized. At low levels of output, average costs tend to be relatively high because of inefficient combination of the factors of production. Similarly, high levels of output lead to costly overtime and other inefficient combinations of the factors of production. The average cost per unit for Brown, Inc. is described by the function

$$f(x) - 3500 - 16x + 0.02x^2$$

where x represents quantity and $f(x)$ represents cost. Determine the quantity that leads to minimum average cost.

The quantity for which costs are minimum is determined by equating the derivative of the average cost function with 0.

$$f'(x) = -16 + 0.04x = 0$$

and

$$x^* = 400$$

Costs are minimized when 400 units are produced per period. The average cost of the 400 units is $f(400) = \$300$. This is a local minimum, since $f(399)$ and $f(401)$ give higher average cost.

10.2.1 HIGHER-ORDER DERIVATIVES

We have shown that the derivative of a function, when evaluated at the point $x = a$, gives the slope of the function at the point $x = a$. It is also possible to differentiate the derivative. The derivative of the derivative is

termed the second derivative. The second derivative, when evaluated at the
point $x = a$, gives the slope of the first derivative at that point. This slope,
as we shall see later, provides useful information in the evaluation of the
original function.

A higher-order derivative is simply the derivative of a derivative. Thus, if

$$y = x^4 + 3x^3$$

is the function, then

$$\frac{dy}{dx} = 4x^3 + 9x^2$$

is the first derivative,

$$\frac{d^2y}{dx^2} = 12x^2 + 18x$$

is the second derivative,

$$\frac{d^3y}{dx^3} = 24x + 18$$

is the third derivative,

$$\frac{d^4y}{dx^4} = 24$$

is the fourth derivative, and

$$\frac{d^5y}{dx^5} = 0$$

is the fifth derivative. Each of these derivatives, when evaluated at the point
$x = a$, gives the slope of the preceding derivative at that point. The first
derivative describes the slope of the function, the second derivative describes
the slope of the first derivative, the third derivative the slope of the second
derivative, etc. In business and economic applications of differential calculus,
we seldom use derivatives other than the first and second derivatives of a
function. Consequently, our discussion of higher-order derivatives is limited
to the first and second derivatives.

10.2.2   THE SECOND DERIVATIVE

The rules for determining second derivatives are the same as those used
to determine the first derivative. The second derivative is merely the derivative of the first derivative; thus, if the first derivative is an algebraic function,
then the appropriate derivative rules are applied to determine the derivative
of the first derivative.

Example: $y = 3x^4 - 6x^3 + 5x^2 - 10x + 20$

$$\frac{dy}{dx} = 12x^3 - 18x^2 + 10x - 10$$

and

$$\frac{d^2y}{dx^2} = 36x^2 - 36x + 10$$

Example:

$$y = \frac{(x+4)^2}{(x^2+2x)}$$

$$\frac{dy}{dx} = \frac{(x^2+2x)[2(x+4)] - (x+4)^2(2x+2)}{(x^2+2x)^2}$$

$$\frac{dy}{dx} = \frac{-6x^2 - 32x - 32}{(x^2+2x)^2}$$

$$\frac{d^2y}{dx^2} = \frac{(x^2+2x)^2(-12x-32) - (-6x^2-32x-32)(2)(x^2+2x)(2x+2)}{(x^2+2x)^4}$$

10.2.3   USE OF THE SECOND DERIVATIVE IN LOCATING LOCAL OPTIMUM

The second derivative is used to identify local maxima and minima. The use of the second derivative in identification of a local minimum is illustrated in Fig. 10.6. Figure 10.6(a) shows that the function is a minimum at $x^* = 3$. The slope of the function is 0 at $x^* = 3$. In addition, the second derivative is positive, as shown in Fig. 10.6(c).

It is not merely coincidence that the second derivative is positive when evaluated at that value of x for which the function is a local minimum. It is always true that a function is a local minimum at $x^* = a$ if the first derivative, when evaluated for $x^* = a$, is 0, and the second derivative when evaluated at $x^* = a$ is positive.

This principle is also illustrated in Fig. 10.7. The function is a local minimum at $x^* = 6$. Since $x^* = 6$ is a critical point, the first derivative is 0 at $x^* = 6$. Also, the second derivative is positive at $x^* = 6$. Consequently, the point $x^* = 6$ satisfies the dual requirement that the first derivative be 0 and the second derivative be positive.

Figure 10.7 also illustrates a local maximum. The function is a local maximum at $x^* = 2$. The first derivative is 0 for $x^* = 2$. As shown in Fig. 10.7(c), the second derivative is negative at $x^* = 2$. This illustrates the use of the second derivative in locating local maxima. A function reaches a local

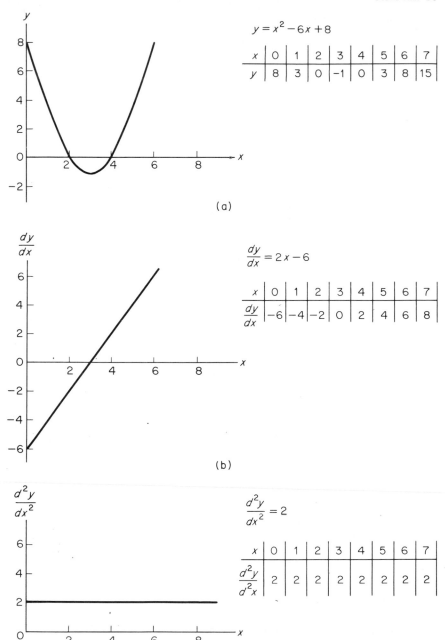

$$y = x^2 - 6x + 8$$

x	0	1	2	3	4	5	6	7
y	8	3	0	-1	0	3	8	15

(a)

$$\frac{dy}{dx} = 2x - 6$$

x	0	1	2	3	4	5	6	7
$\frac{dy}{dx}$	-6	-4	-2	0	2	4	6	8

(b)

$$\frac{d^2y}{dx^2} = 2$$

x	0	1	2	3	4	5	6	7
$\frac{d^2y}{d^2x}$	2	2	2	2	2	2	2	2

(c)

Figure 10.6

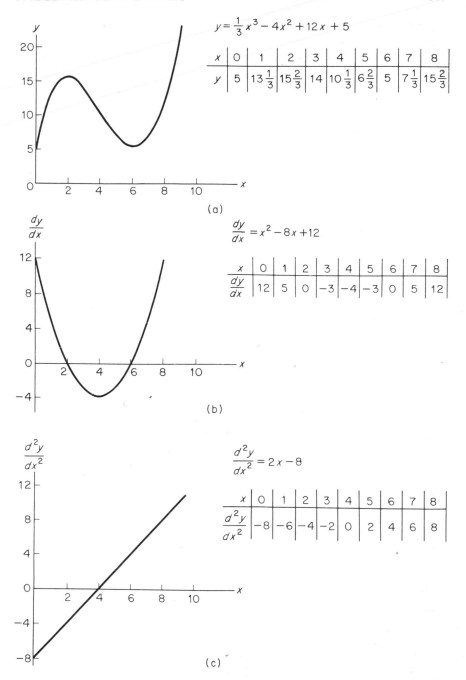

$$y = \frac{1}{3}x^3 - 4x^2 + 12x + 5$$

x	0	1	2	3	4	5	6	7	8
y	5	$13\frac{1}{3}$	$15\frac{2}{3}$	14	$10\frac{1}{3}$	$6\frac{2}{3}$	5	$7\frac{1}{3}$	$15\frac{2}{3}$

(a)

$$\frac{dy}{dx} = x^2 - 8x + 12$$

x	0	1	2	3	4	5	6	7	8
$\frac{dy}{dx}$	12	5	0	-3	-4	-3	0	5	12

(b)

$$\frac{d^2y}{dx^2} = 2x - 8$$

x	0	1	2	3	4	5	6	7	8
$\frac{d^2y}{dx^2}$	-8	-6	-4	-2	0	2	4	6	8

(c)

Figure 10.7

maximum at $x^* = a$ if the first derivative is 0 at $x^* = a$ and the second derivative is negative at $x^* = a$.

In summary, the function $f(x)$ reaches a local:

1. Maximum at $x^* = a$ if $f'(a) = 0$, and $f''(a) < 0$.
2. Minimum at $x^* = a$ if $f'(a) = 0$, and $f''(a) > 0$.

The function reaches an inflection point rather than a local maximum or minimum at $x^* = a$ if both first and second derivatives are 0, i.e., $f'(a) = 0$ and $f''(a) = 0$.

Example: For $f(x) = -5 + 10x - x^2$, determine the critical point. Specify whether the critical point is a maximum, a minimum, or a point of inflection. Determine the value of the function for the critical point.

$$f'(x) = 10 - 2x = 0$$

Thus $x^* = 5$ is a critical point.

$$f''(x) = -2$$

Thus $x^* = 5$ is a local maximum.

$$f(5) = 20$$

is the value of the local maximum.

Example: For $f(x) = x^3 - 7x^2 - 5x + 20$, determine critical points and specify whether the critical points are maxima, minima, or points of inflection. Determine the value of the function for the critical points.

$$f'(x) = 3x^2 - 14x - 5 = 0$$
$$(3x + 1)(x - 5) = 0$$
$$x^* = -\tfrac{1}{3} \quad \text{and} \quad x^* = 5$$

are critical points.

$$f''(x) = 6x - 14$$
$$f''(-\tfrac{1}{3}) = -16$$

Thus $x^* = -\tfrac{1}{3}$ is a maximum.

$$f''(5) = +16$$

Thus $x^* = 5$ is a minimum.

$$f(-\tfrac{1}{3}) = 20.85$$

is the value of the local maximum.

$$f(5) = -55$$

is the value of the local minimum.

Example: For $f(x) = -6x^3 + 2x^2 + 14x - 20$, determine the critical points and specify whether the critical points are maxima, minima, or points of inflection. Determine the value of the function at the critical points.

$$f'(x) = -18x^2 + 4x + 14 = 0$$

$$x = \frac{-4 \pm \sqrt{16 - 4(14)(-18)}}{2(-18)} = \frac{-4 \pm \sqrt{1024}}{-36} = \frac{-4 \pm 32}{-36}$$

and $x^* = 1$ and $x^* = -\frac{7}{9}$ are the critical points.

$$f''(x) = -36x + 4$$

$$f''(1) = -32$$

$x^* = 1$ is a maximum.

$$f''(-\tfrac{7}{9}) = +32$$

$x^* = -\frac{7}{9}$ is a minimum.

$$f(1) = -10$$

is the value of the local maximum.

$$f(-\tfrac{7}{9}) = -26.9$$

is the value of the local minimum.

Example: For $f(x) = 0.01x^3 - 0.15x^2 + 0.50x + 5$, determine critical points and specify whether the critical points are maxima, minima, or points of inflection. Determine the value of the function at the critical points.

$$f'(x) = 0.03x^2 - 0.30x + 0.50 = 0$$

$$x = \frac{-(-0.30) \pm \sqrt{(-0.30)^2 - 4(0.50)(0.03)}}{2(0.03)}$$

$$x = \frac{0.30 \pm \sqrt{0.09 - 0.06}}{0.06} = \frac{0.30 \pm \sqrt{0.03}}{0.06} = \frac{0.30 \pm 0.1732}{0.06}$$

Thus

$$x^* = \frac{0.4732}{0.06} = 7.9$$

and

$$x^* = \frac{0.1268}{0.06} = 2.1$$

are the critical points.

$$f''(x) = 0.06x - 0.30$$

and

$$f''(2.1) = -0.174$$

$x^* = 2.1$ is a maximum.

$$f''(7.9) = +0.174$$

$x^* = 7.9$ is a minimum.

The value of the function at 2.1 and 7.9 is

$$f(2.1) = 5.48 \quad \text{and} \quad f(7.9) = 4.52$$

10.2.4† MEANING OF THE SECOND DERIVATIVE

The second derivative gives the rate of change of the slope of the original function. The reader should note that *the rate of change of the slope of the original function* differs from the *rate of change* (or slope) *of the original function*. The rate of change of the original function is given by the first derivative, whereas the rate of change of the slope of the original function is given by the second derivative.

The relationship between the function, the first derivative, and the second derivative is described below and illustrated in Table 10.1. A knowledge of these relationships is useful in understanding the behavior of functions.

With reference to Table 10.1, a function that is increasing at an increasing rate is a function whose slope is positive and is becoming more positive for larger values of the independent variable. A function that is increasing at a decreasing rate is a function whose slope is positive but is becoming less positive for larger values of x. Similarly, a function that is decreasing at an increasing rate is a function whose slope is negative and becoming more negative for larger values of x. A function that is decreasing at a decreasing rate is a function whose slope is negative and becoming less negative for larger values of x.

A point of inflection of a function occurs when the second derivative equals zero. This occurs at that value of the independent variable for which the function changes from (1) increasing at an increasing rate to increasing at a decreasing rate, (2) increasing at a decreasing rate to increasing at an increasing rate, (3) decreasing at a decreasing rate to decreasing at an increasing rate, and (4) decreasing at an increasing rate to decreasing at a decreasing rate. Illustrations of these four points of inflection are shown in Table 10.1.

As an example of the relationships described in the table, consider again the function graphed in Fig. 10.6(a). The quadratic function decreases as x approaches 3, reaches a minimum at $x^* = 3$, and increases for subsequent increases in x. The slope of the function is plotted in Fig. 10.6(b). This graph shows that the function has a negative slope for values of x less than 3, a slope of 0 at the point $x^* = 3$, and a positive slope for values of x greater than 3.

---

† This section can be omitted without loss of continuity with later sections.

Table 10.1

Value of First Derivative	Value of Second Derivative	Description of the Function	General Appearance
positive	positive	increasing at an increasing rate	$f'(a) = +$ $f''(a) = +$
positive	negative	increasing at a decreasing rate	$f'(a) = +$ $f''(a) = -$
zero	positive	local minimum	$f'(a) = 0$ $f''(a) = +$
zero	negative	local maximum	$f'(a) = 0$ $f''(a) = -$
negative	positive	decreasing at a decreasing rate	$f'(a) = -$ $f''(a) = +$
negative	negative	decreasing at an increasing rate	$f'(a) = -$ $f''(a) = -$

Points of Inflection

1. positive	zero	point of inflection	$f'(a) = +$ $f''(a) = 0$
2. positive	zero	point of inflection	$f'(a) = +$ $f''(a) = 0$
3. negative	zero	point of inflection	$f'(a) = -$ $f''(a) = 0$
4. negative	zero	point of inflection	$f'(a) = -$ $f''(a) = 0$
5.† zero	zero	point of inflection	$f'(a) = 0$ $f''(a) = 0$
			or
			$f'(a) = 0$ $f''(a) = 0$

† Functions of the type $y = x^4$ are special cases since these functions reach a minimum at $x = 0$ (first derivative is zero at $x = 0$) and their second derivative is zero at this minimum, yet the function reaches a minimum rather than a point of inflection at $x = 0$.

The second derivative is graphed in Fig. 10.6(c). This graph shows that the rate of change of the slope of the original function is $\dfrac{d^2y}{dx^2} = 2$. We interpret this to mean that the slope of the original function is becoming less negative as x approaches the critical value $x^* = 3$ and more positive for values of x greater than $x^* = 3$. Alternatively, we can state that the positive second derivative in the example indicates that the function is decreasing but at a decreasing rate for values of x less than 3 and increasing at an increasing rate for values of x greater than 3. The function is a minimum when the first derivative is 0 and the second derivative is positive.

The relationship between the function, the first derivative, and the second derivative is illustrated for a cubic function in Fig. 10.7. The function, graphed in Fig. 10.7(a), is a local maximum at $x^* = 2$ and a local minimum at $x^* = 6$. The slope of the function at both critical points is 0. This is illustrated by the graph of the slope of the function in Fig. 10.7(b).

The second derivative is graphed in Fig. 10.7(c). The second derivative is negative for values of x less than 4 and positive for values of x greater than 4. The function increases, but at a decreasing rate of increase for $0 \leq x < 2$. The function reaches a maximum at $x^* = 2$, and decreases at an increasing rate of decrease for $2 < x \leq 4$. The function reaches a point of inflection at $x = 4$ and for $4 < x < 6$ the function continues to decrease but at a decreasing rate of decrease. The minimum point is reached at $x^* = 6$. For $x > 6$ the function increases at an increasing rate of increase. This description of the behavior of the function is possible either by inspection of the function or by inspection of the first and second derivatives.

Example: Sketch the function whose first and second derivatives are

$$f'(x) = 2x^2 - 12x + 16$$

and

$$f''(x) = 4x - 12$$

The critical points are obtained by setting the first derivative equal to 0 and solving for the roots of the equation.

$$2x^2 - 12x + 16 = 0$$
$$2(x - 4)(x - 2) = 0$$

Thus $x^* = 2$ and $x^* = 4$.

Equating the second derivative with 0 shows the inflection point to be $x = 3$. At $x^* = 2$ the second derivative is negative. Therefore, $x^* = 2$ is a local maximum. Similarly, $x^* = 4$ is a local minimum. Based upon this information, the function can be sketched as follows.

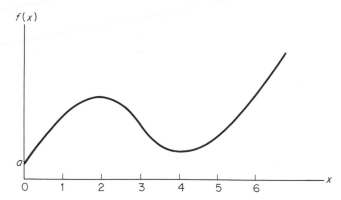

As shown in the preceding example, it is possible to sketch a function if we are given the first derivative of the function. It is not possible, however, to determine the original position of the function on the graph. Thus, in the preceding example we show the value of the function when $x = 0$ as a. The value of the intercept a is lost in taking the derivative of the function, since the derivative of any constant is 0. The general shape of the function can be sketched from the derivatives, but we must have additional information to determine the intercept of the function on the vertical axis.

Example: Verify that the two functions $f_1(x) = x^2 - 6x + 6$ and $f_2(x) = x^2 - 6x + 4$ have the same general shape.

The two functions are identical with the exception that the value of the dependent variable in the first function is two units greater than the value of the dependent variable in the second function. Since both functions have the same shape, the slope of the functions at any value of the independent variable is the same. This is verified by the fact that the derivatives of the functions are identical, i.e., $f'(x) = 2x - 6$.

10.3 Absolute and Local Maximum and Minimum

In the first section of this chapter we discussed both absolute and local maxima and minima. Local maxima and minima can be determined from the first and second derivatives. The absolute maxima and minima can be found only by comparing the local maxima and local minima with the value of the function at the end points, and selecting the absolute maximum and minimum. This procedure is illustrated by the following two examples.

Example: Determine absolute and local maximum and minimum for

$$f(x) = \frac{x^3}{3} - 4x^2 + 12x + 20, \qquad \text{for } 0 \le x \le 10$$

Local maximum and minimum are determined from the first and second derivatives.

$$f'(x) = x^2 - 8x + 12 = 0$$
$$f'(x) = (x - 2)(x - 6) = 0$$
$$x^* = 2 \quad \text{and} \quad x^* = 6$$

are critical points.

$$f''(x) = 2x - 8$$
$$f''(2) = -4$$

Therefore, $x^* = 2$ is a local maximum.

$$f''(6) = +4$$

Therefore, $x^* = 6$ is a local minimum.

The absolute maximum and minimum are found by evaluating the function at the critical points and the end points.

$$f(0) = \frac{(0)^3}{3} - 4(0)^2 + 12(0) + 20 = 20$$

$$f(2) = \frac{(2)^3}{3} - 4(2)^2 + 12(2) + 20 = 30.7$$

$$f(6) = \frac{(6)^3}{3} - 4(6)^2 + 12(6) + 20 = 20$$

$$f(10) = \frac{(10)^3}{3} - 4(10)^2 + 12(10) + 20 = 73.3$$

The absolute minimum is $f(x) = 20$, which occurs when $x = 0$ and $x = 6$, and the absolute maximum is $f(x) = 73.3$, which occurs at $x = 10$.

Example: Determine absolute and local maximum and minimum for

$$f(x) = 10 + 6x - x^2 \qquad \text{for } 0 \le x \le 5$$
$$f'(x) = 6 - 2x = 0$$

Therefore $x^* = 3$ is a critical point.

$$f''(3) = -2$$

and the function is a local maximum, since $f''(3) < 0$.

$$f(0) = 10$$
$$f(3) = 19$$
$$f(5) = 15$$

Thus, $x = 0$ is the absolute minimum and $x = 3$ is the absolute maximum.

Example: Determine the optima for $f(x) = x^3 - 3x^2 - 72x + 16$.

$$f'(x) = 3x^2 - 6x - 72 = 0$$
$$f'(x) = 3(x^2 - 2x - 24) = 0$$
$$f'(x) = 3(x - 6)(x + 4) = 0$$
$$x^* = 6 \quad \text{and} \quad x^* = -4$$

are critical points.

$$f''(x) = 6x - 6$$
$$f''(-4) = -30$$

and the critical point $x^* = -4$ is a local maximum.

$$f''(6) = 30$$

and the critical point $x^* = 6$ is a local minimum.

Since the domain is not specified in this example, it is assumed that $-\infty < x < \infty$. The convention, when the domain is not specified, is to specify only local optima. In this example, the function reaches a local maximum of $f(-4) = 192$ at $x = -4$ and a local minimum of $f(6) = -308$ at $x = 6$.

In summary, a function can reach a local optimum at an interior value of the independent variable, that is, a value of the independent variable that is not an end point of the domain. A function can reach an absolute maximum or minimum at either an interior point or an end point. If, however, the end points are not defined, absolute maximum or minimum are not specified.

PROBLEMS

1. Determine critical points for the following functions and specify whether the function is a maxima, minima, or point of inflection for the critical points.

(a) $f(x) = x^2 - 6x + 3$

(b) $f(x) = 1.5x^2 + 9x + 12$

(c) $f(x) = 2x^2 - 12x + 10$

(d) $f(x) = 3x^2 + 1x - 12$

(e) $f(x) = \dfrac{x^3}{3} - \dfrac{5x^2}{2} + 6x + 12$

(f) $f(x) = \dfrac{2x^3}{3} - \dfrac{x^2}{2} - 6x$

(g) $f(x) = x^3 - 3x + 6$

(h) $f(x) = \dfrac{x^3}{3} + 7x^2 + 40x + 100$

(i) $f(x) = \dfrac{x^3}{3} - \dfrac{3x^2}{2} - 6x + 12$

(j) $f(x) = \dfrac{2x^3}{3} - 3x^2 + 3x + 6$

(k) $f(x) = x^3 + \dfrac{7x^2}{2} - 6x + 10$

(l) $f(x) = \dfrac{2x^3}{3} - x^2 - 24x + 24$

2. Find all local and absolute maxima and minima for the following functions.

(a) $f(x) = x^2 - 8x + 6$ for $0 \leq x \leq 6$

(b) $f(x) = -x^2 + 12x - 10$ for $0 \leq x \leq 10$

(c) $f(x) = \dfrac{-x^3}{3} + 5x^2 - 6x + 10$ for $0 \leq x \leq 20$

(d) $f(x) = 2x + 3$ for $0 \leq x \leq 5$

(e) $f(x) = \dfrac{-2x^3}{3} - 4x^2 + 6x + 10$ for $-10 \leq x \leq 5$

(f) $f(x) = \dfrac{-x^3}{3} - 4x^2 - 4x + 6$ for $-20 \leq x \leq 0$

3. Sketch the function whose first derivative is the following.

(a) $f'(x) = x - 3$ (b) $f'(x) = -2x + 2$

(c) $f'(x) = x^2 - 8x + 4$ (d) $f'(x) = 2x^2 - 20x + 6$

(e) $f'(x) = -x^2 - 12x - 3$ (f) $f'(x) = -3x^2 - 15x + 4$

(g) $f'(x) = 3$ (h) $f'(x) = x^2 - 16x$

4. The profit function of the Clark Manufacturing Company is described by a quadratic function. The following data points were obtained from accounting records at Clark: output of 12,000 units resulted in profit of $120,000; output of 15,000 units resulted in profit of $130,000; output of 18,000 units resulted in profit of $128,000. Determine the quadratic function that describes profit as a function of the number of units produced and determine that level of output which results in maximum profit.

5. The cost of manufacturing a certain assembly at Black Manufacturing has been found to be described by the quadratic function. Three data points from the cost curve are the following: The average cost per unit is 80¢ when output is 600 units; the average cost is 65¢ when output is 700 units; the average cost is 70¢ when output is 800 units. Determine the quadratic function that describes the functional relationship between average cost and output, and determine the output that leads to minimum average cost.

6. It is known that the earning ability of an individual increases as he grows older and then decreases as he approaches retirement age. In a certain profession average earnings at age 40 are $25,000. This figure increases to $32,000 by age 55 and declines to $28,000 at age 60. Determine the quadratic function that describes the relationship between age and earnings, and determine the age that leads to maximum earnings.

7. The average cost per unit of output at Cleveland Electronics is $140 for the 200th unit, $125 for the 250th unit, and $130 for the 275th unit. Determine the quadratic function that describes the relationship between

average cost per unit and output, and find that unit of output which has minimum average cost.

8. Sales of a certain product are known to be related to advertising expenditures. In a controlled experiment in which factors such as price, quality, etc., were held constant and advertising expenditures were varied, the following results were observed: Advertising expenditures of $100 resulted in sales of $10,000; advertising expenditures of $300 resulted in sales of $11,500; advertising expenditures of $500 resulted in sales of $12,500. Determine the function that describes the relationship between advertising and sales, and determine the advertising expenditures that result in maximum sales.

9. The profit of Merril Equipment Company is determined by the number of units of equipment rented by Merril. Profits of $1000 are possible when 75 units are rented; profits of $1200 are possible when 85 units are rented; profits of $1300 are possible with the rental of 100 units of equipment. Determine the functional relationship between profit and the number of units of equipment rented, and determine the number of rental units that make possible maximum profits.

SUGGESTED REFERENCES

The references for this chapter are listed in Chapter 9.

Chapter 11

Single Variable Business
and
Economic Models

Chapters 9 and 10 were concerned primarily with the introduction of the concepts and methodology of differential calculus. Chapter 11 provides some of the more important business and economic applications of differential calculus. In Chapter 11 pricing and output decision of a firm are examined through the use of differential calculus. Two production models are also considered in this chapter, the basic economic order quantity model and the production lot size model.

11.1 Profit Maximization

One of the important applications of differential calculus is in the development of models that describe pricing and output decisions in the business firm. The basic assumption underlying the majority of these models is maximization of profit. The firm is assumed to produce the quantity of output for which profits are maximized. Profits of a firm depend upon both revenue and cost. The revenue of a firm is a function of the price of its products and the quantity sold. The cost of production to the firm depends upon the costs of inputs used in the production process and the quantity of product manufactured. Profits are, of course, the difference between revenue and costs.

We shall consider revenue and cost functions separately in the following two sections. They are then combined into a model of profit maximization.

11.1.1   REVENUE

The revenue generated by a firm from the sale of a product depends upon the price of the product and the quantity of the product sold. The price of the product and the quantity of the product sold are related variables. The relationship between these variables is described by the demand function.

DEMAND FOR A PRODUCT. The demand function for a product is the functional relationship between the quantity of the product offered for sale and the price that the product will bring in the market. The *law of demand* describes this functional relationship: The quantity of a product purchased during a specified period of time will be larger the lower the price, *ceteris paribus*.† The ceteris paribus factors include tastes and incomes of the individuals purchasing the product and the prices of substitute and complementary products. If these ceteris paribus conditions remain constant, then the demand function for a product plots as shown in Fig. 11.1. In this figure

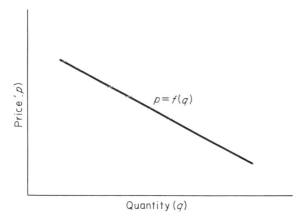

Figure 11.1

we assume that the price is a function of the quantity offered for sale; therefore, price is plotted on the vertical axis and quantity on the horizontal axis. The rationale behind this assumption is that a certain quantity is brought to the market place and the price is established based upon this quantity. The

---

† *Ceteris paribus* means that all other relevant factors remain unchanged.

larger the quantity offered for sale, the smaller the price obtainable for the product.

It is equally justifiable to treat quantity as the independent variable and price as the dependent variable. With a moment's reflection, one realizes that a change in price should result in a change in quantity; or, alternatively, a change in quantity should lead to a change in price. Thus, each variable is dependent upon the other. It is, therefore, the analyst's choice as to which variable is treated as dependent. Most students who have taken an economics course are familiar with diagrams, such as Fig. 11.1, which show price as the dependent variable and quantity the independent variable. We shall follow the custom established by the economist and treat price as the dependent variable in the demand function.

Changes in one or more of the ceteris paribus conditions can be shown by shifts in the demand function. As an example, if incomes of individuals purchasing the product increase, the demand function might shift from position A in Fig. 11.2 to position B. Similarly, should the price of a competing product be reduced, the demand function could possibly shift to position C.

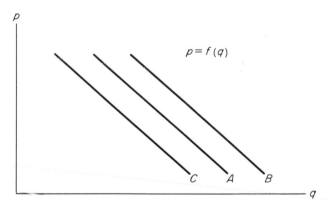

Figure 11.2

DEMAND FOR THE OUTPUT OF A FIRM. The demand function for a product describes the relationship between the demand for various quantities of the product and price of the product. The demand function for the product is not normally equivalent to the demand function for the production of an individual firm. If the individual firm produces only a small segment of the total production of the product, then it might be possible for the firm to increase or decrease the quantity of the product it produces without noticeably affecting the price of the product. As an example, if we assume that the demand function in Fig. 11.3(a) describes the relationship between the price and quantity of wheat grown in the United States, then it is not unreasonable to assume that the demand function for the wheat of an individual farmer is as shown in Fig. 11.3(b). In this figure we again assume that the quantity of wheat determines the price, and therefore quantity is plotted on the horizontal axis with price on the vertical axis. If q^* is produced and brought to market, the price will be p^*. The individual farmer can sell his entire crop for p^*, but he would not be able to raise the price by withholding a portion of his crop. This is the case of *pure competition*.

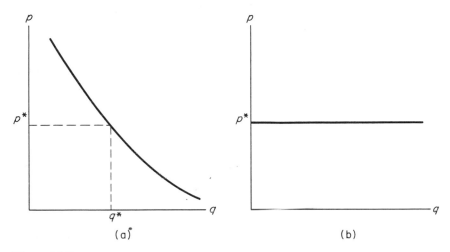

Figure 11.3

There are a number of firms that produce the entire quantity of a product offered for sale in the market. This situation of a single manufacturer occurs when the product is protected, for example, by patents or copyrights, or alternatively, when barriers restrict other firms from engaging in production of the product. The demand for the production of these firms is described by the demand function for the product (Fig. 11.1). This is the case of *pure monopoly*.

The large majority of firms compete in markets that fall in between the extremes of pure competition and pure monopoly. These firms have some control over their price, although if they reduce their price they can expect their competitors to follow with similar reductions. Quantities produced by the individual firms can vary over a limited range without price changes, although again the firm would find it difficult to increase sales significantly without lowering the price of the output. The demand for the production of these firms is described by a negatively sloping demand function similar to that shown in Fig. 11.1. This is the case of *oligopolistic* or *monopolistic competition*.

TOTAL REVENUE. A firm's total revenue that is derived from the sale of a product is given by *price multiplied by quantity*. Expressed mathematically:

$$TR = p \cdot q \tag{11.1}$$

From the demand function we observe that price is a function of quantity:

$$p = f(q) \tag{11.2}$$

Total revenue is thus represented as a function of quantity:

$$TR = f(q) \cdot q \tag{11.3}$$

This function is shown in Fig. 11.4 for two demand functions. It is assumed for both demand functions that the relationship between price and quantity can be approximated by a continuous function.

The total revenue function shown in Fig. 11.4(a) is based upon a negatively sloping demand function. For this case, if additional units of the product are to be offered for sale during the time period, the price of the product could be expected to fall. This results in an increase in total revenue for values of q less than q^*, and a decrease in total revenue for values of q greater than q^*. The total revenue function shown in Fig. 11.4(b) increases as q increases, since p is constant.

MARGINAL REVENUE. One of the useful measures of revenue is marginal revenue. The marginal revenue is the change in total revenue resulting from the sale of an additional unit of the product. Using the symbol Δ to represent change, we can represent the marginal revenue by

$$MR = \frac{\Delta TR}{\Delta q} \tag{11.4}$$

Although products are most often sold in discrete units (i.e., one pound of coffee, one bushel of wheat, etc.), economists customarily use continuous functions rather than discrete functions to describe the relationships between price, quantity, and revenue. For those cases in which the demand and total revenue functions are continuous, the expression $MR = \dfrac{\Delta TR}{\Delta q}$ is replaced by the expression $MR = \lim\limits_{\Delta q \to 0} \dfrac{\Delta TR}{\Delta q}$. The expression $\lim\limits_{\Delta q \to 0} \dfrac{\Delta TR}{\Delta q}$ is the derivative of total revenue with respect to q. The derivative of the total revenue function thus describes the change in total revenue from an additional unit of product sold in the market place. The marginal revenue is given by

$$MR = \frac{d(TR)}{dq} \tag{11.5}$$

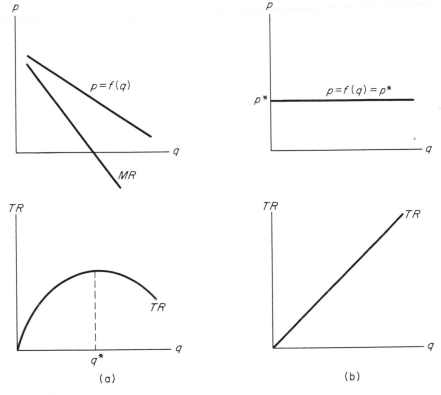

Figure 11.4

To see the relationship between the marginal revenue curve and the demand curve, we must differentiate Formula (11.5). This formula can be rewritten as

$$MR = \frac{d(TR)}{dq} = \frac{d(q \cdot f(q))}{dq} \qquad (11.6)$$

Applying Rule 6 from Chapter 9 for the product of two functions gives

$$MR = f(q) + q[f'(q)] \qquad (11.7)$$

The marginal revenue function is shown to be the sum of the demand function and q times the derivative of the demand function. In Fig. 11.4(a), the demand function has a negative slope, and therefore the derivative of the demand function is negative. Thus, the marginal revenue function plots below the demand function. In Fig. 11.4(b), the derivative of the demand function is 0. Thus, in the case of pure competition the marginal revenue and demand functions are the same.

Again referring to Fig. 11.4(a), note that revenue is a maximum when marginal revenue is 0. The quantity which produces maximum revenue can be determined by equating the marginal revenue function with zero and solving for the critical value q^*.

AVERAGE REVENUE. The average revenue from a product is found by dividing the total revenue by the quantity of the product sold. The function that describes average revenue is the quotient of the total revenue function and the quantity. From (11.8) we see that average revenue function and the demand function are equivalent.

$$AR = \frac{TR}{q} = \frac{f(q)q}{q} = f(q) \qquad (11.8)$$

PRICE ELASTICITY OF DEMAND. The price elasticity of demand provides a measure of the effect of a change in price upon total revenue. If we represent the price elasticity by E, the formula for price elasticity is

$$E = \frac{\frac{\Delta q}{q}}{\frac{\Delta p}{p}} \qquad (11.9)$$

From Formula (11.9) it can be seen that the elasticity of demand is simply the ratio of the percentage change in quantity divided by the percentage change in price. By algebraic manipulation rewrite the formula for elasticity as

$$E = \frac{\Delta q}{\Delta p} \cdot \frac{p}{q} \qquad (11.10)$$

For continuous functions $\Delta p/\Delta q$ is equivalent to dp/dq. Thus, the elasticity for a continuous function is

$$E = \frac{dq}{dp} \cdot \frac{p}{q} \qquad (11.11)$$

The expression $\frac{dq}{dp}$ in Formula (11.11) is the reciprocal of the slope of the demand function. Since the slope of the demand function is negative, the elasticity of demand is negative.

The price elasticity of demand gives the effect upon total revenue of a change in the price of a product. For $0 > E > -1$, a decrease in price results in a less than proportional increase in quantity, and thus a decrease in total revenue. Similarly, for $0 > E > -1$, an increase in price results in an increase in total revenue. For $E = -1$, a change in price results in a proportional change in quantity, and total revenue remains constant. For $E < -1$, a decrease in price results in a greater than proportional increase in quantity demanded and an increase in total revenue. Similarly, an increase in price results in a decrease in total revenue.

It is customary to consider only the absolute value of the elasticity. If we use this convention, for $0 < E < 1$ a decrease in price results in a decrease in total revenue. Similarly, an increase in price results in an increase in total revenue. This is termed *inelastic demand*. For $E = 1$, a decrease in price is counterbalanced by an increase in quantity demanded. This is termed *unit elasticity*. For $E > 1$, a decrease in price is accompanied by an increase in total revenue, and an increase in price is accompanied by a decrease in total revenue. This is termed an *elastic demand*.

Example: As an example of the concepts discussed in Sec. 11.1.1, consider the Hess Electronics Company, which manufactures an electronic module used in television sets. On the basis of market studies, they have found that the quantity of modules sold varies with the price of the module. Specifically, the demand schedule is as follows:

Price per Unit	Units Sold (thousands)
$1.00	5.0
1.25	4.5
1.50	4.0
1.75	3.5
2.00	3.0
2.25	2.5

Based upon the schedule of demand, the demand function is determined. If we assume a continuous linear demand function, the functional relationship between price and quantity is determined by selecting two data points and establishing the linear function.† The demand function is

$$p = f(q) = 3.50 - 0.50q$$

Total revenue is the product of price and quantity:

$$TR = p \cdot q = (3.50 - 0.50q)q = 3.50q - 0.50q^2$$

Marginal revenue is the derivative of total revenue with respect to quantity:

$$MR = \frac{d(TR)}{dq} = 3.50 - 1.00q$$

Average revenue is found by dividing total revenue by q:

$$AR = \frac{TR}{q} = 3.50 - 0.50q$$

---

† The reader may wish to review the method of establishing a linear function through two data points discussed in Chapter 2.

The quantities TR, AR, MR, and demand are plotted in Fig. 11.5. Note that marginal revenue is 0 when total revenue is maximum.

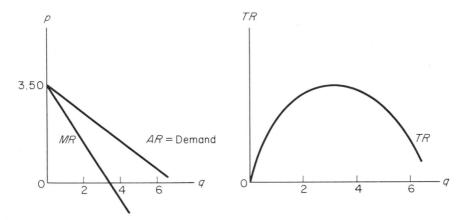

Figure 11.5

The elasticity of demand can be determined for various price levels. Since $p = f(q)$ and $q = h(p)$ are inverse functions, we can determine dq/dp by using the rule for determining derivatives of inverse functions (Rule 10). Thus

$$dq/dp = \frac{1}{dp/dq} = \frac{1}{-0.50} = -2$$

The elasticity, by Formula (11.11), is

$$E = \frac{dq}{dp} \cdot \frac{p}{q} = \frac{-2p}{q}$$

The absolute value of the elasticity is

$$E = \frac{2p}{q}$$

The elasticity of demand for alternative values of p and q is determined in the following manner:

Inelastic:

$$0 < \frac{2p}{q} < 1$$

or, alternatively,

$$0 < 2p < q$$

Unit Elastic:

$$\frac{2p}{q} = 1$$

or, alternatively,

$$2p = q$$

Elastic:

$$\frac{2p}{q} > 1$$

or, alternatively,

$$2p > q$$

The Hess Electronics Company provided an example of the important revenue curves and the elasticity of demand. Although the demand and revenue information was available in the example problem, we did not attempt to make a decision upon pricing and output. The reason, of course, was because the costs involved in the production of the output must also be considered.

In those cases in which all costs are fixed, the price and quantity that lead to both maximum revenue and maximum profits are the same. In this case, we merely maximize total revenue. As an example, consider the following problem.

Example: There are two highways between Kansas City, Missouri, and Wichita, Kansas, the Kansas Turnpike and U.S. Highway 50. The average two-way traffic between these two cities on the turnpike is 1000 cars. It has been estimated that approximately 500 cars travel between these cities using Highway 50. The toll for using the turnpike is $4.00. If we assume that all traffic would use the turnpike if the toll were reduced to $2.00 and furthermore that the demand function for the turnpike is linear, determine the price and quantity that lead to maximum revenue for the turnpike.

We have two points on the demand curve: when $p = \$4.00$, $q = 1000$; and when $p = \$2.00$, $q = 1500$. Following the custom of treating price as the dependent variable and quantity as the independent variable and using the method for establishing a linear function through two points presented in Chapter 2, we obtain the linear demand function

$$p = 8.00 - 0.004q$$

Total revenue is the product of price and quantity:

$$TR = pq = (8.00 - 0.004q)q = 8.00q - 0.004q^2$$

Total revenue is a maximum when marginal revenue is 0:

$$MR = \frac{d(TR)}{dq} = 8.00 - 0.008q = 0$$

$$q^* = 1000$$

From our analysis and on the basis of the assumption of a linear demand curve, we conclude that revenue is a maximum at the current price of $4.00 and quantity of 1000 cars.

11.1.2 COST

The total cost incurred by a firm in the production of a product is customarily assumed to consist of *fixed costs* and *variable costs*. Fixed costs are those costs that remain constant as output varies. These include items such as interest, insurance, property taxes, and salaries of those people necessary even during periods of temporary shutdown of production. Variable costs are those that vary with the volume of output. These include direct labor, cost of materials, and other expenses directly incurred by the production process.

The relationship between total cost and output is described by the cost function. Economists normally consider two cost functions—the *short-run* cost function and the *long-run* cost function. In the short run, certain inputs to the production process are fixed in amount, and a firm can vary production only by increasing or decreasing the variable inputs. In the long run, all inputs are variable. Thus in the short run the production of a product is limited by fixed plant and equipment, whereas in the long run the firm is able to vary the quantity of plant and equipment. We shall consider the short-run cost curves in this text. Long-run curves are quite similar in appearance to short-run cost curves. The student can, therefore, quite easily extend our discussion to incorporate the differences between short-run and long-run curves with the aid of an economics text on price theory.

CONVENTIONAL COST FUNCTIONS. The most common function used to model the total cost for a firm is the cubic function

$$TC = f(q) = a + bq + cq^2 + dq^3 \tag{11.12}$$

where a, b, c, and d are parameters and q is the quantity of output. Since total costs are composed of total fixed costs and total variable costs, the function in (11.12) is often stated as

$$TC = TFC + TVC \tag{11.13}$$

and it can therefore be seen that

$$TFC = a \tag{11.14}$$

and

$$TVC = bq + cq^2 + dq^3 \tag{11.15}$$

The conventional short-run total cost curve is shown in Fig. 11.6(a). Those costs incurred when the quantity produced is 0 are total fixed costs. As output expands, the proportion of variable inputs and fixed inputs to the production process is altered. This alteration in proportions typically results in more efficient and less costly production. This is shown by the *TC* and *TVC* curves. Both curves show that total costs increase; however, the rate of increase in total cost is decreasing. The proportion of variable inputs and fixed inputs continues to change as production expands. After a certain level of production is reached, the proportion of variable and fixed inputs to the production process passes an optimum. This is reflected in the total cost curve by total cost increasing at an increasing rate of increase.

SHORT-RUN COST CURVES. The appropriate functional relationships for the average and marginal cost curves can be determined from the total cost curve. The average cost curves are found by dividing the total cost curves by q. Thus,

$$ATC = \frac{TC}{q} = \frac{a}{q} + b + cq + dq^2 \qquad (11.16)$$

$$AVC = \frac{TVC}{q} = b + cq + dq^2 \qquad (11.17)$$

$$AFC = \frac{TFC}{q} = \frac{a}{q} \qquad (11.18)$$

The marginal cost is given by the change in total cost incurred by the production of an additional unit of product. Marginal cost is thus

$$MC = \frac{\Delta TC}{\Delta q} \qquad (11.19)$$

If we again assume continuous functions, Formula (11.19) is replaced by

$$MC = \frac{d(TC)}{dq} \qquad (11.20)$$

Since total cost is a cubic function, the derivative of total cost (i.e., marginal cost) will be a quadratic function. Formula (11.17) shows that average variable cost is also a quadratic function. Average fixed cost [Formula (11.18)] takes the form of what mathematicians term a rectangular hyperbola.

The sum of average variable cost and average fixed cost equals average total cost. That is,

$$ATC = AVC + AFC \qquad (11.21)$$

The relationship among marginal cost, average variable cost, average total cost, and average fixed cost is diagrammed in Fig. 11.6(b). This figure shows that marginal cost and average variable cost are equal when average variable cost is a minimum. Similarly, marginal cost and average total cost are equal when average total cost is a minimum. This relationship can be demonstrated by determining the minimum value of the AVC curve and the ATC curve. For ATC we observe that

$$ATC = \frac{TC}{q}$$

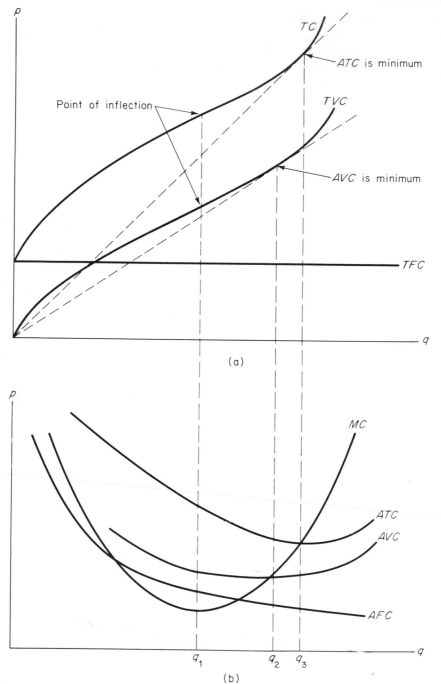

Figure 11.6

Using the derivative formula for a quotient and the fact that the derivative of TC is MC, we obtain

$$\frac{d(ATC)}{dq} = \frac{q(MC) - TC}{q^2} = 0$$

Multiplying both sides of the equation by q^2 gives $q(MC) - TC = 0$. Transposing TC to the right side of the equal sign and dividing the equation by q, we obtain

$$MC = \frac{TC}{q^*}$$

This shows that average total cost is a minimum when marginal cost equals average total cost. Marginal cost equals average total cost at the critical value of q. The same analysis shows that average variable cost is a minimum when marginal cost and average variable cost are equal.

The relationships among the cost curves can be summarized with the aid of Fig. 11.6 as follows:

1. The point of inflection of the total cost curve q_1 corresponds to the minimum point of the marginal cost curve. This occurs when the second derivative of total cost with respect to quantity is 0.
2. The minimum point on the average variable cost curve occurs when marginal cost equals average variable cost and marginal cost is increasing. This point, q_2, is the point of tangency of a line between the origin and the total variable cost curve.
3. Average total cost is a minimum when marginal cost and average total cost are equal and marginal cost is increasing. This point, q_3, is the point of tangency of a line between the origin and the total cost curve.

Example: Total cost of production for the electronic module manufactured by Hess Electronics is (in thousands of dollars)

$$TC = 0.04q^3 - 0.30q^2 + 2q + 1$$

Determine MC, AC, AVC, AFC, and TVC curves, plot these curves, and show the relationships between the curves.

(i) $MC = \dfrac{d(TC)}{dq} = 0.12q^2 - 0.60q + 2.$

MC is a minimum when $\dfrac{d^2(TC)}{dq^2} = 0$. Thus,

$$\frac{d^2(TC)}{dq^2} = \frac{d(MC)}{dq} = 0.24q - 0.60 = 0$$

$$q_1 = 2.5$$

(ii) $TVC = 0.04q^3 - 0.30q^2 + 2q.$

(iii) $AVC = \dfrac{TVC}{q} = 0.04q^2 - 0.30q + 2.$

AVC is a minimum when $\dfrac{d(AVC)}{dq} = 0$ or, alternatively and equivalently, when $MC = AVC.$

$$\frac{d(AVC)}{dq} = 0.08q - 0.30 = 0$$

$$q_2 = 3.75$$

(iv) $AC = \dfrac{TC}{q} = 0.04q^2 - 0.30q + 2 + \dfrac{1}{q}.$

AC is a minimum when $\dfrac{d(AC)}{dq} = 0$ or, alternatively and equivalently, when $MC = AC.$

$$\frac{d(AC)}{dq} = 0.08q - 0.30 - \frac{1}{q^2} = 0$$

$$0.08q^3 - 0.30q^2 - 1 = 0$$

By evaluating this equation for alternative values of q (i.e., solving by trial and error), we find that $q_3 = 4.4$ is the solution.

(v) $AFC = TFC/q = 1/q.$

These curves are plotted in Fig. 11.7.

The total cost function was provided in the Hess Electronics Company illustration. The analyst will very seldom be so fortunate as to have this function readily available. The more common case will be when the cost function must be determined from accounting data. This should present no problem, however, in that the method of fitting nonlinear functions to a set of data points has been explained in Chapter 2.

11.1.3 PROFITS

We are now in a position to combine the revenue and cost functions to obtain an expression for profits. Profits equal total revenue less total cost. Expressed as a function of the quantity produced and sold, profits are

$$P = TR - TC \tag{11.22}$$

The quantity that should be produced to obtain maximum profits can be determined by equating the derivative of the profit function with 0. Thus,

$$\frac{d(P)}{dq} = \frac{d(TR)}{dq} - \frac{d(TC)}{dq} = 0 \tag{11.23}$$

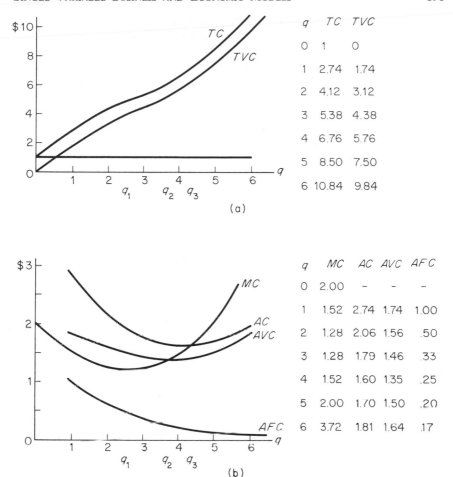

The following data tables appear alongside the graphs in Figure 11.7:

q	TC	TVC
0	1	0
1	2.74	1.74
2	4.12	3.12
3	5.38	4.38
4	6.76	5.76
5	8.50	7.50
6	10.84	9.84

(a)

q	MC	AC	AVC	AFC
0	2.00	–	–	–
1	1.52	2.74	1.74	1.00
2	1.28	2.06	1.56	.50
3	1.28	1.79	1.46	.33
4	1.52	1.60	1.35	.25
5	2.00	1.70	1.50	.20
6	3.72	1.81	1.64	.17

(b)

Figure 11.7

The derivative of total revenue with respect to the variable quantity is marginal revenue [Formula (11.5)], and the derivative of total cost with respect to quantity is marginal cost [Formula (11.20)]. Profits are, therefore, a maximum for the quantity of output for which marginal revenue equals marginal cost.

$$MR = MC \qquad (11.24)$$

The second-order requirement for a maximum is that the second derivative evaluated at the optimum level of output be negative. Thus, for maximum profits, it is also necessary that the second derivative of total revenue minus the second derivative of total cost be negative. That is,

$$\frac{d^2(P)}{dq^2} = \frac{d^2(TR)}{dq^2} - \frac{d^2(TC)}{dq^2} < 0 \qquad (11.25)$$

Example: Determine the optimum quantity and the corresponding price for maximum profits for the Hess Electronics Company.

$$P = TR - TC$$

$$P = (3.50q - 0.50q^2) - (0.04q^3 - 0.30q^2 + 2q + 1)$$

$$\frac{d(P)}{dq} = 3.50 - 1.00q - 0.12q^2 + 0.60q - 2 = 0$$

$$- 0.12q^2 - 0.40q + 1.50 = 0$$

Solving for q by using the quadratic formula gives the positive root, $q^* = 2.25$. The second derivative is

$$\frac{d^2(P)}{dq^2} = -0.24q - 0.40$$

and the second derivative evaluated at $q^* = 2.25$ is less than 0. Therefore, $q^* = 2.25$ is the value of q that leads to maximum profits. Substituting $q^* = 2.25$ in the profit function gives $P = \$910$. From the demand function,

$$p = 3.50 - 0.50q$$

The price which leads to maximum profits is $p = \$2.38$.

The total revenue, total cost, marginal revenue, marginal cost, and profit functions for Hess Electronics are shown in Fig. 11.8.

11.2 Production and Inventory Models

Much effort has been expended in the scientific analysis of production and inventory systems. This area of study is termed *production and inventory theory*. Our purpose in this section is to introduce the basic production and inventory models and to illustrate how these models are employed in the analysis and control of production and inventory systems.

Production models are used for determining the optimal lot size for the production of a product. Typically, the models apply when the ability of the firm to produce the product exceeds the demand for the product. For example, machine shops often produce one product or component for a period of time and, after accumulating sufficient stock, change over to production of a different product or component. Similarly, textile mills produce certain quantities of a specific color and texture of material. Orders for this material are then filled from the stock of inventory, and the textile machinery is used to produce other materials. As another example, toy manufacturers use the same basic

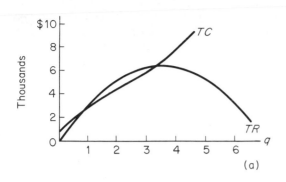

Thousands			
q	TR	TC	Profit
0	0	1.00	−1.00
1	3.00	2.74	.26
2	5.00	4.12	.88
3	6.00	5.38	.62
4	6.00	6.76	−.76
5	5.00	8.50	−3.50
6	3.00	10.84	−7.84

(a)

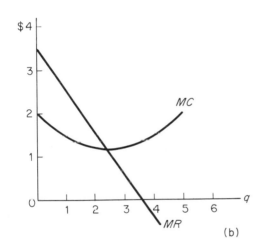

q	MR	MC
0	3.50	2.00
1	2.50	1.52
2	1.50	1.28
3	0.50	1.28
4	−0.50	1.52
5	−1.50	2.00
6	−2.50	3.72

(b)

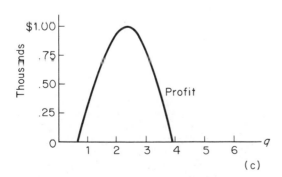

(c)

Figure 11.8

equipment to produce many different toys. A decision must be made, therefore, as to the production lot size of a given toy. After the production run has been completed, the equipment is used to manufacture another toy, and orders for the first toy are filled from inventory.

Inventory models are used to determine the optimal number of units of a product to order and be placed in inventory. Inventory models are quite similar to production models. The major difference between the inventory and production model is in the method of obtaining inventory. In the production model, we assume that inventory is built up gradually during the production run. In an inventory model, the assumption is that the product is purchased from a manufacturer or a wholesaler instead of being manufactured by the firm. It is also assumed that the units ordered from the manufacturer or wholesaler, when received by the firm, are immediately added to inventory. The key difference, then, between the assumptions underlying the two models is that inventory is built up gradually during the production run in the production model, whereas the product becomes inventory immediately upon receipt in the inventory model. Examples of firms that might use an inventory model include book publishing companies and retailers. The publisher receives quantities of books from printing companies and fills orders from this inventory. Similarly, the retailer orders quantities of merchandise from wholesalers and fills customer's orders from this inventory.

The importance of the order quantity or production lot size stems from the economic value of inventory. Inventory is defined as a stock or supply of idle resources that has economic value. This usually refers to physical goods and materials. It can, however, also be interpreted to include control of any idle resources, including labor, cash, capital goods, and storage space.

The inventory or production model is based on the cost of inventory. There are three costs that must be incorporated into the model. These are (1) the cost of carrying inventory, (2) the cost of ordering the inventory or the setup cost associated with a production run, and (3) the cost of the goods produced or ordered.

The cost of carrying inventory includes such items as warehouse expense, insurance, and the cost of the funds that are tied up in inventory. These costs are directly proportional to the quantity of inventory maintained by a firm. The larger the average level of inventory, the greater will be the expense required for storing, insuring, and financing the inventory. If cost of carrying inventory were the only cost considered in the model, inventory costs would be minimized by maintaining no inventory. It is impractical, however, for many firms to maintain no inventory. Sales are lost to competitors who can offer immediate delivery. Consequently, we shall consider only the case in which inventory is maintained.

The cost of ordering the product or the setup cost associated with a production run includes costs such as preparing orders or the down time associated with setting up for a production run. These costs are a function of the number of orders that must be placed or setups that must be made per period of time. For example, placing orders daily would be more costly than placing only one order during the time period. Thus, the cost ordering the product, or the setup cost associated with a production run, also varies with the order quantity or production lot size. If the orders are placed frequently, with resulting small order quantities per order, the cost of ordering is relatively high. Conversely, if the orders are placed infrequently, with resulting large order quantities per order, the cost of ordering is relatively low. Consequently, the cost of ordering is inversely related to the order quantity or the production lot size.

The third cost is the purchase or manufacturing cost of the product. If we assume a given demand and, for the moment, ignore any possible volume discounts for large purchases of the product or the economies that are sometimes associated with long production runs, the total purchase or manufacturing cost is equal merely to the product of the number of units demanded and the cost per unit. Based on this assumption, the total purchase or manufacturing cost is not related to the specific order quantity or production lot size. Instead, the total purchase or manufacturing cost is assumed to be constant. In more complicated models, factors such as volume discounts and economies of scale resulting from long production runs are included.

In developing the production and inventory models, we assume that demand is known and is uniformly distributed throughout the period. This type of model is classified as a *deterministic* inventory model.† An alternative model would be a *stochastic* model in which demand is described by a probability distribution. For the deterministic model, two operating decisions are required:

1. The size of the order (i.e., the number of units to be ordered, or the lot size to be produced).
2. When to place the order or begin production. (We shall assume that this decision is based upon the number of units remaining in inventory.)

We first consider the problem of order size. The objective in to determine production schedules so as to minimize total costs. We shall use the following notation:

a = cost of holding one unit of inventory for one time period
c = setup cost incurred when a new production run is started
or cost of placing an order

---

† The assumption that demand is known simplifies the model. In a good number of practical situations, demand is contractual or is fairly uniform during the period, and this assumption is a reasonable approximation to reality.

k = production rate, in units per time period
r = rate of demand in units per time period
q = quantity ordered or produced at each setup

We shall discuss two basic models in this chapter and two more complicated models in Chapter 13.

11.2.1   ECONOMIC ORDER QUANTITY MODEL

The objective in the economic order quantity model is to determine the order size such that the cost of ordering and carrying inventory is a minimum. It is assumed that shortages are not permitted and that inventory is replenished a fixed number of days following ordering of the inventory.

As an illustration of the model, assume that we have contracted to supply 10,000 assemblies per year. A component of this assembly is ordered from a supplier. The order costs associated with placing an order are $300. The cost of holding a component in inventory (insurance, warehouse space, cost of capital, etc.) is $10 per year. The lead time for inventory delivery is five days. Our problem is to determine the order quantity that minimizes total cost.

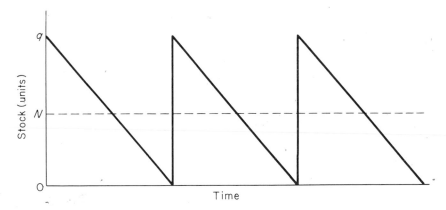

Figure 11.9

In order to develop the model, we have plotted inventory versus time in Fig. 11.9. The order quantity is represented in Fig. 11.9 by q. The letter N represents the inventory level at which an order must be placed to assure delivery before the inventory is exhausted. The objective is to determine the values of q and N such that the cost of maintaining the inventory is a minimum.

The cost of inventory is composed of an order cost C_o, an inventory carrying cost C_c, and the cost of the inventory C_i:

$$TC = f(q) = C_o + C_c + C_i \tag{11.26}$$

Consider first the total order cost per period. The number of units demanded during the period is r and the order quantity is q. Thus, r/q represents the number of orders during the period. Since c represents the cost of placing an order and r/q orders must be placed, the total cost of ordering the inventory is

$$C_o = \frac{rc}{q} \tag{11.27}$$

It can be seen that the order cost C_o decreases as the order size q increases.

The cost of carrying inventory during the period is given by the product of the cost of holding one unit in inventory and the number of units in inventory. Thus

$$C_c = \frac{qa}{2} \tag{11.28}$$

where $q/2$ represents the average inventory level. From (11.28), it can be seen that the cost of carrying inventory increases as the order size increases.

The cost of the inventory based upon a given level of demand is constant and is represented by C_i. The total inventory cost is thus

$$TC = \frac{rc}{q} + \frac{qa}{2} + C_i \tag{11.29}$$

To minimize total cost, we determine the value of q such that $\dfrac{d(TC)}{dq} = 0$. Thus,

$$\frac{d(TC)}{dq} = \frac{-rc}{q^2} + \frac{a}{2} = 0$$

and

$$q^* = \sqrt{\frac{2rc}{a}} \tag{11.30}$$

q^* is termed the economic order quantity.

The fact that q^* is a minimum can be verified by the second derivative. Since the second derivative of TC,

$$\frac{d^2(TC)}{dq^2} = \frac{2rc}{q^3}$$

is positive for all positive values of q, we conclude that q^* gives the order quantity that minimizes total inventory cost.

The order point is determined from the rate of inventory usage and the lead time. Let U represent the inventory consumption rate during lead time, and let L represent the lead time. The inventory consumed is given by

$$N = L \cdot U \tag{11.31}$$

An order of q^* units should be placed when the inventory level falls to N.

Example: For the illustrative problem above, determine the economic order quantity and the order point.

In this problem, $r = 10,000$, $a = \$10$, and $c = \$300$. Therefore,

$$q = \sqrt{\frac{2(10,000)(300)}{10}} = \sqrt{600,000} = 776 \text{ units}$$

We should order 776 components per order.

Assuming that inventory usage is constant and that there are 250 working days per year, we have

$$U = \frac{10,000}{250} = 40 \text{ units per day}$$

If the lead time is five working days, then the order should be placed when the inventory level falls to

$$N = L \cdot U = 5 \cdot 40 = 200 \text{ units}$$

The basic economic order quantity model can be applied directly only if all assumptions underlying the model are reasonably satisfied. One assumption that often is not met is that of constant total cost of the inventory. Quite often volume discounts are allowed, thus encouraging large purchases. These volume discounts must be considered when one is determining the economic order quantity.

The most common form of volume discount is the price break. For example, a manufacturer might offer a price of $20 per unit on orders up to 200 units. For orders exceeding 200 units, the manufacturer reduces the price of all units purchased (including the initial 200) by 5 percent. A price break of $1 per unit is thus offered on purchases of 201 units or more.

This price break or volume discount can be considered in the economic order quantity model. The procedure involves computing the economic order quantity for the several prices. In our example, we would determine the economic order quantity for a price per unit of $20 and the economic order quantity for a price per unit of $19. If the economic order quantity based upon a price per unit of $19 is more than 200 units, the problem is solved. The quantity ordered is merely that quantity calculated from the economic order quantity formula. It often happens, however, that the economic order quantity based upon the $19 price is less than 200 units. Orders of 200 units or less will not be accepted at a price of $19. Using Formula (11.29), we compare the total cost if we assume orders of 201 units and a price of $19 per unit with the total cost if we assume the $20 per unit price and the economic order quantity calculated by using Formula (11.30). The quantity ordered would be that which results in minimum total cost. The calculations are carried out in the following example.

Example: The regional distributor for the southern region of Universal Electric Appliance Company expects to sell 1200 blenders during the coming year. For orders of 200 blenders or less, the price per blender is $20. On orders exceeding 200 blenders, the price is reduced to $19. The cost of storing a blender for one year including possible obsolescence is $4. The ordering cost is $50 per order. Determine the order quantity.

The economic order quantity based on a price of $20 is

$$q_1 = \sqrt{\frac{2(1200)(50)}{4.00}} = 173 \text{ units}$$

It is possible to order quantities of 173 units at the price of $20 per unit. To receive the price of $19 per unit, we must order quantities of 201 units. The order quantity that results in minimum total cost can be determined from (11.29).

$$TC_1 = \frac{1200(50)}{173} + \frac{1}{2}(173)(4.00) + 1200(20)$$

$$= 347.00 + 346.00 + 24,000 = \$24,693$$

$$TC_2 = \frac{1200(50)}{201} + \frac{1}{2}(201)(4.00) + 1200(19)$$

$$- 299.00 \mid 402.00 + 22,800 = \$23,501$$

Order quantities of 173 units results in total inventory cost of $24,693 and order quantities of 201 units results in total inventory cost of $23,501. A saving of $1192 is the result of the order quantity of 201 units. The appropriate order quantity is thus 201 units.

11.2.2 PRODUCTION LOT SIZE MODEL

The objective of the production lot size model is to determine the quantity of product to be manufactured during a production run. The costs incorporated in the model are the cost of setup, the cost of carrying the inventory, and the cost of the inventory. The production lot size is that quantity which results in the minimum total cost. This model differs from the economic order quantity model in that it was assumed in the economic order quantity model that the quantity ordered is, upon arrival, immediately added to inventory. In the production lot size model it is assumed that the product being produced is added to inventory over a period of time.

A plot of inventory versus time for the production lot model is shown in Fig. 11.10. The production rate in this figure is represented by k and the demand rate by r. The length of the production run is q/k. Since k exceeds r, inventory is accumulated during the production run. The rate of accumulation of inventory is $k - r$. The number of units of inventory is given by the product of the rate of accumulation $(k - r)$ and the length of the production run q/k.

As an example, consider the manufacturing situation in which a component is manufactured and stored in inventory. The annual demand is $r = 2000$ units, and the production rate is $k = 12,000$ units. Holding costs are $a = \$10$ per unit per year and the setup cost is $c = \$250$ per setup. Determine the production lot size q.

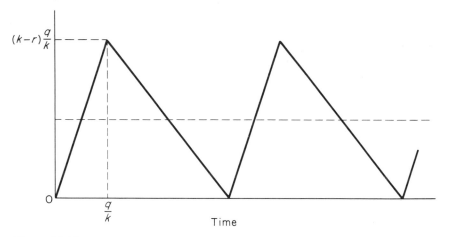

Figure 11.10

Total cost per period is composed of setup cost, carrying cost, and the cost of the inventory:

$$TC = f(q) = C_s + C_c + C_i \qquad (11.32)$$

Since there are r/q setups per period, each costing c, the total setup cost per period is

$$C_s = \frac{rc}{q} \qquad (11.33)$$

The total carrying cost per period is found by multiplying the average number of units of inventory by the cost per period of carrying one unit of inventory. The term q/k gives the proportion of the period required to manufacture q units. Since units are being used concurrently while they are being produced, the net addition to inventory is at the rate of $(k - r)$. The maximum inventory level is thus $(k - r)q/k$. The average inventory is one-half the maximum level of inventory. The total cost of carrying the inventory is

$$C_c = \frac{q(k - r)a}{2k} \tag{11.34}$$

The total cost per period can now be expressed in terms of the production lot size as

$$TC = \frac{rc}{q} + \frac{q(k - r)a}{2k} + C_i \tag{11.35}$$

The minimum cost is found by equating the derivative of the total cost function to 0 and solving for q.

$$\frac{d(TC)}{dq} = \frac{-rc}{q^2} + \frac{a(k - r)}{2k} + 0 = 0$$

$$q^* = \sqrt{\frac{2krc}{a(k - r)}} \tag{11.36}$$

Since the second derivative of TC, $\dfrac{d^2(TC)}{dq^2} = \dfrac{2rc}{q^3}$, is positive for all positive values of q, we conclude that q^* gives the production lot size that minimizes total inventory cost.

Example: For the illustrative problem, determine the production lot size. In this problem, $r = 2000$; $k = 12,000$; $a = \$10$; $c = \$250$.

$$q^* = \sqrt{\frac{2(12,000)(2000)(250)}{10(12,000 - 2000)}} = \sqrt{120,000}$$

$$q^* = 346$$

PROBLEMS

1. The demand function for a particular product is $p = 5.00 - 0.75q$. Determine the total revenue, marginal revenue, average revenue, and elasticity of demand functions. Find the value of q that results in maximum revenue.

2. The demand function for a certain commodity is $p = 1000 - 50q - q^2$ for $0 < q < 25$. Determine the total revenue, marginal revenue, average revenue, and elasticity of demand functions. Find the value of q that results in maximum total revenue, and determine the total revenue, marginal revenue, average revenue, and elasticity of demand for this value of q.

3. The demand function for a product is $p = 2500 - 60q - 0.5q^2$ for $0 < q < 60$. Determine the total revenue, marginal revenue, average revenue,

and elasticity of demand functions. Find the value of q that results in maximum revenue, and determine the total revenue, marginal revenue, average revenue, and elasticity for this value of q.

4. The demand function for a product is $p = 10 - 0.25q - 0.05q^2$ for $0 < q < 10$. Determine the total revenue, marginal revenue, average revenue, and elasticity of demand functions. Find the value of q that results in maximum revenue, and determine the total revenue, marginal revenue, average revenue, and elasticity for this value of q.

5. The ABC Bus Tour Company conducts tours of San Diego. At a price of $5.00 per person, ABC experienced an average demand of 1000 customers per week. Following a price reduction to $4.00 per person, the average demand increased to 1200 customers per week. Assuming a linear demand function, determine the total revenue, marginal revenue, average revenue, and elasticity of demand functions. Find the tour fare and quantity that result in maximum revenue, and determine the total revenue, marginal revenue, average revenue, and elasticity of demand for this value of q.

6. The Santa Catalina Motor Excursion Company offers round trip boat transportation between Santa Catalina Island and the Port of Los Angeles. They are in competition with one other boat line plus an airline. Because of the competition, the owners of the company have not been able to raise their price. The owners of the company believe, however, that an increase in the fare would result in an increase in total revenue. An analysis of the market indicates that a $2.00 increase in the present round trip boat fare of $16.00 would result in a 5 percent decrease in the number of passengers. Using the concept of elasticity of demand, determine if the price increase should be made.

7. The Santa Catalina Motor Excursion Company of the preceding problem was able to convince the other boat line that a $2.00 increase would lead to increased revenue. However, the airline could not raise its fare without governmental approval. If the fare charged by the airline remains unchanged, it is estimated that the proposed $2.00 increase in boat fare would lead to a 15 percent decrease in the number of passengers carried by the boat lines. Should the fare increase be made?

8. The cost function for a particular commodity is a linear function of output for $q \leq 1000$ units. If fixed costs are $1500 and the variable cost of manufacturing is $2.50 per unit, determine total cost, marginal cost, and average cost functions. What value of q gives minimum marginal and average cost?

9. The total cost function for a product is described by the function $TC = 1000 + 20q - q^2 + 0.05q^3$ for $0 \leq q \leq 30$. Determine the marginal cost, average variable cost, and average fixed cost functions. Determine the

value of q for which marginal cost is a minimum. Determine also the values of q for which average cost and average variable cost are minimum. Verify the fact that marginal cost equals average cost when average cost is a minimum and that marginal cost equals average variable cost when average variable cost is a minimum. For what value of q is average fixed cost a minimum?

10. The total cost function for a product is described by the function $TC = 4500 + 1500q - 1.5q^2 + 0.02q^3$. Determine the marginal cost, average cost, average variable cost, and average fixed cost functions. Determine the value of q for which marginal cost is a minimum. Determine also the values of q for which average cost and average variable cost are minimum. Verify the fact that marginal cost equals average cost when average cost is a minimum and that marginal cost equals average variable cost when average variable cost is a minimum.

11. The total revenue function for a particular product is $TR = 600q - 0.5q^2$. The total cost for the product is $TC = 1500 + 150q - 4q^2 + 0.5q^3$. Determine the profit function and the value of q for which profits are maximum. Is this also the value of q for which revenue is a maximum or for which average costs are a minimum?

12. The total revenue function for a product is $TR = q(400 - q)$, and the total cost for the product is $TC = 1000 + 100q - 5q^2 + q^3$. Determine marginal revenue, marginal cost, the output for which profit is a maximum, and the profit at this level of output.

13. The demand for a commodity is $p = 200 - q$. The cost of producing the commodity is composed of a fixed cost of $5000 and a variable cost of $0.50 per unit. Determine the profit function for the commodity and the quantity of output that leads to maximum profit.

14. A tour charges groups of individuals $100 per person less $1 per person for each individual in the group over 25, that is, for example, 26 members of a group would pay $99 each, and the total revenue to the tour organization would be $99(26). The cost of organizing and conducting the tour is $1000 plus $40 per person. What group size leads to maximum profit, and what is this profit?

15. The Western Petroleum Company purchases auto accessories from independent manufacturers and sells these products under the Western Petroleum label through their chain of automobile service stations. Batteries are purchased from the Universal Power Company. The yearly demand for batteries is estimated as 25,000 units. The per year cost of holding the battery, including a charge for possible obsolescence, is $5. The cost of placing an order with Universal Power is $200. Determine the economic order quantity.

16. Western Petroleum expects to sell 100,000 tires during the year. The cost of holding a tire in inventory for one year is $4, and the cost of ordering the tires is $150. Determine the economic order quantity.

17. Universal Power uses the same equipment to manufacture both 6-volt and 12-volt batteries. The yearly demand for 6-volt batteries is 75,000 units, and the demand for 12-volt batteries is 225,000 units. Production capacity is such that 500,000 batteries can be manufactured each year. The cost of changing from the production of one battery to the other is $500; i.e., setup cost is $500. The cost of holding a battery is $6 per year. Determine the production lot sizes for both batteries.

18. The Haynes Music Corporation has leased 300 acres of beach front between San Diego and Los Angeles for a rock music festival. The rock music groups have been engaged, and all other necessary arrangements have been made. Consequently, all major costs have been incurred. Haynes is uncertain what admission price should be charged for the festival. Assuming the objective of maximizing total revenue, determine the appropriate price. The following estimates of demand are available.

Price	Number of Customers
$15	50,000
10	150,000
5	200,000

SUGGESTED REFERENCES

DEAN, BURTON V., et al., *Mathematics for Modern Management* (New York, N.Y.: John Wiley and Sons, Inc., 1963).

GIFFIN, WALTER C., *Introduction to Operations Engineering* (Homewood, Ill.: Richard D. Irwin, Inc., 1971).

HAYNES, W. W., *Managerial Economics: Analysis and Cases*, rev. ed. (Austin, Tex.: Business Publications, Inc., 1969).

NAYLOR, THOMAS H. AND JOHN M. VERNON, *Microeconomics and Decision Models of the Firm* (New York, N.Y.: Harcourt, Brace and World, Inc., 1969).

SCHELLENBERGER, ROBERT E., *Managerial Analysis* (Homewood, Ill.: Richard D. Irwin, Inc., 1969).

SPENCER, MILTON H., *Managerial Economics*, 3rd ed. (Homewood, Ill.: Richard D. Irwin, Inc., 1968).

Calculus of Multivariate, Exponential, and Logarithmic Functions

This chapter continues the discussion of the concepts and techniques that underlie the application of differential calculus in modern managerial decision models. The first section extends the concept of differentiation to multivariate functions. This section is followed by a discussion of the exponential and logarithmic functions. The final section of this chapter introduces a method of optimizing a nonlinear function subject to one or more constraining equations.

12.1 Derivatives of Multivariate Functions

The concept of the derivative extends directly to multivariate functions. In our discussion of derivatives in Chapter 9, we defined the derivative of a function as the instantaneous rate of change of the function with respect to the independent variable. In multivariate functions, the derivative again gives the instantaneous rate of change of the function with respect to an independent variable. Since, however, there is more than one independent variable in multivariate functions, the derivative of the function must be considered separately for each independent variable. The derivative of the function with respect to one of the independent variables is termed the *partial derivative* of the function.

The formula for the partial derivative of the multivariate function with respect to an independent variable is given by (12.1). If $y = f(x_1, x_2)$, the derivative of y with respect to x_1 is

$$\frac{\partial y}{\partial x_1} = \operatorname*{limit}_{\Delta x_1 \to 0} \frac{f(x_1 + \Delta x_1, x_2) - f(x_1, x_2)}{\Delta x_1} \tag{12.1}$$

where the symbol $\dfrac{\partial y}{\partial x_1}$ is read *the partial derivative of y with respect to x_1*. The partial derivative of y with respect to x_2 is expressed as

$$\frac{\partial y}{\partial x_2} = \operatorname*{limit}_{\Delta x_2 \to 0} \frac{f(x_1, x_2 + \Delta x_2) - f(x_1, x_2)}{\Delta x_2} \tag{12.2}$$

where $\dfrac{\partial y}{\partial x_2}$ is read *the partial derivative of y with respect to x_2*.

The partial derivative of a multivariate function gives the instantaneous rate of change of the dependent variable with respect to an independent variable. To determine the effect of a change in a single independent variable upon the dependent variable, the remaining independent variables must not be allowed to change in value. It can be seen that if all independent variables were allowed to vary concurrently, it would be impossible to determine the effect of a change in a single variable. Thus, *all variables other than the variable under consideration are assumed to be constant*.

The partial derivatives of the multivariate function can be determined from the formulas for derivatives of multivariate functions. To illustrate the use of these formulas, we first determine the partial derivatives of $y = 4x_1 + 5x_2$. The partial derivative of y with respect to x_1 is given by Formula (12.1).

$$\frac{\partial y}{\partial x_1} = \operatorname*{limit}_{\Delta x_1 \to 0} \frac{f(x_1 + \Delta x_1, x_2) - f(x_1, x_2)}{\Delta x_1} \tag{12.1}$$

According to the formula, x_1 is replaced by $x_1 + \Delta x_1$. This yields

$$\frac{\partial y}{\partial x_1} = \operatorname*{limit}_{\Delta x_1 \to 0} \left[\frac{4(x_1 + \Delta x_1) + 5x_2 - (4x_1 + 5x_2)}{\Delta x_1} \right]$$

$$\frac{\partial y}{\partial x_1} = \operatorname*{limit}_{\Delta x_1 \to 0} \frac{4 \Delta x_1}{\Delta x_1} = 4$$

The partial derivative of y with respect to x_2 is given by Formula (12.2).

$$\frac{\partial y}{\partial x_2} = \operatorname*{limit}_{\Delta x_2 \to 0} \frac{f(x_1, x_2 + \Delta x_2) - f(x_1, x_2)}{\Delta x_2} \tag{12.2}$$

In the example problem, we replace x_2 by $x_2 + \Delta x_2$.

$$\frac{\partial y}{\partial x_2} = \lim_{\Delta x_2 \to 0} \left[\frac{4x_1 + 5(x_2 + \Delta x_2) - (4x_1 + 5x_2)}{\Delta x_2} \right]$$

$$\frac{\partial y}{\partial x_2} = \lim_{\Delta x_2 \to 0} \frac{5 \, \Delta x_2}{\Delta x_2} = 5$$

Example: Determine the partial derivatives of $f(x_1, x_2) = 3x_1 + 6x_2$.

$$\frac{\partial f}{\partial x_1} = \lim_{\Delta x_1 \to 0} \left[\frac{3(x_1 + \Delta x_1) + 6x_2 - (3x_1 + 6x_2)}{\Delta x_1} \right]$$

$$\frac{\partial f}{\partial x_1} = \lim_{\Delta x_2 \to 0} \frac{3 \, \Delta x_1}{\Delta x_1} = 3$$

$$\frac{\partial f}{\partial x_2} = \lim_{\Delta x_2 \to 0} \left[\frac{3x_1 + 6(x_2 + \Delta x_2) - (3x_1 + 6x_2)}{\Delta x_2} \right]$$

$$\frac{\partial f}{\partial x_2} = \lim_{\Delta x_2 \to 0} \frac{6 \, \Delta x_2}{\Delta x_2} = 6$$

For the multivariate function $y = f(x_1, x_2)$, the symbol for the partial derivative of y with respect to x_1 is $\dfrac{\partial y}{\partial x_1}$, and the symbol for the partial derivative of y with respect to x_2 is $\dfrac{\partial y}{\partial x_2}$. Alternatively, the partial derivative of the function with respect to x_1 can be expressed as f_{x_1} and the partial derivative of the function with respect to x_2 as f_{x_2}.

The derivatives of nonlinear multivariate functions are determined in the same manner as the derivatives of linear multivariate functions. The partial derivative of the function with respect to any independent variable is found by considering the remaining independent variables as constant. This is illustrated by the following examples.

Example: Determine the partial derivatives of $y = x_1^2 + x_1 x_2 + x_2^2$. Application of the formula for the partial derivative of y with respect to x_1 gives

$$\frac{\partial y}{\partial x_1} = \lim_{\Delta x_1 \to 0} \frac{[(x_1 + \Delta x_1)^2 + (x_1 + \Delta x_1)x_2 + x_2^2] - [x_1^2 + x_1 x_2 + x_2^2]}{\Delta x_1}$$

$$\frac{\partial y}{\partial x_1} = \lim_{\Delta x_1 \to 0} \frac{(x_1^2 + 2x_1 \, \Delta x_1 + \Delta x_1^2 + x_1 x_2 + \Delta x_1 x_2 + x_2^2 - x_1^2 - x_1 x_2 - x_2^2)}{\Delta x_1}$$

$$\frac{\partial y}{\partial x_1} = \lim_{\Delta x_1 \to 0} \frac{(2x_1 \, \Delta x_1 + \Delta x_1^2 + \Delta x_1 x_2)}{\Delta x_1} = \lim_{\Delta x_1 \to 0} (2x_1 + \Delta x_1 + x_2)$$

$$\frac{\partial y}{\partial x_1} = 2x_1 + x_2$$

The partial derivative of y with respect to x_2 is

$$\frac{\partial y}{\partial x_2} = \lim_{\Delta x_2 \to 0} \frac{[x_1^2 + x_1(x_2 + \Delta x_2) + (x_2 + \Delta x_2)^2] - [x_1^2 + x_1 x_2 + x_2^2]}{\Delta x_2}$$

$$\frac{\partial y}{\partial x_2} = \lim_{\Delta x_2 \to 0} \frac{(x_1^2 + x_1 x_2 + x_1 \Delta x_2 + x_2^2 + 2x_2 \Delta x_2 + \Delta x_2^2 - x_1^2 - x_1 x_2 - x_2^2)}{\Delta x_2}$$

$$\frac{\partial y}{\partial x_2} = \lim_{\Delta x_2 \to 0} \frac{(x_1 \Delta x_2 + 2x_2 \Delta x_2 + \Delta x_2^2)}{\Delta x_2} = \lim_{\Delta x_2 \to 0} (x_1 + 2x_2 + \Delta x_2)$$

$$\frac{\partial y}{\partial x_2} = x_1 + 2x_2$$

In each of these examples, x_2 is considered a constant when one is determining the partial derivative of y with respect to x_1. Similarly, x_1 is considered a constant when one is determining the partial derivative of y with respect to x_2. Since we are interested in determining the partial derivative of y with respect to only one variable at a time and are treating the other variables as constants, the multivariate function can be considered as a function of only one independent variable for purposes of differentiation. The other variables are treated as constants in applying the rules for finding derivatives.

Consider again the previous examples. If $y = 4x_1 + 5x_2$, and x_2 is treated as a constant, then $\partial y/\partial x_1 = 4$. Similarly, if x_1 is assumed constant, $\partial y/\partial x_2 = 5$. In the third example, treating x_2 as a constant and determining the derivative with respect to x_1, we have $\partial y/\partial x_1 = 2x_1 + x_2$. Again, $\partial y/\partial x_2 = x_1 + 2x_2$.

The rules for determining derivatives of multivariate functions are the same as those for finding derivatives of functions with only one independent variable. In applying these rules, remember that all variables other than the variable for which the derivative is taken with respect to are treated as constants.

Example: Determine the partial derivatives of $y = 3x_1^2 + 2(x_1 x_2)^3 + x_2^4$. To determine the partial derivative of y with respect to x_1, we treat x_2 as a constant.

$$\frac{\partial y}{\partial x_1} = 6x_1 + 6(x_1 x_2)^2 (x_2) + 0$$

Similarly, in determining the partial derivative of y with respect to x_2, we consider x_1 to be constant.

$$\frac{\partial y}{\partial x_2} = 0 + 6(x_1 x_2)^2 (x_1) + 4x_2^3$$

The same procedure is applied in the following examples.

Example: $y = (2x_1 + x_2)^5$.

$$\frac{\partial y}{\partial x_1} = 5(2x_1 + x_2)^4(2)$$

$$\frac{\partial y}{\partial x_2} = 5(2x_1 + x_2)^4(1)$$

Example: $y = 3x_1^2 + 2x_1x_2x_3 + 3x_2x_3^3 + x_3^4$.

$$\frac{\partial y}{\partial x_1} = 6x_1 + 2x_2x_3$$

$$\frac{\partial y}{\partial x_2} = 2x_1x_3 + 3x_3^3$$

$$\frac{\partial y}{\partial x_3} = 2x_1x_2 + 9x_2x_3^2 + 4x_3^3$$

Example: $y = \dfrac{2x_1 - x_2^2}{x_1 + x_2}$.

$$\frac{\partial y}{\partial x_1} = \frac{(x_1 + x_2)(2) - (2x_1 - x_2^2)(1)}{(x_1 + x_2)^2} = \frac{2x_2 + x_2^2}{(x_1 + x_2)^2}$$

$$\frac{\partial y}{\partial x_2} = \frac{(x_1 + x_2)(-2x_2) - (2x_1 - x_2^2)(1)}{(x_1 + x_2)^2}$$

$$\frac{\partial y}{\partial x_2} = \frac{-2x_1x_2 - 2x_1 - x_2^2}{(x_1 + x_2)^2}$$

12.1.1   INTERPRETATION OF PARTIAL DERIVATIVES

The partial derivative of a function with respect to the variable x_1 describes the change in the function for an incremental change in the value of x_1. If the dependent variable is functionally related to two independent variables, this relationship can be shown by a three-dimensional graph. The partial derivative of y with respect to x_1 gives the slope of the function in a plane parallel to the plane formed by the y and x_1 axes. The position of this plane is x_2 units from the origin. We can express the same concept by the statement that the partial derivative of y with respect to x_1 gives the rate of change of the function for an incremental change in x_1, assuming that x_2 is held constant. Similarly, the partial derivative of y with respect to x_2 gives the slope of the function in a plane parallel to the plane formed by the y and x_2 axes. The position of the plane is x_1 units from the origin. An equivalent interpretation is that the partial derivative of y with respect to x_2 gives the rate of change of the function for an incremental change in x_2 with x_1 held constant. The

partial derivatives of a function of two independent variables are illustrated
by the following example.

Example: $f(x_1, x_2) = x_1^2 + 2x_1x_2 + x_2^2$ describes the three-dimensional
surface sketched below. Determine the slope of the function in the plane
parallel to the f and x_1 axes located $x_2 = 2$ units from the origin. Also
determine the slope of the function in the plane parallel to the f and x_2 axes
located $x_1 = 1$ unit from the origin.

The function $f(x_1, x_2) = x_1^2 + 2x_1x_2 + x_2^2$ is sketched in Fig. 12.1(a). The
function that describes the surface in the plane parallel to the f and x_1 axes
that is located $x_2 = 2$ units from the origin is sketched in Fig. 12.1(b). The
function is

$$f(x_1, 2) = x_1^2 + 2x_1(2) + (2)^2$$
$$f(x_1, 2) = x_1^2 + 4x_1 + 4$$

This function is sketched in Fig. 12.1(b). The partial derivative of $f(x_1, 2)$ gives
the slope of the function in this plane.

$$\frac{\partial f}{\partial x_1} = 2x_1 + 4$$

The derivative is sketched in Fig. 12.1(d).

The function that describes the surface in the plane parallel to the f and
x_2 axes that is $x_1 = 1$ unit from the origin is plotted in Fig. 12.1(c) and given
below:

$$f(1, x_2) = 1 + 2x_2 + x_2^2$$

The partial derivative of the function gives the slope of the function in this
plane. The partial derivative, plotted in Fig. 12.1(e), is

$$\frac{\partial f}{\partial x_2} = 2 + 2x_2$$

Functions of more than three dimensions cannot be graphed. Consequently, it is meaningless to speak of the slope of a function of three or more
independent variables. Instead, we are interested in the change in the value
of the function for an incremental change in an independent variable. This
is illustrated by the following examples.

Example: Estimated sales s of a product are related to price p, advertising
a, and the number of salesmen n. The functional relationship is

$$s = (10,000 - 800p)n^{1/2}a^{1/2}$$

If $p = \$5$ and $a = \$10,000$, determine the effect of adding additional salesmen. The marginal effect is described by the partial derivative.

$$\frac{\partial s}{\partial n} = \frac{1}{2} a^{1/2} (10,000 - 800p)n^{-1/2}$$

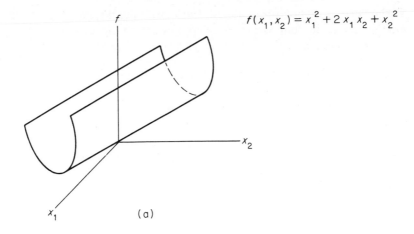

$$f(x_1, x_2) = x_1^2 + 2 x_1 x_2 + x_2^2$$

(a)

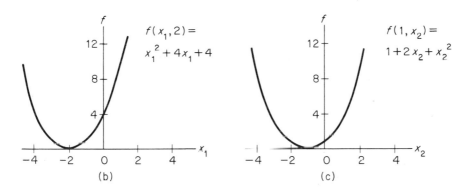

$$f(x_1, 2) = x_1^2 + 4x_1 + 4$$

(b)

$$f(1, x_2) = 1 + 2 x_2 + x_2^2$$

(c)

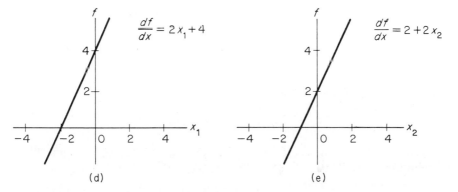

$$\frac{df}{dx} = 2 x_1 + 4$$

(d)

$$\frac{df}{dx} = 2 + 2 x_2$$

(e)

Figure 12.1

If the current number of salesmen is $n = 100$, the marginal increase in sales due to an additional salesman is approximated by evaluating the partial derivative with respect to n for $n = 100$, $p = \$5$, and $a = \$10,000$.

$$\frac{\partial s}{\partial n} = \tfrac{1}{2}(10,000)^{1/2}[10,000 - 800(5)](100)^{-1/2}$$

$$= \tfrac{1}{2}(100)(6000)\tfrac{1}{10} = \$30,000$$

The partial derivative represents the marginal change in s for a marginal change in n. We approximate the total change in s for a one-unit change in n by the product of the partial derivative and the unit change in n. Thus,

$$\Delta s \simeq \Delta n \cdot \frac{\partial s}{\partial n}$$

and

$$\Delta s \simeq (1)(\$30,000) = \$30,000$$

Similarly, if it were possible to add 0.1 of a salesman's time, the addition to total sales would be $3000.

Example: The sales of a manufacturing firm have been found to depend upon the amount invested in research and development R, plant and capital equipment P, and marketing M according to the following function:

$$S = 5R^{0.3}P^{0.5}M^{0.4}, \text{ in millions of dollars}$$

The following amounts are budgeted for investment during the current year: $R = \$12$ million; $P = \$7$ million; and $M = \$10$ million. Determine expected sales, using the predicting function for sales. Determine also the effect of additional expenditures in R, P, and M.

Expected sales can be calculated as follows:†

$$S = 5(12)^{0.3}(7)^{0.5}(10)^{0.4}$$

$$\log S = \log 5 + 0.3 \log 12 + 0.5 \log 7 + 0.4 \log 10$$

$$\log S = 0.6990 + 0.3(1.0792) + 0.5(0.8451) + 0.4(1.0000)$$

$$\log S = 1.8453$$

$$S = 70.03$$

The effect of additional expenditures can be determined from the partial derivatives. The partial derivative of S with respect to R is

$$\frac{\partial S}{\partial R} = 1.5R^{-0.7}P^{0.5}M^{0.4}$$

An additional expenditure of $1 in research and development results in a marginal increase in sales of $\Delta S = \$1 \cdot \partial S/\partial R$,

---

† The student is referred to Appendix B, Sec. B.4, for calculations using logarithms.

$$\frac{\partial S}{\partial R} = 1.5(12)^{-0.7}(7)^{0.5}(10)^{0.4} = \$1.75$$

Using the same procedure, we find that the marginal increase in sales for a \$1 increase in plant and equipment expenditures is \$4.96 and the marginal increase in sales for a \$1 increase in marketing is \$2.78.

12.1.2 HIGHER-ORDER DERIVATIVES OF MULTIVARIATE FUNCTIONS

Higher-order derivatives of multivariate functions are interpreted in the same manner as higher-order derivatives of functions of one independent variable. For $y = f(x_1, x_2)$, the second partial derivative of y with respect to x_1 describes the rate of change of the slope of the original function in planes parallel to the y and x_1 axes. Similarly, the second partial derivative of y with respect to x_2 describes the rate of change of the slope of the original function in the planes parallel to the y and x_2 axes. The symbol for the second partial derivative of y with respect to x_1 is $\partial^2 y / \partial x_1^2$, and the symbol for the second partial of y with respect to x_2 is $\partial^2 y / \partial x_2^2$.

Another higher-order derivative will be of interest in the next section. It is the *cross partial derivative*. For $y = f(x_1, x_2)$, the cross partial derivative is the derivative of $\partial y / \partial x_1$ with respect to x_2. The symbol for this cross partial derivative is $\dfrac{\partial^2 y}{\partial x_1\, \partial x_2}$. Alternatively, a cross partial derivative is the derivative of $\partial y / \partial x_2$ with respect to x_1. This would be written as $\dfrac{\partial^2 y}{\partial x_2\, \partial x_1}$.

An alternative notation is f_{x_1} and $f_{x_1 x_1}$ for the first and second partial derivatives with respect to x_1 and $f_{x_1 x_2}$ or $f_{x_2 x_1}$ for the cross partial derivatives.

The cross partial is a second derivative. It describes the rate of change of the slope of the function in a given plane as the plane is shifted incrementally. Thus, if $\partial y / \partial x_1$ represents the slope of the function in the family of planes that parallel the y and x_1 axes, then $\dfrac{\partial^2 y}{\partial x_1\, \partial x_2}$ describes the rate of change of this slope as the planes are moved marginally along the x_2 axis. The cross partial derivatives are always equal. That is,

$$\frac{\partial^2 y}{\partial x_1\, \partial x_2} = \frac{\partial^2 y}{\partial x_2\, \partial x_1}, \quad \text{or} \quad f_{x_1 x_2} = f_{x_2 x_1}$$

The rules for determining higher-order derivatives of functions of one independent variable apply to multivariate functions. Derivatives of multivariate functions are taken with respect to one independent variable at a time, the remaining independent variables being considered as constants. The same procedure applies in determining higher-order derivatives of multivariate functions. The higher-order derivative is determined with respect to a single independent variable, the remaining independent variables being considered as constants. This procedure is illustrated by the following examples.

Example: Determine all first and second derivatives of

$$y = 3x_1^3 + 2x_1^2x_2 - 3x_1x_2^2 + 4x_2^3$$

$$\frac{\partial y}{\partial x_1} = 9x_1^2 + 4x_1x_2 - 3x_2^2$$

$$\frac{\partial y}{\partial x_2} = 2x_1^2 - 6x_1x_2 + 12x_2^2$$

$$\frac{\partial^2 y}{\partial x_1^2} = 18x_1 + 4x_2 \qquad \frac{\partial^2 y}{\partial x_2^2} = -6x_1 + 24x_2$$

$$\frac{\partial^2 y}{\partial x_1 \, \partial x_2} = 4x_1 - 6x_2 \qquad \frac{\partial^2 y}{\partial x_2 \, \partial x_1} = 4x_1 - 6x_2$$

Example: Determine all first and second derivatives of

$$y = 3x_1^2 + 5x_1x_2 + 4x_2^2$$

$$\frac{\partial y}{\partial x_1} = 6x_1 + 5x_2 \qquad \frac{\partial y}{\partial x_2} = 5x_1 + 8x_2$$

$$\frac{\partial^2 y}{\partial x_1^2} = 6 \qquad \frac{\partial^2 y}{\partial x_2^2} = 8$$

$$\frac{\partial^2 y}{\partial x_1 \, \partial x_2} = 5 \qquad \frac{\partial^2 y}{\partial x_2 \, \partial x_1} = 5$$

Example: Determine all first and second derivatives of

$$y = (x_1^2 + 3x_2^3)^4$$

$$\frac{\partial y}{\partial x_1} = 4(x_1^2 + 3x_2^3)^3(2x_1) = 8x_1(x_1^2 + 3x_2^3)^3$$

$$\frac{\partial y}{\partial x_2} = 4(x_1^2 + 3x_2^3)^3(9x_2^2) = 36x_2^2(x_1^2 + 3x_2^3)^3$$

$$\frac{\partial^2 y}{\partial x_1^2} = 8x_1[3(x_1^2 + 3x_2^3)^2(2x_1)] + 8(x_1^2 + 3x_2^3)^3$$

$$\frac{\partial^2 y}{\partial x_1^2} = 48x_1^2(x_1^2 + 3x_2^3)^2 + 8(x_1^2 + 3x_2^3)^3$$

$$\frac{\partial^2 y}{\partial x_2^2} = 36x_2^2[3(x_1^2 + 3x_2^3)^2(9x_2^2)] + 72x_2(x_1^2 + 3x_2^3)^3$$

$$\frac{\partial^2 y}{\partial x_1\,\partial x_2} = 8x_1[3(x_1^2 + 3x_2^3)^2(9x_2^2)] = 216x_1 x_2^2(x_1^2 + 3x_2^3)^2$$

$$\frac{\partial^2 y}{\partial x_2\,\partial x_1} = 36x_2^2[3(x_1^2 + 3x_2^3)^2(2x_1)] = 216x_1 x_2^2(x_1^2 + 3x_2^3)^2$$

12.1.3   OPTIMUM VALUES OF MULTIVARIATE FUNCTIONS OF TWO INDEPENDENT VARIABLES

We now consider methods for determining the optimum value of functions of two independent variables. Methods of determining optimum values for functions of more than two independent variables are discussed in Sec. 12.1.4.

The optimum value of a multivariate function with two independent variables is determined in much the same manner as the optimum for a function with one independent variable. A function of two independent variables reaches a maximum or minimum when the slope of the function in both the x_1 and x_2 directions is 0. A maximum can be visualized as a hilltop and a minimum as an inverted hilltop.

A maximum or minimum can occur only where the partial derivative of the function with respect to x_1 and the partial derivative of the function with respect to x_2 are simultaneously 0. Symbolically, we state that

$$\frac{\partial f}{\partial x_1}(x_1^* = a, x_2^* = b) = 0$$

and

$$\frac{\partial f}{\partial x_2}(x_1^* = a, x_2^* = b) = 0$$

The two equations are solved simultaneously for $x_1^* = a$ and $x_2^* = b$. The values of x_1 and x_2 for which the derivatives are 0 are again termed *critical points*.[†]

The second derivatives of the function are used to determine if a critical point is a maximum or a minimum. If both second partial derivatives are positive when evaluated at the critical points, the optimum is a minimum. If both second derivatives are negative, the optimum is a maximum.

The condition that the first derivatives equal 0 and the second derivatives be positive or negative is not sufficient to guarantee an optimum for functions of two independent variables. An additional requirement for functions of two independent variables is that the product of the second derivatives

---

† They are also termed *stationary points*.

evaluated at the critical point be greater than the square of the cross partial derivative.† That is,

$$\frac{\partial^2 y}{\partial x_1^2} \cdot \frac{\partial^2 y}{\partial x_2^2} > \left[\frac{\partial^2 y}{\partial x_1 \, \partial x_2}\right]^2$$

The requirements for extrema are illustrated in the following example.

Example: For $z = 4 - x^2 - y^2$, determine the optimum. The function is plotted below. Note that it resembles the upper half of a football. The optimum value of the function is easily seen from the sketch to occur at the critical point $x^* = 0$, $y^* = 0$. Using the methods discussed above, we obtain

$$\frac{\partial z}{\partial x} = -2x = 0 \quad \text{and} \quad \frac{\partial z}{\partial y} = -2y = 0$$

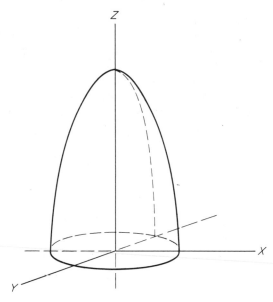

Solving these equations simultaneously gives $x^* = 0$ and $y^* = 0$ as a critical point. We determine the second partial derivatives and the cross partial derivative

$$\frac{\partial^2 z}{\partial x^2} = -2, \quad \frac{\partial^2 z}{\partial y^2} = -2, \quad \text{and} \quad \frac{\partial^2 z}{\partial x \, \partial y} = 0$$

Since

$$(-2)(-2) > (0)^2$$

---

† This requirement is necessary to ensure that the function is an optimum in all directions, rather than just along the principal axes.

the requirement for an optimum that the product of the second derivatives be greater than the square of the cross partial derivative is satisfied. The second derivatives evaluated at the critical points are negative; consequently, the optimum is a maximum. We therefore conclude that the function reaches a maximum value of $z = 4$ when $x^* = 0$ and $y^* = 0$.

In summary, the requirements for an optimum of a function $z = f(x, y)$ of two independent variables are

1. $\dfrac{\partial z}{\partial x}(x^* = a, y^* = b) = 0, \qquad \dfrac{\partial z}{\partial y}(x^* = a, y^* = b) = 0.$

2. $\dfrac{\partial^2 z}{\partial x^2}(x^* = a, y^* = b) \cdot \dfrac{\partial^2 z}{\partial y^2}(x^* = a, y^* = b)$

$$> \left[\dfrac{\partial^2 z}{\partial x\,\partial y}(x^* = a, y^* = b)\right]^2$$

Using the alternative notation for partial derivatives, we can express the requirement as

1. $f_x = 0, f_y = 0 \qquad$ at $(x^* = a, y^* = b)$

2. $f_{xx}f_{yy} > f_{xy}^2 \qquad$ at $(x^* = a, y^* = b)$

If these requirements are met, then for

$$\dfrac{\partial^2 z}{\partial x^2}(x^* = a, y^* = b) > 0$$

the function is a minimum, and for

$$\dfrac{\partial^2 z}{\partial x^2}(x^* = a, y^* = b) < 0$$

the function is a maximum.

It is possible that the product of the second derivatives evaluated at the point $x^* = a$ and $y^* = b$ will equal the square of the cross partial derivative. When this occurs, we cannot state whether or not the function is an optimum from the above conditions. We would instead examine the points around the critical point to determine if an optimum exists.

The following examples illustrate the method for determining extreme values of functions of two independent variables.

Example: Determine the values of x and y for which $z = 4x^2 + 2y^2 + 10x - 6y - 4xy$ is an optimum, specify whether the optimum is a maximum or minimum, and calculate the value of the function at the optimum.

$$\frac{\partial z}{\partial x} = 8x - 4y + 10 = 0$$

$$\frac{\partial z}{\partial y} = -4x + 4y - 6 = 0$$

Solving simultaneously for x and y gives $x^* = -1$ and $y^* = \frac{1}{2}$.

$$\frac{\partial^2 z}{\partial x^2} = 8, \qquad \frac{\partial^2 z}{\partial y^2} = 4, \qquad \frac{\partial^2 z}{\partial x\,\partial y} = -4$$

Since the second derivatives are positive and the product of the second derivatives is greater than the square of the cross partial derivative, the function reaches a minimum at $x^* = -1$ and $y^* = \frac{1}{2}$. The value of the function is

$$z(x^* = -1, y^* = \tfrac{1}{2}) = 4(-1)^2 + 2(\tfrac{1}{2})^2 + 10(-1) - 6(\tfrac{1}{2}) - 4(-1)(\tfrac{1}{2}) = -6\tfrac{1}{2}$$

Example: Determine the values of x and y for which $z = -4x^2 + 4xy - 2y^2 + 16x - 12y$ is an optimum, specify whether the optimum is maximum or minimum, and calculate the value of the function at the optimum.

$$\frac{\partial z}{\partial x} = -8x + 4y + 16 = 0$$

$$\frac{\partial z}{\partial y} = 4x - 4y - 12 = 0$$

Solving simultaneously gives $x^* = 1$ and $y^* = -2$.

$$\frac{\partial^2 z}{\partial x^2} = -8, \qquad \frac{\partial^2 z}{\partial y^2} = -4, \quad \text{and} \quad \frac{\partial^2 z}{\partial x\,\partial y} = 4$$

Since the second derivatives are negative and the product of the second derivatives is greater than the square of the cross partial derivative, the function reaches a maximum at $x^* = 1$ and $y^* = -2$. The value of the function at the optimum is $z = 20$.

Example: Determine the critical points and specify whether the function is a maximum or minimum.

$$f(x, y) = 2x^3 - 2xy + y^2$$

$$\frac{\partial f}{\partial x} = 6x^2 - 2y = 0$$

$$\frac{\partial f}{\partial y} = -2x + 2y = 0$$

Substituting $2x = 2y$ into the first equation, we obtain

$$6x^2 - 2x = 0$$

or, by factoring,

$$2x(3x - 1) = 0$$

The expression $2x(3x - 1)$ equals 0 if $x = 0$ or if $x = \frac{1}{3}$. This means that we must investigate two critical points, namely $x^* = 0$, $y^* = 0$, and $x^* = \frac{1}{3}$,

$y^* = \frac{1}{3}$. We determine which, if either, of these points is an optimum by the requirement

$$\frac{\partial^2 f}{\partial x^2} \cdot \frac{\partial^2 f}{\partial y^2} > \left[\frac{\partial^2 f}{\partial x\, \partial y}\right]^2$$

The second derivatives and the cross partial derivatives are

$$\frac{\partial^2 f}{\partial x^2} = 12x, \qquad \frac{\partial^2 f}{\partial y^2} = 2, \quad \text{and} \quad \frac{\partial^2 f}{\partial x\, \partial y} = -2$$

For an optimum it is required that $(12x)(2) > (-2)^2$.
For $(0, 0)$, $12(0)2 < (-2)^2$, the requirement is not met.
For $(\frac{1}{3}, \frac{1}{3})$, $12(\frac{1}{3})(2) > (-2)^2$, the requirement is met.
For $(\frac{1}{3}, \frac{1}{3})$, determine whether $f(x, y)$ is maximum or minimum.

$$\frac{\partial^2 f}{\partial x^2} = 12x \quad \text{and} \quad 12(\tfrac{1}{3}) > 0$$

the optimum is a minimum.

It is possible for a function of two independent variables to reach a maximum in the plane of one pair of axes and a minimum in another. Figure 12.2 shows a graph of the function $z = x^2 - y^2$. z reaches a maximum

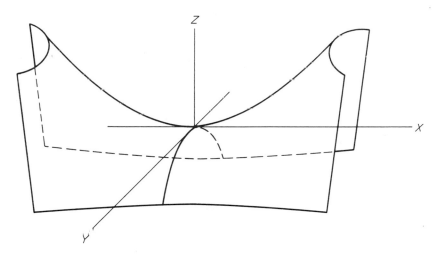

Figure 12.2

at the origin in the plane formed by the z and y axes but simultaneously reaches a minimum in the plane formed by the z and x axes. The point $(x^* = 0, y^* = 0)$ for this function is neither a maximum nor a minimum; it is called a *saddle point*.

We have stated that an optimum occurs at the critical points, provided that the product of the second derivatives exceeds the square of the cross partial derivative. If, as in the example illustrated in Fig. 12.2, the square of the cross partial derivative exceeds the product of the second derivatives, the function has reached a saddle point.

A function that reaches a saddle point is illustrated by the following example.

Example: Verify that a saddle point exists for the function

$$z = 18x^2 - 6y^2 - 36x - 48y$$

We first determine the critical values of x and y by equating the partial derivative of z with respect to x and the partial derivative of z with respect to y with 0 and solving the pair of equations simultaneously for x^* and y^*.

$$\frac{\partial z}{\partial x} = 36x - 36 = 0$$

$$\frac{\partial z}{\partial y} = -12y - 48 = 0$$

Solving for the critical values of x and y gives $x^* = 1$ and $y^* = -4$. The second derivatives of z are

$$\frac{\partial^2 z}{\partial x^2} = 36, \qquad \frac{\partial^2 z}{\partial y^2} = -12, \quad \text{and} \quad \frac{\partial^2 z}{\partial x\, \partial y} = 0$$

The second derivatives are of different sign. The requirement for a saddle point that

$$\frac{\partial^2 z}{\partial x^2} \cdot \frac{\partial^2 z}{\partial^2 y} < \left[\frac{\partial^2 z}{\partial x\, \partial y} \right]^2$$

is satisfied. We therefore conclude that the function reaches a saddle point at $x^* = 1$ and $y^* = -4$.

Multivariate functions are used in business and economics to describe problems in which two or more independent variables are functionally related to a single dependent variable. Chapter 13 introduces several of the more important models that are based upon the calculus of multivariate functions. We shall delay the development of the models until that chapter. At this point, however, we can illustrate the process of optimizing multivariate functions that describe business and economic problems.

Example: The Corola Typewriter Company manufactures two types of typewriters. Both typewriters are machines of fine quality, one being an electric portable and the second a manual portable. The revenue function that describes total revenue from sales of the two machines is rather complicated because of the interaction between sales of the competing types of machines. The revenue function, in units of thousands, is

$$R = 8E + 5M + 2EM - E^2 - 2M^2 + 20$$

where E represents the units of electric in thousands and M represents units of manual in thousands. Determine the quantity of electric and manual typewriters which leads to maximum revenue.

To determine the critical values of E and M we equate the partial derivatives of the revenue function with 0.

$$\frac{\partial R}{\partial E} = 8 + 2M - 2E = 0$$

$$\frac{\partial R}{\partial M} = 5 - 4M + 2E = 0$$

Solving simultaneously for E and M gives

$$E^* = 10.5 \quad \text{and} \quad M^* = 6.5$$

The second derivatives are negative and their product exceeds the square of the cross partial derivatives. This indicates that the critical values of the variables represent a maximum rather than a minimum or an inflection point. Consequently, revenue is maximum when 10,500 electric and 6500 manual typewriters are manufactured. The maximum revenue is $R = \$78,250$.

The Corola Typewriter Company offers an example in which two products have a degree of substitutability between each other. The example that follows is based upon the same principle, with the exception that factors used in the production process are considered rather than the output from the production process. The objective is to maximize output by using the appropriate quantities of factors of production.

Example: Western Valley Farms, Inc. has developed a sales revenue function that treats farm equipment E and labor L as independent variables with sales revenue as the dependent variable. The revenue function is

$$R = 18L + 24E + 10LE - 5E^2 - 8L^2$$

where L represents units of labor and E represents units of farm equipment. Determine the amount of labor and equipment that gives maximum revenue.

To determine the critical values of L and E, we equate the partial derivatives of the revenue function with 0.

$$\frac{\partial R}{\partial L} = 18 + 10E - 16L = 0$$

$$\frac{\partial R}{\partial E} = 24 - 10E + 10L = 0$$

Solving the two equations simultaneously gives

$$L^* = 7 \quad \text{and} \quad E^* = 9.4$$

The sales revenue generated from using these factors in the production process is $R = \$176$. Since the second derivatives are negative and since the product of the second derivatives exceeds the square of the cross partial derivative, the revenue is maximum when $L = 7$ units of labor and $E = 9.4$ units of equipment are used in the production process.

12.1.4 OPTIMUM VALUES OF MULTIVARIATE FUNCTIONS OF MORE THAN TWO INDEPENDENT VARIABLES†

Optimum values of functions of more than two variables are determined by equating the partial derivatives to zero and solving the resulting equations simultaneously for the critical values of each independent variable. Thus, for $y = f(x_1, x_2, \ldots, x_n)$, the procedure for determining the critical points is to equate the partial derivatives to zero and to solve the n partial derivative equations simultaneously for the critical points. That is,

$$\frac{\partial f}{\partial x_1}(x_1^*, x_2^*, \ldots, x_n^*) = 0$$

$$\frac{\partial f}{\partial x_2}(x_1^*, x_2^*, \ldots, x_n^*) = 0$$

$$\frac{\partial f}{\partial x_n}(x_1^*, x_2^*, \ldots, x_n^*) = 0$$

The rules for specifying if a critical point is a maximum, a minimum, or some combination of maximum and minimum are beyond the scope of this text.‡ We can, however, determine if the function is a maximum or minimum in specific planes.

Consider the function $y = f(x_1, x_2, x_3)$. The critical points are determined by solving the following equations simultaneously.

$$\frac{\partial f}{\partial x_1} = 0, \quad \frac{\partial f}{\partial x_2} = 0, \quad \frac{\partial f}{\partial x_3} = 0$$

For the critical point (x_1^*, x_2^*, x_3^*), we can state that the function is a maximum in the planes parallel to the y and x_1 axes, the y and x_2 axes, and the y and x_3 axes provided that the second derivatives are negative. The function is a minimum in these planes if the second derivatives are positive.

---

† This section can be omitted without loss of continuity.

‡ The technique is discussed by Alpha C. Chiang, *Fundamental Methods of Mathematical Economics* (McGraw-Hill, 1967), pp. 232–235.

To summarize, for a multivariate function to achieve an optimum it is necessary that all first derivatives be zero and all second derivatives (excluding cross partials) have the same sign. These conditions, however, are not sufficient. A third requirement, which is beyond the scope of this text, must be met. The student who takes advanced studies will be introduced to the requirement. The student who is not interested in advanced mathematics need only remember that functions of three or more independent variables cannot be treated with the same ease as those with only two independent variables.

The method of determining optimum values of a multivariate function along the major axes is illustrated in the following example.

Example: Determine the critical points for the following functions. Specify whether the function reaches a maximum or minimum along the major axes at the critical point.

$$f(x_1, x_2, x_3) = 2x_1^2 + x_1x_2 + 4x_2^2 + x_1x_3 + x_3^2 + 2$$

$$\frac{\partial f}{\partial x_1} = 4x_1 + x_2 + x_3 = 0$$

$$\frac{\partial f}{\partial x_2} = x_1 + 8x_2 = 0$$

$$\frac{\partial f}{\partial x_3} = x_1 + 2x_3 = 0$$

Solving simultaneously gives $x_1^* = 0$, $x_2^* = 0$, and $x_3^* = 0$ as the critical point. The second derivatives are

$$\frac{\partial^2 f}{\partial x_1^2} = 4, \qquad \frac{\partial^2 f}{\partial x_2^2} = 8, \qquad \frac{\partial^2 f}{\partial x_3^2} = 2$$

Since each of the second derivatives is positive, we conclude that the function reaches a minimum in the major axes at the critical point $x_1^* = 0$, $x_2^* = 0$, and $x_3^* = 0$.

12.2 Exponential and Logarithmic Functions

Up to this point in the text, our discussion of functions has been limited to algebraic functions. Another important class of functions that are widely used in business and economic models is the *exponential* and *logarithmic functions*. The exponential and logarithmic functions are subclassifications of a more general classification of functions termed *transcendental functions*. Since the exponential and logarithmic are the most widely applicable of the transcendental functions to business and economic problems, we shall limit our discussion to these two functional forms.

12.2.1 EXPONENTIAL FUNCTION

An exponential function is a function of the form

$$f(x) = k(a)^{cx} \qquad\qquad (12.3)$$

where $f(x)$ is the dependent variable, x is the independent variable, a is a constant termed the *base* that is greater than 0 and not equal to 1, and k and c are parameters that position the function on the coordinate axes. The possible forms of the exponential function are graphed in Fig. 12.3.

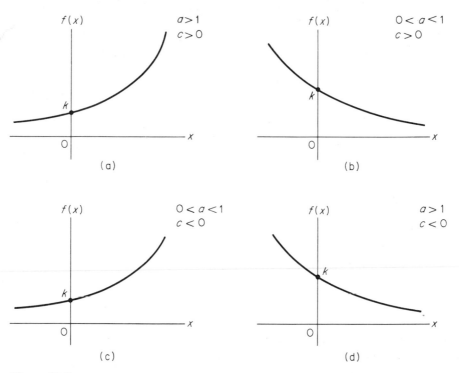

Figure 12.3

Figures 12.3(a) and 12.3(c) both show an exponential function in which the dependent variable increases as the independent variable increases. It can be shown that by proper selection of the constants a and c in Fig. 12.3(a), the functional relationship in Fig. 12.3(c) is completely described. Similarly, the functional relationships shown in Figs. 12.3(b) and 12.3(d) are interchangeable. The most commonly used forms of the exponential function are those in which a is greater than 1, the functions shown in Figs. 12.3(a) and 12.3(d). Those shown in Figs. 12.3(b) and 12.3(c) are, however, also applied in business and economic problems.

One of the characteristics of the exponential function that finds widespread application in business and economic problems is that the function describes constant rates of growth. As the independent variable increases by a constant amount in the exponential function, the dependent variable increases or decreases by a constant percentage. Thus, the value of an investment that increases by a constant percentage each period, the sales of a company that increase at a constant rate each period, and the value of an asset that declines at a constant rate each period are examples of functional relationships that are described by the exponential function. This important application of exponential functions is discussed in the section entitled "Mathematics of Finance" in Chapter 13.

12.2.2   THE NUMBER e

One of the most commonly used bases in the exponential function is the number e. e is defined as the limit as n approaches infinity of $(1 + 1/n)^n$. That is,

$$e = \operatorname*{limit}_{n \to \infty} \left(1 + \frac{1}{n}\right)^n \tag{12.4}$$

The numerical value of e with accuracy of nine places is 2.718281828. Replacing a by e in the formula for an exponential function, we obtain

$$f(x) = ke^{cx} \tag{12.5}$$

where k and c are again parameters, $f(x)$ and x are the dependent and independent variables, and e is the base of the function and has the value 2.718281828.

12.2.3   DERIVATIVES OF EXPONENTIAL FUNCTIONS

The derivative of an exponential function gives the slope or rate of change of the function for values of the independent variables. There are two rules for determining the derivatives of exponential functions. One applies to exponential functions in base e, the other to exponential functions with base other than e.

Rule 11. If $y = e^{g(x)}$, where $g(x)$ is a differentiable function of x, then

$$\frac{dy}{dx} = e^{g(x)}g'(x)$$

Rule 12. If $y = a^{g(x)}$, where $g(x)$ is a differentiable function of x, then

$$\frac{dy}{dx} = a^{g(x)}g'(x) \ln a$$

where "ln a" refers to the natural logarithm of a.
These two rules are illustrated by the following examples.

Example: $y = e^x$; determine dy/dx.

$$\frac{dy}{dx} = e^x(1)$$

The derivative of e^x is also e^x.

Example: $y = 2^x$; determine dy/dx.

$$\frac{dy}{dx} = 2^x(1) \ln 2 = 2^x(0.69315)$$

The value 0.69315 for ln 2 was obtained from Table IV.

Example: $y = 0.3e^{5x}$; determine the slope of the function at $x = 0.5$.

$$\frac{dy}{dx} = 0.3e^{5x}(5) = 1.5e^{5x}$$

$$\frac{dy}{dx}(x = 0.5) = 1.5e^{2.5} = 1.5(12.18) = 18.27$$

The slope of the function at $x = 0.5$ is 18.27. The value of 12.18 for $e^{2.5}$ was obtained from Table V.

Example: $y = 4(3)^{-0.1x}$; determine dy/dx.

$$\frac{dy}{dx} = 4(3)^{-0.1x}(-0.1) \ln 3$$

Example: $y = xe^{x^2}$; determine dy/dx.

$$\frac{dy}{dx} = x[e^{x^2}(2x)] + e^{x^2}(1)$$

This result comes from the product rule and the exponential rule.

Example: $y = xe^{2x}$; determine the optimum value of the function and specify whether the function is a maximum or a minimum.

$$\frac{dy}{dx} = xe^{2x}(2) + e^{2x}(1) = 0$$

$$e^{2x}(2x + 1) = 0$$

In order for the expression to equal 0, either e^{2x} equals 0 or $(2x + 1)$ equals

0. There is no value for x such that e^{2x} equals 0. Therefore, the critical value of x is determined by equating $2x + 1$ with 0. The function is an optimum when $x^* = -\frac{1}{2}$. To determine if the critical point is a maximum or minimum, we evaluate the second derivative at the critical point.

$$\frac{d^2y}{dx^2} = e^{2x}(2) + (2x + 1)e^{2x}(2)$$

$$= 2e^{2x}(2x + 2)$$

$$\frac{d^2y}{dx^2}\left(x^* = -\frac{1}{2}\right) = 2e^{-1}(-1 + 2) = \frac{2}{e} > 0$$

Since the second derivative is positive, the function is a minimum at $x^* = -\frac{1}{2}$.

Example: $y = e^{(2x_1 + x_2)}$; determine the first derivatives.

$$\frac{\partial y}{\partial x_1} = e^{(2x_1 + x_2)}(2) = 2e^{(2x_1 + x_2)}$$

$$\frac{\partial y}{\partial x_2} = e^{(2x_1 + x_2)}(1) = e^{(2x_1 + x_2)}$$

Example: $y = x_1 x_2 e^{(x_1 + 2x_2)}$; determine the first derivatives.

$$\frac{\partial y}{\partial x_1} = (x_1 x_2)e^{(x_1 + 2x_2)}(1) + e^{(x_1 + 2x_2)}(x_2)$$

$$\frac{\partial y}{\partial x_1} = e^{x_1 + 2x_2}(x_1 x_2 + x_2)$$

$$\frac{\partial y}{\partial x_2} = x_1 x_2 e^{(x_1 + 2x_2)}(2) + e^{x_1 + 2x_2}(x_1)$$

$$\frac{\partial y}{\partial x_2} = e^{(x_1 + 2x_2)}(2x_1 x_2 + x_1)$$

Example: The demand function for a commodity is described by the exponential function $p = 7.50e^{-0.001q}$. Determine the quantity and price for which total revenue is a maximum.

$$TR = q \cdot p = q(7.50e^{-0.001q})$$

$$\frac{d(TR)}{dq} = 7.50q[e^{-0.001q}(-0.001)] + 7.50e^{-0.001q} = 0$$

$$= 7.50e^{-0.001q}(-0.001q + 1) = 0$$

We must determine the value of q for which the expression is 0. For the expression to equal 0, either $7.50e^{-0.001q}$ equals 0 or $-0.001q + 1$ equals 0. Since $7.50e^{-0.001q}$ is not equal to 0 for any value of q, we equate $-0.001q + 1$ with 0 and obtain $q = 1000$.

12.2.4 LOGARITHMIC FUNCTIONS

A logarithmic function is a function of the form

$$y = \log_a x \tag{12.6}$$

where y is the dependent variable, x is the independent variable, and a is a constant termed the base that is greater than 0 and not equal to 1.

The relationship among the dependent variable y, the independent variable x, and the base a can be explained as follows. In a logarithmic function, the dependent variable is equal to the exponent (or power) to which the base a must be raised to give x. To illustrate, assume that we have the logarithmic function

$$y = \log_{10} x$$

Remembering that y is the exponent to which 10 must be raised to equal x, we can calculate the value of y for any value of x in the domain $0 < x < \infty$. For instance, if $x = 100$ and $10^2 = 100$, then $y = 2$. Similarly, if $x = 1000$, then $y = 3$.

Example: Plot the logarithmic function

$$y = \log_{10} x \qquad \text{for } 0 < x \le 50$$

The graph of this function is shown below. The values of the dependent variable y are obtained from Table III, Common Logarithms.

x	0.1	1	5	10	25	50
y	−1	0	0.6990	1	1.3979	1.6990

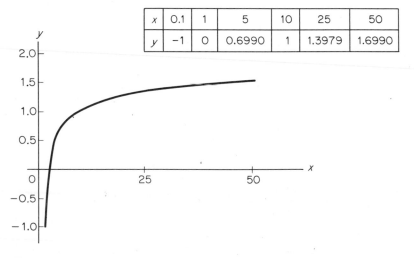

Figure 12.4

The logarithmic function is the inverse of the exponential function. In Sec. 9.4.4, we explained that if $y = f(x)$ is solved for x to give $x = g(y)$, then $y = f(x)$ and $x = g(y)$ are inverse functions.

It can be shown that the logarithmic and exponential functions are inverse functions. Given the exponential function $y = e^x$, one can obtain $x = \ln y$ by taking the logarithm of both sides of the exponential function.† Thus, $y = e^x$ and $x = \ln y$ are inverse functions.

The logarithmic function describes a functional relationship in which constant proportional changes in the independent variable result in constant arithmetic changes in the dependent variable. As an example, consider the exponential function $y = 10^x$. As x increases in units of 1, y increases by a factor of 10. If this exponential function is rewritten as a logarithmic function, we obtain $x = \log_{10} y$. In the logarithmic form, for $y = 10$, $x = 1$, for $y = 100$, $x = 2$, etc. We conclude that constant proportional changes in the independent variable y result in constant arithmetic changes in x. Fig. 12.5 provides a graph of the logarithmic function.

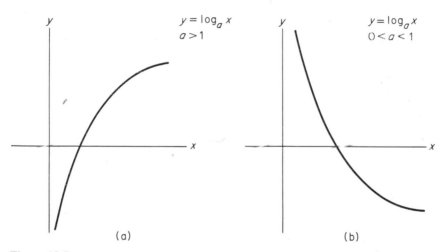

$$y = \log_a x$$
$$a > 1$$

$$y = \log_a x$$
$$0 < a < 1$$

(a)

(b)

Figure 12.5

There are two rules for determining the derivatives of logarithmic functions. Rule 13 applies to logarithmic functions of the base e, and Rule 14 applies to all bases other than e.

Rule 13. If $y = \ln g(x)$, where $g(x)$ is a differentiable function of x, then

$$\frac{dy}{dx} = \frac{g'(x)}{g(x)}$$

---

† Logarithms are explained in Appendix B.

Rule 14. If $y = \log_a g(x)$, where $g(x)$ is a differentiable function of x, then

$$\frac{dy}{dx} = \frac{g'(x)}{g(x)} \cdot \log_a e$$

Example: $y = \ln x$; determine dy/dx.

$$\frac{dy}{dx} = \frac{1}{x}$$

Example: $y = \log_{10} x$; determine dy/dx.

$$\frac{dy}{dx} = \frac{1}{x} \log_{10} e$$

and since $\log_{10} e = 0.4343$,

$$\frac{dy}{dx} = \frac{0.4343}{x}$$

Example: $y = \ln (x^2 + 2x + 10)$; determine dy/dx.

$$\frac{dy}{dx} = \frac{2x + 2}{x^2 + 2x + 10}$$

Example: $y = [\ln (x + 3)]^2$; determine dy/dx.

$$\frac{dy}{dx} = 2[\ln (x + 3)]^1 \frac{1}{x + 3}$$

Example: Determine the slope of the function in the preceding example at $x = 2$.

$$\frac{dy}{dx}(x = 2) = \frac{2 \ln 5}{5} = \frac{2}{5}(1.60944) = 0.6438$$

Example: $y = \ln (x^2 - 4x + 6)$; determine the critical point. Specify whether the function is a maximum or minimum at the critical point.

$$\frac{dy}{dx} = \frac{2x - 4}{x^2 - 4x + 6} = 0$$

Multiplying both sides by $x^2 - 4x + 6$ gives

$$2x - 4 = 0$$

and $x^* = 2$ is a critical point. Evaluating the second derivative at $x^* = 2$ gives

$$\frac{d^2y}{dx^2} = \frac{(x^2 - 4x + 6)(2) - (2x - 4)(2x - 4)}{(x^2 - 4x + 6)^2}$$

$$\frac{d^2y}{dx^2}(x = 2) = 1 > 0$$

Thus, the function reaches a minimum at $x^* = 2$.

Example: $y = \ln(x_1^2 + 2x_1x_2)$; determine the partial derivatives.

$$\frac{\partial y}{\partial x_1} = \frac{2x_1 + 2x_2}{x_1^2 + 2x_1x_2}$$

$$\frac{\partial y}{\partial x_2} = \frac{2x_1}{x_1^2 + 2x_1x_2}$$

12.2.5 LOGARITHMIC DIFFERENTIATION†

The derivatives of certain functions can be most easily determined if the function is expressed in logarithms. This technique, termed *logarithmic differentiation*, is especially applicable to functions of the form

$$y = f(x)^{g(x)} \tag{12.7}$$

As an example, consider the function $y = x^x$. Neither the rule for the derivative of a variable raised to a constant, Rule 3, nor that for a constant raised to a variable, Rule 12, applies to this function. Consequently, we must develop a formula that gives the derivative of a variable raised to a variable power.

In developing the rule, we first take the natural logarithm of both sides of (12.7).

$$\ln y = g(x)\ln f(x)$$

To illustrate the derivative process for this function, we shall consider the derivative of the term on the left side of the equal sign and that on the right side separately. The derivative of $\ln y$ with respect to x is found by using Rule 13:

$$\frac{d(\ln y)}{dx} = \frac{1}{y}\frac{dy}{dx}$$

By the application of the product rule, Rule 6, and Rule 13, the derivative of $g(x) \ln f(x)$ is

$$\frac{d[g(x)\ln f(x)]}{dx} = g(x)\frac{f'(x)}{f(x)} + g'(x)\ln f(x)$$

The derivative of

$$\ln y = g(x)\ln f(x)$$

is thus

$$\frac{1}{y}\frac{dy}{dx} = g(x)\frac{f'(x)}{f(x)} + g'(x)\ln f(x)$$

Multiplying both sides by y gives

$$\frac{dy}{dx} = y\left[g(x)\frac{f'(x)}{f(x)} + g'(x)\ln f(x)\right]$$

---

† This section can be omitted without loss of continuity.

If we now replace y by $f(x)^{g(x)}$, we obtain a formula for determining the derivative of a variable raised to a variable power.

Rule 15. If $y = f(x)^{g(x)}$, the derivative is given by

$$\frac{dy}{dx} = f(x)^{g(x)} \left[g(x)\frac{f'(x)}{f(x)} + g'(x) \ln f(x) \right]$$

Example: $y = x^x$; determine dy/dx.

$$\frac{dy}{dx} = x^x \left[\frac{x(1)}{x} + (1) \ln x \right] = x^x(1 + \ln x)$$

Example: $y = (x + 4)^{x^2}$; determine dy/dx.

$$\frac{dy}{dx} = (x + 4)^{x^2} \left[x^2 \frac{1}{(x + 4)} + 2x \ln (x + 4) \right]$$

12.3 Constrained Optima

The techniques of differential calculus can be used to determine the optimum value of a nonlinear function that is subject to one or more constraints. For instance, one fairly common problem involves maximizing a nonlinear profit function subject to constraints on factory capacity and working capital. Alternatively, another problem might involve the maximization of sales revenue subject to constraints on advertising and on production capacity. This type of problem involves determining the optimum value of a *nonlinear objective function* subject to one or more *linear or nonlinear constraints*.

The reader will recognize the similarity between the problems described in the preceding paragraph and the linear programming problem discussed in Chapter 5. The important difference between the two types of problems is in the form of the objective function and constraints. Both the objective function and the constraints must be linear in the linear programming problem. The problem presented in this section differs in that the objective function must be nonlinear. The constraining equations, however, can be linear or nonlinear.

The general problem is that of finding the extreme points of the multivariate function $z = f(x, y)$ subject to equalities of the form $g(x, y) = 0$. The function $z = f(x, y)$ is termed the *objective function* and $g(x, y) = 0$ is the *constraining equation*. If we represent the critical values of x and y by x^* and y^*, the problem becomes that of determining critical points such that

$$\frac{\partial f}{\partial x}(x^*, y^*) = 0, \qquad \frac{\partial f}{\partial y}(x^*, y^*) = 0, \quad \text{and} \quad g(x^*, y^*) = 0$$

12.3.1   SUBSTITUTION

One method of solving the constrained optima problem is that of substitution. This involves solving the constraining equation for one variable in terms of the other and substituting this expression in the objective function. The objective function is then optimized by using the procedure discussed in the preceding section. This is illustrated by the following example.

Example: Determine the local optima of $z = x^2 + 3xy + y^2$, subject to the constraining equation $x + y = 100$.

Solving the constraining equation for y and substituting into the objective function, we obtain

$$z = x^2 + 3x(100 - x) + (100 - x)^2$$

To determine the critical point we equate dz/dx with 0 and solve for x.

$$\frac{dz}{dx} = 2x + 3x(-1) + 3(100 - x) + 2(100 - x)(-1) = 0$$

and

$$x^* = 50, \qquad y^* = 100 - x^* = 50$$

are the critical points.

Applying the second derivative test to the function gives

$$\frac{d^2z}{dx^2} = -2$$

This indicates that the constrained function has a local maximum in the plane parallel to the z and x axes. If the objective function is written in terms of y rather than x, we find that the function also reaches a local maximum in the plane parallel to the z and y axes. The value of the constrained maximum is $z = 12,500$.

The method of substitution in solving constrained optima problems becomes quite difficult if the constraint is at all complex. For example, the method of substitution is rather cumbersome because of the difficulty of solving for x in terms of y for constraints of the form $ax^2 + bxy + y^2 = 0$. An alternative method of solution is, however, available. This is the method of Lagrangian multipliers.

12.3.2   LAGRANGIAN MULTIPLIERS

Lagrangian multipliers provide a method of determining the optimum value of a differentiable nonlinear function subject to linear or nonlinear constraints. The method of solution is to form the Lagrangian expression,

$$F(x, y, \lambda) = f(x, y) + \lambda g(x, y) \tag{12.8}$$

This expression consists of the objective function $f(x, y)$, and the product of λ (lambda) and the constraining equation $g(x, y)$. The coefficient of the constraining equation, λ, is termed the Lagrangian multiplier. Since the constraining equation is equal to 0, the addition of the term $\lambda g(x, y)$ to the objective function $f(x, y)$ does not change the value of the function. The necessary condition for an extreme point is that the partial derivatives of the function with respect to x, y, and λ equal 0; that is,

$$\frac{\partial F}{\partial x} = 0, \quad \frac{\partial F}{\partial y} = 0, \quad \text{and} \quad \frac{\partial F}{\partial \lambda} = 0$$

The three equations with variables x, y, and λ are solved simultaneously for the critical values of x^*, y^*, and λ^*.

Since the Lagrangian expression has more than two independent variables, the second derivative test does not apply in specifying whether the function has reached a local maximum or minimum. Rather, we can determine if the function is a local maximum or minimum by evaluating the function at points adjacent to the critical points in the constraining plane. The procedure is summarized as follows. Let x^* and y^* represent critical points in the Lagrangian expression. $f(x^*, y^*)$ is a constrained maximum if

$$f(x^*, y^*) > f(x^* - \Delta x, y^* + \Delta y)$$

and

$$f(x^*, y^*) > f(x^* + \Delta x, y^* - \Delta y)$$

It is a constrained minimum if

$$f(x^*, y^*) < f(x^* - \Delta x, y^* + \Delta y)$$

and

$$f(x^*, y^*) < f(x^* + \Delta x, y^* - \Delta y)$$

where both $g(x^* + \Delta x, y^* - \Delta y) = 0$, and $g(x^* - \Delta x, y^* + \Delta y) = 0$. In selecting Δx and Δy, Δx need not equal Δy. However, it is important to remember that Δx and Δy must be selected such that the constraining equation remains equal to 0.

The method of Lagrangian multipliers is illustrated by the following examples.

Example: Determine the critical points and the constrained optima for

$$z = x^2 + 3xy + y^2$$

subject to

$$x + y = 100$$

The Lagrangian expression is

$$F(x, y, \lambda) = x^2 + 3xy + y^2 + \lambda(x + y - 100)$$

and the partial derivatives are

$$\frac{\partial F}{\partial x} = 2x + 3y + \lambda = 0$$

$$\frac{\partial F}{\partial y} = 3x + 2y + \lambda = 0$$

$$\frac{\partial F}{\partial \lambda} = x + y - 100 = 0$$

The three equations are solved simultaneously for

$$x^* = 50, \quad y^* = 50, \quad \text{and} \quad \lambda^* = -250$$

The critical values of x, y, and λ can be substituted into the Lagrangian expression to obtain the constrained optimum. For the example problem,

$$f(50, 50, -250) = (50)^2 + 3(50)(50) + (50)^2 - 250(50 + 50 - 100)$$
$$= 2500 + 7500 + 2500 - 250(0)$$
$$= 12,500$$

To determine if the function reaches a maximum or a minimum, we evaluate the function at points adjacent to $x = 50$ and $y = 50$. The function is a constrained maximum, since

$$F(49, 51, -250) = 12,499$$

and

$$F(51, 49, -250) = 12,499$$

Example: Determine the critical points and the constrained optima for

$$z = x^2 - xy + y^2$$

subject to

$$x + y = 100$$

The Lagrangian expression is

$$F(x, y, \lambda) = x^2 - xy + y^2 + \lambda(x + y \quad 100)$$

and the partial derivatives are

$$\frac{\partial F}{\partial x} = 2x - y + \lambda = 0$$

$$\frac{\partial F}{\partial y} = -x + 2y + \lambda = 0$$

$$\frac{\partial F}{\partial \lambda} = x + y - 100 = 0$$

The three equations are solved simultaneously for

$$x^* = 50, \quad y^* = 50, \quad \text{and} \quad \lambda = -50$$

The constrained optimum is

$$F(50, 50, -50) = (50)^2 - (50)(50) + (50)^2 - 50(0)$$
$$= 2500$$

This constrained optimum is a local minimum, since

$$F(49, 51, -50) = 2503$$

and

$$F(51, 49, -50) = 2503$$

In this and the preceding example the constraining equation $g(x, y) = x + y - 100 = 0$ is satisfied by the data points $(49, 51)$ and $(51, 49)$. We have, therefore, been able to use these points to determine if the optimum was a constrained maximum or constrained minimum.

As stated previously, the second derivative test does not apply in Lagrangian multipliers. This is illustrated by the preceding two examples. In both examples the second derivatives were positive. On the basis of the second derivatives we would conclude that the functions were constrained minima. This conclusion is incorrect for the first example, since the constrained optimum was a maximum.

The Lagrangian multiplier λ provides an indication of the effect of a unit change in the constant in the constraining equation on the objective function. From the Lagrangian expression in (12.8), note that the expression $\lambda g(x, y)$ is added to the objective function to obtain the constrained optimum. Since $g(x, y)$ equals 0, the product of $\lambda g(x, y)$ is 0. If, however, we alter the constraining equation to allow $g(x, y)$ to assume an incrementally negative or positive value, then, provided λ is not 0, $\lambda g(x, y)$ is no longer 0. This term now adds to or subtracts from the objective function. The effect of $\lambda g(x, y)$ depends upon the magnitude of λ and the signs of both λ and $g(x, y)$.

To illustrate, assume that the constant in the constraining equation in the first example problem is increased by 1. The constraint is now $x + y = 101$, and $g(x, y) = x + y - 101$. For $x = 50$, $y = 50$, and $\lambda = -250$, $g(x, y) = -1$ and $\lambda g(x, y) = +250$. The effect of increasing the constant in the constraint by 1 results in the addition of a positive term to the constrained optimum. Similarly, if the constant in the constraining equation had been reduced by 1, the constrained optimum would be reduced.

The objective function is not linear. Therefore, the exact effect upon the objective function of a unit or incremental change in the constraint is not specified by the multiplier. The multiplier does, however, show the approximate effect of a one-unit change on the constrained optimum. This is shown by the following example.

Example: Determine the effect of increasing the constant in the constraint of the preceding problem to 101.

$$z = x^2 + 3xy + y^2$$

subject to

$$x + y = 101$$

$$F(x, y, \lambda) = x^2 + 3xy + y^2 + \lambda(x + y - 101)$$

Solving for the critical value gives

$$x^* = 50.5, \qquad y^* = 50.5, \quad \text{and} \quad \lambda^* = -252.5$$

and

$$F(50.5, 50.5, -252.5) = 12{,}751.25$$

The constrained optimum for $g(x, y) = x + y - 101$ exceeds the constrained optimum for $g(x, y) = x + y - 100$ by 251.25. This is approximately equal to $\lambda g(x, y) = -250(-1) = 250$.

Interpretation of the Lagrangian multiplier can be summarized as follows. The magnitude of the Lagrangian multiplier indicates the approximate change in the objective function for a unit change in the constant in the constraint. If λ is positive, the constrained optimum will increase if the constant in the constraint is decreased and decrease if the constant in the constraint is increased. If λ is negative, the constrained optimum will increase if the constant in the constraint is increased and decrease if the constant in the constraint is decreased.

Example: Determine the constrained optimum and the approximate effect of a unit change in the constraining equation for the function

$$z = 3x^2 - 6xy + y^2$$

subject to the constraint that

$$2x + y = 150$$

The Lagrangian expression and the partial derivatives are

$$F(x, y, \lambda) = 3x^2 - 6xy + y^2 + \lambda(2x + y - 150)$$

$$\frac{\partial F}{\partial x} = 6x - 6y + 2\lambda = 0$$

$$\frac{\partial F}{\partial y} = -6x + 2y + \lambda = 0$$

$$\frac{\partial F}{\partial \lambda} = 2x + y - 150 = 0$$

The critical values of x, y, and λ are $x^* = 39.5$, $y^* = 71$, and $\lambda^* = 95$. Substitution of these critical values into the Lagrangian expression gives a constrained optimum of $F(39.5, 71, 95) = -7105.25$. We can determine if this optimum is a maximum or minimum by application of the method established above. Since $g(x, y) = 2x + y - 150$, we select $\Delta x = 0.5$ and

$\Delta y = 1.0$ so that $g(x^* + \Delta x, y^* - \Delta y) = 0$ and $g(x^* - \Delta x, y^* + \Delta y) = 0$. This gives

$$F(40, 70, 95) = -7100$$

and

$$F(39, 72, 95) = -7100$$

The critical values of $x^* = 39.5$ and $y^* = 71$ yield a value, $F(39.5, 71, 95) = -7105.25$, which is smaller than at any adjacent point on the constraining plane. Consequently, the function reaches a constrained minimum at the critical points.

The method of determining constrained optima by Lagrangian multipliers can be expanded to include more than one constraint. For the general case of m variables with n constraints, a multiplier is introduced for each of the n constraints. The $(m + n)$ partial derivatives of the Lagrangian expression are equated to 0 and solved simultaneously for the m critical values and the n Lagrangian multipliers. This procedure is illustrated in the following example for three variables and two constraints.

Example: Determine the critical points of the function

$$z = f(x, y, w) = 2xy - 3w^2$$

subject to

$$x + y + w = 15 \quad \text{and} \quad x - w = 4$$

Since we have two constraining equations, we must use two Lagrangian multipliers. The method of solution involves forming the Lagrangian expression and equating the partial derivatives with 0.

$$F(x, y, w, \lambda_1\lambda_2) = 2xy - 3w^2 + \lambda_1(x + y + w - 15) + \lambda_2(x - w - 4)$$

$$\frac{\partial F}{\partial x} = 2y + \lambda_1 + \lambda_2 = 0$$

$$\frac{\partial F}{\partial y} = 2x + \lambda_1 = 0$$

$$\frac{\partial F}{\partial w} = -6w + \lambda_1 - \lambda_2 = 0$$

$$\frac{\partial F}{\partial \lambda_1} = x + y + w - 15 = 0$$

$$\frac{\partial F}{\partial \lambda_2} = x - w - 4 = 0$$

Solving the five equations simultaneously gives

$$x^* = 4.43, \quad y^* = 10.14, \quad w^* = 0.43, \quad \lambda_1 = -8.86, \quad \lambda_2 = -11.42$$

The Lagrangian multipliers have the same signs, λ_1 and λ_2 being negative. This indicates that if the constant in the first constraining equation is incrementally increased, the objective function will increase. Similarly, if the constant in the second constraining equation is increased, the objective function will also increase. The value of the function at the constrained optimum is

$$F(4.43, 10.14, 0.43, -8.86, -11.42) = 89.5$$

The procedure for determining if the optimum is a constrained maximum or minimum is conceptually the same as expressed above. Given the function $F(x, y, w)$ and the constraining equations $g(x, y, w)$ and $h(x, y, w)$, we investigate points adjacent to the critical values of x, y, and w. For the case of three independent variables, 12 adjacent points must be evaluated. This set of 12 points is

$$
\begin{aligned}
S = \{ & (x + \Delta x, y - \Delta y, w), (x + \Delta x, y - \Delta y, w + \Delta w), \\
& (x + \Delta x, y - \Delta y, w - \Delta w), (x + \Delta x, y + \Delta y, w - \Delta w), \\
& (x - \Delta x, y + \Delta y, w), (x - \Delta x, y + \Delta y, w + \Delta w), \\
& (x - \Delta x, y + \Delta y, w - \Delta w), (x - \Delta x, y - \Delta y, w - \Delta w), \\
& (x, y + \Delta y, w - \Delta w), (x, y - \Delta y, w + \Delta w), \\
& (x + \Delta x, y, w - \Delta w), (x - \Delta x, y, w + \Delta w) \}
\end{aligned}
$$

Each adjacent point must be selected so that the constraining equations are satisfied, i.e., $g(x, y, w) = 0$ and $h(x, y, w) = 0$. If these conditions are met, the function is a constrained minimum when the function evaluated at the critical point is less than it is when it is evaluated at any adjacent point. It is a constrained maximum when the reverse is the case.

The method of Lagrangian multipliers is useful in allocating scarce resources between alternative uses. The following two example problems illustrate the Lagrangian technique for optimizing revenue functions subject to budgetary constraints.

Example: Revenue of the Rinehart Distributing Company is related to advertising A and the quantity Q produced and sold according to the function

$$R = 520 - 5A^2 + 21A + 15QA - 4.5Q^2 + 15Q$$

The budgetary constraint for advertising and production is

$$2Q + A = 10$$

Determine the values of Q and A that maximize revenue subject to the budgetary constraint.

To determine the critical values of the variables, we form the Lagrangian expression

$$
\begin{aligned}
F(A, Q, \lambda) = {} & 520 - 5A^2 + 21A + 15QA - 4.5Q^2 \\
& + 15Q + \lambda(2Q + A - 10)
\end{aligned}
$$

The partial derivatives of this expression are equated to 0.

$$\frac{\partial F}{\partial A} = -10A + 15Q + \lambda + 21 = 0$$

$$\frac{\partial F}{\partial Q} = 15A - 9Q + 2\lambda + 15 = 0$$

$$\frac{\partial F}{\partial \lambda} = A + 2Q - 10 = 0$$

The three equations are solved simultaneously, giving

$$A^* = 4.08, \qquad Q^* = 2.96, \quad \text{and} \quad \lambda = -24.6$$

The maximum revenue subject to the budgetary constraint is

$$F(4.08, 2.96, -24.6) = 708.8$$

From the value of the Lagrangian multiplier, we see that a one-unit increase in the budgetary constraint, i.e.,

$$2Q + A = 11$$

would result in approximately 24.6 units of additional revenue.

Example: The revenue function for the Corola Typewriter Company was given on p. 422 as

$$R = 8E + 5M + 2EM - E^2 - 2M^2 + 20$$

where E represented the number of electric portables and M the number of manual portable typewriters. The maximum revenue occurred when $E^* = 10,500$ and $M^* = 6500$. The maximum revenue was $R = \$78,250$. Assume that capacity constraints limit production to 14,000 typewriters. Determine the optimum product mix based upon this constraint.

The constraint can be incorporated into the problem through the method of Lagrangian multipliers. The Lagrangian expression is

$$F(E, M, \lambda) = 8E + 5M + 2EM - E^2 - 2M^2 + 20 + \lambda(E + M - 14)$$

The partial derivatives of the Lagrangian expression are equated to 0.

$$\frac{\partial F}{\partial E} = 8 + 2M - 2E + \lambda = 0$$

$$\frac{\partial F}{\partial M} = 5 - 4M + 2E + \lambda = 0$$

$$\frac{\partial F}{\partial \lambda} = -14 + M + E = 0$$

Solving the three equations simultaneously gives

$$M^* = 5.3, \qquad E^* = 8.7, \quad \text{and} \quad \lambda^* = -1.2$$

The sales revenue resulting from 5300 manual typewriters and 8700 electric typewriters is $R = \$76,500$.

12.3.3 INEQUALITY RESTRICTIONS

The method of Lagrangian multipliers can be modified to incorporate constraints that take the form of inequalities rather than equalities. The problem now becomes that of determining the extreme points of the multivariate function $z = f(x, y)$ subject to the inequality $g(x, y) \leq 0$ or $g(x, y) \geq 0$. We shall show a relatively simple extension of the Lagrangian technique which provides a solution to the problem of optimizing an objective function subject to a single constraining inequality. The general problem of optimizing an objective function subject to n inequalities is not considered in this text.

In the problem of optimizing an objective function subject to a constraining inequality, two cases must be considered. First, the constraining equality may act as an upper or lower bound on the function. Consider the objective function $z = x^2 + 3xy + y^2$ and the constraining inequality $x + y \leq 100$. In this problem, it is obvious that $x + y \leq 100$ acts as an upper bound on the values of x and y. Were it not for the constraining inequality, the function would reach a maximum at infinity.

The constraining inequality need not, however, act as an upper or lower bound on the function. As a second case, consider the function

$$z = -4x^2 + 4xy - 2y^2 + 16x - 12y$$

The critical values of x and y were determined on p. 420 to be $x^* = 1$ and $y^* = -2$. If this problem were modified by the addition of the constraining inequality $x + y \leq 10$, the solution to the problem is not changed. Consequently, the constraining inequality does not act as an upper or lower bound on the function in this case.

For either of the above cases, the method of optimizing a function subject to a constraining inequality is to assume that the constraining inequality is an equality; that is, assume $g(x, y) = 0$. The inequality is changed to an equality and the critical values and constrained optimum are obtained by using the method of Lagrangian multipliers. The sign of the Lagrangian multiplier is used to determine whether the constraint is actually limiting the optimum value of the objective function. The procedure is as follows:

For maximizing the objective function subject to the constraining inequality $g(x, y) \leq 0$:

1. If $\lambda > 0$, the restriction is not a limitation; we resolve the problem, ignoring the restriction, to obtain the optimum.
2. If $\lambda \leq 0$, the restriction acts as an upper bound, and the result obtained by assuming that $g(x, y) = 0$ is the constrained optimum.

For minimizing the objective function subject to $g(x, y) \leq 0$:

1. If $\lambda > 0$, the restriction is a limitation, and the constrained optimum is obtained by assuming that $g(x, y) = 0$.
2. If $\lambda \leq 0$, the restriction is not a limitation; we resolve the problem, ignoring the restriction, to obtain the optimum.

For maximizing the objective function subject to $g(x, y) \geq 0$:

1. If $\lambda > 0$, the restriction is a limitation, and the constrained optimum is obtained by assuming that $g(x, y) = 0$.
2. If $\lambda \leq 0$, the restriction is not a limitation; we resolve the problem, ignoring the restriction, to obtain the optimum.

For minimizing the objective function subject to $g(x, y) \geq 0$:

1. If $\lambda > 0$, the restriction is not a limitation; we resolve the problem, ignoring the restriction, to obtain the optimum.
2. If $\lambda \leq 0$, the restriction is a limitation, and the constrained optimum is obtained by assuming that $g(x, y) = 0$.

The following examples illustrate this technique.

Example: Determine the maximum of the function $z = x^2 + 3xy + y^2$ subject to

$$x + y \leq 100$$

To determine the solution we treat the inequality as an equality and form the Lagrangian expression

$$F(x, y, \lambda) = x^2 + 3xy + y^2 + \lambda(x + y - 100)$$

The problem was solved on p. 436. The critical values are

$$x^* = 50, \qquad y^* = 50, \quad \text{and} \quad \lambda^* = -250$$

Since the Lagrangian multiplier is negative, we conclude that the value of the function is limited by the constraint. This is also obvious from inspection, since z approaches infinity as x or y approaches infinity. The extreme value of the function thus occurs as x and y reach the limiting value of the inequality.

Example: Consider again the Corola Typewriter Company. The revenue function for Corola, expressed in terms of electric E and manual M typewriters, was

$$R = 8E + 5M + 2EM - E^2 - 2M^2 + 20$$

This function was optimized without limitations on production on p. 422. The maximum revenue of $78,250 occurred when $E^* = 10.5$ and $M^* = 6.5$.

On p. 442, we revised the problem by assuming that production capacity constraints limited production to $E + M = 14$. This problem was solved to give $E^* = 8.7$, $M^* = 5.3$, and $\lambda^* = -1.2$ with revenue of $76,500. The negative coefficient of the Lagrangian multiplier shows that the maximum is limited by the equality $E + M = 14$.

Assume that we reformulate the problem again by incorporating the inequality that production be less than or equal to 20,000 units; i.e.,

$$E + M \leq 20$$

Considering the inequality as an equality, we form the Lagrangian expression

$$F(E, M, \lambda) = 8E + 5M + 2EM - E^2 - 2M^2 + 20 + \lambda(E + M - 20)$$

The partial derivatives are equated to 0, and the resulting three equations are solved simultaneously to give

$$E^* = 12.3, \qquad M^* = 7.7, \quad \text{and} \quad \lambda^* = 1.2$$

Total revenue for the constrained optimum is $R = 76,400. This represents a reduction of $1850 from the unconstrained optimum. The positive value of λ indicates that the unconstrained maximum is greater than the constrained maximum. The solution to the unconstrained problem satisfies the inequality; that is, $10.5 + 6.5 \leq 20.0$.

Example: The operations research department of the Cost Plus Defense Corporation has established a function that describes profits from army and navy contracts. This function is

$$P = 600 - 4A^2 + 20N + 2AN - 6N^2 + 12A$$

where A = army contracts in millions of dollars.

N = navy contracts in millions of dollars.

P = profits in thousands of dollars.

Company policy dictates that a minimum of $5 million in contracts will be accepted during the coming year. Thus, the constraining inequality is

$$A + N \geq 5$$

Determine the values of A and N that maximize profits.

To obtain the solution, we form the Lagrangian expression

$$F(A, N, \lambda) = 600 - 4A^2 + 20N + 2AN - 6N^2 + 12A + \lambda(A + N - 5)$$

The partial derivatives of this expression are equated to 0:

$$\frac{\partial F}{\partial A} = -8A + 2N + 12 + \lambda = 0$$

$$\frac{\partial F}{\partial N} = 2A - 12N + 20 + \lambda = 0$$

$$\frac{\partial F}{\partial \lambda} = A + N - 5 = 0$$

The three equations are solved simultaneously to give

$$A = 2.584, \qquad N = 2.416, \quad \text{and} \quad \lambda = 3.834$$

Profits are $630,080.

Profits, subject to the constraint, are maximized when army contracts of $2,584,000 and navy contracts of $2,416,000 are obtained. Since λ is positive, we recognize that the constraint acts as a lower bound on the values of the variables. This can be shown by determining the optimum value of A and N without the constraint. The optimum value of P, if we assume no constraining inequality, is found by equating the partial derivatives $\partial F/\partial A$ and $\partial F/\partial N$ to 0 and solving the resulting two equations simultaneously. The solution of these two equations gives $A = 2$ and $N = 2$. The profit from this level of sales is $656,000.

PROBLEMS

1. Find the first and second partial derivatives and the cross partial derivative of the following multivariate functions.
   (a) $f(x, y) = 2x + 3xy + 3y$
   (b) $f(x, y) = 2x^2 + 3xy^2 - 5y^2 + y^3$
   (c) $f(x, y) = (x^2 + y^2)^3$
   (d) $f(x, y) = (2x + xy^2 + y^2)^3$
   (e) $f(x, y) = (x + y)(x + y)^2$
   (f) $f(x, y) = (2x + 3y^2)^2/(x^2 + y^2)$
   (g) $f(x, y) = x/y$
   (h) $f(x, y) = \ln(x^2 + xy + y^2)$
   (i) $f(x, y) = \ln(2x + y)^2$
   (j) $f(x, y) = x^2 \ln(x^2 + y^2)$
   (k) $f(x, y) = (x + 2y) \ln(x + 3y)$
   (l) $f(x, y) = e^{x+y}$
   (m) $f(x, y) = e^{(x^2+y^2)}$
   (n) $f(x, y) = 3^{2x}$
   (o) $f(x, y) = e^e$

2. Find the first and second partial derivatives for the following functions.
   (a) $f(x, y, z) = 2x + 3y + 4z$
   (b) $f(x, y, z) = (x^2 + xy + y^2 + xyz + yz^2)$
   (c) $f(x, y, z) = (x^3 + xyz + y^2 + z^4)^{1/2}$
   (d) $f(x, y, z) = \ln(x + y + z)$
   (e) $f(x, y, z) = x + yz + (1/xyz)$
   (f) $f(x, y, z) = e^{(2x+y+z)}$
   (g) $f(x, y, z) = e^{x^2+y^2+z^2}$

3. Find the maxima, minima, or saddle point for the following functions.
   (a) $f(x, y) = 2x^2 - 4x + y^2 - 4y + 4$
   (b) $f(x, y) = 3x^2 + 24x + 16y^2 - 32y - 10$
   (c) $f(x, y) = -x^2 + y^2 + 10x - 15y - 5$
   (d) $f(x, y) = -8x^2 - 3y^2 - 144x - 12y + 15$
   (e) $f(x, y) = 2x^2 - 8xy - 8x + 12y^2 - 48y + 24$
   (f) $f(x, y) = 2x^2 - 4x + 8y^2 + 80y + 50xy + 100$
   (g) $f(x, y) = -12x^2 + 24x + 10xy + 4y^2 - 200y + 16$
   (h) $f(x, y) = 8x^2 + 16x + 3y^2 - 12y - 24xy - 15$
   (i) $f(x, y) = \ln(x^2 - 4x + 2y^2 - 16y)$
   (j) $f(x, y) = \ln(2x^2 - 24x - 6y^2 + 12y)$
   (k) $f(x, y) = e^{(x^2 - 8x - y^2 - 4y)}$
   (l) $f(x, y) = e^{(4x^2 + 40x - 3y^2 + 12y)}$
   (m) $f(x, y) = 4^{(2x^2 - 8xy - 8x + 12y^2 - 48y)}$

4. Find the maxima and minima of the following functions subject to the constraining equations or inequalities.
   (a) $f(x, y) = 5x^2 + 6xy - 3y^2 + 10$ and $x + 2y = 24$
   (b) $f(x, y) = 12xy - 3y^2 - x^2$ and $x + y = 16$
   (c) $f(x, y) = x^2 + 2y^2 - xy$ and $x + y = 8$
   (d) $f(x, y) = 3x^2 + 4y^2 - xy$ and $2x + y = 21$
   (e) $f(x, y) = -3x^2 - 4y^2 + 6xy$ and $3x + y = 19$
   (f) Minimize $f(x, y) = 4x^2 + 5y^2 - 6y$ and $x + 2y \geq 20$.
   (g) Maximize $f(x, y) = 10xy - 5x^2 - 7y^2 + 40x$ and $x + y \leq 12$.
   (h) Minimize $f(x, y) = 12x^2 + 4y^2 - 8xy - 32x$ and $x + y \leq 2$.

5. The sales of the Ace Novelty Company are a function of price p, advertising a, and the number of salesmen n. The functional relationship is

$$S = (10{,}000 - 700p)n^{2/3}a^{1/2}$$

The current price is $5.00, advertising is $10,000, and 99 salesmen are employed. Using the partial derivatives, determine
   (a) The effect of an additional $1 of advertising expenditure upon sales.
   (b) The effect of raising price by $0.01 upon sales.
   (c) The effect of employing one additional salesman.

6. The cost of construction of a project depends upon the number of skilled workers x and unskilled workers y. If the cost is given by

$$c(x, y) = 40{,}000 + 9x^3 - 72xy + 9y^2$$

   (a) Determine the number of skilled and unskilled workers that results in minimum cost.
   (b) Determine the minimum cost.

7. The yearly profits of a small service organization, Bill Athy, Inc., are dependent upon the number of workers x and the number of units of

advertising y, according to the function

$$p(x, y) = 412x + 806y - x^2 - 5y^2 - xy$$

(a) Determine the number of workers and the number of units in advertising that results in maximum profit.

(b) Determine maximum profits.

8. Suppose that the quantity q sold of a certain item is a function of the price p and advertising a, given by

$$q = [10,000 + 200(1 - e^{-0.50a})]e^{-0.25p}$$

Use the partial derivatives to determine the effect of

(a) A price change from \$2.00 to \$2.01 when advertising is \$10.

(b) An increase in advertising from \$10 to \$11 when price is \$2.00.

9. Assume in Problem 6 that union contracts require that each skilled laborer must direct five or more unskilled laborers. Determine the number of skilled and unskilled laborers that leads to minimum cost.

10. In Problem 7, assume that workers receive \$100 per week (\$5200 per year) and that advertising cost is \$100 per unit. Working capital is such that a maximum total expenditure of \$100,000 can be made on labor and advertising.

(a) Determine the optimum allocation of these funds.

(b) Determine profits based upon this allocation.

SUGGESTED REFERENCES

The references for this chapter are listed in Chapter 9.

Chapter 13

Multivariate and Exponential Business Models

This chapter introduces additional managerial decision models that utilize the techniques of differential calculus. These include production and inventory models, the least squares model for establishing functions, and a model describing continuous growth in a firm or an economy. The chapter also includes a discussion of mathematics of finance.

13.1 Production and Inventory Models

In the discussion of production and inventory models in Sec. 11.2, we illustrated the economic order quantity model and the production lot size model. These two basic models are now modified to permit shortages in inventory. It is assumed in the following two models that orders that cannot be filled from inventory are backlogged. However, since there is a certain inconvenience to the customer in being required to wait for delivery of his purchase, a cost of shortage is incorporated in the model. This cost of shortage could be in the form of a discount to the customer, such as is typical in purchasing from a catalogue. It also could be a penalty to the merchant, which represents his estimate of the economic value of the ill will created by being unable to fill a customer's order.

The notation used for the following two models is the same as that in Sec. 11.2 with the exception that we include the cost of shortage. The inventory costs and decision variables are represented as follows:

a = cost of holding one unit of inventory for one time period.
b = cost of shortage of one unit for one time period.
c = setup cost incurred when a new production run is started or cost of placing an order.
k = production rate, in units per time period.
r = rate of demand, in units per time period.
q = quantity ordered or produced at each setup.

13.1.1 ECONOMIC ORDER QUANTITY WITH SHORTAGES

In the economic order quantity model with shortages we assume that shortages are permitted and backlogged. We also assume that inventory is purchased and arrives a fixed number of days following the placing of the order. As an example, consider a retailer who purchases stock from wholesalers. His customers place an order that is either immediately filled or filled from the next order received from the wholesaler. His annual demand is $r = 5000$ per year. Holding cost is $a = \$10$ per unit per year. Shortage cost, which includes a penalty for customer dissatisfaction and possible loss of the customer, is $b = \$15$ per unit short per year. Order cost is $c = \$300$ per order. Orders are received 10 days after the order is placed. Determine the order quantity that results in minimum total cost. Also determine the inventory level at which an order should be placed.

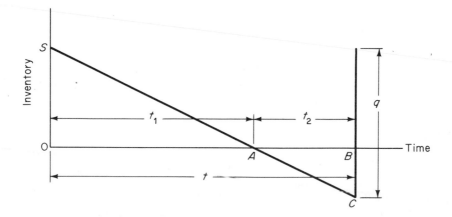

Figure 13.1

A graph of the inventory level over time for a single inventory cycle is shown in Fig. 13.1. As in the economic order quantity model in Chapter 11, the rate of demand for the product is assumed to be constant. This is shown by the constant rate of decrease of inventory during the cycle.

In Fig. 13.1, q represents the order quantity. Since orders from customers for the product are backlogged and filled from the next shipment, the maximum level of inventory is represented in the figure by S. The length of the inventory cycle, normally expressed as a fraction of a year, is represented by t. Similarly, t_1 represents the time during which orders are filled from inventory and t_2 represents the time during which orders are backlogged. Both t_1 and t_2 are also expressed as fractions of the total period, rather than of the inventory cycle.

From similar triangles, it can be seen that

$$\frac{t_1}{S} = \frac{t}{q} \tag{13.1}$$

thus

$$t_1 = \frac{S \cdot t}{q} \tag{13.2}$$

and since

$$t_1 + t_2 = t \tag{13.3}$$

we obtain

$$t_2 = t - t_1 = t - \frac{S \cdot t}{q} = (q - S)\frac{t}{q} \tag{13.4}$$

The total cost of an inventory policy can be separated into the cost of holding or carrying inventory C_c, the cost of shortage C_s, the cost of ordering the product C_o, and the purchase cost of the product C_I. Thus,

$$TC = C_c + C_s + C_o + C_I$$

Each of the separate costs are first determined for a single inventory cycle. The cost of holding the inventory during a cycle is equal to the product of the holding cost per unit a, the average number of units in inventory $\frac{S}{2}$, and the length of time the units are held in inventory t_1. The cost of holding the inventory for one cycle is thus

$$C_c/\text{cycle} = \frac{a(St_1)}{2} = \frac{aS^2t}{2q} \tag{13.5}$$

The cost of shortage during a cycle is given by the product of the cost of shortage per unit b, the average number of units short $\frac{q - S}{2}$, and the length of time the units are short, t_2. This cost, again for a single inventory cycle, is

$$C_s/\text{cycle} = \frac{b(q - S)t_2}{2} = \frac{b(q - S)^2t}{2q} \tag{13.6}$$

Since the cost of placing an order is c, the inventory cost per cycle is

$$TC/\text{cycle} = c + \frac{aS^2t}{2q} + \frac{b(q - S)^2t}{2q} \tag{13.7}$$

The number of cycles per period is r/q. The total cost per period for inventory is thus

$$TC = \left[c + \frac{aS^2t}{2q} + \frac{b(q - S)^2t}{2q} \right] \frac{r}{q} + C_I \tag{13.8}$$

and since r/q is equal to $1/t$, the expression reduces to

$$TC = \frac{cr}{q} + \frac{aS^2}{2q} + \frac{b(q - S)^2}{2q} + C_I \tag{13.9}$$

The variables in the total cost equation, Eq. (13.9), are the order quantity q and the maximum inventory level S. In order to determine the values of q and S that minimize total cost, we equate the partial derivative of total cost with respect to q and the partial derivative of total cost with respect to S to 0. These two equations are solved simultaneously for q and S. Thus,

$$\frac{\partial(TC)}{\partial q} = \frac{-cr}{q^2} - \frac{aS^2}{2q^2} - \frac{b(q - S)^2}{2q^2} + \frac{b(q - S)}{q} = 0$$

$$\frac{\partial(TC)}{\partial S} = \frac{2aS}{2q} - \frac{2b(q - S)}{2q} = 0$$

From the second equation we obtain

$$aS - bq + bS = 0$$

or

$$S = \frac{bq}{(a + b)} \tag{13.10}$$

To solve for q, we multiply both sides of the first equation by $-2q^2$. This gives

$$2cr + aS^2 + b(q - S)^2 - 2bq(q - S) = 0$$

Substituting $S = \dfrac{bq}{(a + b)}$ into this equation and solving for q gives

$$2cr + \frac{a(bq)^2}{(a + b)^2} + b\left[q - \frac{bq}{a + b} \right]^2 - 2bq\left[q - \frac{bq}{a + b} \right] = 0$$

$$2cr + \frac{a(bq)^2}{(a + b)^2} + \frac{b(aq)^2}{(a + b)^2} - \frac{2bq(aq)}{(a + b)} = 0$$

$$2cr(a + b)^2 + abq^2(a + b) - 2abq^2(a + b) = 0$$

$$abq^2(a + b) - 2abq^2(a + b) = -2cr(a + b)^2$$

$$q^2 = \frac{-2cr(a + b)^2}{-ab(a + b)} = \frac{2cr(a + b)}{ab}$$

$$q = \sqrt{\frac{2cr(a + b)}{ab}} \tag{13.11}$$

Example: Determine the order quantity, the maximum level of inventory, the maximum number of units short, and the reorder point for the illustrative problem.

The order quantity is

$$q = \sqrt{\frac{2cr(a+b)}{ab}} = \sqrt{\frac{2(300)(5000)(10+15)}{10(15)}} = \sqrt{500,000} = 707 \text{ units}$$

and the maximum level of inventory is

$$S = \frac{bq}{a+b} = \frac{15(707)}{25} = 424 \text{ units}$$

The maximum number of units short is

$$S - q = 707 - 424 = 283 \text{ units}$$

If we assume 250 working days per year and a yearly demand of $r = 5000$ units, the daily demand is 20 units. Since 10 days are required between the time an order is placed and received, the order should be placed when the inventory level reaches

$$N = 200 - 283 = -83 \text{ units}$$

The reorder point therefore occurs when the shortage has reached 83 units.

Example: Pemco, Inc., is the sole distributor of a patented industrial component. The sales of this component are evenly distributed throughout the year and total 10,000 units. Pemco's current inventory policy is to maintain sufficient inventory to ensure that all orders can be immediately filled from inventory. Although this inventory policy has worked satisfactorily in the past, the management of Pemco believes that inventory costs can be reduced by modifying the inventory policy to permit shortage and backlogging of orders during a portion of the inventory cycle. The costs of the current inventory policy are (1) holding cost, $8 per unit per year and (2) order cost, $400 per order. Management estimates that under the proposed inventory policy the cost of shortage would be $12 per unit per year.

Determine the difference in cost between the proposed inventory policy and the current policy. Calculate the maximum inventory, the maximum shortage, the length of the inventory cycle, and the length of time during each cycle that orders would be backlogged for the proposed inventory plan.

The difference in cost between the two inventory policies can be found by calculating the total inventory cost for each policy. The economic order quantity for the current policy is given by Formula (11.30).

$$q = \sqrt{\frac{2rc}{a}} = \sqrt{\frac{2(10,000)(400)}{8}} = \sqrt{1,000,000} = 1000 \text{ units}$$

The total inventory cost under the current policy is given by Formula (11.29) and is

$$TC = \frac{rc}{q} + \frac{qa}{2} + C_I$$

$$TC = \frac{10,000(400)}{1000} + \frac{1000(8)}{2} + C_I$$

$$TC = \$8000 + C_I$$

The economic order quantity under the proposed inventory policy is given by Formula (13.11) and is

$$q = \sqrt{\frac{2cr(a+b)}{ab}} = \sqrt{\frac{2(400)(10,000)(8+12)}{8(12)}} = 1290 \text{ units}$$

The maximum inventory level under this policy, from Formula (13.10), is

$$S = \frac{bq}{a+b} = \frac{12(1290)}{20} = 774 \text{ units}$$

The total cost of this policy, given by Formula (13.9), is

$$TC = \frac{cr}{q} + \frac{aS^2}{2q} + \frac{b(q-S)^2}{2q} + C_I$$

$$TC = \frac{400(10,000)}{1290} + \frac{8(774)^2}{2(1290)} + \frac{12(1290-774)^2}{2(1290)} + C_I$$

$$TC = 3101 + 1858 + 1238 + C_I$$

$$TC = \$6197 + C_I$$

Subtracting the total cost of the proposed policy from the total cost of the present policy gives the difference in cost of the two inventory policies. This difference is

$$\Delta TC = \$8000 + C_I - (\$6197 + C_I) = \$1803$$

The total cost of the proposed inventory policy is $1803 less than total cost of the current policy.

The maximum shortage under the proposed policy is given by

$$q - S = 1290 - 774 = 516 \text{ units}$$

The length of the inventory cycle is

$$t = \frac{q}{r} = \frac{1290}{10,000} = 0.129 \text{ years}$$

and the length of time during the cycle in which orders are backlogged is

$$t_2 = (q - S)\frac{t}{q} = 516\frac{(0.129)}{1290} = 0.0516 \text{ years}$$

The proportion of the cycle during which orders must be backlogged is

$$\frac{t_2}{t} = \frac{0.0516}{0.129} = 0.40$$

Orders are thus backlogged during 40 percent of each inventory cycle.

13.1.2 PRODUCTION LOT QUANTITY WITH SHORTAGES

The production model described in this section is quite similar to the production model described in Chapter 11. The two models differ only in that shortages are permitted and are backlogged. Once again, the production lot quantity model describes the case in which more than one production run is made during the period (i.e., the year) considered.

The costs included in the model are the cost of holding the inventory a, the cost of shortage b, the cost of setting up for a new production run c, and the cost of manufacturing the product C_I. The production rate is represented by k and the usage rate by r. Since continual production is not required, k is greater than r.

A production and inventory usage cycle is shown in Fig. 13.2. It is important to note that t and t_i again represent fractions of the time period rather than a fraction of the production cycle.

We shall first consider a single production-inventory cycle. Setup cost for the cycle is c. Shortage cost is the product of the average number of units short $EF/2$, the length of time the shortage exists ($t_1 + t_4$), and the cost per unit short b. Since OA and EF are equal,

$$C_s = \frac{EF(t_1 + t_4)b}{2} \tag{13.12}$$

and since $EF = rt_4$, formula (13.12) can be expressed as

$$C_s = \frac{brt_4(t_1 + t_4)}{2} \tag{13.13}$$

Inventory carrying cost during the cycle is given by the average number of units in inventory $BC/2$, the length of time the units are in inventory ($t_2 + t_3$), and the cost per unit per time period of inventory a. Since BC is equal to rt_3, inventory carrying costs for a single cycle are

$$C_c = \frac{art_3(t_2 + t_3)}{2} \tag{13.14}$$

The total cost of carrying the inventory, shortages, and the setup cost for one cycle is

$$TC/\text{cycle} = \tfrac{1}{2}[art_3(t_2 + t_3) + brt_4(t_1 + t_4)] + c \tag{13.15}$$

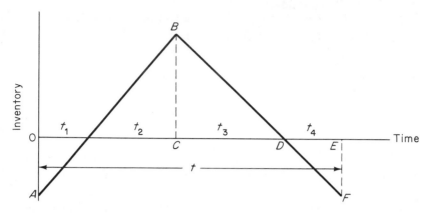

Figure 13.2

The total cost for the period is determined by multiplying the cost per cycle by the number of cycles per period r/q and adding the cost of the product manufactured.

$$TC = \frac{r}{2q} \left[art_3(t_2 + t_3) + brt_4(t_1 + t_4) \right] + \frac{cr}{q} + C_I$$

$$= \frac{1}{2q} (art_3rt_2 + art_3rt_3 + brt_4rt_1 + brt_4rt_4 + 2cr) + C_I$$

(13.16)

Total cost can be expressed as a function of t_1 and t_2 by using the following equalities:

    (i) $q = k(t_1 + t_2)$.
    (ii) peak inventory $= t_2(k - r) = rt_3$.
    (iii) peak shortage $= t_1(k - r) = rt_4$.

Substitution of these equalities yields

$$TC = \frac{1}{2k(t_1 + t_2)} \left[at_2^2(k - r)r + at_2^2(k - r)^2 + bt_1^2(k - r)r \right.$$

$$\left. + bt_1^2(k - r)^2 + 2cr \right] + C_I$$

(13.17)

By equating the partial derivatives of total cost with respect to t_1 and t_2 to 0, we find the optimal values of t_1 and t_2. These are

$$t_1 = \sqrt{\frac{2acr}{bk(k - r)(a + b)}}$$

(13.18)

$$t_2 = \sqrt{\frac{2bcr}{ak(k - r)(a + b)}}$$

(13.19)

By substitution for t_1 and t_2, we can obtain t and q.

$$t = \sqrt{\frac{2ck(a+b)}{arb(k-r)}} \tag{13.20}$$

$$q = \sqrt{\frac{2crk(a+b)}{ab(k-r)}} \tag{13.21}$$

Example: Assume that yearly demand is $r = 1000$ units and the production rate is $k = 6000$ units. Inventory holding costs are $a = \$10$ per unit per year, shortage costs are $b = \$15$ per unit per year, and setup costs are $c = \$200$ per setup. Determine q, t_1, t_2, t_3, t_4, and t.

$$q = \sqrt{\frac{2(200)(1000)(6000)(25)}{10(15)(5000)}} = \sqrt{80,000} = 283$$

$$t_1 = \sqrt{\frac{2(10)(200)(1000)}{15(6000)(5000)(25)}} - \sqrt{0.0003555} = 0.01886 \text{ year}$$

$$t_2 = \sqrt{\frac{2(15)(200)(1000)}{10(6000)(5000)(25)}} = \sqrt{0.0008} = 0.02828 \text{ year}$$

$$t_3 = \frac{t_2(k-r)}{r} = \frac{0.02828(5000)}{1000} = 0.14140 \text{ year}$$

$$t_4 = \frac{t_1(k-r)}{r} = \frac{0.01886(5000)}{1000} = 0.09430 \text{ year}$$

and the inventory-production cycle is

$$t = t_1 + t_2 + t_3 + t_4 = 0.28284$$

As a check on the calculations, we note that

$$t = \frac{q}{r} = \frac{283}{1000} = 0.283$$

13.2 The Method of Least Squares

An important application of differential calculus is the least squares model. The least squares model provides a method of fitting curves to data points. As an example, assume that an electric company is faced with the problem of forecasting sales of electricity for the next five years. Assume that sales of electricity are dependent upon the number of homes in the area served by the company. From records for the past ten years, total sales and the total number of homes were obtained. These totals are plotted in Fig. 13.3.

The plot of data in Fig. 13.3 is termed a *scatter diagram*. Each point is

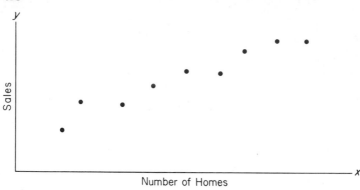

Figure 13.3

called a *data point*. The least squares model provides a method of establishing a function that describes the relationship between the variables, i.e., between the values on the vertical and horizontal axes. In terms of our example, this means establishing a functional relationship between electricity sales and the number of homes. The relationship would be used to forecast electricity sales on the basis of estimates of the number of homes in the area during the coming five-year period.

The scatter diagram in Fig. 13.3 shows that the relationship between the sales and the number of homes is approximately linear. Only two data points are necessary to establish a linear function. Since there are more than two data points, the analyst must either arbitrarily select two points that he believes are representative, or he must position the linear function through the data based upon some criterion of *best fit*. The method of least squares provides this criterion of best fit.

The method of least squares is commonly used to determine the relationship between variables in the regression model, the time series model, and econometric forecasting models. These subjects are beyond the design of this text. However, an understanding of the least squares model will be quite useful in the development of simple forecasting equations and in the study of these additional topics at a later date.

13.2.1 LINEAR FIT

The least squares model provides a method of establishing a curve through data points. The initial task is to determine the appropriate form of the curve, i.e., linear, exponential, quadratic, etc. A common method of determining the appropriate function is to plot the data points on a scatter diagram and determine by inspection the most appropriate form of the function. Fig. 13.4 provides an example in which the linear function applies.

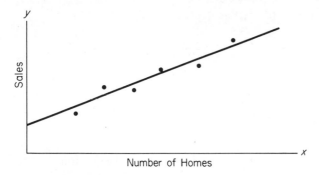

Figure 13.4

Each data point (x_i, y_i) in Fig. 13.4 represents an actual observation of the relationship between the number of homes x_i and sales y_i. The objective is to establish a linear function through the data points based on the criterion of least squares. The general form of this linear function is

$$\tilde{y}_i = a + bx_i \qquad (13.22)$$

$\tilde{y}_i$ (read "y tilde") represents the value of y predicted for (based upon) a given value of x_i. The criterion of least squares is expressed mathematically as

$$\text{Minimize} \sum_{i=1}^{n} (y_i - \tilde{y}_i)^2 \qquad (13.23)$$

where the symbol $\sum_{i=1}^{n}$, read "the summation as i is incremented in units of 1 from 1 to n," represents the summation of n terms.

Formula (13.23) states that the function is to be positioned on the scatter diagram in such a way that the summation of the squared deviations between the observed y values y_i and the predicted y values $\tilde{y}_i$ is a minimum. We can substitute (13.22) in (13.23) and obtain

$$\text{Minimize} \sum_{i=1}^{n} [y_i - (a + bx_i)]^2 \qquad (13.24)$$

Since we must determine a and b such that the sum of the squared deviations is a minimum, the function should be differentiated with respect to a, then with respect to b, and these partial derivatives should be equated with 0. Thus,

$$\frac{\partial F}{\partial a} = 2 \sum_{i=1}^{n} [y_i - (a + bx_i)]^1 (-1) = 0$$

and

$$\frac{\partial F}{\partial b} = 2 \sum_{i=1}^{n} [y_i - (a + bx_i)]^1 (-x_i) = 0$$

These two equations can be expressed as

$$\sum_{i=1}^{n} y_i = na + b \sum_{i=1}^{n} x_i \qquad (13.25)$$

and

$$\sum_{i=1}^{n} y_i x_i = a \sum_{i=1}^{n} x_i + b \sum_{i=1}^{n} x_i^2 \qquad (13.26)$$

Equations (13.25) and (13.26) are termed the "normal equations." The normal equations are a set of two linear equations with unknown values a and b. Their simultaneous solution gives the values of a and b for the function that provides the best fit to the data according to the criterion of least squares.

The method of least squares is the most widely used technique for establishing the coefficients in the regression and econometric models. The least squares predicting function has certain characteristics that analysts find desirable. These characteristics are discussed in detail in statistical textbooks. The characteristics are

1. The least squares criterion provides the best fit to the data in the sense that the sum of the squared deviations, $\sum (y - \tilde{y})^2$, of the observed values from the predicted values is a minimum.
2. The deviations above the line equal those below the line; that is, $\sum (y - \tilde{y}) = 0$.
3. The linear function passes through the overall mean (average) of the data $(\bar{x}, \bar{y})$.
4. In those cases in which the data points are obtained by sampling from a larger population, the least squares estimates are "best estimates" of the population parameters in terms of the statistical properties of "unbiasedness and efficiency."

The following example illustrates use of the method of least squares in the development of a function for the forecasting of sales.

Example: Mr. Leonard Barker of the Finance Department of United Machinery Company wishes to determine the relationship between productivity and the profits of United Machinery. The productivity and profit data for six years are given in the following table. These data are plotted on the accompanying scatter diagram. Determine the predicting function for profit as a function of productivity.

	Profit (millions of dollars)	Productivity (output per man-week)
1968	3.1	145
1969	3.5	160
1970	3.6	185
1971	3.6	190
1972	3.8	200
1973	4.2	220

On the basis of the scatter diagram of profit versus productivity, we conclude that the relationship between the two variables is best described by a linear function. The general form of this function is given by (13.22). Calculation of the coefficients a and b is illustrated as follows.

x	y	$y \cdot x$	x^2
145	3.1	449.5	21,025
160	3.5	560.0	25,600
185	3.6	666.0	34,225
190	3.6	684.0	36,100
200	3.8	760.0	40,000
220	4.2	924.0	48,400
1100	21.8	4043.5	205,350

From these calculations $n = 6$, $\sum x = 1100$, $\sum y = 21.8$, $\sum y \cdot x = 4043.5$, and $\sum x^2 = 205,350$. The values are substituted into the normal equations giving

$$21.8 = 6a + 1100b$$

$$4043.5 = 1100a + 205,350b$$

These two equations are solved simultaneously for a and b. The resulting least squares estimates of a and b are $a = 1.298$ and $b = 0.01274$. The predicting equation is

$$\tilde{y}_i = 1.298 + 0.01274x_i$$

Mr. Barker, after consulting with the engineering staff of United Machinery, prepared estimates of productivity for 1974 through 1978. If we assume that the past relationship between productivity and profits remains the same, we can estimate profits using the predicting function. These estimates are shown in the following table.

Year	Productivity	Profits
1974	230	$4.23
1975	240	4.36
1976	245	4.42
1977	250	4.48
1978	255	4.55

Example: A personnel manager is interested in determining a possible linear relationship between an employment screening examination and an employee's performance rating. In order to determine this relationship, the the manager selected the records of ten recently hired employees. The scores of each employee on the screening examination, together with the performance ratings of the employee, are shown below.

Employee	Screening Exam	Performance Rating
1	70	80
2	65	70
3	80	80
4	85	90
5	70	75
6	50	60
7	60	70
8	80	90
9	90	95
10	75	80

Use the least squares criterion to determine the linear function that relates scores on the screening exam to the performance rating.

A scatter diagram showing screening exam scores and performance rating scores is given below. This diagram shows that the relationship between the scores on the screening exam and the scores on the performance rating can, at least for the sample of ten employees, be described as being linearly related.

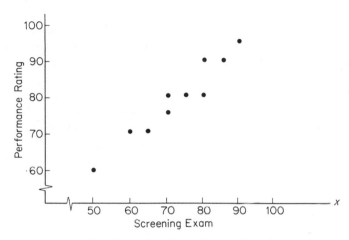

The intercept a and the slope b of the linear function are determined as follows:

x	y	$x \cdot y$	x^2
70	80	5600	4900
65	70	4550	4225
80	80	6400	6400
85	90	7650	7225
70	75	5250	4900
50	60	3000	2500
60	70	4200	3600
80	90	7200	6400
90	95	8550	8100
75	80	6000	5625
725	790	58,400	53,875

Substituting these values into the normal equations gives

$$790 = 10a + 725b$$

$$58,400 = 725a + 53,875b$$

Simultaneous solution of the two equations for a and b gives $a = 16.87$ and $b = 0.86$. The linear function is thus

$$\bar{y} = 16.87 + 0.86x$$

13.2.2 CURVILINEAR FIT

The least squares model can also be used to establish nonlinear functions through data points. Ideally, a nonlinear function should be positioned so that the sum of the squared deviations between the observed y values and the predicted $\bar{y}$ values is a minimum. In practice, however, the function is normally positioned so that the sum of the squared deviations between the logarithms of the observed values and the logarithms of the predicted values is a minimum. This modification is made to permit direct application of the normal equations [Equations (13.25) and (13.26)].

Two of the more commonly applied curvilinear functions are used to illustrate the method for establishing nonlinear functions through data points. These functions are the exponential function and the power function.

EXPONENTIAL FUNCTION. The exponential function was introduced in Chapter 12. One form of the exponential function is given by (12.5) as

$$y = ke^{cx} \tag{12.5}$$

To apply the linear fit model to the exponential function, we express (12.5) in terms of natural logarithms.

$$\ln(y) = \ln(k) + cx \tag{13.27}$$

Equation (13.27) is linear when expressed in terms of $\ln(y)$ and x. This is apparent from the fact that the intercept $\ln(k)$ and the slope c are constants. The dependent variable $\ln(y)$ and the independent variable x are thus linearly related. The values of $\ln(k)$ and c can be determined by solving Equations (13.25) and (13.26) for $\ln(k)$ and c. In Equations (13.25) and (13.26) we replace y_i by $\ln(y_i)$, a by $\ln(k)$, and b by c. The normal equations then become

$$\sum_{i=1}^{n} \ln(y_i) = n \cdot \ln(k) + c \sum_{i=1}^{n} x_i \tag{13.28}$$

and

$$\sum_{i=1}^{n} \ln(y_i)x_i = \ln(k) \sum_{i=1}^{n} x_i + C \sum_{i=1}^{n} x_i^2 \tag{13.29}$$

Simultaneous solution of the two equations gives $\ln(k)$ and c.

Example: Bob's Family Restaurants operates restaurants throughout the southeastern United States. Sales for a five-year period are shown in the table and plotted in the scatter diagram below. If the growth trend that has been established over the five years were to continue for three more years, determine the predicting equation for sales and use this equation to predict sales for these three years.

Sales—Bob's Family Restaurants

Year	1969	1970	1971	1972	1973
Sales	$200,000	$238,000	$290,000	$348,000	$430,000

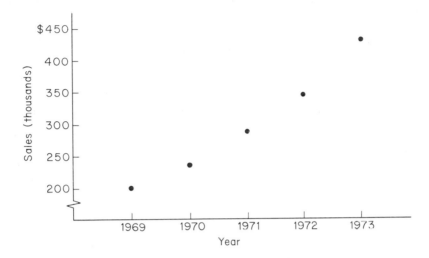

From the scatter diagram, we conclude that an exponential predicting equation best describes the growth in sales. Calculation of the coefficients k and c is illustrated below. To simplify the calculations, the five years are represented by $x = 1, 2, 3, 4, 5$.

x	y	$\ln(y)$	$x \ln(y)$	x^2
1	200	5.29832	5.29832	1
2	238	5.47227	10.94454	4
3	290	5.66988	17.00964	9
4	348	5.85220	23.40880	16
5	430	6.06379	30.31895	25
15		28.35646	86.07989	55

These values are substituted into Equations (13.28) and (13.29) giving

$$28.35646 = 5 \ln(k) + 15c$$
$$86.97989 = 15 \ln(k) + 55c$$

Solving the two equations simultaneously gives

$$\ln(k) = 5.09229 \quad \text{and} \quad c = 0.193$$

The predicting function, written in terms of natural logarithms, is thus

$$\ln(y) = 5.09229 + 0.193x$$

The sales forecast, obtained by substituting the values 6, 7, and 8 for x in the predicting function is \$518,000 for 1974, \$628,000 for 1975, and \$762,000 for 1976.

The logarithmic function can be expressed as an exponential function. This is

$$y = 163e^{0.193x}$$

POWER FUNCTION. The power function is a special case of the algebraic functions introduced in Chapter 2. The general form of the power function is

$$y = ax^b \tag{13.30}$$

The logarithmic form of the power function is

$$\ln(y) = \ln(a) + b\ln(x) \tag{13.31}$$

Since the intercept $\ln(a)$ and the slope b are constant, it can be seen that the logarithmic form of the power function is a linear function.

We can apply the linear fit least squares model to Equation (13.31) by replacing y_i by $\ln(y_i)$, a by $\ln(a)$, and x_i by $\ln(x_i)$ in Equations (13.25) and (13.26). This gives

$$\sum_{i=1}^{n} \ln(y_i) = n\ln(a) + b\sum_{i=1}^{n}\ln(x_i) \tag{13.32}$$

and

$$\sum_{i=1}^{n} \ln(y_i)\ln(x_i) = \ln(a)\sum_{i=1}^{n}\ln(x_i) + b\sum_{i=1}^{n}\ln(x_i)^2 \tag{13.33}$$

These two equations are solved simultaneously for $\ln(a)$ and b.

Example: Fit a power function of the form $y = ax^b$ to the sales data for Bob's Family Restaurants.

The values of $\ln(x)$ and $\ln(y)$ are obtained from the Table of Natural Logarithms, Table IV, and are shown in the following table.

x	y	$\ln(x)$	$\ln(y)$	$\ln(x)\ln(y)$	$\ln(x)^2$
1	200	0	5.29832	0	0
2	238	0.69315	5.47227	3.79310	0.48046
3	290	1.09861	5.66988	6.22897	1.20694
4	348	1.38629	5.85220	8.11285	1.92180
5	430	1.60944	6.06379	9.75930	2.59030
		4.78749	28.35646	27.89422	6.19952

These values are substituted into the normal equations, giving

$$28.35646 = 5 \ln (a) + 4.78749b$$

$$27.89422 = 4.78749 \ln (a) + 6.19952b$$

Solving these two equations simultaneously gives $\ln (a) = 5.23094$ and $b = 0.45989$. The predicting equation written in terms of natural logarithms is $\ln (y) = 5.23094 + 0.45989 \ln (x)$. The sales forecast based upon this predicting equation is \$426,000 for 1974, \$457,000 for 1975, and \$487,000 for 1976.

Rewriting this predicting equation in the form of a power function gives

$$y = 187x^{0.45989}$$

13.2.3 DETERMINING THE BEST FIT

It is not always possible to determine by inspection the appropriate form of the predicting equation. In many cases, the data when plotted on a scatter diagram could be described by either the exponential or power function. It is also possible that the analyst is uncertain as to selection between a linear or power function, linear or exponential function, or perhaps among all three forms considered in this section.

The selection of the general form of the predicting equation should be based upon a careful study of the economic or physical relationship between the variables. It is possible, for example, that the data may appear to be best described by an exponential function. After careful analysis of the variables, however, the analyst may conclude that future values of the dependent variable are more likely to be linearly related to the independent variable. In this case, the linear function should be used instead of the exponential.

If the analyst believes that the selection of the general form of the predicting equation should be based solely upon the observed historical relationship between the variables, then that functional relationship which gives the minimum sum of the squared derivations is the appropriate function. The sum of the squared derivations is given by (13.23).

$$\sum_{i=1}^{n} (y_i - \tilde{y}_i)^2 \tag{13.23}$$

An alternative measure of the sum of the squared deviations is the *variance of the estimate*. This term is represented by s_e^2. The formula for the variance of the estimate is

$$s_e^2 = \frac{1}{n-2} \sum_{i=1}^{n} (y_i - \tilde{y}_i)^2 \tag{13.34}$$

The variance of the estimate is simply the sum of the squared deviations divided by $n - 2$. The functional relationship that gives the smallest variance of the estimate best describes the historical data.

Example: Determine the variance of the estimate for both the exponential and power function for the sales data for Bob's Family Restaurants.

Year	Sales	Exponential Estimate	Difference Squared	Power Estimate	Difference Squared
1969	200	197	9	187	169
1970	238	239	1	257	361
1971	290	290	0	310	400
1972	348	352	16	354	36
1973	430	428	4	392	1444
			30		2410

The variance of the estimate, s_e^2, for the exponential function is

$$s_e^2 = \frac{1}{n - 2} \sum_{i=1}^{n} (y_i - \tilde{y}_i)^2 = \frac{1}{3} (30) = 10$$

The variance of the estimate for the power function is

$$s_e^2 = \frac{1}{n - 2} \sum_{i=1}^{n} (y_i - y_i)^2 = \frac{1}{3} (2410) = 803$$

Since the variance of the estimate for the exponential function is smaller than that for the power function, we conclude that the exponential function best describes the historical data for Bob's Family Restaurants.

13.3 Mathematics of Finance

The interest rate is one of the important parameters in those business decisions that involve capital investments, the issuance of bonds, the return on invested capital, savings, etc. Consequently, it is important to understand the effect of the interest rate on a sum of money invested for a period of time. This section introduces and illustrates the important financial formulas that are used to calculate the time value of money. These formulas are customarily classified under the heading of the mathematics of finance.

13.3.1 SIMPLE AND COMPOUND INTEREST

As the reader is aware, the term *interest* refers to the price that one must pay to borrow a sum of money. This price is normally expressed as a percentage of the sum borrowed. Customarily, this percentage, or, alternatively, the *interest rate*, is based on a time period of one year. To illustrate, an interest rate of 6 percent would, unless specified differently, mean that the price of borrowing a sum of money for one year is equal to 6 percent of the sum of money borrowed. Since the base period for calculating the price of borrowing the sum of money is one year, the interest rate is referred to as the *annual interest rate*.

The term *simple interest* is used when interest is paid only on the original sum of money borrowed. This is contrasted to *compound interest,* in which the interest owed is added to the sum borrowed and subsequent interest is paid on the total of the original sum borrowed and the accumulated interest owed. To illustrate these two forms of interest payment, consider the following two examples.

Example: Bob Short has arranged a loan from a local bank. The sum of money borrowed is $1000 and the simple interest rate is 6 percent. Determine the interest that Bob must pay during the first year.

The interest is calculated by multiplying the interest rate by the sum borrowed. The interest is $1000(0.06), or $60. The total that must be paid the bank at the end of the one-year period is $1060.

Example: Bill Smith also arranged a one-year loan from a local bank. The terms of Bill's loan specify that the interest be calculated at the end of each six-month period, however, rather than at the end of the year. The sum borrowed is again $1000 and the annual interest rate is 6 percent. Assuming Bill repays both the original sum borrowed and the accumulated interest at the end of the one year period, determine the interest and the total amount that must be paid.

The interest for the first six months is $1000(0.03), or $30. Since Bill does not pay the $30 interest at the end of the first six months, the interest is added to the original sum borrowed. Bill thus owes $1030 during the second six months. The interest charges during the second six months are $1030(0.03) = $30.90. The total owed the bank at the second year is

$$A = \$1000 + \$30 + \$30.90 = \$1060.90$$

These two examples illustrate the effect of compounding: Although the interest rate quoted to both borrowers was 6 percent, Bill Smith found that the loan cost more than 6 percent of the sum borrowed. In both Bob and Bill's loan, the named or *nominal annual interest rate* was 6 percent. Because the interest was compounded on Bill's loan, the *effective annual interest rate* was greater than the nominal rate.

13.3.2 COMPOUND AMOUNT FORMULA

One of the more common financial transactions involves the deposit of a certain sum of money in a savings institution. The money deposited earns interest and the interest earned is allowed to accumulate over time. The sum of money deposited is normally referred to as the *principal* and is denoted by P. The principal plus the interest that has accumulated over the period of time is referred to as the *amount* and is represented by A. Since the interest earned during a certain time period is added to the original principal invested rather than withdrawn, the interest is compounded. The amount of money, then, depends on the original principal P, the nominal interest rate i, the number of years duration of the investment t, and the number of compounding periods per year n.

The formula for the amount can be most easily developed if we initially assume that the interest is compounded only one time per year. Based on this assumption of annual compounding, the interest earned on the principal P during the first year is iP. The sum of the principal and interest at the end of the first year is

$$P + iP = P(1 + i)$$

During the second year, interest will be earned on $P(1 + i)$ dollars. The sum of the interest and principal at the end of the second year would thus be

$$P(1 + i) + iP(1 + i) = P(1 + i)^2$$

Continuing in this fashion, we find that the amount that results from the annual compounding of P dollars at interest rate i is

$$A = P(1 + i)^t \qquad (13.35)$$

This formula is termed the *compound amount* formula.

To illustrate the compound amount formula, assume that an investment of $1000 in securities will grow at an annual rate of 8 percent for 10 years. The amount of the investment at the end of the tth year is given by $A = \$1000(1.08)^t$. The amount at the end of each of the ten years is shown in Table 13.1.

Table 13.1

t	1	2	3	4	5	6	7	8	9	10
A	$1080	1166	1259	1360	1469	1586	1713	1850	1998	2158

The values of A in Table 13.1 are calculated from Formula (13.35) with the use of logarithms.† To illustrate, the calculations for the amount at the end of 5 years are

$$A = \$1000(1.08)^5$$
$$\log A = \log (1000) + 5 \log (1.08)$$
$$\log A = 3.0000 + 5(0.0334)$$
$$\log A = 3.1670$$
$$A = \$1469$$

The following examples illustrate typical applications of the compound amount formula.

Example: Determine the interest and amount if \$1000 is invested at 6 percent for 8 years.

$$A = 1000(1.06)^8$$
$$\log A = \log (1000) + 8 \log (1.06)$$
$$\log A - 3.0000 + 8(0.0253)$$
$$\log A = 3.2024$$
$$A = \$1593$$

The interest during the eight-year period totals \$593, and the amount at the end of the eight-year period is \$1593.

Example: The cost of sending a student to a certain school has been increasing at an average rate of 6 percent per year. If the present cost is \$1500, determine the cost in 10 years.

The functional relationship is

$$A = 1500(1.06)^{10}$$
$$\log A = \log (1500) + 10 \log (1.06)$$
$$\log A = 3.1761 + 10(0.0253)$$
$$\log A = 3.4291$$

Thus,

$$A = \$2686$$

Example: An individual has the option of investing \$5000 in a time deposit account that returns 5.50 percent interest or in mutual funds. The fee for purchasing the mutual funds is a one-time fee of $8\frac{1}{2}$ percent paid at the

† The use of logarithms is explained in Appendix B.

time of purchase of the fund. If the investment has a five-year duration, what interest rate must the fund yield to match that of the time deposit?

The amount at the end of the five-year period from the time deposit is

$$A = \$5000(1.055)^5 = \$6535$$

The principal invested in the mutual fund is

$$P = \$5000 - 0.085(\$5000) = \$4575$$

The interest rate is determined by solving the compound amount formula for i.

$$A = P(1 + i)^t$$
$$\$6535 = \$4575(1 + i)^5$$
$$\log (6535) = \log (4575) + 5 \log (1 + i)$$
$$3.81525 - 3.66040 = 5 \log (1 + i)$$
$$0.15485 = 5 \log (1 + i)$$
$$\log (1 + i) = \frac{0.15485}{5} = 0.03097$$
$$1 + i = 1.072$$
$$i = 0.072 \text{ or } 7.2\%$$

Example: A company that manufactures copying equipment has determined that the life of a copier is limited by obsolescence. If a copier whose initial value is $10,000 decreases in value by 20 percent of its value at the beginning of the preceding year, determine the value of the machine at the end of each of the five years.

The functional relationship is

$$A = \$10,000(0.80)^t$$

The copier's value at the end of the first year is

$$A = \$10,000(\tfrac{8}{10})^1 = \$8000$$

The copier's value at the end of the second year is

$$A = \$10,000(\tfrac{8}{10})^2$$
$$\log A = \log (10,000) + 2[\log 8 - \log 10]$$
$$\log A = 4.000 + 2[0.9031 - 1.0000]$$
$$\log A = 3.8062$$
$$A = \$6400$$

The value at the end of each year is given in the following table. The calculations were made by using Formula (13.35) and logarithms.

t	1	2	3	4	5
A	$8000	6400	5120	4096	3277

Example: The purchasing power of the dollar declined at an average annual rate of 4 percent during 1968, 1969, and 1970. If we assume a continuation of this rate of decline, what will a 1970 dollar be worth in 1990?

$$A = \$1(1 - 0.04)^{20}$$

$$A = \$1(0.96)^{20}$$

$$\log A = \log(1) + 20[\log(96) - \log(100)]$$

$$\log A = 0.0000 + 20[1.9823 - 2.0000]$$

$$\log A = -0.3540$$

Since -0.3540 is equal to $0.6460 - 1.0000$,

$$\log A = 0.6460 - 1.0000$$

and

$$A = \frac{4.426}{10} = \$0.4426 \text{ or } 44.26\cancel{c}$$

13.3.3 PRESENT VALUE OF AN AMOUNT

The term *present value* refers to the present or current value of an amount of money that will be received at some time in the future. To illustrate, assume that one will receive $1000 ten years hence. The present value of this $1000 is equal to the principal that, if currently invested, would amount to $1000 at the end of the ten-year period.

The present value of an amount of money received t years in the future with interest of i per year can be determined by solving Formula (13.35) for P. The present value of an amount of money is

$$P = A(1 + i)^{-t} \tag{13.36}$$

In the present value formula, an amount A received t years from now has a present value of P. If we assume a positive interest rate, P will be less than A. P can be invested at rate i and will increase according to (13.35) to equal A. Consequently, P is termed the present value of A.

Example: Mr. Smith has an obligation of $500 due five years from now. If interest is assumed to be 7 percent and is compounded yearly, what is the present value of the obligation?

$$P = \$500(1 + 0.07)^{-5} = 500(1.07)^{-5}$$

$$\log P = \log 500 - 5 \log(1.07)$$

$$\log P = 2.6990 - 5(0.0294)$$
$$\log P = 2.5520$$
$$P = \$356.$$

Example: Determine the present value of receiving $1000 ten years hence, assuming annual compounding and an interest rate of 6 percent.

$$P = \$1000(1.06)^{-10}$$
$$\log P = \log(1000) - 10\log(1.06)$$
$$\log P = 3.0000 - 10(0.0253)$$
$$\log P = 2.7470$$
$$P = \$558.50$$

Example: An individual has the alternatives of selling an asset for a current cash price of $6000 or accepting a promissory note for $10,000 due 8 years hence. Determine the interest rate that equates the two alternatives.

$$\$6000 = 10,000(1+i)^{-8}$$
$$\log(6000) = \log(10,000) - 8\log(1+i)$$
$$3.7782 = 4.0000 - 8\log(1+i)$$
$$\log(1+i) = \frac{(4.0000 - 3.7782)}{8}$$
$$\log(1+i) = 0.0277$$
$$1+i = 1.066$$
$$i = 0.066 \text{ or } 6.6\%$$

The calculation of the amount and the present value has been made in the preceding examples by using logarithms. These calculations can be simplified through the use of tables of the compound interest and present value factors. A table of compound interest factors gives the value of the expression $(1+i)^n$ for various values of i and n. Table I in the Appendix gives these factors. A table of present value factors gives the value of the expression $(1+i)^{-n}$. These factors are given in Table II in the Appendix. The following examples illustrate the use of the compound interest and present value tables.

Example: If $1500 is invested at 4 percent annual interest for 8 years, determine the amount of interest and principal.

$$A = \$1500(1.04)^8$$
$$A = \$1500(1.368569)$$
$$A = \$2052.85$$

Example: Determine the present value of $10,000 received 5 years from now, assuming 6 percent annual interest.

$$P = \$10,000(1.06)^{-5}$$
$$P = \$10,000(0.747258)$$
$$P = \$7472.58$$

Example: If $500 is deposited in a savings account paying 4 percent interest, compounded annually, what amount will be on deposit in the account at the end of 16 years?

$$A = \$500(1.04)^{16}$$
$$A = \$500(1.872981)$$
$$A = \$936.49$$

Example: In a moment of weakness, John Smith promised his wife Mary that they would take an "around the world" vacation trip beginning on their tenth wedding anniversary. John anticipates the trip's costing $5000. John and Mary were married last month. What amount would John need to deposit now at interest of 5 percent in order to pay for the trip?

$$P = \$5000(1.05)^{-10}$$
$$P = \$5000(0.613913)$$
$$P = \$3069.57$$

Example: The long-term growth rate of the United States gross national product has been 4 percent per year. If the gross national product in 1971 was $1 trillion and assuming a continuation in the growth rate, in what year will the gross national product reach $2 trillion?

$$A = P(1.04)^{t}$$
$$\$2 = \$1(1.04)^{t}$$

or

$$(1.04)^{t} = 2$$

From the table for amount at compound interest, we find that t is approximately 18 years. Consequently, we predict that the gross national product will reach $2 trillion in 1989.

13.3.4 MULTIPLE COMPOUNDING

Although the interest rate is normally expressed on an annual basis, it is quite common to compound the interest more than one time per year. For instance, many savings institutions compound interest quarterly or monthly.

Quarterly compounding means that the interest earned during a quarter is added to the accumulated investment at the end of each quarter, rather than at the end of the year. Similarly, monthly compounding means that the interest earned during the month is added to the accumulated investment at the end of each month.

In the case of multiple compounding, the nominal annual interest rate does not give the percentage of change in the amount invested during the year. If, for example, $1 is invested at $i = 4$ percent and compounded quarterly, the percentage of change from the beginning to the end of the year will be 4.06 percent. This rate exceeds the 4 percent nominal rate because the interest is added to the principal four times during the year rather than only at the end of the year. Interest is thus earned on interest during the year. This rate incorporates the effect of compounding f times during the year and is termed the *effective annual interest rate*.

The compound amount formula can be modified to include more than one compounding period per year. If i is the nominal annual interest rate, t is the number of years, and f is the frequency per year of compounding, the compound interest formula becomes

$$A = P\left(1 + \frac{i}{f}\right)^{ft} \tag{13.37}$$

where i/f is the interest rate per compounding period and ft is the number of periods.

The present value formula, multiple compounding again being assumed, becomes

$$P = A\left(1 + \frac{i}{f}\right)^{-ft} \tag{13.38}$$

Example: Calculate the amount of a $1000 investment compounded quarterly at 6 percent interest for 5 years. Compare this amount to the amount if interest is compounded annually.

Compounding quarterly gives

$$A = \$1000\left(1 + \frac{0.06}{4}\right)^{4(5)}$$

$$A = \$1000(1.015)^{20}$$

$$A = \$1346.86$$

Compounding yearly gives

$$A = \$1000(1.06)^5$$

$$A = \$1338.23$$

Example: Determine the present value of $1000 received in 5 years if the nominal interest rate is 6 percent and interest is compounded twice per year. Compare this to the present value if interest is compounded yearly.

Compounding twice a year gives

$$P = \$1000 \left(1 + \frac{0.06}{2}\right)^{-2(5)}$$

$$P = \$1000(1.03)^{-10}$$

$$P = \$744.09$$

Compounding annually gives

$$P = \$1000(1.06)^{-5}$$

$$P = \$747.26$$

The effective annual interest rate for multiple compounding can easily be determined. If we represent the effective annual interest rate by r and the nominal annual interest rate by i, we can solve for r as follows:

$$P(1 + r)^t = P\left(1 + \frac{i}{f}\right)^{ft}$$

or, by taking the tth root of the expression,

$$(1 + r) = \left(1 + \frac{i}{f}\right)^f$$

Thus,

$$r = \left(1 + \frac{i}{j}\right)^f - 1 \tag{13.39}$$

Example: Determine the effective annual interest rate if the nominal annual rate of 4 percent is compounded quarterly.

$$r = \left(1 + \frac{0.04}{4}\right)^4 - 1$$

$$r = (1.01)^4 - 1$$

$$r = 0.040604 = 4.0604\%$$

The effective annual rate is 4.0604%.

Example: Determine the effective annual interest rate if the nominal annual rate of 6 percent is compounded monthly.

$$r = \left(1 + \frac{0.06}{12}\right)^{12} - 1$$

$$r = (1.005)^{12} - 1$$

$$r = 0.061678 = 6.1678\%$$

13.3.5 CONTINUOUS COMPOUNDING

Certain assets earn on a continuous basis throughout the year. As an example, capital equipment that is used in the manufacture of daily output contributes a continuous flow of earnings to the firm. The rate of return of this equipment is somewhat distorted if it is assumed that the return occurs only once each year. The actual rate of return should be calculated by incorporating the continuous flow of earnings into the formula.

The compound amount formula can be modified to incorporate continuous compounding. We can determine the effect of continuous compounding by letting the number of compounding periods f approach infinity in the compound amount formula. Thus,

$$A = \lim_{f \to \infty} \left[P \left(1 + \frac{i}{f} \right)^{ft} \right]$$

which can be rewritten as

$$A = P \lim_{f \to \infty} \left[\left(1 + \frac{1}{f/i} \right)^{f/i} \right]^{it}$$

From Formula (12.4), we recognize that the expression within the brackets equals e. Therefore, the formula for the amount for continuous compounding reduces to

$$A = Pe^{it} \tag{13.40}$$

The present value of an amount A received t years hence if the principal is compounded continuously is determined by solving formula (13.40) for P.

$$P = Ae^{-it} \tag{13.41}$$

Values of the term e^x and e^{-x} for $x = it$ are given in Table V in the Appendix.

The use of Table V in determining the present value and amount if continuous compounding is assumed is illustrated by the following examples.

Example: Assume that \$10,000 is invested for 10 years at a nominal annual interest rate of 6 percent. Compare the amount if the principal is compounded continuously and if it is compounded annually.

For continuous compounding:

$$A = 10{,}000e^{(0.06)10} = 10{,}000e^{0.6}$$
$$A = 10{,}000(1.822) = \$18{,}220$$

For annual compounding:

$$A = 10{,}000(1.06)^{10}$$
$$A = 10{,}000(1.7908) = \$17{,}908$$

The difference over the 10-year period between continuous compounding and annual compounding is $312.

Example: One share of Miller Growth Fund, a mutual fund, was valued at $10.00 on January 1 and $11.00 on December 31. Determine the rate of return, assuming continuous appreciation throughout the year.

The method of determining i involves solving the continuous compounding formula for i. Thus,

$$11.00 = 10.00e^{i(1)}$$

$$\ln(11.00) = \ln(10.00) + i$$

$$i = 2.39790 - 2.30259$$

$$i = 0.09531 = 9.53\%$$

Example: Determine the present value of a note of $5000 that matures in 10 years. Assume continuous compounding and a nominal annual interest rate of 7 percent.

$$P = 5000e^{-(0.07)10}$$

$$P = 5000e^{-0.7}$$

$$P = 5000(0.497)$$

$$P = \$2485$$

Example: Company A has stated that it expects sales to increase from the present level of $1,000,000 by $50,000 per year. Company B expects a continuous growth rate of 6 percent and it reported sales of $900,000 during the past accounting period. Develop predicting formulas for both companies and calculate anticipated sales 10 years hence.

For Company A:

$$S = 1,000,000 + 50,000t$$

and

$$S(10) = 1,000,000 + 500,000 = \$1,500,000$$

For Company B:

$$S = \$900,000e^{(0.06)t}$$

and

$$S(10) = \$900,000e^{0.60}$$

$$S(10) = \$1,639,800$$

The effective annual interest rate from continuous compounding can be determined by equating Formulas (13.35) and (13.40). Again representing the effective annual interest rate by r and the nominal annual rate by i, we obtain

$$P(1 + r)^t = Pe^{it}$$

or

$$(1 + r)^t = e^{it}$$

Taking the tth root of the equation gives

$$1 + r = e^i$$

and the effective annual interest rate is

$$r = e^i - 1 \qquad\qquad (13.42)$$

Example: Determine the effective annual interest rate, assuming continuous compounding and a nominal interest rate of $i = 0.10$.

$$r = e^{0.10} - 1$$
$$r = 1.105 - 1$$
$$r = 0.105 = 10.5\%$$

Example: Determine the effective annual interest rate assuming continuous compounding and a nominal interest rate of $i = 0.05$.

$$r = e^{0.05} - 1$$
$$r = 1.0513 - 1$$
$$r = 0.0513 = 5.13\%$$

Example: Western Pacific Savings and Loan advertises that money deposited in a savings account earns interest of 6 percent compounded daily. Determine the effective annual interest rate paid by Western Pacific.

An exact solution for the effective annual interest rate requires solving Formula (13.39) for r. That is,

$$r = \left(1 + \frac{0.06}{365}\right)^{365} - 1$$

Solving this equation requires very detailed tables of logarithms. Since these tables are not readily available, we can determine the approximate effective annual interest rate by assuming continuous compounding. The rate is

$$r = e^{0.06} - 1$$
$$r = 1.0618 - 1$$
$$r = 0.0618 = 6.18\%$$

13.3.6 ANNUITIES

Another common form of a financial agreement is the *annuity*. Annuities consist of a series of equal payments, each payment normally being made at the end of a designated period of time. A common example of an annuity is the mortgage payments on a home loan. Other examples include installment payments on an automobile loan, payments on life insurance, payments for retirement plans, sinking fund payments, and payments to a savings plan.

The *amount of an annuity* is the sum of the periodic payments and the interest earned from these payments. As an example, assume that an individual plans to establish a trust fund for his children's education. The trust fund will be established by annually depositing $1000 in a savings institution. Each payment, or deposit, to the trust fund will be made at the *end* of the year, rather than at the beginning of the year. The payments deposited will earn 6 percent interest compounded annually.

If we let i represent the annual interest rate, t the number of years, p the annuity payment made at the *end of each year*, and A the sum of the annuity payments and interest, we can calculate A as follows:

$$A = p + p(1 + i) + p(1 + i)^2 + \ldots + p(1 + i)^{t-1} \qquad (13.43)$$

The final annuity payment is written first and earns no interest, the next to final payment earns interest for one year, and the first payment, which is made at the end of the first year, earns interest for $t - 1$ years. This expression can be solved for A by a simple algebraic maneuver. This consists of multiplying the expression in (13.43) by $(1 + i)$. This gives

$$A(1 + i) = p(1 + i) + p(1 + i)^2 + p(1 + i)^3 + \ldots + p(1 + i)^t \qquad (13.44)$$

subtracting (13.43) from (13.44) gives

$$A(1 + i) - A = p(1 + i)^t - p$$

which when solved for A gives the formula for the amount of an annuity.

$$A = p \left[\frac{(1 + i)^t - 1}{i} \right] \qquad (13.45)$$

We can also determine the periodic payment necessary, when invested at interest rate i for t years, to sum to the amount A. This is found by solving (13.45) for p.

$$p = A \left[\frac{i}{(1 + i)^t - 1} \right] \qquad (13.46)$$

It is important to remember that these formulas were developed based on the assumption that the payments are made at the end of each year, rather than at the beginning. The formula for the amount of an annuity can be modified to account for payments made at the beginning of the year by replacing p in (13.45) by $p(1 + i)$. The formula for the periodic payment, Formula (13.46), is modified by replacing A by $A(1 + i)^{-1}$. The use of these formulas is illustrated by the following examples. In these examples, Table I in the Appendix is used to determine the value of $(1 + i)^t$. The annuity factor is then calculated by simple arithmetic.

Example: Determine the amount of the trust fund mentioned above if payments are made for ten years. The interest rate is $i = 0.06$, the time period is $t = 10$ years, and the amount is

$$A = p\left[\frac{(1 + i)^t - 1}{i}\right]$$

$$A = \$1000\left[\frac{(1.06)^{10} - 1}{0.06}\right]$$

$$A = \$1000\left[\frac{1.790848 - 1}{0.06}\right]$$

$$A = \$13,180.80$$

Example: Determine the amount in the preceding example if the payments are made at the beginning, rather than the end, of each year.

$$A = p(1 + i)\left[\frac{(1 + i)^t - 1}{i}\right]$$

$$A = \$13,180.80(1.06)$$

$$A = \$13,971.65$$

Example: Mr. Clark plans on investing $1000 per year in a savings plan that earns 5 percent interest compounded annually. Determine the sum of the annuity payments and interest at the end of 10 years.

$$A = \$1000\left[\frac{(1.05)^{10} - 1}{0.05}\right]$$

$$A = \$1000\left[\frac{1.628895 - 1}{0.05}\right]$$

$$A = \$1000(12.5779) = \$12,577.90$$

Example: Sioux Falls Steel Company recently placed a $5 million bond issue with a group of private investors. One of the requirements of the investors was that the firm establish a sinking fund that will retire the bonds in 15 years. If Sioux Falls Steel plans to invest the sinking fund payments in government bonds that earn 6 percent interest, what yearly payment is necessary to retire the $5 million bond issue at the end of 15 years?

$$p = \$5,000,000\left[\frac{0.06}{(1.06)^{15} - 1}\right]$$

$$p = \$5,000,000\left[\frac{0.06}{2.396558 - 1}\right]$$

$$p = \$5,000,000(0.04296) = \$214,800$$

Formulas (13.45) and (13.46) were developed based on the assumption of annual annuity payments. In many cases, however, annuity payments are made more than one time per year. For example, most installment loans require monthly payments. Similarly, some savings plans are based on quarterly payments. These cases can be taken into account by replacing i by i/f and t by ft in Formulas (13.45) and (13.46), where f represents the frequency or number of annuity payments per year. The term i/f is the interest rate per period, and ft is the number of periods or number of annuity payments. We again assume that payments are made at the end of each period and that interest is compounded at the time of the payment.

Example: Determine the amount of an annuity if payments of $100 are deposited quarterly at 4 percent nominal annual interest for 5 years. Assume that the payments are deposited at the end of each quarter and interest is compounded quarterly.

$$A = p \left[\frac{\left(1 + \frac{i}{f}\right)^{ft} - 1}{i/f} \right]$$

$$A = \$100 \left[\frac{(1.01)^{20} - 1}{0.01} \right]$$

$$A = \$100 \left[\frac{1.220190 - 1}{0.01} \right]$$

$$A = \$2201.90$$

It is often necessary to calculate the current value of receiving p per year for t years, assuming an interest rate of i. This sum is termed the *present value of an annuity*. As an example, one might be required to determine the present value of receiving $1000 per year for 10 years if interest is 6 percent per year. Since money has a value over time, the present value of $1000 per year for 10 years will be less than the value of receiving the $10,000 during the first year. The present value of this series of annuity payments will, of course, depend upon the interest rate.

The formula for the present value of an annuity is derived by summing the present values of each of the individual annuity payments. The present value of the first payment, made one year from now, is $p(1 + i)^{-1}$. The present value of the second payment, made two years from the present, is $p(1 + i)^{-2}$. The present value of all payments is the geometric series

$$PV = p(1 + i)^{-1} + p(1 + i)^{-2} + \ldots + p(1 + i)^{-t} \qquad (13.47)$$

If all terms in (13.47) are multiplied by $(1 + i)$, we obtain

$$PV(1 + i) = p + p(1 + i)^{-1} + \ldots + p(1 + i)^{-(t-1)} \qquad (13.48)$$

Subtracting (13.47) from (13.48) gives:

$$PV(1 + i) - PV = p - p(1 + i)^{-t}$$

which, when solved for PV, gives

$$PV = p \left[\frac{1 - (1 + i)^{-t}}{i} \right] \tag{13.49}$$

The periodic payment that has a present value of PV is given by solving (13.49) for p:

$$p = PV \left[\frac{i}{1 - (1 + i)^{-t}} \right] \tag{13.50}$$

The value of $(1 + i)^{-t}$ is given in Table II in the Appendix. The use of the present value and periodic payment formulas is illustrated by the following examples.

Example: Determine the present value of receiving \$1000 per year for 10 years if interest is 6 percent per annum.

$$PV = \$1000 \left[\frac{1 - (1.06)^{-10}}{0.06} \right]$$

$$PV = \$1000 \left[\frac{1 - 0.558395}{0.06} \right]$$

$$PV = \$1000(7.3601) = \$7360$$

Example: Mr. Miller is considering the purchase of a home. The loan balance is \$20,000 and the interest rate is 7 percent. Determine the yearly payments for a 25-year loan.

$$p = \$20,000 \left[\frac{0.07}{1 - (1.07)^{-25}} \right]$$

$$p = \$20,000 \left[\frac{0.07}{1 - 0.184249} \right]$$

$$p = \$20,000(0.08581) = \$1716$$

Example: An individual purchases \$1940 of furniture, which is to be paid for in 12 monthly payments of \$176 per month. What is the nominal annual rate of interest?

The nominal rate of interest can be determined by solving the present value of an annuity formula for i.

$$1940 = 176 \left[\frac{1 - (1 + i)^{-12}}{i} \right]$$

$$11.02(i) = 1 - (1 + i)^{-12}$$

We now must solve for i by trial and error. This is accomplished by selecting alternative values of i until a satisfactory value of i is determined. Assume

$$i = 0.01$$

then

$$11.02(0.01) = 1 - (1.01)^{-12}$$
$$0.1102 < 0.1126$$

Assume

$$i = 0.0125$$

then

$$11.02(0.0125) = 1 - (1.0125)^{-12}$$
$$0.1378 < 0.1385$$

Assume

$$i = 0.0150$$

then

$$11.02(0.0150) = 1 - (1.0150)^{-12}$$
$$0.1653 > 0.1636$$

From these calculations, it is apparent that i is between 0.0125 and 0.0150. Further refinement yields a monthly rate of approximately 0.013 or 1.3 percent. This corresponds to a yearly rate of 15.6 percent.

13.4 Growth Rate of Functions

The *growth rate*, or rate of growth, of a function gives the percentage change in the value of the dependent variable during a specified period of time. The growth rate is normally expressed on an annual basis. Thus, one might read that the sales of a company have been increasing at 10 percent per year. Similarly, an economist might state that the long-term growth rate of the gross national product is 4 percent.

The growth rate of a function that grows periodically, rather than continuously, is given by

$$r = \frac{\dfrac{\Delta f(t)}{\Delta t}}{f(t)} \tag{13.51}$$

where r represents the growth rate, $f(t)$ represents the value of the dependent variable at time t, $\Delta f(t)$ represents the change in the value of the dependent variable, and Δt represents the change in time. To illustrate, assume that $2 is earned quarterly on an investment of $100. The growth rate of the investment, expressed on an annual basis, is

$$r = \frac{\frac{\Delta f(t)}{\Delta t}}{f(t)} = \frac{2}{\frac{1/4}{1000}} = 0.08 \text{ or } 8\%$$

The growth rate of the investment, or, alternatively, the interest rate on the investment is 8 percent.

The growth rate for continuous functions is given by the ratio of the derivative of the function and the function. Thus,

$$r = \frac{f'(t)}{f(t)} \tag{13.52}$$

The derivative of the function gives the change in the function for an instantaneous change in time, whereas the denominator represents the base value of the function at time t. It can be seen that Formulas (13.52) and (13.51) are quite similar, (13.52) being the limiting value of (13.51) as Δt approaches zero.

We stated in Chapter 12 that one of the important properties of the exponential function was that the function describes constant rates of growth. To illustrate this property, consider the formula for the amount assuming continuous compounding. This formula is

$$A = Pe^{it} \tag{13.40}$$

The formula is, of course, an exponential function with base e and exponent it. The growth rate of this function is given by (13.52) and is

$$r = \frac{f'(t)}{f(t)} = \frac{Pe^{it}(i)}{Pe^{it}} = i$$

The growth rate of the function is thus equal to the nominal rate of interest. Furthermore, the growth rate is constant.

The growth rate for the amount, if annual compounding is assumed, is also constant. The formula for the amount if annual compounding is assumed is an exponential function with base $(1 + i)$ and exponent t. The formula is

$$A = P(1 + i)^t \tag{13.35}$$

The continuous growth rate of the exponential function is

$$r = \frac{f'(t)}{f(t)} = \frac{P(1 + i)^t \ln (1 + i)}{P(1 + i)^t} = \ln (1 + i)$$

The result that $r = \ln (1 + i)$ for the compound amount formula can be interpreted as follows: The nominal annual rate of growth of the compound amount formula is i. This nominal rate of growth is equivalent to a continuous rate of growth of $\ln (1 + i)$. As we would expect, the rate of growth for the the function is constant.

One of the important results of the preceding discussion has been to show that the exponential function can be used to model a situation in which constant growth rates are required. Constant growth rates sometimes apply to sales, profits, income, savings, etc. In these cases, the analyst should certainly consider an exponential function for describing the dependent variable as a function of time.

Growth rates are not constant for functions other than exponential functions. Although we do not intend to offer proof of this statement, we can demonstrate the statement for selected functions.

Example: The sales of a new product are described by the following function:

$$S(t) = 10,000t^2$$

Determine the growth rate of sales.

$$r = \frac{S'(t)}{S(t)} = \frac{20,000t}{10,000t^2} = \frac{2}{t}$$

The growth rate for this simple power function depends upon the value of the independent variable. This characteristic of varying growth rates is also true for more complicated algebraic functions.

Example: Following an advertising campaign, the sales of a product increase and then begin to decline. The sales are forecast by the quadratic function

$$S(t) = 1000 + 100t - 10t^2$$

where t represents months. Determine the growth rate at the conclusion of the campaign ($t = 0$) and for each of the nine months following the conclusion. Determine sales during each of these months. The growth rate is

$$r = \frac{S'(t)}{S(t)} = \frac{100 - 20t}{1000 + 100t - 10t^2}$$

The growth rate and the sales are given for $t = 0$ through $t = 9$ in the following table.

t	0	1	2	3	4	5	6	7	8	9
r	10%	7.34%	5.16%	3.30%	1.61%	0%	−1.61%	−3.30%	−5.16%	−7.34%
S	1000	1090	1160	1210	1240	1250	1240	1210	1160	1090

Example: The Beck Investment Company found that the value of their land investments could be described by

$$V = 1,000,000(1.50)^{\sqrt{t}}$$

Determine the growth rate of this function.

$$r = \dfrac{\dfrac{dV}{dt}}{V} = \dfrac{1{,}000{,}000(1.50)^{\sqrt{t}}(1/2)t^{-1/2}\ln(1.50)}{1{,}000{,}000(1.50)^{\sqrt{t}}}$$

$$r = \dfrac{\ln(1.50)}{2\sqrt{t}} = \dfrac{0.40547}{2\sqrt{t}}$$

The growth rate for Beck Investments depends upon the value of t. For $t = 1$, the growth rate is 20.27 percent. When $t = 4$ the rate has fallen to 10.14 percent.

13.5 Functions Defined over Time

In this section we consider the problem of when to sell an asset whose value is changing over time. The objective is to maximize the present value of the asset. To illustrate the technique, consider the following example.

Example: The Beck Investment Company has extensive land holdings in Florida. They have estimated that the value of these holdings is given by the functional relationship:

$$V = \$1{,}000{,}000(1.50)^{\sqrt{t}}$$

Assuming a cost of capital of r percent (on a continuous basis) and disregarding any costs of upkeep of the undeveloped land, determine the optimal period of holding the land.

The present value of the land is given by Formula (13.41). Thus,

$$P = Ve^{-rt} = 1{,}000{,}000(1.50)^{\sqrt{t}}e^{-rt}$$

This can be written in its logarithmic form as

$$\ln P = \ln(1{,}000{,}000) + \sqrt{t}\ln(1.50) - rt$$

To determine the optimal value of t, we equate the derivative of the function with 0 and solve for t. Since the function is logarithmic, Rule 15 is used for the derivative of logarithmic functions.

$$\frac{1}{P}\frac{dP}{dt} = \frac{1}{2}\ln(1.50)t^{-1/2} - r = 0$$

$$\frac{dP}{dt} = P\left[\frac{1}{2}\ln(1.50)t^{-1/2} - r\right] = 0$$

Since P is not equal to 0, $\frac{1}{2} \ln (1.50)t^{-1/2} - r = 0$. Thus,

$$t^{-1/2} = \frac{2r}{\ln (1.50)}$$

and

$$t = \left[\frac{\ln (1.50)}{2r} \right]^2$$

If we assume that $r = 8$ percent, then the optimum holding period for the land is

$$t = \left[\frac{\ln (1.50)}{0.16} \right]^2 = \left[\frac{0.40547}{0.16} \right]^2 = 6.4 \text{ years}$$

The present value of the land is

$\ln (P) = \ln (1,000,000) + \sqrt{6.4} \ln (1.50) - 0.08(6.4)$

$\ln (P) = \ln (1,000,000) + 2.53(0.40547) - 0.08(6.4)$

$\ln (P) = \ln (1,000,000) + 1.0258 - 0.512 = \ln (1,000,000) + 0.51384$

$P = \$1,000,000(1.672) = \$1,672,000$

The technique utilized in determining the optimum time span for holding the land in the Beck Investment Company problem can be applied to any similar problem. The procedure involves expressing the value of the function in terms of its present value and maximizing the present value of the function. This involves equating the derivative of the present value function with 0 and solving for the optimum time interval. As a second illustration, consider the following example.

Example: Coleman and Associates are engaged in the business of importing, aging, and distributing premium table wines. The value of the wine increases over time according to the following function:

$$V = 4.00(1.35)^{t^{2/3}}$$

The present value of the wine, if we assume continuous appreciation and cost of capital r, is

$$P = 4.00(1.35)^{t^{2/3}}e^{-rt}$$

The optimum value of P is determined by equating the derivative of the present value function with 0.

$$\ln P = \ln (4.00) + t^{2/3} \ln (1.35) - rt$$

$$\frac{1}{P} \frac{dP}{dt} = \frac{2 \ln (1.35)}{3} t^{-1/3} - r = 0$$

Solving for t gives

$$t^{1/3} = \frac{2 \ln (1.35)}{3r}$$

$$t = \left[\frac{2 \ln (1.35)}{3r}\right]^3$$

If we assume $r = 10$ percent, the optimum time for aging the wine is

$$t = \left[\frac{2(0.300)}{0.30}\right]^3 = (2.00)^3 = 8.0 \text{ years}$$

The present value of the wine is

$$P = 4.00(1.35)^{(8.0)^{2/3}}[e^{-0.10(8.0)}]$$

$$P = 4.00(1.35)^4[e^{-0.8}] = \$5.98 \text{ per bottle}$$

The price per bottle obtainable at the end of the 8-year holding period is

$$P = 4.00(1.35)^{t^{2/3}} = 4.00(1.35)^4 = \$13.35$$

The wine will thus be worth \$13.35 per bottle at the end of the 8-year period. Assuming that the cost of capital is 10 percent and assuming continuous compounding, we find that the present value of the wine is \$5.98 per bottle.

PROBLEMS

1. An automobile transmission shop advertises 48-hour automatic transmission exchange. In order to guarantee 48-hour turn-around time, the shop maintains a relatively large stock of remanufactured transmissions. The cost of storing a transmission is \$25 per year. If the shop does not have the proper transmission in stock, it must immediately request one from its supplier and pay for a rush delivery. The shortage cost averages \$30. The shop expects a demand for 500 of a common type of transmission during the year. The normal cost of ordering the transmissions is \$200 per order. Orders are received 5 working days after the order is placed. Determine the economic order quantity, the maximum inventory of this particular transmission, and the inventory cycle.

2. A television sales store can afford to stock only a limited number of color television sets because of the high cost of the inventory. A certain store expects to sell 250 of a particular set during the year. The holding cost is \$30 per set per year. If the set is out of stock, the customer is offered a discount for the inconvenience of waiting for delivery of the set from the factory. The discount averages \$50 per set short per year. The cost of placing an order for the particular set is \$100. Determine the

economic order quantity and the maximum quantity of this type of set in inventory at any one time.

3. The yearly demand for a particular product is 2000 units, and the production rate for this product is 8000 units per year. Inventory holding costs are $10 per unit per year, shortage costs are $15 per unit short per year, and setup costs are $500. Determine q, t_1, t_2, t_3, t_4, t, and the maximum inventory.

4. Ace Manufacturing Company manufactures various types of small electric motors, which are sold to various other firms as components. The yearly demand for a particular motor is 3000 units, and the production rate is 12,000 units per year. Inventory holding costs are $5.00 and the shortage costs are $2.00 per unit short. Setup costs are $350. Determine the production lot size, the maximum inventory, and the length of the production cycle.

5. An automobile retailer must determine the optimum number of a particular model of a new car to hold in inventory. Sales of the model are forecast as 30 units during the month of interest. The holding cost of the car is $60 per month. The shortage cost, which includes a discount to the customer, is $150 per unit short, and the cost of placing an order is $300. Determine the number of this model of car that should be ordered, the maximum number of cars in inventory during the month, and the portion of the month in which customers cannot be promised immediate delivery.

6. The data that follow represent the results of a study of the relationship between productivity of workers and scores on selected aptitude tests.

Employee No.	Y Productivity	X Aptitude Score
1	41.5	110
2	43.0	125
3	42.5	140
4	44.0	150
5	44.5	165
6	45.2	170
7	46.0	190
8	47.5	200
9	49.0	210
10	48.0	215
Total	451.2	1675

Sum of squares:

$$\sum x^2 = 292{,}375, \qquad \sum xy = 76{,}558$$

(a) Plot these data on a scatter diagram and verify that a linear function can be used to predict productivity based upon aptitude score.

(b) Determine the slope and intercept of the predicting function, using the method of least squares.

(c) Determine the variance of the estimate.

7. The number of revenue passengers flown by Trans World Airlines for the period from 1959 through 1968 is given in the following table.

Year	Year No.	Revenue Passengers (millions)
1959	0	5.9
1960	1	5.8
1961	2	5.5
1962	3	5.9
1963	4	6.8
1964	5	8.2
1965	6	9.7
1966	7	9.9
1967	8	12.9
1968	9	13.9
Total	45	84.5

(a) Plot these data on a scatter diagram and verify that an exponential function can be used to describe the growth in revenue passengers.

(b) Determine the exponential function, using the method of least squares.

(c) Use the exponential function to predict the number of revenue passengers in 1969 and 1970.

8. From 1940 to 1970 the price of eggs rose from $0.23 per dozen to $0.70. If this rate of price increase continues, what would be the price of one dozen eggs in 2000?

9. Harvey Smith received a spendable income of $84.50 per week in 1960. His spendable income has increased to $124.00 per week in 1970. In 1960 the consumer price index was 103.1. The index was 134.0 in 1970. Determine the annual rate of increase in real spendable income for Mr. Smith.

10. Gross national product in the United States increased from $284.8 billion in 1950 to $503.7 billion in 1960. Determine the growth rate during this period.

11. An investor has the opportunity to purchase a second mortgage on a home for $3500. The mortgage matures in five years and returns $5000. Assuming annual compounding, determine the rate of interest on the investment.

12. A company's sales grew from $10 million to $15 million in eight years. Determine the continuous rate of growth in sales.

13. Great Southern Savings and Loan currently pays 5 percent interest compounded quarterly on passbook savings accounts. If the compounding period is changed from quarterly to daily, determine the difference in the effective annual rate of interest. (*Hint:* Assume that daily compounding can be approximated by continuous compounding.)

14. A share of Investors Mutual declined in value from $10 per share to $9 per share between July 1 and December 31, 1969. Assuming continuous interest, determine the effective annual rate of decline for this mutual fund.

15. An individual can borrow $1000 at 8 percent interest, or he can discount a note for $1000 at 7.5 percent interest. Determine the effective annual interest for both alternatives.

16. If $100 is deposited at the beginning of every year for 5 years and compounded semiannually at a rate of 6 percent per year, determine the principal at the end of the five-year period.

17. Determine the yearly payments for a $30,000 home loan amortized over 30 years at 8 percent interest.

SUGGESTED REFERENCES

The references for this chapter are listed in Chapter 11.

Chapter 14

Integral
Calculus

We have discussed differential calculus and business and economic models based upon differential calculus. In the study of calculus, it is customary to consider differential calculus and integral calculus as two related branches of calculus. The relationship between differential and integral calculus is established in Sec. 14.2 of this chapter. Before we turn to this relationship, it is important to understand the concept and meaning of integral calculus. Consequently, our first task in this chapter is to present the meaning of integral calculus. After this concept is clearly established, we can then discuss the relationship between the two branches of calculus and methods of determining integrals.

14.1 The Integral

The task of defining the integral is best accomplished by illustrating the procedure for determining the area of an irregular figure. Assume that we wish to determine the area in Fig. 14.1 between the horizontal axis and the function $y = f(x)$ between $x = a$ and $x = b$. There is no geometrical formula that gives the area for this irregularly shaped function. We can, how-

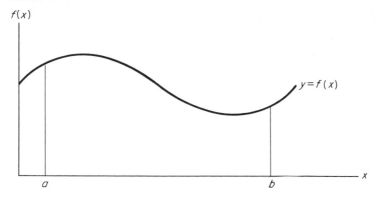

Figure 14.1

ever, determine the approximate area of the figure by subdividing the area into rectangles and calculating the sum of the areas of the rectangles. The area has been subdivided into rectangles in Fig. 14.2.

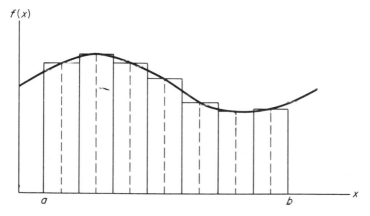

Figure 14.2

The area of each rectangle in Fig. 14.2 is determined by dividing the distance from a to b into small intervals and treating each interval as the base of a rectangle. If n rectangles with equal base are constructed, the base of each rectangle is $\Delta x = (b - a)/n$. The height of the rectangle is given by the value of the function at the midpoint of each rectangle, $f(x_i)$. The approximate area can now be found by summing the areas of the n rectangles. Since the area of a rectangle is given by the product of the base and height, $f(x_i)\,\Delta x$, the approximate area is given by

$$A \simeq f(x_1) \, \Delta x_n + f(x_2) \, \Delta x_n + \ldots + f(x_n) \, \Delta x_n \tag{14.1}$$

where

$$\Delta x_n = \frac{b - a}{n}$$

Using the Greek letter $\sum$ to represent the summation of a series of terms, we can rewrite (14.1) as

$$A \simeq \sum_{i=1}^{n} f(x_i) \, \Delta x_n \tag{14.2}$$

where

$$\Delta x_n = \frac{b - a}{n}$$

From Fig. 14.2, it can be seen that the accuracy of the approximation is dependent upon the number of rectangles in the interval from a to b. As the number of rectangles increases, the effect of the "gaps" between the function and the top of the rectangles on the area becomes of less importance. In other words, as Δx grows smaller or alternatively as n becomes larger, the approximation becomes more accurate. The area is given as the limit as Δx approaches 0 (i.e., as n approaches ∞) of the summation of the products of $f(x)$ and Δx. Expressed mathematically, the area is given by

$$A = \operatorname*{limit}_{n \to \infty} \sum_{i=1}^{n} f(x_i) \, \Delta x_n \tag{14.3}$$

where

$$\Delta x_n = \frac{b - a}{n}$$

The magnitude of each term in the summation of (14.3) decreases as the number of terms increases. For instance, if Δx decreases from $\Delta x = 1.0$ to $\Delta x = 0.1$, then the magnitude of $f(x) \, \Delta x$ decreases by a factor of ten and the number of terms increases by a factor of ten. It can thus be seen that the *area is given by the limiting value of the sum of a number of terms, when the number of terms increases infinitely as the value of each term approaches zero.* This limiting value A is termed the *definite integral* of the function $f(x)$ for the interval $x = a$ to $x = b$.

The definite integral of a function $f(x)$ over the interval a to b is a numerical value that is associated with the function and the interval. More specifically, this numerical value is the limiting value of the sum of a number of terms when the number of terms increases infinitely as the numerical value of each term approaches zero. The symbolism commonly used for the definite integral of the function is

$$\int_a^b f(x) \, dx \tag{14.4}$$

The summation sign is replaced by the elongated S or integral sign. a and b are, respectively, the lower and upper limits of integration. $f(x)$ represents the function for which the definite integral is being determined, and dx replaces Δx and is termed the *differential*.

14.2 Relationship between Integral and Differential Calculus

In order to determine the value of a definite integral, the reader must understand two additional concepts. These are the relationship between the *indefinite integral* and the derivative, and the *fundamental theorem* of calculus. We shall first show the relationship between the indefinite integral and the derivative. Sec. 14.3 introduces the fundamental theorem of calculus.

The indefinite integral and the derivative are related in the following manner. Assume that we want to find the indefinite integral of a function $f(x)$. *The indefinite integral of the function $f(x)$ is another function $g(x)$ rather than a numerical value.* The procedure for determining the indefinite integral of $f(x)$ is to find the function $g(x)$ such that the derivative of $g(x)$ is $f(x)$. If the derivative of $g(x)$ is $f(x)$, then $g(x)$ is the indefinite integral of $f(x)$. Thus,

$$\int f(x)\,dx = g(x) \qquad\qquad (14.5)$$

provided that

$$g'(x) = f(x)$$

If $f(x)$ is thought of as the derivative of some function, then the indefinite integral of $f(x)$ is the function $g(x)$ such that $g'(x) = f(x)$. Integration is thus sometimes thought of as the reverse process of differentiation. For this reason, the term *antiderivative* is often used synonymously with the term *indefinite integral*. If we can find a function $g(x)$ and verify that the derivative of $g(x)$ equals $f(x)$, then we can conclude that $g(x)$ is the antiderivative or the indefinite integral of $f(x)$. These concepts are illustrated by the following examples.

Example. Determine $\int x\,dx$. Note that $f(x) = x$.

If $g(x) = \frac{1}{2}x^2 + c$, then $g'(x) = x$. By (14.5), we conclude that $\int x\,dx = \frac{1}{2}x^2 + c$, where c is any constant including 0.

Example. Find $\int (x^2 + 4x)\,dx$. In this example $f(x) = x^2 + 4x$.

If $g(x) = \frac{1}{3}x^3 + 2x^2 + c$, then $g'(x) = x^2 + 4x$, and, therefore,

$$\int (x^2 + 4x)\, dx = \frac{1}{3}x^3 + 2x^2 + c$$

Example. Determine $\int 2x\, dx$.

For $f(x) = 2x$,

$$g(x) = x^2 + c \quad \text{and} \quad \int 2x\, dx = x^2 + c$$

The constant term in each of the preceding examples is necessary to illustrate that the indefinite integral of $f(x)$ is an entire family of functions. There are an infinite number of functions, each differing only by the value of the constant c, which have the derivative $f(x)$. In the preceding example, $g(x) = x^2 + c$ was the integral of $f(x) = 2x$. For any of the infinite number of possible values of the constant c, the derivative of $g(x)$ continues to be $f(x) = 2x$. In determining the indefinite integral, therefore, c is included as a part of the integral. This constant is termed the *constant of integration*. Figure 14.3 illustrates several possible values of the constant of integration for the preceding example.

The constant of integration can only be determined if additional information is given. For example, if it is known that the value of the dependent

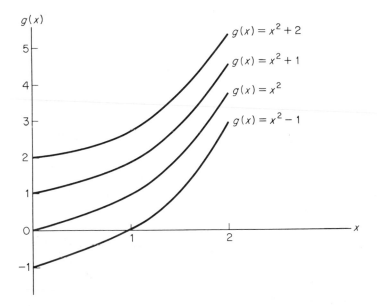

Figure 14.3

variable for $x = 0$ is $g(0) = 1$ for the function sketched in Fig. 14.3, we know that the indefinite integral is $g(x) = x^2 + 1$. This additional information is called the *initial condition*.

Example. Determine $\int x^{-1/2}\, dx$, given the initial condition that $g(4) = 6$.

$$g(x) = 2x^{1/2} + c \quad \text{and} \quad g(4) = 6$$

thus

$$g(4) = 2(4)^{1/2} + c = 6 \quad \text{and} \quad c = 6 - 4 = 2$$

Therefore,

$$g(x) = 2x^{1/2} + 2$$

Example. Determine $\int (3x + 4)\, dx$, when $g(0) = 10$.

$$g(x) = \frac{3x^2}{2} + 4x + 10$$

Methods of determining the indefinite integral of a function are discussed in Sec. 14.4.

14.3   Fundamental Theorem of Calculus

The fundamental theorem of calculus relates the concepts of the definite integral and the indefinite integral. The definite integral of the function $f(x)$ between the limits $x = a$ and $x = b$ is a numerical value that can be determined from the indefinite integral. If we again represent the indefinite integral of the function $f(x)$ by $g(x)$, then the fundamental theorem of calculus states that

$$\int_a^b f(x)\, dx = g(b) - g(a). \tag{14.6}$$

This theorem states that the value of the definite integral of the function $f(x)$ between the limits a and b is given by the indefinite integral $g(x)$ evaluated at the upper limit of integration b minus the indefinite integral evaluated at the lower limit of integration a. Determining the definite integral thus merely involves evaluating the indefinite integral at $x = b$ and at $x = a$ and subtracting the two values. The constant of integration, which was required for the indefinite integral, is not included in the definite integral. This theorem is illustrated by the following examples.

Example. Using the methods of integral calculus, determine the area between the function $f(x) = x$ and the horizontal axis from $x = 0$ to $x = 5$.

The function $f(x) = x$ plots as shown below as a 45-degree triangle with base 5 and height 5. Using the formula for the area of a triangle $A = \frac{1}{2}bh$,

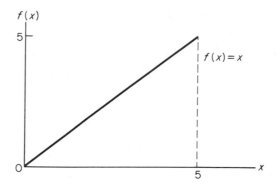

we observe that the area is 12.5. Using the concept of the definite integral, we have

$$A = \int_0^5 x\,dx = \frac{x^2}{2}\Big|_0^5 = \frac{25}{2} - \frac{0}{2} = 12.5$$

This example introduces the symbolism commonly employed in the evaluation of definite integrals. The integral of $\int x\,dx$ is $g(x) = x^2/2$. Since the limits of integration are $x = 0$ and $x = 5$, we determine $g(5)$ and $g(0)$.

$$g(5) = \frac{5^2}{2} = 12.5$$

$$g(0) = \frac{0^2}{2} = 0$$

and

$$g(5) - g(0) = 12.5$$

An alternative way of expressing the same mathematical operation is to use the vertical line: $\Big|_a^b$. Thus we write

$$\int_0^5 x\,dx = \frac{x^2}{2}\Big|_0^5$$

The integral is then evaluated at the upper and lower limits, and the difference between these two values is the value of the definite integral.

This example can also be used to illustrate the reason that we do not include the constant of integration in the definite integral. Including the constant of integration for the integral evaluated at the upper limit gives

$$g(5) = 12.5 + c$$

Similarly, evaluating the function at the lower limit gives

$$g(0) = 0 + c$$

Subtracting the two gives

$$g(5) - g(0) = 12.5 + c - 0 - c = 12.5$$

Since the constant is eliminated by the subtraction, it is unnecessary to include the constant. The symbolism and the concepts are further illustrated by the following examples:

Example. Evaluate the definite integral $\int_2^6 (x + 2) \, dx$.

$$\int_2^6 (x + 2) \, dx = \frac{x^2}{2} + 2x \Big|_2^6 = \left(\frac{36}{2} + 12\right) - \left(\frac{4}{2} + 4\right) = 30 - 6 = 24$$

Example. Evaluate the definite integral $\int_3^8 (2x + 4) \, dx$.

$$\int_3^8 (2x + 4) \, dx = x^2 + 4x \Big|_3^8 = (64 + 32) - (9 + 12) = 75$$

A function must be continuous in the interval of integration in order to have a definite integral. This is illustrated by Fig. 14.4. Figure 14.4 shows a

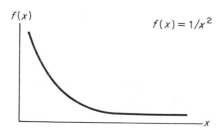

$f(x)$

$f(x) = 1/x^2$

Figure 14.4

plot of the function $f(x) = 1/x^2$. This function cannot be evaluated for a lower limit of integration of $x = 0$. The function $f(x) = 1/x^2$ is not defined for $x = 0$ and is discontinuous at $x = 0$. The function does have a definite integral, however, for $x > 0$. This is illustrated by the following example.

Example. Evaluate the definite integral $\int_1^5 \frac{dx}{x^2}$.

$$\int_1^5 \frac{dx}{x^2} = \frac{-1}{x} \Big|_1^5 = \frac{-1}{5} - \frac{(-1)}{1} = \frac{4}{5}$$

14.3.1 PROPERTIES OF DEFINITE INTEGRALS

Important properties of the definite integral can be illustrated with the aid of Fig. 14.5. This figure shows a continuous function $f(x)$ with values a, b,

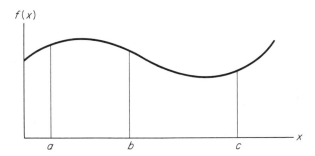

Figure 14.5

and c of x. One of the properties of the definite integral is that the definite integral of the function $f(x)$ with lower and upper limits of integration a and c, respectively, has the same value as the sum of the definite integrals of $f(x)$ with limits a and b and limits b and c. This is expressed as

$$\int_a^c f(x)\, dx = \int_a^b f(x)\, dx + \int_b^c f(x)\, dx \qquad (14.7)$$

This relation holds no matter what the relative values of a, b, and c.

A second property of the definite integral of the function $f(x)$ is that the value of the definite integral that has the same upper and lower limit is zero. Thus

$$\int_a^a f(x)\, dx = 0 \qquad (14.8)$$

In the discussion of the definite integral, it was assumed that the lower limit of integration was numerically smaller than the upper limit of integration. It is also possible to determine the definite integral of a function over an interval in which the upper limit is numerically smaller than the lower limit. A third property of definite integrals is that the definite integral of the function $f(x)$ with limits of integration a to b is equal to the negative of the definite integral of the function with limits of integration b to a. This is expressed as

$$\int_a^b f(x)\, dx = -\int_b^a f(x)\, dx \qquad (14.9)$$

These properties are illustrated by the following examples.

Example. Show that $\int_1^5 x \, dx = \int_1^3 x \, dx + \int_3^5 x \, dx$.

$$\int_1^5 x \, dx = \frac{x^2}{2} \Big|_1^5 = \frac{25}{2} - \frac{1}{2} = 12$$

$$\int_1^3 x \, dx = \frac{x^2}{2} \Big|_1^3 = \frac{9}{2} - \frac{1}{2} = 4$$

and

$$\int_3^5 x \, dx = \frac{x^2}{2} \Big|_3^5 = \frac{25}{2} - \frac{9}{2} = 8$$

Therefore,

$$\int_1^5 x \, dx = \int_1^3 x \, dx + \int_3^5 x \, dx = 12$$

Example. Show that $\int_3^6 x^2 \, dx = - \int_6^3 x^2 \, dx$.

$$\int_3^6 x^2 \, dx = \frac{x^3}{3} \Big|_3^6 = \frac{(6)^3}{3} - \frac{(3)^3}{3} = 72 - 9 = 63$$

$$\int_6^3 x^2 \, dx = \frac{x^3}{3} \Big|_6^3 = \frac{(3)^3}{3} - \frac{(6)^3}{3} = 9 - 72 = -63$$

Therefore,

$$\int_3^6 x^2 \, dx = - \int_6^3 x^2 \, dx$$

Example. Show that $\int_1^3 x \, dx = \int_1^5 x \, dx + \int_5^3 x \, dx$.

$$\int_1^3 x \, dx = \frac{x^2}{2} \Big|_1^3 = \frac{9}{2} - \frac{1}{2} = 4$$

$$\int_1^5 x \, dx = \frac{x^2}{2} \Big|_1^5 = \frac{25}{2} - \frac{1}{2} = 12$$

$$\int_5^3 x \, dx = \frac{x^2}{2} \Big|_5^3 = \frac{9}{2} - \frac{25}{2} = -8$$

Therefore, the definite integral is 4. This illustrates that (14.7) holds no matter what the relative values of *a*, *b*, and *c*.

Example. Evaluate $\int_2^2 x \, dx$.

$$\int_2^2 x \, dx = \frac{x^2}{2} \Big|_2^2 = \frac{4}{2} - \frac{4}{2} = 0$$

14.4   Methods of Integration

In comparing integral and differential calculus, most mathematicians would agree that the integration of functions is a more complicated process than differentiation of functions. Functions can be differentiated through application of a number of relatively straightforward rules. This is not true in determining the integrals of functions. Integration is much less straightforward and often requires considerable ingenuity. Certain functions can be integrated quite simply by applying rules of integration. Other functions require more complicated techniques, such as integration by substitution and integration by parts. Certain other functions can most readily be integrated through the use of tables of integrals. Still other rather ordinary-appearing functions cannot be integrated.

We shall discuss four methods of integration in this section. These are rules of integration, integration by substitution, integration by parts, and integration by use of tables of integrals. Additional methods of integration are available but are not included in this text.

14.4.1   RULES OF INTEGRATION

This section introduces rules for determining the integrals of certain functions and illustrates the rules through examples. When one is applying the rules for determining indefinite integrals, it is necessary to add the constant c to the integral. As stated earlier, the value of this constant is determined from additional information concerning the integral. The constant may be nonzero or zero and is necessary, since the derivative $f(x)$ of the integral $g(x)$ does not retain any constant.

Rule 1. The integral of a constant is

$$\int k \, dx = kx + c$$

The constant c is necessary, since the derivative of $kx + c$ is k. For $k = 1$, the rule states that $\int dx = x + c$.

Example. Determine $\int 10 \, dx$.

$$\int 10 \, dx = 10x + c$$

Example. Determine $\int -20\,dx$.

$$\int -20\,dx = -20x + c$$

Example. Determine $\int e\,dx$.

$$\int e\,dx = ex + c$$

Rule 2. The integral of a variable to a constant power (power function) is

$$\int x^n\,dx = \frac{x^{n+1}}{n+1} + c \qquad \text{for } n \neq -1$$

The student should carefully note that this rule applies for all power functions with the important exception of x^{-1}. A rule for determining the integral of x^{-1} is introduced as Rule 7 on p. 508.

The integral of a power function is verified by noting that the derivative of $\frac{x^{n+1}}{n+1} + c$ is x^n.

Example. $\int x^3\,dx = \frac{x^4}{4} + c.$

Exumple. $\int x^{3/2}\,dx = \frac{2x^{5/2}}{5} + c.$

Example. $\int x^{-2}\,dx = -x^{-1} + c.$

Example. $\int x^k\,dx = \frac{x^{k+1}}{k+1} + c.$

Example. $\int x^{-3/4}\,dx = 4x^{1/4} + c.$

Rule 3. The integral of a constant times a function is the constant times the integral of the function.

$$\int kf(x)\,dx = k\int f(x)\,dx$$

Example. $\int 12x^2\,dx = 12\int x^2\,dx = \frac{12x^3}{3} + c = 4x^3 + c.$

Example. $\int \frac{1}{4}x \, dx = \frac{1}{4} \int x \, dx = \frac{x^2}{8} + c.$

Example. $\int 6x^{-5} \, dx = 6 \int x^{-5} \, dx = -\frac{6}{4} x^{-4} + c.$

Example. $\int -3x^{-1/2} \, dx = -3 \int x^{-1/2} \, dx = -6x^{1/2} + c.$

Example. Evaluate the definite integral $\int_0^6 2x \, dx.$

$$\int_0^6 2x \, dx = x^2 \Big|_0^6 = (36 - 0) = 36$$

Rule 4. The integral of the sum or difference of two or more functions is the sum or difference of their integrals. That is,

$$\int [f(x) + g(x)] \, dx = \int f(x) \, dx + \int g(x) \, dx$$

and

$$\int [f(x) - g(x)] \, dx = \int f(x) \, dx - \int g(x) \, dx$$

Example. $\int (x + 4) \, dx = \int x \, dx + \int 4 \, dx = \frac{x^2}{2} + 4x + c.$

Example. $\int (x^2 - 3x + 6) \, dx = \int x^2 \, dx - \int 3x \, dx + \int 6 \, dx$

$$= \frac{x^3}{3} - \frac{3x^2}{2} + 6x + c.$$

Example. $\int (x^{-3} - x^{-2}) \, dx = \int x^{-3} \, dx - \int x^{-2} \, dx$

$$= -\frac{x^{-2}}{2} + x^{-1} + c.$$

Example. Evaluate the definite integral $\int_{-1}^5 (3x - 4) \, dx.$

$$\int_{-1}^5 (3x - 4) \, dx = \int_{-1}^5 3x \, dx - \int_{-1}^5 4 \, dx = \frac{3x^2}{2} \Big|_{-1}^5 - 4x \Big|_{-1}^5$$

$$= \tfrac{3}{2}[(5)^2 - (-1)^2] - 4[5 - (-1)]$$

$$= \tfrac{3}{2}(25 - 1) - 4(5 + 1)$$

$$= 36 - 24 = 12$$

Rule 5. The integral of the exponential function is

$$\int a^{kx} \, dx = \frac{a^{kx}}{k \ln (a)} + c$$

Example. $\displaystyle \int 3^{2x} \, dx = \frac{3^{2x}}{2 \ln 3} + c.$

Example. $\displaystyle \int 10^{0.5x} \, dx = \frac{10^{0.5x}}{0.5 \ln 10} + c.$

Example. $\displaystyle \int e^x \, dx = \frac{e^x}{\ln e} + c = e^x + c,$ since $\ln e = 1.$

Example. $\displaystyle \int 4e^{2x} \, dx = \frac{4e^{2x}}{2} + c = 2e^{2x} + c.$

Example. Evaluate the definite integral $\displaystyle \int_0^1 4e^{-4x} \, dx.$

$$\int_0^1 4e^{-4x} \, dx = -e^{-4x} \Big|_0^1 = (-e^{-4}) - (-e^0)$$

$$= -0.018 + e^0 = -0.018 + 1 = 0.982$$

Rule 6. The integral of the logarithmic function is

$$\int \log_a (kx) \, dx = x \log_a (kx) - x \log_a e + c$$

This rule is verified by determining the derivative of the integral.

$$\frac{d}{dx} [x \log_a (kx) - x \log_a e + c]$$

$$= x \left(\frac{k}{kx} \right) \log_a e + \log_a (kx) - \log_a e = \log_a (kx)$$

If the function is $\ln (kx)$, the integral becomes

$$\int \ln (kx) \, dx = x \ln (kx) - x + c$$

Example. Evaluate the definite integral $\displaystyle \int_1^4 \ln x \, dx.$

$$\int_1^4 \ln x \, dx = x \ln x - x \Big|_1^4 = (4 \ln 4 - 4) - (\ln 1 - 1)$$

$$= 4 \ln 4 - \ln 1 - 3 = 4(1.38629) - 0 - 3$$

$$= 2.54516$$

Example. Determine $\int \ln (3x)\, dx.$

$$\int \ln (3x)\, dx = x \ln (3x) - x + c$$

Rule 7. The integral of $f(x) = x^{-1}$ is

$$\int \frac{dx}{x} = \int x^{-1}\, dx = \ln x + c$$

The rule applies when the variable x is raised to the constant power $n = -1$. For x raised to any power other than $n = -1$, Rule 2 rather than Rule 7 applies. Note that $f(x) = x^{-1}$ is equivalent to $f(x) = \dfrac{1}{x}.$

14.4.2 VARIABLE LIMITS OF INTEGRATION

The limits of integration (a, b) in the preceding examples have been assumed to have fixed numerical values. In certain problems, such as those often encountered in the study of probability, one of the limits of integration is a variable rather than a constant. As an illustration, let us consider the integral of $f(x)$ over the variable interval (a, x). The value of the definite integral is a function of the upper limit of integration. The definite integral is

$$F(x) = \int_a^x f(t)\, dt \tag{14.10}$$

We have denoted the variable of integration as t rather than x to avoid confusion with the upper limit of integration.

    Example. Determine the definite integral $F(x)$ for the function $f(x) = x$ for the variable limits of integration $(0, x)$.

    This problem is the same as that discussed on p. 500 for the area of the triangle. The hypotenuse of the triangle is described by the function $f(x) = x$ and the base of the triangle is given by the interval $(0, x)$. The area of the triangle is a function of the length of the base $(0, x)$. For a variable base the area is

$$F(x) = \int_0^x t\, dt = \frac{t^2}{2}\bigg|_0^x = \frac{x^2}{2}$$

In the earlier example the base was $x = 5$.

$$F(5) = \frac{5^2}{2} = 12.5$$

This agrees with the earlier calculations.

Example. Determine the definite integral $F(x)$ for the function $f(x) = x^2 - 2x$ for $(2, x)$. We replace x by t as the variable in the function to obtain

$$F(x) = \int_2^x (t^2 - 2t)\, dt = \frac{t^3}{3} - t^2 \bigg|_2^x = \frac{x^3}{3} - x^2 - \frac{8}{3} + 4$$

$$F(x) = \tfrac{1}{3}(x^3 - 3x^2 + 4)$$

The relationship that exists between the function $F(x)$ and $f(x)$ is that the derivative of $F(x)$ is $f(x)$. The function $F(x)$ defined by (14.10) has the derivative given by

$$F'(x) = f(x) \tag{14.11}$$

This can also be expressed as

$$\frac{d}{dx} \int_a^x f(t)\, dt = f(x)$$

This formula states that the derivative of a definite integral with respect to an upper limit of integration is equal to the function evaluated at the upper limit.

Example. Determine the derivative of $F(x) = \int_0^x t\, dt$.

$$F'(x) = \frac{d}{dx} \int_0^x t\, dt = x$$

Example. Determine the derivative of $F(x) = \int_2^x (t^3 + 2t^2)\, dt$.

$$F'(x) = \frac{d}{dx} \int_2^x (t^3 + 2t^2)\, dt = x^3 + 2x^2$$

14.4.3 INTEGRATION BY SUBSTITUTION

The rules of integration discussed in the preceding sections directly apply only to a limited number of types of functions. The use of the rules of integration can be extended by the technique of substitution. Certain functions can be integrated by substituting a variable for a function and integrating with respect to the substituted variable. If the function to be integrated is of the form

$$\int f(x)\, dx = \int h(g(x))g'(x)\, dx$$

the integration can be performed by substitution. The function $h(g(x))$ is a composite function. The function $f(x)$ is equal to the product of the composite function $h(g(x))$ and the derivative of $g(x)$, $g'(x)$. The method of substitution involves first substituting the variable u for the function $g(x)$, i.e., $u = g(x)$.

The function $h(g(x))$ now becomes $h(u)$. The next step is to determine the derivative of u with respect to x. This is

$$\frac{du}{dx} = g'(x)$$

In the study of differential calculus, du/dx was considered as a single symbol. In integral calculus we give separate meanings to du and dx, so that du/dx is now considered as a quotient: the differential of u divided by the differential of x. Since we are now considering du and dx as separate terms, we can write

$$du = g'(x)\, dx$$

or alternatively,

$$dx = \frac{du}{g'(x)}$$

provided

$$g'(x) \neq 0$$

The substitution can now be made in the integral

$$\int f(x)\, dx = \int h(g(x))g'(x)\, dx$$

We substitute $u = g(x)$ and $dx = \dfrac{du}{g'(x)}$ to obtain

$$\int h(g(x))g'(x)\, dx = \int h(u)\, du \qquad (14.12)$$

The integral of $h(u)$ is now determined.

The method of integration by substitution is illustrated by the following examples.

Example. Determine $\displaystyle\int \frac{2x\, dx}{(x^2 + 3)}$.

This integral can be expressed as $\displaystyle\int (x^2 + 3)^{-1} 2x\, dx$. If

$$h(g(x)) = (x^2 + 3)^{-1}$$

then

$$g(x) = x^2 + 3$$

If we substitute u for $g(x)$, then

$$u = x^2 + 3$$

and

$$\frac{du}{dx} = 2x \quad \text{or} \quad du = 2x\, dx$$

Substituting these values gives

$$\int u^{-1}\, du = \ln u + c$$

By resubstituting $g(x)$ for u, we obtain

$$\int (x^2 + 3)^{-1} 2x\, dx = \ln (x^2 + 3) + c$$

Example. Determine $\int xe^{x^2+9}\, dx$.

Let

$$g(x) = u = x^2 + 9$$

The derivative of u with respect to x gives

$$\frac{du}{dx} = 2x$$

or, since we are considering du and dx as separate symbols,

$$\frac{du}{2x} = dx$$

Based upon the substitution, the integral becomes

$$\tfrac{1}{2} \int e^u\, du = \tfrac{1}{2}e^u + c$$

By resubstituting $g(x)$ for u, we obtain

$$\int xe^{x^2+9}\, dx = \tfrac{1}{2}e^{x^2+9} + c$$

Example. Evaluate $\int_1^4 3x^2(4x^3 + 5)\, dx$.

We can determine the indefinite integral by substitution.

$$u = 4x^3 + 5, \qquad du = 12x^2\, dx, \quad \text{and} \quad dx = \frac{du}{12x^2}$$

The indefinite integral written in terms of u is

$$\frac{1}{4} \int u\, du = \frac{u^2}{8} + c$$

The definite integral written in terms of x is

$$\int_1^4 3x^2(4x^3 + 5)\, dx = \frac{(4x^3 + 5)^2}{8} \Big|_1^4$$

$$= \frac{[4(64) + 5]^2 - [4(1) + 5]^2}{8} = \frac{68{,}121 - 81}{8}$$

$$= 8505$$

This result can also be obtained by expressing the limits of integration in terms of u. Since $u = 4x^3 + 5$, the lower limit is $u = 4(1)^3 + 5 = 9$ and the upper limit is $u = 4(4)^3 + 5 = 261$. The definite integral is

$$\frac{1}{4} \int_9^{261} u \, du = \frac{u^2}{8} \Big|_4^{261} = 8505$$

Example. Evaluate $\int_0^3 \frac{xe^{x^2}}{3} \, dx$.

By substituting $u = x^2$ and $\frac{du}{2x} = dx$, and expressing the limits of integration as $u = 0$ and $u = 9$, we obtain

$$\frac{1}{6} \int_0^9 e^u \, du = \frac{e^u}{6} \Big|_0^9 = \frac{e^9 - e^0}{6} = \frac{8103.1 - 1}{6} = 1350.3$$

14.4.4 INTEGRATION BY PARTS

The formula for the derivative of the product of two functions is often useful in evaluating integrals. In differential calculus, the derivative of the product of the two functions, $f(x)$ and $g(x)$, is

$$\frac{d}{dx} [f(x) \cdot g(x)] = f(x)g'(x) + g(x)f'(x)$$

The integral of this derivative gives

$$f(x) \cdot g(x) = \int f(x)g'(x) \, dx + \int g(x)f'(x) \, dx$$

which can be written as

$$\int f(x)g'(x) \, dx = f(x)g(x) - \int g(x)f'(x) \, dx \qquad (14.13)$$

This relationship can be used to determine integrals of certain functions. The following examples illustrate the technique of integration by parts.

Example. Determine $\int xe^{-x} \, dx$.

The method of integration by parts involves separating the function to be integrated into $f(x)$ and $g'(x)$. In this example we let $f(x) = x$ and $g'(x) = e^{-x}$. Formula (14.13) requires $f'(x)$ and $g(x)$. These functions are obtained by differentiating $f(x)$ and integrating $g'(x)$. Thus,

$$f(x) = x \quad \text{and} \quad f'(x) = 1$$
$$g'(x) = e^{-x} \quad \text{and} \quad g(x) = \int e^{-x} \, dx = -e^{-x}$$

Substituting these terms in Formula (14.13) gives

$$\int xe^{-x} \, dx = -xe^{-x} - \int -e^{-x} \, dx$$

which can be integrated to yield

$$\int xe^{-x} \, dx = -xe^{-x} - e^{-x} + c = -e^{-x}(x + 1) + c$$

Example. Determine $\int x^2 e^{-x} \, dx$.

Let

$$f(x) = x^2 \quad \text{and} \quad g'(x) = e^{-x}$$

The derivative of $f(x)$ and the integral of $g'(x)$ are

$$f'(x) = 2x, \qquad g(x) = -e^{-x}$$

From Formula (14.13) we obtain

$$\int x^2 e^{-x} \, dx = -x^2 e^{-x} - \int -2x e^{-x} \, dx$$

$$= -x^2 e^{-x} + 2 \int x e^{-x} \, dx$$

The integral $\int x e^{-x} \, dx$ was determined in the preceding example. The integral is thus

$$\int x^2 e^{-x} \, dx = -x^2 e^{-x} - 2 e^{-x}(x+1) + c$$

One of the difficult problems in applying the formula for integration by parts is to determine the terms to substitute for $f(x)$ and $g'(x)$ in the integral $\int f(x) g'(x) \, dx$. Although there are no specific rules for determining the appropriate substitution, a guide line that the student can follow is to select $g'(x)$ so that $g(x)$ can be determined and $f(x)$ so that $f'(x)$ can be determined. In Formula (14.13), the integral of $\int f(x) g'(x) \, dx$ is given by the $f(x) g(x)$ minus $\int g(x) f'(x) \, dx$. If the substitutions have been correctly made, $f(x) g(x)$ can be determined and $\int g(x) f'(x) \, dx$ can be integrated. If $\int g(x) f'(x) \, dx$ appears to be a more complicated integral than $\int f(x) g'(x) \, dx$, there is a good chance that the original substitutions for $f(x)$ and $g'(x)$ were incorrect.

Example. Determine $\int x \ln x \, dx$.

Let

$$f(x) = \ln x \quad \text{and} \quad g'(x) = x$$

Then

$$f'(x) = \frac{1}{x} \quad \text{and} \quad g(x) = \frac{x^2}{2}$$

$$\int x \ln x \, dx = \frac{x^2 \ln x}{2} - \int \frac{x}{2} \, dx$$

$$= \frac{x^2 \ln x}{2} - \frac{x^2}{4} + c$$

Example. Determine $\int_1^3 x^2 \ln x \, dx$.

Let
$$f(x) = \ln x \quad \text{and} \quad g'(x) = x^2$$
Then
$$f'(x) = \frac{1}{x} \quad \text{and} \quad g(x) = \frac{x^3}{3}$$

$$\int x^2 \ln x \, dx = \frac{x^3 \ln x}{3} - \int \frac{x^2}{3} \, dx$$

$$= \frac{x^3 \ln x}{3} - \frac{x^3}{9} = \frac{x^3}{3} \left(\ln x - \frac{1}{3} \right)$$

Evaluating the definite integral for the limits of integration gives

$$\int_1^3 x^2 \ln x \, dx = \frac{x^3}{3} \left(\ln x - \frac{1}{3} \right) \Big|_1^3 = 9 \left(\ln 3 - \frac{1}{3} \right) - \frac{1}{3} \left(\ln 1 - \frac{1}{3} \right)$$

$$9(1.09861 - 0.33333) - \tfrac{1}{3}(0.0 - 0.33333) = 6.88752 + 0.11111 = 6.99863$$

14.4.5 INTEGRATION BY TABLES OF INTEGRALS

Determining the integrals of certain functions requires techniques of integration that are beyond the scope of business-oriented texts. Such integrals can most readily be determined by the use of tables of integrals. Table VI contains those integrals most commonly used in business and economic models. For a more extensive reference, the reader is referred to *Standard Mathematical Tables*, a widely used mathematical reference, which contains more than 500 integrals.

Example. Determine $\int \frac{dx}{x \ln x}$.

Referring to integral no. 23, Table VI, we see that

$$\int \frac{dx}{x \ln x} = \ln (\ln x) + c$$

Example. Determine $\int (3 + 4x)^5 \, dx$.

This integral has the form of integral no. 7, Table VI. The integral is

$$\int (3 + 4x)^5 \, dx = \frac{1}{4(6)} (3 + 4x)^6 + c$$

$$= \frac{(3 + 4x)^6}{24} + c$$

Example. Determine $\int x\sqrt{9 - x^2}\, dx$.

From integral no. 15, Table VI, we see that

$$\int x\sqrt{9 - x^2}\, dx = -\tfrac{1}{3}\sqrt{(9 - x^2)^3} + c$$

Example. Determine $\int \dfrac{dx}{16 - 4x^2}$.

Factoring a 4 from the denominator of the function gives

$$\int \frac{dx}{16 - 4x^2} = \frac{1}{4} \int \frac{dx}{4 - x^2}$$

The function can now be integrated by using integral no. 9.

$$\frac{1}{4} \int \frac{dx}{4 - x^2} = \frac{1}{4} \cdot \frac{1}{2(2)} \ln \left(\frac{2 + x}{2 - x} \right) + c$$

Example. Determine $\int (4x^2 - 9)^{-1/2}\, dx$.

The integral is determined by first expressing $(4x^2 - 9)^{-1/2}$ as $[4(x^2 - \tfrac{9}{4})]^{-1/2}$. The expression can now be simplified to

$$[4(x^2 - \tfrac{9}{4})]^{-1/2} = \tfrac{1}{2}(x^2 - \tfrac{9}{4})^{-1/2}$$

The integral, from no. 12 in Table VI, is

$$\int (4x^2 - 9)^{-1/2}\, dx = \tfrac{1}{2} \int (x^2 - \tfrac{9}{4})^{-1/2}\, dx$$

$$= \tfrac{1}{2} \ln [x + (x^2 - \tfrac{9}{4})^{1/2}] + c$$

14.5 Applications of Integral Calculus

Several important applications of integral calculus are introduced in this section. One of these applications is the evaluation of the area underlying a function. As shown earlier in this chapter, integral calculus enables one to determine the area bounded by the function and the horizontal axis between the limits of integration. Although the business analyst may rarely have occasion to utilize this specific application, an understanding of this concept will be quite useful in the study of probability models.

Economic applications of integral calculus are also discussed in this section. These applications include marginal and total cost functions, capital formation, and continuous financial processes.

14.5.1 AREA

The definite integral $\int_a^b f(x)\, dx$ gives the area bounded by $f(x)$, the horizontal axis, $x = a$, and $x = b$. As an example, consider the area of the rectangle shown in Fig. 14.6.

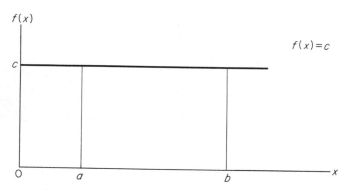

Figure 14.6

The formula for the area of a rectangle is $A =$ base·height. The rectangle shown in the figure has base $(b - a)$ and height c. The area of the rectangle is thus $A = (b - a)c$.

The same result can be obtained through calculus. The function that describes the upper base of the rectangle is $f(x) = c$. The definite integral, evaluated between $x = a$ and $x = b$, gives the area bounded by $x = a$, $x = b$, the horizontal axis, and the function $f(x) = c$. This integral is $\int_a^b c\, dx = cx \Big|_a^b = (b - a)c$.

In using the definite integral for the calculation of areas, area above the horizontal axis is positive and that below the horizontal axis is negative. The process of integration automatically subtracts negative area from positive area, thus giving the net area. Consequently, if the analyst is concerned with the absolute area rather than the net area, he must determine the area beneath the horizontal axis and that above the horizontal axis separately, and then sum the absolute values of the areas. This is illustrated by the following example.

Example. Determine both the net area and the absolute area between the function $f(x) = -4 + x$ and the horizontal axis between $x = 2$ and $x = 6$. This function is shown in Fig. 14.7. From the symmetry of the diagram, it can

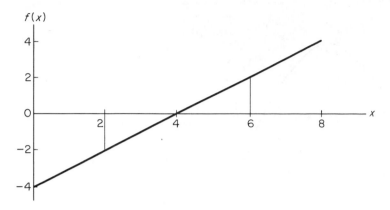

Figure 14.7

be seen that the net area is zero. This is also shown by the integral.

$$A = \int_2^6 (-4 + x)\, dx = -4x + \frac{x^2}{2} \bigg|_2^6 = -6 + 6 = 0$$

The absolute area can be found by considering the triangles separately.

$$A_1 = \int_2^4 (-4 + x)\, dx - -4x + \frac{x^2}{2} \bigg|_2^4 = -2$$

$$A_2 = \int_4^6 (-4 + x)\, dx = -4x + \frac{x^2}{2} \bigg|_4^6 = +2$$

The absolute value of the area is, therefore, $A = |A_1| + |A_2| = 4$.

The preceding two examples dealt with the area underlying linear functions. The definite integral can, of course, also be used to determine the area bounded by the horizontal axis and curvilinear functions. The procedure for determining the area is the same as that for linear functions, in that we again evaluate the definite integral, using the upper and lower boundaries of the area as the limits of integration. This is illustrated by the following example.

Example. Determine the area bounded by the horizontal axis, the function $f(x) = -x^2 + 12x - 20$, $x = 2$, and $x = 10$. This function is shown in Fig. 14.8. The area is

$$A = \int_2^{10} (-x^2 + 12x - 20)\, dx = \frac{-x^3}{3} + 6x^2 - 20x \bigg|_2^{10} = 85\tfrac{1}{3}$$

Example. Determine the net area between $x = 0$ and $x = 12$ for the function in the preceding example.

The definite integral is

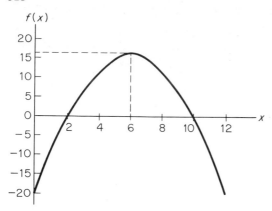

Figure 14.8

$$A = \int_0^{12} (-x^2 + 12x - 20)\, dx = \frac{-x^3}{3} + 6x^2 - 20x \Big|_0^{12} = 48$$

The net area between $x = 0$ and $x = 12$ is $A = 48$. Since, from the preceding example, the area between $x = 2$ and $x = 10$ is $85\frac{1}{3}$, we conclude that the negative area is $85\frac{1}{3} - 48 = 37\frac{1}{3}$. This can be verified by determining the integral between $x = 0$ and $x = 2$ and between $x = 10$ and $x = 12$.

$$A = \int_0^2 (-x^2 + 12x - 20)\, dx + \int_{10}^{12} (-x^2 + 12x - 20)\, dx$$

$$A = \frac{-x^3}{3} + 6x^2 - 20x \Big|_0^2 + \frac{-x^3}{3} + 6x^2 - 20x \Big|_{10}^{12}$$

$$A = -18\tfrac{2}{3} - 18\tfrac{2}{3} = -37\tfrac{1}{3}$$

AREA BETWEEN FUNCTIONS. The definite integral can be used to determine the area between two functions. Given two functions $f(x)$ and $g(x)$, the area bounded by these functions and $x = a$ and $x = b$ is given by

$$A = \int_a^b [f(x) - g(x)]\, dx \tag{14.14}$$

The integral gives the total area bounded by the two functions between $x = a$ and $x = b$ in Fig. 14.9. Since $f(x)$ is greater than $g(x)$ between the limits of integration, the area will be positive. This area is indicated by the shading in Fig. 14.9. Had we chosen to evaluate $\int_a^b [g(x) - f(x)]\, dx$, the sign prefacing the area would have been negative. The absolute value of the two areas is, however, the same.

Example. Determine the area between the two functions $g(x) = x - 3$ and $h(x) = 3x - x^2$. These functions are shown in Fig. 14.10.

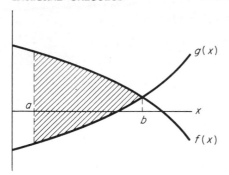

Figure 14.9

The limits of integration are those values of x for which the two functions are equal. If we equate the two functions, we obtain

$$3x - x^2 = x - 3$$
$$x^2 - 2x - 3 = 0$$
$$x = -1, \quad x = 3$$

The area between the two functions for $x = -1$ to $x = 3$ is

$$A = \int_{-1}^{3} [(3x - x^2) - (x - 3)] \, dx$$

$$A = \frac{3x^2}{2} - \frac{x^3}{3} - \frac{x^2}{2} + 3x \bigg|_{-1}^{3} = 10\tfrac{2}{3}$$

We can also determine the area bounded by more than two functions. In the following example the problem is to determine the area between the function $f(x)$ and the function $h(x)$.

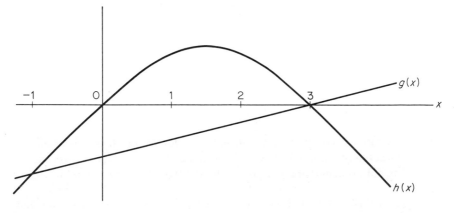

Figure 14.10

Example. Determine the area bounded by $f(x)$ and $h(x)$.

$$f(x) = 4 + 5x - x^2 \qquad \text{for } 0 \leq x \leq 6$$

$$h(x) = \begin{cases} 4 + \tfrac{2}{3}x & \text{for } 0 \leq x \leq 3 \\ 12 - 2x & \text{for } 3 \leq x \leq 6 \end{cases}$$

These functions are shown in Fig. 14.11.

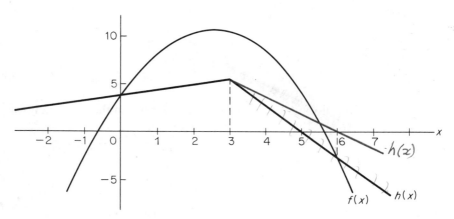

Figure 14.11

The limits of integration were determined by equating the functions and solving for $x = 0$, $x = 3$, and $x = 6$. The area is found by evaluating the definite integrals

$$A = \int_0^3 [(4 + 5x - x^2) - (4 + \tfrac{2}{3}x)]\, dx$$

$$+ \int_3^6 [(4 + 5x - x^2) - (12 - 2x)]\, dx$$

$$A = 4x + \frac{5x^2}{2} - \frac{x^3}{3} - 4x - \frac{x^2}{3} \Big|_0^3 + 4x + \frac{5x^2}{2} - \frac{x^3}{3} - 12x + x^2 \Big|_3^6$$

$$A = 10.5 + 7.8 = 18.0$$

14.5.2 ECONOMIC APPLICATIONS

The relationship between the integral and the derivative forms the basis for numerous economic models. We shall illustrate basic models in cost theory, capital formulation, and mathematics of finance which utilize this relationship.

MARGINAL AND TOTAL COST FUNCTIONS. We have shown that the marginal cost function is given by the derivative of the total cost function. This relationship is reversible; i.e., the integral of the marginal cost function is the total cost function.

$$TC = \int MC = TVC + FC \qquad (14.15)$$

The integral of marginal cost gives total variable cost plus the constant. The constant represents fixed cost, which does not vary with changes in the quantity of goods produced. The relationship between total cost and marginal cost is illustrated by the following examples.

Example. The cost of producing a simple component is composed of a fixed cost and a variable cost. If the fixed cost is k and the variable cost is c per unit, show that the total cost is given by $TC = cq + k$.

The cost of producing an additional unit is c. The marginal cost function is, therefore, $MC = c$. The integral of this function is

$$TC = \int c \, dq = cq + k$$

Total cost is composed of total variable cost plus fixed cost.

Example. The cost of manufacturing an item consists of $1000 fixed cost and $1.50 per unit variable cost. Determine the total cost function and the cost of manufacturing 2000 units.

$$TC = \int 1.50 \, dq = \$1.50q + k$$

where $k = \$1000$.

The cost of manufacturing 2000 units is

$$TC = \$1.50(2000) + \$1000 = \$4000$$

Example. Assume that the marginal cost of manufacturing a product is the following function of output:

$$MC = 3q^2 - 4q + 5$$

Determine the total cost function.

$$TC = \int (3q^2 - 4q + 5) \, dq = q^3 - 2q^2 + 5q + k$$

where k is the fixed cost of manufacturing the product.

The definite integral can be used to determine the total variable cost of producing some quantity of output. The total variable cost of producing the ath to the bth units is given by $\int_{(a-1)}^{b} MC$. This is illustrated by the following examples.

Example. The marginal cost function for a certain commodity is $MC = 3q^2 - 4q + 5$. Determine the cost of producing the eleventh through the fifteenth units, inclusive.

$$TVC = \int_{10}^{15} (3q^2 - 4q + 5)\, dq = \left| q^3 - 2q^2 + 5q \right|_{10}^{15}$$

$$= (15)^3 - 2(15)^2 + 5(15) - (10)^3 + 2(10)^2 - 5(10) = \$2150$$

The evaluation of the definite integral gives the same result that would be obtained by evaluating the total cost function for b units and the total cost function for $(a - 1)$ units and subtracting the two costs. This difference is the incremental cost of producing the ath to the bth units of output. For the above example,

$$TC(15 \text{ units}) = (15)^3 - 2(15)^2 + 5(15) + k = \$3000$$
$$TC(10 \text{ units}) = (10)^3 - 2(10)^2 + 5(10) + k = \$850$$
$$TC(15) - TC(10) = \$2150$$

The cost of producing units 11 through 15 is \$2150.

CAPITAL FORMATION AND CONTINUOUS FINANCIAL PROCESSES. The growth rate for a continuous function was discussed in Chapter 13. The formula for the rate of growth was

$$r = \frac{f'(t)}{f(t)} \tag{13.52}$$

If the rate of growth is known rather than the function, we can determine the function by integrating Formula (13.52). To demonstrate the methodology, assume that \$1 is invested at r percent and the interest is compounded continuously for a period of t years. Determine the function that describes the continuous growth.

We rewrite (13.52) to obtain

$$f'(t) = rf(t) \tag{14.16}$$

Since r is a constant, the integral of the function is

$$\int f'(t)\, dt = r \int f(t)\, dt \tag{14.17}$$

From (14.16) we observe that the derivative of the function $f(t)$ is equal to r times the function $f(t)$. The only function that has the property that the derivative of the function is equal to a constant times the function is the exponential function. We thus conclude that $f(t) = e^{rt}$. Integrating both sides of (14.17) gives

$$\int f'(t)\, dt = r \int e^{rt}\, dt$$

or

$$f(t) = e^{rt} \tag{14.18}$$

We can verify that $f(t) = e^{rt}$ by differentiating $f(t)$.

$$f'(t) = re^{rt}$$

which can be expressed as

$$f'(t) = rf(t)$$

since $f(t) = e^{rt}$. The function that describes the amount of the investment at the end of period t is thus

$$A = Pe^{rt} \tag{14.19}$$

where P is the value of the function when $t = 0$, i.e., the initial investment. This formula is derived in Chapter 13 based upon the concept of the limit and is given by (13.40) as $A = Pe^{it}$. The use of this formula in models of capital formation is illustrated by the following examples.

Example. An investment of \$100 at 10 percent is compounded continuously for five years. Determine the amount of the investment at the end of the fifth year.

$$A = \$100e^{0.10t} = \$100e^{0.5} = \$164.90$$

Example. Determine the amount which must be invested at 8 percent to accumulate to \$500 at the end of ten years.

$$P = \$Ae^{-0.08t} = \$500e^{-0.08(10)} = \$224$$

$$P = Ae^{-Rt}$$

Formula (14.19) is used to determine the amount A to which an investment of P will accumulate when invested at interest rate r and compounded continuously for t years. An interesting extension of (14.19) occurs when we determine the integral of the function. This integral gives the amount to which a series of n equal payments of p invested at interest r and compounded continuously will accumulate. The reader will recognize this as the amount of an annuity assuming continuous compounding. The integral of (14.19) is

$$A_n = \int_0^t pe^{rx}\, dx$$

where x is substituted for t as the variable of integration. Performing the integration gives

$$A_n = \frac{pe^{rx}}{r}\bigg|_0^t$$

$$A_n = \frac{p(e^{rt} - 1)}{r} \tag{14.20}$$

This formula is similar to (13.45), in which periodic compounding rather than continuous compounding is assumed. The only difference between the periodic and the continuous case is that $(1 + i)^t$ is replaced by e^{rt}. Applications of the formula are provided by the following examples.

Example. The president of a certain country has called for sacrifices in the consumption of consumer goods in order to shift production to capital goods. The current level of investment in capital goods is $10 billion. The plan proposed by the president involves increasing this level of investment by 10 per cent per year for ten years. Assuming continuous compounding, determine the amount that must be invested during the ten-year period.

The total investment during the ten years can be determined from (14.20).

$$A_n = \frac{\$10(e^{0.10(10)} - 1)}{0.10} = \frac{10(2.718 - 1)}{0.10}$$

$$A_n = \$171.8 \text{ billion}$$

Example. For the preceding example, determine the investment required during the ninth and tenth years.

The total investment during the final two years can be determined from (14.19).

$$A_n = \int_8^{10} 10e^{0.10t}\, dt = \frac{10e^{0.10t}}{0.10}\bigg|_8^{10}$$

$$A_n = 100(e^{1.0} - e^{0.8}) = 100(2.718 - 2.226)$$

$$A_n = \$49.2 \text{ billion}$$

Example. An individual deposits $1000 at the end of each year in a savings account. The account pays 6 percent interest compounded daily. Determine the amount on deposit at the end of 10 years.

In working this problem, we shall assume continuous compounding rather than 365 compounding periods per year. Based upon this approximation, the amount is

$$A_n = \frac{P\,(e^{rt} - 1)}{r}$$

$$A_n = \frac{1000}{0.06}\,(e^{0.6} - 1) = \frac{1000}{0.06}\,(1.822 - 1)$$

$$A_n = \$13,700$$

Example. In the preceding example, calculate the amount, assuming that the $1000 was deposited at the first of the year rather than the last of the year. We can determine the amount by multiplying the formula by e^r.

$$A_n = \frac{P}{r}\,(e^{rt} - 1)(e^r)$$

$$A_n = 13,700\,(e^{0.06}) = 13,700(1.0618)$$

$$A_n = \$14,640$$

The formula for the amount of an annuity if periodic compounding is assumed can be modified to account for payment at the first of the period rather than the last by multiplying the formula by $(1 + i)$. Thus, the formula for periodic compounding is

$$A_n = \frac{p}{i}[(1 + i)^t - 1)](1 + i) \tag{14.21}$$

and the formula for the amount of an annuity with the payment at the first of the period if continuous compounding is assumed is

$$A_n = \frac{p}{r}[e^{rt} - 1]e^r \tag{14.22}$$

Integral calculus can be used to derive the formula for the present value of an annuity. Based upon continuous compounding, the present value of an amount A received t years hence is

$$P = Ae^{-rt} \tag{13.41}$$

Instead of receiving a single payment of A at the end of t years, consider the case in which we receive a payment of A at the end of each year. The present value of this series of payments is found by integrating (13.41) between the limits of integration of $t = 0$ and $t = t$. Evaluation of this definite integral gives the present value of an annuity if continuous compounding is assumed. The formula is

$$PV = \int_0^t Ae^{-rx}\, dx$$

$$PV = \frac{A}{r}[-e^{-rt}\Big|_0^t$$

$$PV = \frac{A}{r}[-e^{-rt} - (-e^0)]$$

$$PV = A\frac{(1 - e^{-rt})}{r} \tag{14.23}$$

This formula is similar to the formula for the present value of an annuity if a discrete number of compounding periods is assumed,

$$PV = p\left[\frac{1 - (1 + i)^{-t}}{i}\right] \tag{13.49}$$

Example. A research and development firm has patented a new photocopy process, which they have licensed to a major manufacturer. The agreement calls for the annual payment of $50,000 for 10 years. If the discount rate is 8 percent and continuous compounding is assumed, determine the present value of this agreement.

$$PV = 50,000 \frac{(1 - e^{-0.08(10)})}{0.08}$$

$$PV = \frac{50,000(1 - 0.4493)}{0.08}$$

$$PV = \$344,200$$

PROBLEMS

1. Determine the integrals.

(a) $\int 6\, dx$ (b) $\int k\, dx$

(c) $\int \frac{1}{3}\, dx$ (d) $\int -2\, dx$

(e) $\int (a + b)\, dx$ (f) $\int k^2\, dx$

(g) $\int dx$ (h) $\int \frac{k^2}{k + 1}\, dx$

(i) $\int x\, dx$ (j) $\int x^2\, dx$

(k) $\int x^{-2}\, dx$ (l) $\int x^{1/2}\, dx$

(m) $\int \frac{dx}{x^3}$ (n) $\int x^{2k}\, dx$

(o) $\int \sqrt{x}\, dx$ (p) $\int \frac{k^2}{x^2}\, dx$

(q) $\int_0^2 x^2\, dx$ (r) $\int_1^3 \frac{1}{\sqrt{x}}\, dx$

(s) $\int_0^5 dx$ (t) $\int_2^6 x^{-1/2}\, dx$

(u) $\int_1^{10} 3x^3\, dx$ (v) $\int_0^5 x^{2.2}\, dx$

2. Determine the integrals.

(a) $\int (3x + 4)\, dx$ (b) $\int (x^2 + 2x + 6)\, dx$

(c) $\int (2x^3 + x^2)\, dx$ (d) $\int (x^{1/2} + x^{1/3})\, dx$

(e) $\int (x + 1)(x + 1)\, dx$ (f) $\int (x - 2)(x + 3)\, dx$

(g) $\int_1^4 (x^2 - 2x + 3)\, dx$

(h) $\int_2^5 (x - 3)(x + 1)\, dx$

(i) $\int 3e^{3x}\, dx$

(j) $\int 5^x\, dx$

(k) $\int \frac{1}{2}e^{2x}\, dx$

(l) $\int 2^{2x}\, dx$

(m) $\int_0^{10} 3e^{-3x}\, dx$

(n) $\int_0^\infty e^{-x}\, dx$

(o) $\int \ln (x)\, dx$

(p) $\int \ln (2x)\, dx$

(q) $\int \log (3x)\, dx$

(r) $\int \log_5 (x)\, dx$

(s) $\int_1^{10} \ln (x)\, dx$

(t) $\int \log_{10} (2x)\, dx$

3. Determine the integrals. Use the method of substitution.

(a) $\int (x - 1)^{5/2}\, dx$

(b) $\int x(x^2 - 2)^2\, dx$

(c) $\int x^2(4x^3 - 2)^9\, dx$

(d) $\int xe^{x^2}\, dx$

(e) $\int x^2 e^{x^3}\, dx$

(f) $\int 4xe^{-x^2-4}\, dx$

(g) $\int \frac{1}{5x - 6}\, dx$

(h) $\int \frac{dx}{4 - 3x}$

(i) $\int_2^5 \frac{2x}{x^2 - 2}\, dx$

(j) $\int_{-2}^2 xe^{x^2}\, dx$

(k) $\int x \ln (x^2 - 6)\, dx$

(l) $\int x \ln x^2\, dx$

(m) $\int_0^1 \frac{2x}{x^2 + 1}\, dx$

(n) $\int_0^2 x(x^2 + 1)^2\, dx$

4. Determine the integrals. Use the method of integration by parts.

(a) $\int x^2 \ln x\, dx$

(h) $\int xe^{-x}\, dx$

(c) $\int x^3 e^{2x}\, dx$

(d) $\int x^{1/2} \ln x\, dx$

(e) $\int (\ln x)^2\, dx$

(f) $\int xe^{2x}\, dx$

(g) $\int x3^x\, dx$

5. Calculate the area between the horizontal axis and the function $f(x) = x^2$ between $x = 0$ and $x = 5$, using Formula (14.1) with $n = 5$ and $f(x)$

equal to the value of the function at the midpoint of each interval Δx. Recalculate, using $n = 10$. Calculate the area, using Formula (14.4) and compare the result with that obtained by using Formula (14.1).

6. Calculate the value of the definite integral $\int_0^{10} x^3 \, dx$, using Formula (14.1) with $n = 5$. Compare this value to that obtained from Formula (14.4).

7. Determine the area bounded by the horizontal axis and the following functions for the specified domain of x.
(a) $f(x) = 2x$ for $0 \leq x \leq 6$
(b) $f(x) = 6 - x$ for $0 \leq x \leq 4$
(c) $f(x) = 3 + 2x$ for $-1 \leq x \leq 10$
(d) $f(x) = 10 - x$ for $-10 \leq x \leq 0$
(e) $f(x) = 3x + 4$ for $0 \leq x \leq 8$
(f) $f(x) = -6 + x$ for $0 \leq x \leq 6$
(g) $f(x) = 0.5 + 0.5x$ for $5 \leq x \leq 10$
(h) $f(x) = 15 - 15x$ for $0 \leq x \leq 1$
(i) $f(x) = 15 - 15x$ for $1 \leq x \leq 2$
(j) $f(x) = -60 + 10x$ for $0 \leq x \leq 10$
(k) $f(x) = 6$ for $5 \leq x \leq 10$
(l) $f(x) = x^2$ for $0 \leq x \leq 5$
(m) $f(x) = -x^2$ for $0 \leq x \leq 5$
(n) $f(x) = 2 + 3x + x^2$ for $0 \leq x \leq 6$
(o) $f(x) = 3 + 6x - x^2$ for $0 \leq x \leq 6$
(p) $f(x) = 100 - 15x + x^2$ for $5 \leq x \leq 10$
(q) $f(x) = 1 + x + x^2 + x^3$ for $0 \leq x \leq 10$
(r) $f(x) = 6 - x + x^2 - x^3$ for $0 \leq x \leq 5$
(s) $f(x) = 3 + x + x^2 - x^3$ for $0 \leq x \leq 5$
(t) $f(x) = 0.5e^{-0.5x}$ for $0 \leq x \leq 20$
(u) $f(x) = 4e^{6x}$ for $0 \leq x \leq 10$
(v) $f(x) = 6e^{-6x}$ for $0 \leq x \leq 1000$
(w) $f(x) = 10e^{-10x}$ for $0 \leq x \leq 0.05$
(x) $f(x) = 0.5e^{-0.5x}$ for $1 \leq x \leq 2$
(y) $f(x) = 0.3e^{-0.3x}$ for $0.5 \leq x \leq 1.5$
(z) $f(x) = 0.1e^{-0.1x}$ for $0.1 \leq x \leq 0.2$

8. Determine the area between $f(x)$ and $g(x)$ for the specified domain of x.
(a) $f(x) = 3 + 2x$, $g(x) = x$ for $0 \leq x \leq 5$
(b) $f(x) = 6$, $g(x) = 3$ for $0 \leq x \leq 10$
(c) $f(x) = 10 - x$, $g(x) = 2 + x$ for $0 \leq x \leq 4$
(d) $f(x) = 25 - 2x$, $g(x) = -25 + 3x$ for $0 \leq x \leq 5$
(e) $f(x) = 5 + x$, $g(x) = 5 - x$ for $0 \leq x \leq 10$
(f) $f(x) = 6 + 5x - x^2$, $g(x) = 6 - 5x + x^2$ for $0 \leq x \leq 10$

(g) $f(x) = 10 - 2x + x^2$, $g(x) = 10 + 2x - x^2$ for $0 \le x \le 4$
(h) $f(x) = 15 - 3x + x^2$, $g(x) = 5 - 3x + x^2$ for $0 \le x \le 6$
(i) $f(x) = 6 + x + x^2$, $g(x) = 5 + x + x^2$ for $0 \le x \le 10$
(j) $f(x) = 15 - 2x + x^2$, $g(x) = -15 + 2x - x^2$ for $5 \le x \le 10$
(k) $f(x) = 20 - 2x + x^2$, $g(x) = 10$ for $2 \le x \le 6$
(l) $f(x) = 10 + 4x - x^2$, $g(x) = 4$ for $3 \le x \le 6$
(m) $f(x) = 20 - x + x^2$, $g(x) = -20 + x - x^2$ for $-4 \le x \le 10$

SUGGESTED REFERENCES

The references for this chapter are listed in Chapter 9.

Chapter 15

Probability

The final two chapters of this text deal with probability and probability functions. Most students have an intuitive notion of the meaning of probability. For instance, the majority of students have placed a bet on the outcome of a football or basketball game. In placing this bet, the student either consciously or subconsciously calculated the probability or the "chances" of his team winning. If the teams were evenly matched, one might say that the chance of winning the bet was fifty-fifty. An alternative statement is that the probability of winning was 0.5.

Most readers would agree that an understanding of probability would be quite useful when placing bets in games of chance. In fact, the theory of probability was first developed to model the outcomes of games of chance (e.g., dice, roulette, cards, etc.). If, however, our purpose is not to study the intricacies of a gambling casino or the operation of a football betting pool, the question then arises as to the usefulness of probability in making business decisions. The answer to this question is really quite simple. Many business decisions must be based on predictions of future events. These predictions inevitably carry with them uncertainty and probable error. An understanding of the theory of probability better enables the administrator or decision maker to recognize the importance of the uncertainty in the prediction. It also enables him to take into account the effect of the probable error in making the business decision.

Probability, like linear programming and calculus, is based on certain definitions and assumptions. Our task is to introduce the important definitions and assumptions and to provide realistic and important examples. We shall begin with the concept of a sample space.

15.1 Sample Space

The term *sample space* refers to the set of all possible outcomes of a statistical process or experiment. To illustrate, assume that the experiment involves tossing two coins and recording the outcome of the tosses. The sample space, or *set of possible outcomes*, is

$$S = \{(HH), (HT), (TH), (TT)\}$$

where (HH) means a head on the toss of the first coin and a head on the toss of the second coin, (HT) means a head on the toss of the first coin and a tail on the toss of the second, etc. The sample space is the set of possible outcomes of the tosses of the two coins.

The elements within the sample space are often referred to as *sample points*. If we designate the sample points by o_i, the sample space can be written as

$$S = \{o_1, o_2, o_3, \ldots, o_n\} \qquad (15.1)$$

where the sample points $o_1, o_2, o_3, \ldots, o_n$ describe all possible outcomes of the statistical process or experiment. In the example of tossing two coins, the sample space was

$$S = \{(HH), (HT), (TH), (TT)\}$$

and the sample points were (HH), (HT), (TH), and (TT).

In defining the term sample space, we have used the terms statistical process and experiment. The terms statistical process and experiment are used interchangeably. Both terms refer to a process that leads to some well-defined outcome. In the preceding example the process of tossing the two coins was the experiment or statistical process. The outcome of the experiment consisted of recording a head or a tail on the toss of the first coin and a head or a tail on the toss of the second coin. Other examples of experiments include observing the outcome from the roll of a pair of dice, recording the number of automobiles sold by a car dealer during a given week, observing whether the closing price of a particular stock is up, down, or unchanged from the previous day, and asking an individual his brand preference in color television. In each of these examples we only require that the experiment leads to some well-defined outcome and that the set of all possible outcomes can be specified.

Example: Specify the sample space that results from the experiment of tossing three coins and recording the outcome of the toss of each coin as a head *H* or tail *T*.

The sample space is

$$S = \{(HHH), (HHT), (HTH), (THH), (HTT), (THT), (TTH), (TTT)\}$$

Example: In designing a questionnaire, the income of the respondent is often important. Could the following income categories be considered as a sample space?

Your Income (Check the appropriate box)?

☐ $0-$4999 ☐ $5000-$9999 ☐ $10,000-$14,999 ☐ Over $15,000

The income categories would be a sample space. The sample space is

$$S = \{(0-4999), (5000-9999), (10,000-14,999), (\text{Over } 15,000)\}$$

Example: Assume that the income categories in the questionnaire in the preceding example were as follows:

Your Income (Check one box)?

☐ $0-$4999 ☐ $4000-$9999 ☐ $7500-$14,999 ☐ Above $12,500

Do these categories represent a sample space?

These categories would not be a sample space. A sample space is defined as the *set* of all possible outcomes of the experiment. Since the income categories shown overlap, one of the requirements for a set is not satisfied. This requirement, it will be remembered, is that the objects in the set (i.e., the income categories) must be distinct. The categories in this example do not meet this requirement.

15.1.1 SAMPLE SPACE AND CARTESIAN PRODUCT

The concept of the Cartesian product of two or more sets was introduced in Chapter 1. The reader will remember that the Cartesian product of the sets A and B, designated as $A \times B$, was the set containing all ordered pairs (a, b) such that $a \in A$ and $b \in B$. The concept of the Cartesian product can sometimes be quite useful in specifying the sample space of an experiment.

To illustrate, assume that our experiment again involves the toss of two coins. Suppose we let $A = \{H, T\}$ represent the sample space for the first toss and $B = \{H, T\}$ represent the sample space for the second toss. The sample space for the experiment of tossing the two coins is given by the Cartesian product of A and B. Thus

$$S = A \times B = \{(HH), (HT), (TH), (TT)\}$$

It can be seen that the sample space formed by the Cartesian product of A and B is the same as that specified earlier for the toss of the two coins.

Example: Assume that an experiment involves surveying individuals with regard to their political affiliation. As a part of the survey, it is important that the following questions are answered.

1. Age
   ☐ Under 39   ☐ Over 39
2. Sex
   ☐ Male   ☐ Female
3. Political affiliation
   ☐ Democrat   ☐ Republican   ☐ Other

Use the Cartesian product of the three sets to form the sample space of the experiment.

Let A represent the set of possible ages, B the set of possible sex, and C the set of political affiliation. Then

$$A = \{\text{Under 39, Over 39}\}$$
$$B = \{M, F\}$$

and

$$C = \{D, R, O\}$$

The sample space for the experiment is $S = A \times B \times C$, or

$S = \{$(Under 39, M, D), (Under 39, M, R), (Under 39, M, O), (Under 39, F, D), (Under 39, F, R), (Under 39, F, O), (Over 39, M, D), (Over 39, M, R), (Over 39, M, O), (Over 39, F, D), (Over 39, F, R), (Over 39, F, O)$\}$

Example: Assume that an experiment involves rolling a pair of dice. Use the Cartesian product of the outcome on each die to specify the sample space.

Let $A = \{1, 2, 3, 4, 5, 6\}$ represent the six sides on the first die and $B = \{1, 2, 3, 4, 5, 6\}$ represent the six sides on the second die. The possible outcomes of rolling the pair of dice are given by the Cartesian product of A and B. The Cartesian product is shown by the box diagram in Table 15.1.

Table 15.1

A	B	*Second Die*					
		1	*2*	*3*	*4*	*5*	*6*
1		(1, 1)	(1, 2)	(1, 3)	(1, 4)	(1, 5)	(1, 6)
2		(2, 1)	(2, 2)	(2, 3)	(2, 4)	(2, 5)	(2, 6)
First 3		(3, 1)	(3, 2)	(3, 3)	(3, 4)	(3, 5)	(3, 6)
Die 4		(4, 1)	(4, 2)	(4, 3)	(4, 4)	(4, 5)	(4, 6)
5		(5, 1)	(5, 2)	(5, 3)	(5, 4)	(5, 5)	(5, 6)
6		(6, 1)	(6, 2)	(6, 3)	(6, 4)	(6, 5)	(6, 6)

15.1.2 TREE DIAGRAMS

Rather than listing the elements in the sample space, it is sometimes helpful to show the possible outcomes of an experiment by the use of a *tree diagram*. To illustrate the construction of a tree diagram, consider again the experiment of surveying individuals with regard to their age, sex, and political affiliation. The possible outcomes of the questions concerning age, sex, and political affiliation were described by the sets

$$A = \{\text{Under 39, Over 39}\}$$

$$B = \{\text{M, F}\}$$

and

$$C = \{\text{D, R, O}\}$$

The tree diagram for the sample space of this experiment is constructed by showing sets A, B, and C as branches on a tree. Starting from a common point, we draw two lines to represent the elements in set A. From each line, two additional lines are drawn to represent the two elements in set B. To complete the tree diagram, three lines are drawn to represent the elements in set C. The completed diagram is shown in Fig. 15.1. The branches on the tree represent the sample space of the experiment. The sample space is, of course, the same as that given in the earlier example.

Example: An experiment consists of observing the purchases of stock by a brokerage house. If stock is purchased on the New York Stock Exchange,

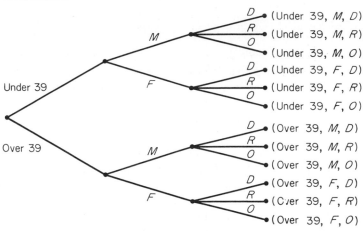

Figure 15.1

the American Stock Exchange, or "over-the-counter" in "round lots" (i.e., 100 share blocks of stock) or "odd lots" (i.e., less than 100 shares), determine the sample space for the experiment.

The sample space is shown by the tree diagram in Fig. 15.2.

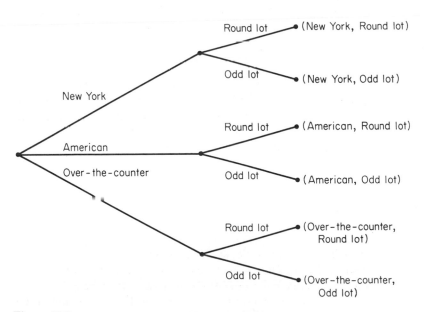

Figure 15.2

15.1.3 FINITE AND INFINITE SAMPLE SPACE

The term "sample space" has been defined as the set of all possible outcomes of an experiment. In illustrating this definition, we have used experiments in which the sample space contained a finite number of possible outcomes. For instance, the sample space shown in Fig. 15.1 contains 12 sample points or possible outcomes, and the sample space in Fig. 15.2 contains six possible outcomes.

It is important to point out that in many experiments the sample space contains an infinite, rather than a finite, number of sample points. As a simple illustration, consider the experiment of determining the time required to complete a long-distance telephone call. The sample space for this experiment consists of all points along a line that begins at zero and continues to infinity. Since there are an infinite number of points on any continuous line, the sample space for this experiment would contain an infinite number of sample points. This would also be true if the upper bound on the time limit were 3 minutes rather than infinity. Again, we recognize that there are an infinite number of points along a continuous line that begins at 0 and continues to 3. We shall consider experiments in which the sample space is described by points along a continuum in Chapter 16.

It is also possible for the sample space to contain a countable, although infinite, number of sample points. To illustrate what is meant by a *countably infinite* number of sample points, consider an experiment that involves determining the number of defects in a bolt of cloth. The sample space for this experiment is

$$S = \{0, 1, 2, 3, \ldots\}$$

The first sample point in the sample space is 0, implying no defects in the bolt of cloth. The second sample point is 1, meaning one defect is a possible outcome from inspecting the cloth. Similarly, two defects are possible. In fact, there is no maximum number of defects in the bolt of cloth. Consequently, the sample space contains a countable, although infinite, number of sample points.

Example: Specify whether the sample space for the following experiments would contain a finite, countably infinite, or an infinite number of sample points.

1. Drawing the winning ticket for a sweepstakes.
   Answer: Finite. The sample space would be a list of all entrants.
2. Tossing a single coin until the occurrence of a "head."
   Answer: Countably infinite. The tosses of the coin are countable, and there is no upper limit on how many times the coin must be tossed until the occurrence of a head.

3. Determining the time before the arrival of the next car at a toll booth.
Answer: Infinite. The sample space consists of the continuum between zero and infinity.

15.2   Subsets of a Sample Space: Events

We have defined the term "sample space" as the set of all possible outcomes of a statistical process or experiment. Assume, for the moment, that we are specifically interested in one or more of the possible outcomes of the experiment. These possible outcomes would be described by a subset of the sample space. In the theory of probability, a subset of a sample space is defined as an *event*.

To illustrate an event as a subset of a sample space, consider the experiment of tossing two coins. The sample space for this experiment is

$$S = \{(HH), (HT), (TH), (TT)\}$$

Assume that we are interested in the event "two heads." The subset of the sample space that defines this event is

$$A = \{(HH)\}$$

Similarly, if our interest were in the event "at least one head," the appropriate subset would be

$$B = \{(HH), (HT), (TH)\}$$

The subsets A and B of the sample space S are examples of events.

Example: Consider the experiment of rolling a pair of dice. The sample space for this experiment was given by Table 15.1 and is repeated in Table 15.2.

Table 15.2

A \ B	1	2	3	4	5	6
			Second Die			
1	(1, 1)	(1, 2)	(1, 3)	(1, 4)	(1, 5)	(1, 6)
2	(2, 1)	(2, 2)	(2, 3)	(2, 4)	(2, 5)	(2, 6)
First 3	(3, 1)	(3, 2)	(3, 3)	(3, 4)	(3, 5)	(3, 6)
Die 4	(4, 1)	(4, 2)	(4, 3)	(4, 4)	(4, 5)	(4, 6)
5	(5, 1)	(5, 2)	(5, 3)	(5, 4)	(5, 5)	(5, 6)
6	(6, 1)	(6, 2)	(6, 3)	(6, 4)	(6, 5)	(6, 6)

Using this table, determine the sample points in the following events.

1. The sum of the spots on the pair of dice is 7.

$$\{(6, 1), (5, 2), (4, 3), (3, 4), (2, 5), (1, 6)\}$$

2. The sum of the spots on the pair of dice is 12.

$$\{(6, 6)\}$$

3. The number on the second die is larger than that on the first die.

$$\{(1, 2), (1, 3), (1, 4), (1, 5), (1, 6), (2, 3), (2, 4), (2, 5), (2, 6), (3, 4),$$
$$(3, 5), (3, 6), (4, 5), (4, 6), (5, 6)\}$$

4. The sum of the spots on the pair of dice is 2

$$\{(1, 1)\}$$

The subset or events in the preceding example can be used to introduce the terms *simple* and *compound* events. When a subset or event contains only one sample point, the event is termed a simple event. Thus, the events describing statements 2 and 4 in the preceding example are simple events. If the subset contains more than one sample point, the event is referred to as a compound (or *joint*) event. The events described by statements 1 and 3 are compound events.

Example: Using the sample space in Table 15.2, specify the following events and state whether the events are simple or compound.

1. The sum of the spots on the pair of dice is 3.

$$\{(1, 2), (2, 1)\} \qquad \text{Compound.}$$

2. Five plus the number on the first die equals the number on the second die.

$$\{(1, 6)\} \qquad \text{Simple.}$$

3. The number on the first die is 1.

$$\{(1, 1), (1, 2), (1, 3), (1, 4), (1, 5), (1, 6)\} \qquad \text{Compound.}$$

15.2.1 SET OPERATIONS AND EVENTS

The set operations of intersection, union, and complementation are quite useful in specifying events.† To illustrate, consider again the experiment of rolling a pair of dice. The sample space for this experiment was given by Tables 15.1 and 15.2 and is repeated in Table 15.3. Assume that the following events are defined on this sample space.

---

† The set operations of intersection, union, and complementation were introduced in Chapter 1.

Table 15.3

A \ B		1	2	*Second Die* 3	4	5	6
First Die	1	(1, 1)	(1, 2)	(1, 3)	(1, 4)	(1, 5)	(1, 6)
	2	(2, 1)	(2, 2)	(2, 3)	(2, 4)	(2, 5)	(2, 6)
	3	(3, 1)	(3, 2)	(3, 3)	(3, 4)	(3, 5)	(3, 6)
	4	(4, 1)	(4, 2)	(4, 3)	(4, 4)	(4, 5)	(4, 6)
	5	(5, 1)	(5, 2)	(5, 3)	(5 4)	(5, 5)	(5, 6)
	6	(6, 1)	(6, 2)	(6, 3)	(6, 4)	(6, 5)	(6, 6)

1. The event "The sum of the spots on the two dice is 7."

$$A = \{(6, 1), (5, 2), (4, 3), (3, 4), (2, 5), (1, 6)\}$$

2. The event "The number on the second die is 3."

$$B = \{(1, 3), (2, 3), (3, 3), (4, 3), (5, 3), (6, 3)\}$$

3. The event "the sum of the spots on the two dice is 11."

$$C = \{(6, 5), (5, 6)\}$$

Given the events A, B, and C, we can specify additional events, using the operations of intersection, union, and complementation. For instance, consider the event "seven or eleven." In combining events the conjunction *or* represents the union of the events. The subset of sample points included in the event "seven or eleven" is thus

$$A \cup C = \{(6, 1), (5, 2), (4, 3), (3, 4), (2, 5), (1, 6), (6, 5), (5, 6)\}$$

Similarly, the event "The sum of the spots on the two dice is 7 and the number on the second die is 3" can be described by using set operations. The conjunction *and* refers to the intersection of the events. The subset of sample points included in the event is

$$A \cap B = \{(4, 3)\}$$

Finally, the event "The sum of the spots on the two dice is not 7 or 11" is described by the complement of the union of events A and C. The event is

$$(A \cup C)' = \{(1, 1), (1, 2), (1, 3), (1, 4), (1, 5), (2, 1), (2, 2), (2, 3), (2, 4), (2, 6),$$
$$(3, 1), (3, 2), (3, 3), (3, 5), (3, 6), (4, 1), (4, 2), (4, 4), (4, 5), (4, 6),$$
$$(5, 1), (5, 3), (5, 4), (5, 5), (6, 2), (6, 3), (6, 4), (6, 6)\}.$$

Example: Consider the experiment of tossing three coins and recording the outcome of each toss as a "head" or a "tail." The sample space for this experiment is

$$S = \{(HHH), (HHT), (HTH), (THH), (HTT), (THT), (TTH), (TTT)\}$$

Assume that the following events are defined in this sample space.

1. The event "three heads."
$$A = \{(HHH)\}$$

2. The event "two heads and one tail."
$$B = \{(HHT), (HTH), (THH)\}$$

3. The event "three tails."
$$C = \{(TTT)\}$$

Given the events A, B, and C, specify the following events, using the operations of intersection, union, and complementation.

1. The event "at least two heads." This event is
$$A \cup B = \{(HHH), (HHT), (HTH), (THH)\}$$

2. The event "at least two tails." The event is
$$(A \cup B)' = \{(HTT), (THT), (TTH), (TTT)\}$$

3. The event "three heads" and "two heads and one tail." The event "three heads" cannot occur simultaneously with the event "two heads and one tail." The subset is thus
$$A \cap B = \{\ \}$$
or, alternatively,
$$A \cap B = \phi$$

4. The event "three heads or not three heads." This event is given by the union of A and A' and is
$$A \cup A' = \{(HHH), (HHT), (HTH), (THH),$$
$$(HTT), (THT), (TTH), (TTT)\}$$
or, alternatively,
$$A \cup A' = S$$

The events described by statements 3 and 4 illustrate events that are, respectively, *mutually exclusive* and *exhaustive*. To illustrate, consider two events H and G that are defined on the sample space S. The events H and G are termed mutually exclusive if the subset defined by the intersection of H and G is null, i.e., if H and G have no common sample points. Symbolically, the two events H and G are mutually exclusive if

$$H \cap G = \phi \tag{15.2}$$

The events "three heads" and "two heads and one tail" are clearly mutually exclusive, i.e., $A \cap B = \phi$. One cannot toss three coins and get both "three heads" and "two heads and one tail" as the outcome of the tosses.

To define events that are exhaustive, consider the events I and J that are defined on a sample space S. The events I and J are termed exhaustive if the subset formed by the union of I and J is equal to the sample space S. Thus, the events I and J are exhaustive if

$$I \cup J = S \qquad (15.3)$$

The events A and A' in the preceding example are, of course, exhaustive; i.e., $A \cup A' = S$. This merely says the result of the toss of three coins will be "three heads" or "not three heads." In fact, we know from Chapter 1 that the subset formed by the union of any event A and its complement A' is equal to the universal set U. It therefore follows that, for any event A defined on the sample space S, the events A and A' are exhaustive. Similarly, from the definition of complementary events, we know that the intersection of two complementary events is null and the events must, therefore, also be mutually exclusive.

Example: Consider an experiment of drawing a card from a deck of playing cards. Indicate if the following events are mutually exclusive, exhaustive, or both.

1. $A = \{\text{black card}\}$, $B = \{\text{red card}\}$.
   Answer. $A \cap B = \phi$, $A \cup B = S$. The events are both mutually exclusive and exhaustive.
2. $A = \{\text{heart}\}$, $B = \{\text{spade}\}$, $C = \{\text{club}\}$.
   Answer. $A \cap B = \phi$, $A \cap C = \phi$, $B \cap C = \phi$, $A \cup B \cup C \neq S$. The events are mutually exclusive but not exhaustive.
3. $A = \{\text{black card}\}$, $B = \{\text{club}\}$.
   Answer. $A \cap B \neq \phi$, $A \cup B \neq S$. The events are neither mutually exclusive nor exhaustive.
4. $A = \{\text{red card}\}$, $B = \{\text{spade}\}$.
   Answer. $A \cap B = \phi$, $A \cup B \neq S$. The events are mutually exclusive but not exhaustive.
5. $A = \{\text{face card}\}$, $B = \{\text{not a face card}\}$.
   Answer. $A \cap B = \phi$, $A \cup B = S$. The events are both mutually exclusive and exhaustive.

15.2.2 PARTITION OF THE SAMPLE SPACE

Events defined on a sample space that are both mutually exclusive and exhaustive form a *partition* of the sample space. Thus, in the simplest case, given an event A defined on the sample space S, the events A and A' form a *twofold* partition of the sample space. Similarly, if the n events $A_1, A_2, \ldots, A_n$ are both mutually exclusive and exhaustive, the n events form an n-fold partition of the sample space.

As an example of partitioning a sample space, consider an experiment of selecting a panel of jurors from a list of registered voters. If we let A represent the event "male" and B the event "female," then $A \cap B = \phi$ and $A \cup B = S$, and the events A and B are mutually exclusive and exhaustive. The events A and B form a twofold partition of the sample space.

Continuing with the example, assume that C represents the event "under twenty-five years of age," D the event "twenty-five to forty years of age," and E the event "over forty years of age." The events C, D, and E are mutually exclusive, since no two of the events can occur at once. Since one of the events must occur, they are also exhaustive. Thus, the events C, D, and E form a threefold partition of the sample space.

Example: Explain why the events A, B, and C in the preceding illustration concerning voter selection do not form a partition of the sample space.

The requirements for partitioning a sample space are that the events be both mutually exclusive and exhaustive. Given the events A, B, and C in the preceding example, we note that $A \cap B = \phi$ but that $A \cap C \neq \phi$ and $B \cap C \neq \phi$. Therefore, the events A, B, and C do not form a partition of the sample space.

15.3 Counting the Possible Outcomes

Constructing a complete list of the possible outcomes (i.e., sample points) of an experiment can be a time-consuming and burdensome task. Fortunately, there are many problems in probability in which a listing of the possible outcomes or sample points in an experiment is not required. In these types of problems we need know only the total number of sample points in the sample space of the experiment and the number of sample points in a specific event. Our task in this section is to introduce the rules and formulas that are used in counting these possible outcomes. The relationship between counting the number of possible outcomes and the probability of the event is discussed in Sec. 15.4.

15.3.1 A FUNDAMENTAL RULE FOR COUNTING

The number of possible outcomes of an experiment can often be determined by applying the following fundamental rule of counting.

If an experiment consists of k separate steps and the first step can occur in n_1 different ways, the second in n_2 different ways, . . . , and the kth in n_k different ways, then the number of different possible outcomes of the experiment is

$$n_1 \cdot n_2 \cdot \ldots \cdot n_k \qquad (15.4)$$

To illustrate the fundamental rule, consider an experiment of selecting a committee composed of one labor leader, one governmental official, and one community leader from a group of three labor leaders, four governmental officials, and five community leaders. There are three ways in which the selection of the labor leader can be made, four ways in which the selection of the governmental official can be made, and five ways in which the selection of the community leader can be made. The total possible number of ways of selecting the committee is thus $3 \cdot 4 \cdot 5 = 60$.

Example: Use the fundamental rule of counting to determine the number of possible outcomes in the experiment of tossing a coin three times and recording the outcomes of each toss as a head or a tail.

The experiment of tossing a coin three times consists of three steps. There are two possible outcomes on the first toss (i.e., heads and tails), two possible outcomes on the second toss, and two possible outcomes on the third toss. The total number of possible outcomes is thus $2 \cdot 2 \cdot 2 = 8$.

It is interesting to notice that the fundamental rule of counting is the same as the rule introduced in Chapter 1 for determining the number of elements in the Cartesian product of two or more sets. The reader will remember that the Cartesian product of sets $A_1, A_2, \ldots, A_k$ containing, respectively, n_1, $n_2, \ldots, n_k$ elements has $n_1 \cdot n_2 \cdot \ldots \cdot n_k$ elements.

As an illustration of the relationship between the rule for determining the number of elements in the Cartesian product of k sets and the number of possible outcomes of an experiment consisting of k steps, consider the example discussed above. If $A = \{H, T\}$ represents the set of possible outcomes on the first toss of the coin, $B = \{H, T\}$ the possible outcomes on the second toss, and $C = \{H, T\}$ the possible outcomes on the third toss, the Cartesian product of A, B, and C is

$$\{(HHH), (HHT), (HTH), (THH), (HTT), (THT), (TTH), (TTT)\}$$

The Cartesian product contains $2 \cdot 2 \cdot 2 = 8$ elements and represents the sample space of the experiment. The number of possible outcomes is, of course, the same as was found by using the fundamental rule of counting.

Example: In a marketing survey, an individual is asked to taste four different types of wine and rank them in order of preference. How many different rankings are possible?

There are four possible choices for the first preference. After this selection is made, there are three possible choices for the second preference. Similarly, there are two choices for the third preference and one choice for the least preferred wine. The number of possible rankings is thus $4 \cdot 3 \cdot 2 \cdot 1$, or 24.

Example: An experiment consists of drawing five cards from a well shuffled deck of 52 playing cards. If it is assumed that a card once drawn from the deck is not replaced, in how many possible ways may the selection of five cards be made?

There are 52 possible outcomes from the first draw, 51 possible outcomes from the second draw, etc. The total number of possible outcomes is $52 \cdot 51 \cdot 50 \cdot 49 \cdot 48$, or 311,875,200.

15.3.2 PERMUTATIONS

The fundamental rule of counting was used in the previous example to determine the number of possible ways of selecting five cards from a deck of 52 playing cards. In counting the number of ways of making the selection, we assumed that the order of selection of an individual card was of importance. This means that the selection of the ace of hearts followed by the king of diamonds differs from the selection of the king of diamonds followed by the ace of hearts. This difference occurs because the order in which the cards were selected in the two hands differs.

The selection of r objects from a set of n objects in which the order of selection of the r objects is of importance is referred to as a *permutation* of the objects.

The concept of a permutation of objects can perhaps be illustrated by considering the number of possible ways of arranging the first three letters of the alphabet. Given the set of letters $\{a, b, c\}$, there are three possible choices for the first letter, two possible choices for the second letter, and one choice for the third letter. The number of possible arrangements of the letters, again if it is assumed that the order of the selection of the letters is of importance, is $3 \cdot 2 \cdot 1$, or 6 possible arrangements. These are

$$abc \qquad bac \qquad cab$$
$$acb \qquad bca \qquad cba$$

Each of these possible arrangements is referred to as a permutation of the set of letters.

The formula for the number of possible permutations of the n objects in a set is derived from the fundamental rule of counting. Since there are n objects in the set, there are n ways in which the first item can be selected. Similarly, there are $(n - 1)$ ways of selecting the second item, $(n - 2)$ ways of selecting the third item, and finally, $(n - n + 1)$ ways of selecting the final item. The formula for counting the number of possible permutations of n objects in a set of n objects is thus

$$_nP_n = n! \qquad (15.5)$$

where $n! = n(n - 1)(n - 2) \ldots 1$.

Example: A management consulting firm has four contracts that must be assigned to four separate consultants. In how many possible ways can the assignments be made?

From Formula (15.5), the number of possible assignments is

$$_4P_4 = 4! = 24$$

Formula (15.5) applies when one is counting the total number of ways the n objects in a set can be selected or arranged. Quite often, we are interested in counting the total number of possible ways of selecting a subset of r of the n objects in a set. To illustrate, consider the problem of selecting a president, vice-president, secretary, and treasurer from a group of ten candidates. Although there are ten objects (i.e., candidates) in the set, only four of the ten are to be selected. Again by using the fundamental rule of counting, the number of possible permutations is computed by recognizing that there are ten ways of selecting the president, nine ways of selecting the vice-president, eight ways of selecting the secretary, and seven ways of selecting the treasurer. The total number of possible permutations is thus $10 \cdot 9 \cdot 8 \cdot 7$, or 5040. There are 5040 possible ways of selecting four objects from a set of ten objects when the order of selection of the objects is of importance.

The formula for determining the number of possible permutations of r objects selected from a set of n objects is

$$n(n - 1)(n - 2) \ldots (n - r + 1) \qquad (15.6)$$

Formula (15.6) can be rewritten as

$$_nP_r = \frac{n!}{(n - r)!} \qquad (15.7)$$

Example: Using the preceding example of selecting four officers from a group of ten candidates, show that Formulas (15.6) and (15.7) are equivalent.

From Formula (15.6), the total number of possible ways of selecting a president, a vice-president, a secretary, and a treasurer from a group of ten candidates is

$$10 \cdot 9 \cdot 8 \cdot 7 = 5040$$

By Formula (15.7), the total number of possible ways is

$$_{10}P_4 = \frac{10!}{6!} = 10 \cdot 9 \cdot 8 \cdot 7 = 5040$$

It can be seen that the two formulas are equivalent. Formula (15.7) is merely an alternative expression of Formula (15.6).

Example: A retailing chain has 20 stores. Three of the stores are to be selected for testing three different products. If it is assumed that only one product will be tested at a specific store, how many different arrangements are possible?

$$_{20}P_3 = \frac{20!}{17!} = 20 \cdot 19 \cdot 18 = 6840$$

Example: An experiment consists of selecting two numbers from the set of digits, $D = \{0, 1, 2, 3, 4, 5, 6, 7, 8, 9\}$. If it is assumed that the first number selected is not replaced, how many different arrangements of the two numbers are possible?

$$_{10}P_2 = \frac{10!}{8!} = 90$$

If the first number is replaced before selecting the second number, how many different arrangements are possible?

If the fundamental rule of counting is used, there are ten ways of selecting the first digit and ten ways of selecting the second digit. Consequently, there are one hundred ways of arranging the two numbers.

15.3.3   COMBINATIONS

In determining the number of possible outcomes of an experiment that involves the selection of r objects from a set of n objects, the order in which the r objects are selected is often not important. For instance, consider an experiment that involves selecting three committee members from a list of ten candidates. In this case we are interested in the number of different groups of three objects selected from a set of ten objects. Since a committee consisting of, say, Jones, Smith, and Rogers is the same as a committee consisting of Smith, Jones, and Rogers, the order of selecting the individual members of the committee is of no importance.

A group of r objects selected from a set of n objects in which the order of selection is of no importance is referred to as a *combination* of the objects. To develop the formula for the number of combinations, consider the preceding illustration of selecting three committee members from a list of ten candidates. From Formula (15.7), the number of different permutations of three objects selected *with regard to order* from a set of ten objects is

$$_{10}P_3 = \frac{10!}{7!} = 720$$

Assume that Smith, Jones, and Rogers are included in the list of ten candidates. The permutations that include Smith, Jones, and Rogers are $\{(S, J, R), (S, R, J), (R, S, J), (R, J, S), (J, R, S), (J, S, R)\}$.

The six permutations that include Smith, Jones, and Rogers represent, however, only one combination (i.e., committee). Since the same result would be true for any combination of three of the ten objects, the number of combinations is only one-sixth as large as the number of permutations. The number of different three-man committees is, therefore, $\frac{720}{6}$, or 120.

This result can be generalized by recognizing that the number of permutations per combination is $r!$. Thus, in the preceding problem there are $3! = 3 \cdot 2 \cdot 1$, or 6 permutations per combination. The number of possible combinations is therefore given by dividing the number of permutations by the number of permutations per combination. The formula is

$$_nC_r = \frac{n!}{(n-r)!r!} \tag{15.8}$$

Example: Use Formula (15.8) to determine the number of ways of selecting three committee members from a list of ten candidates.

$$_{10}C_3 = \frac{10!}{7!3!} = \frac{10 \cdot 9 \cdot 8}{3 \cdot 2 \cdot 1} = 120$$

In certain formulas that are used in probability, the notation $_nC_r$ is cumbersome. For this reason, an alternative notation is often used. The notation is $\binom{n}{r}$. The number of combinations of r objects selected from a set of n objects can thus also be expressed as

$$\binom{n}{r} = \frac{n!}{(n-r)!r!} \tag{15.9}$$

Formulas (15.8) and (15.9) are, of course, the same.

Example: Seven students wish to select four of their group to do research on a team project. How many different combinations are possible?

$$_7C_4 = \frac{7!}{3!4!} = 35$$

Example: A grocery chain has fourteen stores. Management decides to try out a new product in four of the stores. How many different possible four-store samples are there?

$$_{14}C_4 = \frac{14!}{10!4!} = 1001$$

Example: From a list of twelve applicants for a job, five are to be selected. How many different combinations of five selected from twelve are possible?

$$_{12}C_5 = \frac{12!}{7!5!} = 792$$

15.4 Assigning Probabilities

A probability is a numerical measure of the likelihood of the occurrence of an event. To illustrate, consider the experiment of tossing two coins and recording the outcome of the toss of each coin as a head H or tail T. The sample space for the experiment is

$$S = \{(HH), (HT), (TH), (TT)\}$$

Assume that we want to determine the likelihood of obtaining two heads as the outcome of this experiment. Since each of the four sample points in S is equally likely, the probability of the event "two heads" is 0.25. Similarly, the probability of each of the remaining sample points in S is also 0.25.

Probabilities are assigned on the basis of either *objective* or *subjective* estimates of the likelihood of occurrence of an event. *Objective probabilities* are determined from theoretical analysis of the experiment or from actual observation of numerous repetitions of the experiment. In the experiment involving the toss of two coins, for instance, we reasoned that each of the outcomes is equally likely. Since there are four possible outcomes and each outcome is equally likely, the probability of two heads is 0.25. This probability was determined by theoretical analysis rather than from actually observing repeated tosses of two coins. The probability of the event was given by

$$P(E) = \frac{n(E)}{n(S)} \qquad\qquad (15.10)$$

where $n(E)$ represents the number of sample points in event E and $n(S)$ represents the total number of equally likely sample points in the sample space S. In the example, $n(E)$ is 1, $n(S)$ is 4, and $P(HH) = \frac{1}{4}$ or 0.25.

Objective probabilities determined from actually observing numerous repetitions of the experiment are calculated by the formula

$$P(E) = \frac{f}{n} \qquad\qquad (15.11)$$

where f is the frequency or number of occurrences of event E and n is the total number of times the experiment has been observed. For example, consider the manufacture of a certain component. If, during a long production run, 100,000 components have been manufactured and 5000 of these have been defective, an empirical estimate of the probability of a defective unit is $P(E) = 5000/100,000 = 0.05$.

Example: In the game of roulette, a ball is dropped on a revolving disk. The disk contains 38 slots—18 red, 18 black, and 2 green. Assuming that the ball is equally likely to drop in any slot, determine the probability of the event "black."

There are a total of 38 equally likely slots. Since 18 of these are black, we reason that the probability of the event black is

$$P(\text{black}) = \frac{n(\text{black})}{n(S)} = \frac{18}{38}$$

Example: A manufacturer of transistors must determine the probability of a defective transistor. One thousand transistors are randomly selected from the production line and tested. Forty are found to be defective. Determine the probability of a defective transistor.

A total of 1000 transistors were tested and 40 were defective. The probability of a defective is

$$P(\text{defective}) = \frac{f}{n} = \frac{40}{1000} = 0.04$$

Subjective probabilities differ from objective probabilities in that a subjective probability is a measure of an individual's belief in the likelihood of the occurrence of some event. Subjective probabilities are normally based on the opinion of a "reasonable" or "knowledgeable" person. Most often, subjective probabilities are used when it is impossible or impractical to determine the objective probability of the event. To illustrate, consider the case of introducing a new product. A major food company plans to market a new breakfast product. In order for the product to earn a reasonable return on investment, sales must be at least $10 million. The probability of sales reaching this figure is subjectively estimated by the vice-president of marketing as $P(\text{sales} \geq \$10 \text{ million}) = 0.80$.

In determining the subjective probability, the vice-president called on the combined expertise of people in sales, finance, and marketing. It was, of course, impossible to theoretically determine this probability or to introduce this particular product a large number of times and thereby determine the frequency of occurrence of the event "sales $\geq$ \$10 million." The probability, instead, represents a subjective estimate of a knowledgeable individual of the likelihood of the sales of the product exceeding $10 million.

Example: In a recent panel discussion of the prospects for developing a cure for cancer, Dr. Jefferies stated that he believed the probability of finding a cure for cancer before the year 2000 was 0.80. Dr. Baxter, however, estimated that the probability was only 0.50. Explain how two probabilities for the same event can differ.

The probabilities differ because they are subjective estimates based on the personal belief of the two doctors. One could not expect two or more knowledgeable individuals to have the same subjective belief about a highly uncertain event.

15.4.1   AXIOMS OF PROBABILITY

In assigning either objective or subjective probabilities to a set of events, three requirements must be satisfied. First, the probability of any event must be equal to or greater than zero and less than or equal to one. A probability of zero means that the event cannot occur, and a probability of one means that the event must occur. Second, the sum of the probabilities of the events must equal one. This means that the total of the probabilities of the events in the sample space is one, or, alternatively, that one of the events must occur as a result of the experiment. Finally, the probability of the occurrence of two or more mutually exclusive events is equal to the sum of the probabilities assigned to the individual events. Thus, in the example above, the probability of two heads or two tails is equal to the sum of the probabilities assigned to the events two heads and two tails, that is:

$$P((HH) \cup (TT)) = P(HH) + P(TT) = 0.50$$

The requirements for assigning probabilities are termed the three axioms of probability. For a sample space S with the exhaustive events $A_1, A_2, \ldots, A_n$, the axioms can be summarized as:

Axiom 1. $0 \leq P(A_j) \leq 1$ (15.12)

Axiom 2. $P(S) = 1$ (15.13)

Axiom 3. $P(A_i \cup A_j) = P(A_i) + P(A_j)$ for $A_i \cap A_j = \phi$ (15.14)

Axiom 1 states that the probability assigned to an event must be a number between 0 and 1, inclusive. Axiom 2 requires that the sum of the probabilities must equal 1. Axiom 3 states that the probability of the occurrence of the mutually exclusive events A_i or A_j is given by the sum of the probabilities assigned to A_i and A_j.

Axioms 2 and 3 can be used to determine the probabilities of complementary events. The events A and A' being given, it follows from Axiom 2 and from the definition of complementary events that

$$P(A) + P(A') = 1$$ (15.15)

Alternatively, this relationship can be expressed as

$$P(A') = 1 - P(A)$$ (15.16)

Example: At a recent stockholders' meeting of a large corporation, the president stated that the probability of an increase in profits during the coming year was 0.60. He also stated that the probability of profits remaining unchanged was 0.30 and that the probability of a decline in profit was 0.10. Are these statements consistent with the axioms of probability?

Assume that I represents the event that profits increase, U the event that profits remain unchanged, and D the event that profits decline. Then, $P(I) = 0.60$, $P(U) = 0.30$, and $P(D) = 0.10$. The probabilities of I, U, and D are all represented by numbers between 0 and 1. Furthermore, the sum of the probabilities for the three mutually exclusive and exhaustive events are $P(I) + P(U) + P(D) = 0.60 + 0.30 + 0.10 = 1.0$. Finally, the probability of the union of the two events I and D is $P(I \cup D) = P(I) + P(D) = 0.60 + 0.10 = 0.70$.

Example: Determine the probability that profits will not increase in the preceding example.

The event that profits will not increase is described as I'. The probability of I' is $P(I') = 1 - P(I) = 1 - 0.60 = 0.40$.

The third axiom of probability was defined for events that are mutually exclusive. Quite often, however, probabilities must be assigned to events that are not mutually exclusive. In these cases, Axioms 1 and 2 still apply. Axiom 3 must be modified to take into account the fact that $A_i \cap A_j \neq \phi$. The modification is

$$\text{Axiom 3'.} \quad P(A_i \cup A_j) = P(A_i) + P(A_j) - P(A_i \cap A_j) \quad (15.17)$$

where $A_i \cap A_j \neq \phi$.

To illustrate Axiom 3', consider the events A and B defined on the sample space S in Fig. 15.3. The events A and B in Fig. 15.3(a) are mutually exclusive,

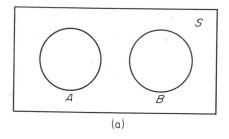

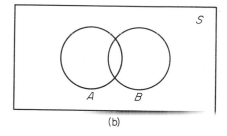

(a)   (b)

Figure 15.3

i.e., $A \cap B = \phi$. The probability of the events A or B occurring is thus $P(A \cup B) = P(A) + P(B)$. The events A and B in Fig. 15.3(b), however, are not mutually exclusive, i.e., $A \cap B \neq \phi$. The event $A \cap B$ is a subset of both the event A and the event B. Since the $P(A \cap B)$ is included in both $P(A)$ and $P(B)$, it must be subtracted from $P(A) + P(B)$ in order to avoid double counting. The probability of the events A or B occurring is, therefore, $P(A \cup B) = P(A) + P(B) - P(A \cap B)$.

As an example of assigning probabilities to events, consider the experiment of drawing a card from a well-shuffled deck of 52 playing cards. The sample space for this experiment consists of the 52 cards in the deck. Assume that the events "R = red card," "C = club," and "S = spade" are defined on the sample space. The events R, C, and S are mutually exclusive and exhaustive. Since there are 26 red cards, 13 spades, and 13 clubs in the deck and the selection of any one of the 52 cards is equally likely, the probabilities of the events R, S, and C are $P(R) = \frac{26}{52}$, $P(C) = \frac{13}{52}$, and $P(S) = \frac{13}{52}$. Notice that the probabilities assigned to the events are numbers between 0 and 1 and that $P(R) + P(C) + P(S) = \frac{26}{52} + \frac{13}{52} + \frac{13}{52} = 1$, indicating that one of the events must occur. Also note that the probability of a red card or a spade is given by $P(R \cup S) = P(R) + P(S) = \frac{26}{52} + \frac{13}{52} = \frac{39}{52}$; that is, the probability is given by the sum of the probabilities of the two mutually exclusive events.

To illustrate Axiom 3', assume that event "F = face card" is defined on the sample space. A deck of cards has four face cards (jack, queen, king, ace) in each suit and four suits (clubs, diamonds, hearts, spades). The probability of the event F is therefore $P(F) = \frac{16}{52}$.

Assume that we wish to determine the probability of the event "red card" or "face card." Since the events red card and face card are not mutually exclusive, the probability is $P(R \cup F) = P(R) + P(F) - P(R \cap F) = \frac{26}{52} + \frac{16}{52} - \frac{8}{52} = \frac{34}{52}$. The probability is found by subtracting the probability of a red face card from the sum of the probabilities of the events red card and face card.

Example: Consider two events A and B defined on a sample space S. If $P(A) = 0.40$, $P(B) = 0.30$, $P(A \cap B) = 0.10$, and $P(S) = 1.00$, determine $P(A \cup B)$, $P(A' \cap B')$, and $P((A' \cap B) \cup (A \cap B'))$. Is the assignment of probabilities consistent with the three axioms of probability?

The events are shown in the Venn diagram in Fig. 15.4. To verify that the probabilities are consistent with the axioms of probability, notice that

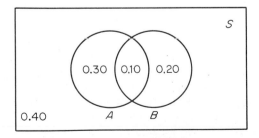

Figure 15.4

the probabilities assigned are numbers between 0 and 1, that $P(S) = 1$, and that $P(A \cup B) = P(A) + P(B) - P(A \cap B)$.

The probabilities of $(A \cup B)$, $(A' \cap B')$, and $(A' \cap B) \cup (A \cap B')$ are $P(A \cup B) = 0.60$, $P(A' \cap B') = 0.40$, and $P((A' \cap B) \cup (A \cap B')) = 0.50$.

15.4.2 MARGINAL AND JOINT PROBABILITIES

One of the more useful methods of summarizing the probabilities of inter-related events is through the use of a *probability table*. To illustrate, consider an experiment recently conducted by a large department store. The purpose of this experiment was to determine the buying patterns of men and women shoppers in the department store. The experiment consisted of noting the proportion of men and women shoppers who made at least one purchase before leaving the store. The results of this experiment are shown as probabilities in Table 15.4.

Table 15.4   Buying Patterns of Men and Women Shoppers

	Men (M)	Women (W)	Total
Buyer (B)	0.10	0.10	0.20
Nonbuyer (B')	0.20	0.60	0.80
Total	0.30	0.70	1.00

Four events are defined in the table. These are "Buyer (B)," "Nonbuyer (B')," "Men (M)," and "Women (W)." The probability of each of the events is read from the margins of the probability table. The probability of a buyer is thus $P(B) = 0.20$. Similarly, the probability of a nonbuyer is $P(B') = 0.80$, the probability of a man shopper is $P(M) = 0.30$, and the probability of a woman shopper is $P(W) = 0.70$. Probabilities such as $P(B)$, $P(B')$, $P(M)$, and $P(W)$ that appear in the margin of a probability table are referred to as *marginal* probabilities. The term marginal probability comes from the fact that the probabilities are shown in the margins of the table.

The probability of the joint occurrence of two events is shown in the body of the probability table. For example, the probability that a randomly selected individual entering the store is both a man and a buyer is $P(M \cap B) = 0.10$. The probability of the joint occurrence of the events man and buyer is termed a *joint* probability.

Joint probabilities of two or more events are denoted as the probability of the intersection of the events. In the example, the probability of an individual's being a man and a buyer is $P(M \cap B) = 0.10$. Similarly, the probability of an individual's being a woman and a nonbuyer is $P(W \cap B') = 0.60$. The probability statement, $P(M \cap B)$, is read "the probability of the joint occurrence of the events M *and* B," or, alternatively, "the probability of M *and* B."

Example: For the example shown in Table 15.4, determine the probabilities of the events $(M \cap B')$, $(B \cap W)$, $(W \cup B)$, and $(M \cup B')$.

$P(M \cap B') = 0.20$

$P(B \cap W) = 0.10$

$P(W \cup B) = P(W) + P(B) - P(W \cap B) = 0.70 + 0.20 - 0.10 = 0.80$

$P(M \cup B') = P(M) + P(B') - P(M \cap B') = 0.30 + 0.80 - 0.20 = 0.90$

Example: The dean of a school of business recently authorized a study of the relative performance of individuals in the school's masters program. Specifically, the study classified individuals by undergraduate degree in business (B) and nonbusiness (B'), and by honor student (H) and nonhonor student (H'). The results of the study are shown in Table 15.5. Based on the entries

Table 15.5

	Honor (H)	Nonhonor (H')	Total
Undergraduate Degree			
Business (B)	0.10	0.30	0.40
Nonbusiness (B')	0.20	0.40	0.60
Total	0.30	0.70	1.00

in the probability table, determine the following:

1. Marginal probability of an honor student.

$$P(H) = 0.30$$

2. Joint probability of nonbusiness and nonhonor.

$$P(B' \cap H') = 0.40$$

3. The probability of nonbusiness or honor.

$$P(B' \cup H) = P(B') + P(H) - P(B' \cap H)$$
$$= 0.60 + 0.30 - 0.20 = 0.70$$

4. Are the events H and B mutually exclusive?

$$\text{No, } P(H \cap B) \neq 0$$

5. Do the events H and H' form a partition of the sample space?
   Yes, complementary events are both mutually exclusive
   and exhaustive. Thus, they partition the sample space.

$$P(H \cap H') = 0 \quad \text{and} \quad P(H \cup H') = P(H) + P(H') = P(S)$$

15.4.3 CONDITIONAL PROBABILITY

In studying the buying patterns of men and women shoppers, we were able to determine the marginal probability of the event "man" and the joint probability of the event "buyer and man." Referring to Table 15.4 and our earlier cáclulations, we find that these probabilities were $P(M) = 0.30$ and $P(B \cap M) = 0.10$. Based on a relative frequency interpretation of probability, this means that 30 percent of the shoppers are men and 10 percent of the shoppers are both men and buyers.

Suppose that, instead of determining the probability of a customer's being both a man and a buyer, we were asked to determine the probability of a man customer making a purchase. Referring to Table 15.4, we note that 10 percent of the customers are buyers and men and 30 percent of the customers are men. The proportion of the men customers who are buyers is $0.10/0.30 = 0.33$. Thus, 33 percent of the men customers are buyers or, alternatively, the probability of a buyer, given the fact that the customer is a man, is 0.33.

The probability of the event "buyer" given the event "man" is referred to as a conditional probability. The *conditional probability* of an event is the probability of the event's occurrence, given some specified condition. In our example, the probability of the event "buyer," given the condition that the event "man" has occurred, was 0.33. This is written as $P(B \mid M) = 0.33$, where the symbol "|" is read as "given." †

The formula for the conditional probability of the event A given the occurrence of the event B is

$$P(A \mid B) = \frac{P(A \cap B)}{P(B)} \tag{15.18}$$

where $P(A \cap B)$ represents the joint probability of the events A and B and $P(B)$ the marginal probability of B. The conditional probability, $P(A \mid B)$, is defined only if $P(B) > 0$.

Formula (15.18) is illustrated by Fig. 15.5. The crosshatched area in the figure represents the event $A \cap B$. Since we are given the condition that event B has occurred, the sample space is reduced to the sample points in B. The probability of A given B, $P(A \mid B)$, is thus equal to the proportion of the

---

† Notice that the symbol "|" is read "such that" when used to define a set and "given" when used to express conditional probability.

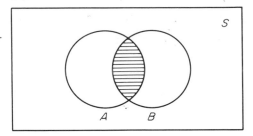

Figure 15.5

sample points in A that are in the reduced sample space represented by B. Consequently, $P(A \mid B) = P(A \cap B)/P(B)$.

By multiplying both sides of Formula (15.18) by $P(B)$, the joint probability of $A \cap B$ can be expressed as

$$P(A \cap B) = P(A \mid B)P(B) \qquad (15.19)$$

Since $P(A \cap B) = P(B \cap A)$, it also follows that

$$P(A \mid B)P(B) = P(B \mid A)P(A) \qquad (15.20)$$

Example: A large midwestern corporation has a savings plan available to its employees. Although there are no restrictions that limit participation, only 50 percent of the employees who are under 35 years of age have elected to participate in the plan. Of the total work force, 40 percent are under 35 years of age and 70 percent participate in the savings plan. Let U denote the event that an employee is "under 35 years of age" and P denote the event that an employee "participates" in the plan. Determine the following probabilities: $P(P' \mid U)$; $P(P \mid U')$; $P(P' \mid U')$; and $P(U \mid P)$.

In determining the probabilities, it will be helpful to construct a probability table. From the data we know that $P(U) = 0.40$, $P(U') = 1 - P(U) = 0.60$, $P(P) = 0.70$, $P(P') = 1 - P(P) = 0.30$, and $P(P \mid U) = 0.50$. From the formula for conditional probability, we see that

$$P(P \cap U) = P(P \mid U)P(U) = 0.50(0.40) = 0.20$$

These probabilities are entered in Table 15.6.

Table 15.6

	U	U'	Total
P	0.20	0.50	0.70
P'	0.20	0.10	0.30
Total	0.40	0.60	1.00

The table can be completed by recognizing that

$$P(P' \cap U) = P(U) - P(P \cap U) = 0.40 - 0.20 = 0.20$$
$$P(P \cap U') = P(P) - P(P \cap U) = 0.70 - 0.20 = 0.50$$
$$P(P' \cap U') = P(P') - P(P' \cap U) = 0.30 - 0.20 = 0.10$$

These probabilities are shown in Table 15.7.

Table 15.7

	U	U'	Total
P	0.20	0.50	0.70
P'	0.20	0.10	0.30
Total	0.40	0.60	1.00

The conditional probabilities, $P(P' \mid U)$, $P(P \mid U')$, $P(P' \mid U')$, and $P(U \mid P)$, are

$$P(P' \mid U) = \frac{P(P' \cap U)}{P(U)} = \frac{0.20}{0.40} = 0.50$$

$$P(P \mid U') = \frac{P(P \cap U')}{P(U')} = \frac{0.50}{0.60} = 0.834$$

$$P(P' \mid U') = \frac{P(P' \cap U')}{P(U')} = \frac{0.10}{0.60} = 0.166$$

$$P(U \mid P) = \frac{P(U \cap P)}{P(P)} = \frac{0.20}{0.70} = 0.286$$

The probability table in the preceding example can be used to illustrate an additional probability formula. Notice in Table 15.7 that

$$P(U) = P(P \cap U) + P(P' \cap U) = 0.40$$

A similar probability statement can be written for any row or column in the table. Although this relationship might seem obvious, it is worthwhile formally to state that the marginal probability of an event A can be determined from the fact that

$$P(A) = P(A \cap B) + P(A \cap B') \tag{15.21}$$

or alternatively,

$$P(A) = P(A \mid B)P(B) + P(A \mid B')P(B') \tag{15.22}$$

where A and B are events defined on the sample space S.

15.4.4 INDEPENDENCE

In illustrating the concept of the conditional probability of an event A given the occurrence of an event B, we used examples in which the conditional probability of A given B differed from the marginal probability of A, i.e.,

$$P(A \mid B) = \frac{P(A \cap B)}{P(B)} \neq P(A)$$

In problems of this type, the events A and B are termed *dependent events*.

When knowledge of the occurrence or nonoccurrence of an event B does not alter the probability of occurrence of the event A, the two events are said to be *independent*. In the case of independent events, the conditional probability of an event A given the occurrence of an event B is equal to the marginal probability of A. Thus, for independent events,

$$P(A \mid B) = P(A) \tag{15.23}$$

For the special case in which A and B are independent, Formula (15.19) for the joint probability of events,

$$P(A \cap B) = P(A \mid B)P(B) \tag{15.19}$$

can be revised to read

$$P(A \cap B) = P(A)P(B) \tag{15.24}$$

The joint probability of two independent events is thus equal to the product of the marginal probabilities of the events.

Example: An experiment consists of drawing a card from a deck of 52 playing cards. Let A represent the event the card is an ace and C the event the card is a club. Verify that the events A and C are independent.

Since there are four aces and thirteen clubs in the deck of cards, the marginal probability of an ace is

$$P(A) = \tfrac{4}{52} = \tfrac{1}{13}$$

and the marginal probability of a club is

$$P(C) = \tfrac{13}{52}$$

The joint probability of the "ace of clubs" is

$$P(A \cap C) = P(A)P(C) = \tfrac{1}{13} \cdot \tfrac{13}{52} = \tfrac{1}{52}$$

The conditional probability of an "ace" given a "club" is

$$P(A \mid C) = \frac{P(A \cap C)}{P(C)} = \frac{1/52}{13/52} = \frac{1}{13}$$

The marginal probability of an "ace" equals the conditional probability of an "ace given a club"; i.e.,

$$P(A) = P(A \mid C) = \tfrac{1}{13}$$

Consequently, the events are independent.

Example: Verify that the events given in the following probability table are independent.

	A	A'	Total
B	0.18	0.12	0.30
B'	0.42	0.28	0.70
Total	0.60	0.40	1.00

From Formula (15.24), we know that the events A and B are independent if $P(A \cap B) = P(A)P(B)$. We can verify that the events in the probability table are independent by observing that

$$P(A \cap B) = P(A)P(B) = (0.60)(0.30) = 0.18$$

$$P(A' \cap B) = P(A')P(B) = (0.40)(0.30) = 0.12$$

$$P(A \cap B') = P(A)P(B') = (0.60)(0.70) = 0.42$$

and

$$P(A' \cap B') = P(A')P(B') = (0.40)(0.70) = 0.28$$

The requirement for independence is given by Formula (15.23). For any two events A and B to be independent, it is necessary that the events not be mutually exclusive, i.e., $A \cap B \neq \phi$. Although this is a necessary condition for independence, the fact that both events share some common sample points is not a sufficient condition. To illustrate, consider the events in Tables 15.8 and 15.9. The events are not mutually exclusive in either table. However, the

Table 15.8

	A	A'	Total
B	0.12	0.28	0.40
B'	0.18	0.42	0.60
Total	0.30	0.70	1.00

A and B are not mutually exclusive but they are independent; i.e.,

$$P(A \mid B) = P(A) = 0.30$$

Table 15.9

	F	F'	Total
U	0.20	0.20	0.40
U'	0.10	0.50	0.60
Total	0.30	0.70	1.00

F and U are not mutually exclusive but they are dependent; i.e.,

$$P(F \mid U) = \frac{P(F \cap U)}{P(U)} = 0.50$$

$$\neq P(F)$$

events A and B in Table 15.8 are independent while the events F and U in Table 15.9 are dependent.

An example of events that are mutually exclusive is shown in Table 15.10. The events A and B in the table have no common sample points; i.e., $A \cap B = \phi$.

Table 15.10

	A	A'	Total
B	0	0.40	0.40
B'	0.30	0.30	0.60
Total	0.30	0.70	1.00

A and B are mutually exclusive and are therefore dependent; i.e.,

$$P(A \mid B) = 0 \neq P(A)$$

Consequently, the occurrence of event B eliminates any possibility of the occurrence of event A and $P(A \mid B) = 0$. Since the probability of A given B does not equal the marginal probability of event A, the events are dependent.

In summary, if two events are mutually exclusive, they are dependent. If the events are not mutually exclusive, they can be independent or dependent. They are independent if

$$P(A \mid B) = P(A)$$

and dependent if

$$P(A \mid B) = \frac{P(A \cap B)}{P(B)} \neq P(A)$$

15.5 Bayes' Theorem

One of the more important formulas for calculating the conditional probability of events was developed in 1763 by the Reverend Thomas Bayes. This formula, widely known as *Bayes' theorem*, forms the cornerstone of the branch of statistics referred to as *decision theory* or, alternatively, *Bayesian statistics*. Although a thorough discussion of decision theory is beyond the scope of this text, we can illustrate the formula known as Bayes' theorem and offer some preliminary remarks regarding the importance of the formula.

15.5.1 AN ILLUSTRATION

As an illustration of Bayes' theorem, consider the following problem. A manufacturing company has three factories that produce the same item. The items are shipped from the three factories and stored in a central warehouse. Production at the three factories differs both in terms of quantity and reliability. Factory A produces 35 percent of the items, Factory B produces 20 percent, and Factory C produces the remaining 45 percent. From past records we know that one percent of the items from Factory A are defective, two percent of the items from Factory B are defective, and four percent of the items from Factory C are defective. If we assume that the items in the warehouse are placed in a common inventory and are not separated by origin, determine the probability that an item selected from the central warehouse and found to be defective came from Factory A.

The probability that the defective item came from Factory A is found by using Bayes' theorem. To apply the theorem, we define the following events:

$$A = \text{item selected came from Factory A}$$

$$B = \text{item selected came from Factory B}$$

$$C = \text{item selected came from Factory C}$$

$$D = \text{item selected is defective}$$

The events A, B, and C are mutually exclusive and exhaustive. The probabilities of the events are $P(A) = 0.35$, $P(B) = 0.20$, and $P(C) = 0.45$. The conditional probabilities of a defective item given each of the factories are known from past records. These are $P(D \mid A) = 0.01$, $P(D \mid B) = 0.02$, and $P(D \mid C) = 0.04$. On the basis of this information, we are asked to determine the probability that an item selected from the pooled inventory and found to be defective came from Factory A; i.e., $P(A \mid D)$.

From Formula (15.18), we know that

$$P(A \mid D) = \frac{P(A \cap D)}{P(D)}$$

The joint probability of A and D can be expressed as

$$P(A \cap D) = P(D \cap A) = P(D \mid A)P(A)$$

and the marginal probability of D can be written as

$$P(D) = P(D \cap A) + P(D \cap B) + P(D \cap C)$$

or, alternatively,

$$P(D) = P(D \mid A)P(A) + P(D \mid B)P(B) + P(D \mid C)P(C)$$

The conditional probability of A given D can, therefore, be expressed as

$$P(A \mid D) = \frac{P(D \mid A)P(A)}{P(D \mid A)P(A) + P(D \mid B)P(B) + P(D \mid C)P(C)}$$

This expression illustrates Bayes' theorem as applied to our example problem. The conditional probability of A given D is

$$P(A \mid D) = \frac{0.01(0.35)}{0.01(0.35) + 0.02(0.20) + 0.04(0.45)} = \frac{0.0035}{0.0255} = 0.137$$

The conditional probability that the defective item came from Factory B can be found by using Bayes' theorem. The probability is

$$P(B \mid D) = \frac{P(D \mid B)P(B)}{P(D \mid A)P(A) + P(D \mid B)P(B) + P(D \mid C)P(C)} = \frac{0.0040}{0.0255} = 0.157$$

Similarly, the probability of C given D is

$$P(C \mid D) = \frac{P(D \mid C)P(C)}{P(D \mid A)P(A) + P(D \mid B)P(B) + P(D \mid C)P(C)} = \frac{0.0180}{0.0255} = 0.706$$

The marginal probabilities of events such as A, B, and C in this example are often referred to as *prior probabilities*. The term comes from the fact that they represent the probabilities of an item coming from the three factories *prior* to any additional information concerning the process, i.e., whether the item is good or defective. The conditional probabilities that were calculated by using Bayes' theorem are termed *posterior* probabilities. This term comes from the fact that the probabilities are calculated by using information that became available after examining the process, i.e., after it was known that the item was defective. Note that both the prior and the posterior probabilities partition the sample space. Thus, the sum of the prior probabilities is 1 and the sum of the posterior probabilities is 1, i.e.:

$$P(A) + P(B) + P(C) = 0.35 + 0.20 + 0.45 = 1.00$$

and

$$P(A \mid D) + P(B \mid D) + P(C \mid D) = 0.137 + 0.157 + 0.706 = 1.00$$

It is sometimes convenient in applying Bayes' theorem to use a tabular format. The calculations for the preceding problem are shown in Table 15.11. Columns (1), (2), and (3) are given in the statement of the problem.

Table 15.11   Computation of Posterior Probabilities

(1) Factory	(2) Prior Probability	(3) Conditional Probability	(4) Joint Probability	(5) Posterior Probability
A	$P(A) = 0.35$	$P(D \mid A) = 0.01$	$P(A \cap D) = 0.0035$	$P(A \mid D) = 0.137$
B	$P(B) = 0.20$	$P(D \mid B) = 0.02$	$P(B \cap D) = 0.0040$	$P(B \mid D) = 0.157$
C	$P(C) = 0.45$	$P(D \mid C) = 0.04$	$P(C \cap D) = 0.0180$	$P(C \mid D) = 0.706$
	$\overline{1.00}$		$\overline{P(D) = 0.0255}$	$\overline{1.000}$

Column (4) gives the joint probability and is given by the product of the corresponding entries in columns (2) and (3). The sum of the joint probabilities, column (4), is the marginal probability of a defective. The posterior probabilities in column (5) are found by dividing the joint probabilities in column (4) by their sum.

Example: In a recent report, a doctor stated that 0.20 percent of the population of a certain area have tuberculosis. In discussing methods of identifying the disease, he stated that a person with tuberculosis will get a positive reading from the tuberculin skin test 98 percent of the time. Unfortunately, people without tuberculosis will show positive on the test 10 percent of the time. Determine the probability that a person randomly selected from the population has tuberculosis, given that he has a positive reading on the tuberculin skin test.

To determine the probability of tuberculosis, given a positive reading, let T = "person has tuberculosis," T' = "person does not have tuberculosis," and P = "positive reading from skin test." The prior probabilities of T and T' are $P(T) = 0.002$ and $P(T') = 0.998$. The conditional probabilities of P given T and P given T' are $P(P \mid T) = 0.98$ and $P(P \mid T') = 0.10$. The posterior probability of T given P is

$$P(T \mid P) = \frac{P(P \mid T)P(T)}{P(P \mid T)P(T) + P(P \mid T')P(T')}$$

$$= \frac{0.90(0.002)}{0.98(0.002) + 0.10(0.998)} = 0.0193$$

15.5.2   FORMAL STATEMENT OF BAYES' THEOREM

We may now offer a formal statement of Bayes' theorem. Let $E_1, E_2, \ldots, E_n$ be n mutually exclusive and exhaustive events with probabilities $P(E_1), P(E_2), \ldots, P(E_n)$. Let C be an event for which the conditional probabilities of C given E_1, C given $E_2, \ldots, C$ given E_n are $P(C \mid E_1), P(C \mid E_2), \ldots, P(C \mid E_n)$.

Assume that both the marginal probabilities $P(E_i)$ and the conditional probabilities $P(C \mid E_i)$, for $i = 1, 2, \ldots, n$, are known. The conditional probabilities of an event E_j given C are found by

$$P(E_j \mid C) = \frac{P(C \cap E_j)}{P(E_j)} = \frac{P(C \mid E_j)P(E_j)}{\sum_{i=1}^{n} P(C \mid E_i)P(E_i)} \tag{15.25}$$

Formula 15.25 is Bayes' theorem.

Example: The Federated Broadcasting Company is considering introducing a new television series. On the basis of the opinion of knowledgeable individuals in the company, management of Federated estimates that the probability that the show will be a "success" is $P(S) = 0.80$. Before actually filming the show, however, management has scheduled a viewing of the pilot film before a selected audience. The members of the audience will be asked to indicate whether they believe the show is "excellent."

From past records, management knows that 90 percent of shows that proved to be successful had been rated as excellent by a preview audience. Conversely, only 30 percent of the shows that later proved to be unsuccessful had received an excellent rating from the preview audience. Determine the posterior probability of a successful show, given that the preview audience rated the show as excellent.

To apply Formula (15.25), we let

$$S = \text{event show is a success}$$

$$S' = \text{event show is not a success}$$

$$E = \text{event show is rated as excellent}$$

The prior probabilities are $P(S) = 0.80$ and $P(S') = 0.20$. The conditional probabilities are $P(E \mid S) = 0.90$ and $P(E \mid S') = 0.30$. The posterior probability of S given E is

$$P(S \mid E) = \frac{P(E \mid S)P(S)}{P(E \mid S)P(S) + P(E \mid S')P(S')}$$

$$P(S \mid E) = \frac{0.90(0.80)}{0.90(0.80) + 0.30(0.20)} = 0.923$$

15.5.3 IMPORTANCE OF BAYES' THEOREM

Bayesian statistics, or decision theory, has become one of the more important and widely employed mathematical techniques for managerial decisions. The reason for this importance comes from the modern interpretation of prior and posterior probabilities. Prior probabilities often come from subjective estimates of knowledgeable individuals. Using Bayes' theorem, we can revise these probabilities to include available sample information. The revised, or posterior probabilities, provide the basis for business decisions.

The major difference between modern decision theory and the older classical statistical techniques is that the classical techniques do not permit subjective probability estimates. Using classical techniques, one would make a decision based entirely upon sample evidence. Many decision makers, however, believe that *all evidence* should be included in the decision process. Bayesian statistics provide a method of including both the subjective (prior) probabilities and the objective (sample) probabilities.

It is important to remember that we have provided only a brief introduction to the topic of decision theory. The reader interested in pursuing the topic should refer to one of the texts suggested at the end of this chapter.

PROBLEMS

1. Specify the sample space for each of the following experiments.
   (a) In a survey of families with two children, the sex of each child in the family is recorded with the oldest child listed first.
   (b) In an experiment comparing purchasing habits, a respondent is asked whether he agrees (*a*) or disagrees (*d*) with each of three statements.
   (c) In a test of laundry soap, a housewife is asked to rank brands *x*, *y*, and *z* by order of preference.
   (d) Four parts turned out by a machine are inspected and denoted as good (*g*) or defective (*d*).

2. Three well-known economists are questioned regarding a proposed tax increase. Let *F* represent favoring the increase and *N* represent not favoring the increase. Specify the sample space of possible replies.

3. Ajax Corporation recently hired three new salesmen—Smith, Jones, and Clark. Each salesman must be assigned a sales territory. The territories are California, Oregon, and Arizona. If the salesmen are thought of as a set *A* and the territories a set *B*, the possible assignments can be viewed as a sample space with $S = A \times B$. List the elements in *S*.

4. Beginning next Monday, three commercials will be broadcast on a local radio station—one commercial on Monday, a second on Tuesday, and a third on Wednesday. If it is assumed that set *A* is the set of days and set *B* the set of different commercials, the possible assignments of commercials to days can be viewed as a sample space with $S = A \times B$. List the elements in *S*.

5. The proud father of a new baby boy anticipates his son's attending either Harvard or Yale and studying either law or medicine. Treating the possi-

ble choice of schools and course of study as an experiment, construct a tree diagram that represents the sample space of the experiment.

6. A shipment of parts must be sent from a manufacturer in London to a distributor in Kent, Ohio. The parts can be air freighted to New York, Philadelphia, or Chicago and then shipped by rail or truck to Kent. Treating the possible routes as an experiment, construct a tree diagram that represents the sample space of the experiment.

7. An individual has three coins in his pocket—a penny, a nickel, and a dime. He takes two coins from his pocket.
   (a) Construct a tree diagram that shows the possible order of selection, assuming that the first coin is replaced before the second coin is selected.
   (b) Construct a tree diagram that shows the possible order of selection, assuming that the first coin is not replaced before the second coin is selected.

8. An office manager must make work assignments to three girls—Alice, Betty, and Carol. One of the girls will take dictation, a second will type, and the third will file. Only Alice and Betty can take dictation, whereas all three girls can type and file. Treating the assignment of the jobs to the girls as an experiment, construct a tree diagram that represents the sample space of the experiment.

9. An experiment consists of counting the number of customers in a queue at a checkout counter. Give the sample space for this experiment. Does the sample space contain a finite or a countably infinite number of sample points?

10. For each of the following experiments, give the sample space and specify whether the sample space contains a finite or a countably infinite number of sample points.
    (a) The number of defects in a lot of five parts.
    (b) The number of flaws in a bolt of cloth.
    (c) The number of misspelled words in a list of ten words.

11. An experiment consists of inspecting three parts to determine if the parts are good (g) or defective (d). Give the sample points in each of the following events and state whether the event is a simple or a composite event.
    (a) None of the parts is defective.
    (b) No more than one part is defective.
    (c) At least one part is defective.
    (d) All three parts are defective.
    (e) The first part inspected is defective.

12. An avionics manufacturer produces an aircraft radio that gives both the distance to a station and the ground speed of the aircraft. Before deliver-

ing a radio, it must be inspected for accuracy with respect to both distance and ground speed. If the radio is within tolerance for both distance and ground speed, it is shipped to the customer. If not, it is retained for repairs. Defining the event D as in tolerance with respect to distance and G as in tolerance with respect to ground speed, interpret the following events:

(a) $D' \cap G'$
(b) $D \cap G'$
(c) $D' \cap G$
(d) $D \cap G$

13. The personnel manager of the Rogers Manufacturing Company has gathered the following statistics:

Department	Experience (years)			
	Less than 3 (D)	3 to 5 (E)	More than 5 (F)	Total
Production (A)	8	9	33	50
Engineering (B)	2	6	12	20
Sales (C)	5	10	15	30
Total	15	25	60	100

Determine the number of employees in each of the following categories:
(a) $A \cap D$ (b) $A \cup D$
(c) $B \cup F$ (d) $B' \cap F'$
(e) $A \cup B \cup C$ (f) $A \cap B \cap C$
(g) $A \cap (D \cup E)$ (h) F'

14. Referring to the table in Problem 13, specify if the following events are mutually exclusive, exhaustive, both mutually exclusive and exhaustive, or neither mutually exclusive nor exhaustive.
(a) A, B
(b) A, A'
(c) C, F

15. Four assignments must be made to four equally qualified employees. In how many different possible ways can the four assignments be made?

16. Six company presidents have been asked to speak at an upcoming conference. How many different speaking orders are possible?

17. A personnel officer plans on visiting ten college campuses during the coming year to interview for prospective employees. In how many ways can the ten trips be ordered?

18. In a marketing survey, a respondent is asked to compare five different brands of coffee and to state her first three choices in order of preference. How many different rankings are possible?

19. The license plates of the state of California consist of three letters followed by three digits. If each letter and each digit is used only once on a specific license plate, how many different license plates are possible?

20. With reference to Problem 19, how many different license plates are possible if the letters and numbers can be repeated?

21. From a group of fifteen products, five are to be selected for a special promotional sale. How many different combinations of five products are possible?

22. Three products will be selected from a group of ten products for market testing. How many different three product combinations are possible?

23. A group of twelve men attend a conference. If each man shakes hands with every other man, how many handshakes are exchanged?

24. An airline serves twenty major cities. If the airline provides direct service between every possible pair of cities, how many different routes does the airline fly?

25. Mark Thomas, president of Thomas Ford, Inc., has gathered the following historical data on car sales:

		Method of Payment		
		Cash (C)	Time (T)	Total
Type of car purchased (percent)	New (N)	12	30	42%
	Used (U)	20	38	58%
	Total	32%	68%	100%

(a) Would the numbers in the table be classified as objective or subjective probabilities?

(b) What is the marginal probability of an individual's paying cash for a car?

(c) Determine the joint probability of a purchase's being a new car and on time.

(d) Determine the conditional probability that a used car purchaser will pay cash.

(e) Are the events "new car purchase" and "payment by cash" statistically independent?

26. On the basis of his knowledge of Ajax Corporation and the stock market, an investment counselor developed the following probability table:

**Expected Performance of Ajax Corporation Stock
and the Dow-Jones Industrial Index**

		Stock Price—Ajax Corp.		
		Increase (A)	Decrease (A')	Total
Dow-Jones Index	Increase (B)	0.50	0.10	0.60
	Decrease (B')	0.20	0.20	0.40
	Total	0.70	0.30	1.00

(a) Would the probabilities be classed as objective or subjective?

(b) What is the marginal probability of an increase in the price of Ajax Corporation stock?

(c) What is the probability of an increase in Ajax Corporation stock, given an increase in the Dow-Jones Index?

(d) What is the probability of a decrease in Ajax Corporation stock, given an increase in the Dow-Jones Index?

(e) Are the events "increase in Ajax Corporation stock" and "increase in the Dow-Jones Index" statistically independent?

27. Suppose that 40 percent of companies in a certain industry have at least one lawyer on the board of directors and 70 percent have at least one banker. Assuming that the events "lawyer" (L) and "banker" (B) are independent, determine the proportion of companies in the industry that have both a lawyer and a banker on their board of directors.

28. An analysis of the sales of Brooks Brothers Clothiers shows that 20 percent of the purchases were made by men and that 40 percent of the purchases were over $20 in value. The records also show that 40 percent of the purchases of over $20 were made by men.

(a) What is the probability of a purchase of over $20, given that the customer is a man?

(b) What is the probability of a purchase of less than $20, given that the customer is a woman?

(c) Are the events "man" and "over $20" dependent?

29. Management of the Caroline Corporation is currently negotiating for two separate contracts, A and B. Management believes that the probability of winning contract A is 0.60. Contract A will be decided before contract B, and the winner of A will have a definite advantage in the negotiations

for B. Management believes that the probability of winning B is 0.70 if they first win A. However, if they fail to win A, the probability of winning B drops to 0.10.

(a) What is the marginal probability of winning contract B?

(b) What is the joint probability of losing both contracts?

30. The Sutton Corporation must raise funds for expansion. Management believes that there is a 70 percent chance of raising the funds through a private offering of common stock. They also believe that there is a 90 percent chance of obtaining the necessary capital through a public offering, provided that a financial report on Sutton is favorable. If, however, the report is not favorable, the probability of raising the necessary funds from a public stock offering drops to 0.50. Management estimates that the probability of receiving a favorable report is 0.60. Should a public or a private stock offering be made?

31. The Omaha Beef Packing Company processes beef products, such as sirloin, round, roasts, etc. The beef products are classified by inspectors as Prime, Choice, and Grade A. The manager of the company estimates that 20 percent of the beef is actually Prime, 50 percent is Choice, and 30 percent is Grade A. Although the inspectors are well trained, there is a certain probability that they will classify a side of beef in one category when, in reality, it should be classified in another. The manager's estimates of the probabilities that a side of beef will be correctly classified are shown in the following table.

Conditional Probabilities for Correct Classification			
If actual quality is:	Probability that it will be actually classified:		
	Prime	Choice	Grade A
Prime	0.50	0.30	0.20
Choice	0.20	0.50	0.30
Grade A	0.10	0.30	0.60

(a) If an inspector classifies a side of beef as Prime, what is the probability that it is actually Prime? Choice? Grade A?

(b) If an inspector classifies a side of beef as Choice, what is the probability that it is actually Prime? Choice? Grade A?

32. Airport Rent-A-Car, Inc. is considering placing a fleet of rental cars at a major airport. Because of competition, the management is uncertain whether the demand for the cars will be high (H), moderate (M), or low (L). Their prior probability assessment of each of these possible levels

of demand is $P(H) = 0.60$, $P(M) = 0.20$, and $P(L) = 0.20$. A survey of potential customers indicates that the demand will be moderate. However, from previous experience with such surveys, management has found that the survey will indicate moderate demand 30 percent of the time when, in reality, demand turns out to be high. Similarly, the survey will indicate moderate demand 40 percent of the time when demand is low and 70 percent of the time when demand is really moderate. On the basis of this information, determine the posterior probabilities that demand is high, moderate, and low.

SUGGESTED REFERENCES

CHOU, YA-LUN, *Statistical Analysis* (New York, N.Y.: Holt, Rinehart and Winston, Inc., 1969).

DEGROOT, MORRIS H., *Optimal Statistical Decisions* (New York, N.Y.: McGraw-Hill Book Company, Inc., 1970).

EWART, P., J. FORD, AND C. LIN, *Probability for Statistical Decision Making* (Englewood Cliffs, N.J.: Prentice-Hall, Inc., 1974).

HAMBURG, MORRIS, *Statistical Analysis for Decision Making* (New York, N.Y.: Harcourt, Brace & World, Inc., 1970).

HAYS, WILLIAM L. AND ROBERT L. WINKLER, *Statistics* (New York, N.Y.: Holt, Rinehart and Winston, Inc., 1971).

LAPIN, LAWRENCE L., *Statistics for Modern Business Decisions* (New York, N.Y.: Harcourt Brace Jovanovich, Inc., 1973).

PETERS, W. S. AND G. W. SUMMERS, *Statistical Analysis for Business Decisions* (Englewood Cliffs, N.J.: Prentice-Hall, Inc., 1968).

RAIFFA, HOWARD, *Decision Analysis* (Reading, Mass.: Addison-Wesley Publishing Company, Inc., 1968).

SCHLAIFER, ROBERT, *Analysis of Decisions Under Uncertainty* (New York, N.Y.: McGraw-Hill Book Company, Inc., 1969).

SPURR, W. A. AND C. P. BONINI, *Statistical Analysis for Business Decisions* (Homewood, Ill.: Richard D. Irwin, Inc., 1973).

Chapter 16

Probability
Functions

In Chapter 15 we introduced the basic concepts of probability. In this chapter we extend these concepts to include selected probability functions or, alternatively, probability distributions.† The reader will remember from Chapter 2 that a function is a mathematical relationship in which the values of a single dependent variable are determined from the values of one or more independent variables. A probability function, like the functions discussed in Chapter 2, describes the relationship between the dependent and the independent variable. In a probability function the dependent variable is the probability of the occurrence of an event and the independent variable is a numerical value that is assigned to the event. A probability function is thus used to determine the probability of an event that is defined in terms of a numerical value.

As a simple illustration of a probability function, consider the problem of determining the probability of the weights of infants at birth. In this example a probability function could be used to calculate the probability that a randomly selected infant would weigh, say, between six and seven pounds at birth. The dependent variable would be the probability, and the

---

† The terms "probability function" and "probability distribution" can be used interchangeably.

independent variable would be the weight of the infant. The relationship between weight and probability would be given by the probability function.

Certain probability functions are widely used as an aid in making managerial decisions. One of the most commonly used functions is the *normal probability function*. The reason for the widespread use of the normal probability function is that it accurately describes the probabilities associated with the occurrence of a large number of probabilistic events. The normal probability function, along with selected additional functions, is discussed later in this chapter.

16.1 Random Variables

We have stated that a probability function describes the relationship between the probability of an event and a numerical value that represents the event. In our introductory remarks the probability of the event was referred to as the dependent variable. Although the properties of functions that were discussed in Chapter 2 apply to probability functions, the terminology commonly used in probability functions differs from that introduced in Chapter 2. One of the major differences is in the use of the term "random variable" instead of the term "independent variable."

The term *random variable* refers to a rule for assigning numerical values to the outcome of a random process or an experiment. As an example, consider the experiment of tossing two coins and recording the outcome of each toss as a head or a tail. The sample space for this experiment is

$$S = \{(HH), (HT), (TH), (TT)\}$$

Suppose that we are interested in the number of heads resulting from the toss of the coins. If this were the case, we could define X as equalling the number of heads resulting from the random process. In this experiment X would be the random variable. The rule for determining the value of the random variable would be to assign to X the number of heads resulting from the toss of the coins. The possible values of the random variable would thus be $x = 0$, $x = 1$, and $x = 2$. Notice that the random variable X is designated by an uppercase letter and values of the random variable are designated by lowercase letters.

It is important to note that a random variable refers to a rule that links the elements in the sample space with the set of real numbers. In the preceding example the sample space consisted of the set $S = \{(HH), (HT), (TH), (TT)\}$. The rule involved assigning as values of the random variable the number of heads resulting from the toss of the coins. The real numbers were the values of the random variable, namely, $x = 0$, $x = 1$, and $x = 2$.

Example. An experiment consists of inspecting three randomly selected parts. Each part is classified as good, *g*, or defective, *d*. The sample space for this experiment is

$$S = \{(ggg), (ggd), (gdg), (dgg), (gdd), (dgd), (ddg), (ddd)\}$$

Suppose we are interested in the number of defective parts in our sample. The sample points can be converted into numerical values by defining the random variable X as equalling the number of defectives. The values of the random variable are thus $x = 0$, $x = 1$, $x = 2$, and $x = 3$.

Example. An experiment consists of rolling a pair of dice. The sample space for this experiment is

$$\begin{aligned} S = \{&(1, 1), (1, 2), (1, 3), (1, 4), (1, 5), (1, 6), (2, 1), (2, 2), (2, 3), (2, 4), (2, 5),\\ &(2, 6), (3, 1), (3, 2), (3, 3), (3, 4), (3, 5), (3, 6), (4, 1), (4, 2), (4, 3), (4, 4),\\ &(4, 5), (4, 6), (5, 1), (5, 2), (5, 3), (5, 4), (5, 5), (5, 6), (6, 1), (6, 2), (6, 3),\\ &(6, 4), (6, 5), (6, 6)\} \end{aligned}$$

Suppose that we are interested in the sum of the numbers on the dice resulting from the roll. The random variable X would then be defined as the sum of the numbers on the pair of dice. The possible values of the random variable would be $x = 2, 3, \ldots, 12$.

16.1.1 TRADITIONAL DEFINITION OF A RANDOM VARIABLE

On the basis of the example in the preceding section, we can offer a formal definition of the term "random variable." A random variable is traditionally defined as a function (or rule) whose domain is the sample space of an experiment and whose range is the set of real numbers. The rule of the function links (or maps) the elements in the domain (i.e., sample space) with the set of real numbers. In other words, the random variable is a rule that assigns numerical values to the elements of the sample space. The values that are assigned to the elements of the sample space are the values of the random variable.

The term "random" comes from the fact that the value of the variable is determined from the outcome of an experiment or a random process. Thus, the particular numerical value of the random variable is not known until the outcome of the experiment has been observed. The possible numerical values are, however, known before the experiment, and the probability of the occurrence of alternative values of the random variable can be calculated from the probability function.

16.1.2 DISCRETE AND CONTINUOUS
RANDOM VARIABLES

Random variables are commonly classified as being either *discrete* or *continuous*. A random variable is termed discrete if the set of possible values of the random variable contains a finite or a countably infinite number of numerical values. For instance, in the experiment involving tossing two coins and recording the number of heads, the set of possible values of the random variable was

$$\{x \mid x = 0, 1, 2\}$$

This set contains a finite number of numerical values. Consequently, the random variable is discrete.

As another example of a discrete random variable, consider an experiment that involves counting the number of defects in a bolt of cloth. If the random variable X represents the number of defects, the set of possible values of X is

$$\{x \mid x = 0, 1, 2, 3, \ldots\}$$

Since this set has a countably infinite number of numerical values, the random variable is discrete.

A random variable is termed continuous if the random variable can have any value along a continuum between specific limits. To illustrate, consider an experiment that involves measuring the time between arrivals of customers at a service facility. If the random variable X represents time between arrivals of customers, the possible values of the random variable are described by the set

$$\{x \mid 0 \leq x \leq \infty\}$$

In this example the random variable can have any one of the infinite number of values along the continuum from $x = 0$ to $x = \infty$.

Suppose that we modify the above example by assuming that the maximum possible time between arrivals of customers is 60 minutes. Based on this assumption, the possible values of the random variable are described by the set

$$\{x \mid 0 \leq x \leq 60\}$$

Since the random variable can have any of the infinite number of possible values along the continuum between $x = 0$ and $x = 60$, the random variable X is again continuous.

In the preceding example we assume that the random variable can have any value in the continuum between 0 and 60. Quite obviously, however, the values of the random variable will be limited by the accuracy with which we measure the outcome of the experiment. In the example, for instance, we may measure time between arrivals to the nearest one-tenth of a second. Although as a practical matter, a measurement to the nearest one-tenth second may be quite satisfactory, it would be theoretically possible to measure the time between arrivals to any desired precision. For this reason, random variables that can theoretically have any value in a continuum are classified as continuous rather than discrete.

To distinguish between the random variable and the value of the random variable, we customarily denote the random variable by a capital letter and possible values of the random variable by a lowercase letter. Thus, the random variable X represented time between arrivals in the preceding example and the possible values of the random variable were $0 \leq x \leq 60$.

Example. In testing automobile head lamps, an inspector selects a lot of five lamps and determines (1) if the lamps function properly and (2) the length of time the lamps will burn before failure. If we define the random variable D as the number of defectives in the sample, D is a discrete random variable. The possible values of D are $\{d \mid d = 0, 1, 2, 3, 4, 5\}$. If we define L as the length of time a randomly selected lamp will burn before failure, L is a continuous random variable. The possible values of L are $\{l \mid 0 \leq l \leq \infty\}$.

16.2 Probability Functions

To introduce the basic concept of a probability function, consider again the experiment of tossing two coins and recording the outcome of each toss as a head or a tail. The sample space for this experiment was

$$S = \{(HH), (HT), (TH), (TT)\}$$

Assume that we define the random variable X as representing the number of heads resulting from the toss of the coins. The possible values of X were $x = 0, 1, 2$. Since each outcome in the sample space is equally likely, the probability of no heads is $P(X = 0) = \frac{1}{4}$, the probability of one head is $P(X = 1) = \frac{1}{2}$, and the probability of two heads is $P(X = 2) = \frac{1}{4}$. The functional relationship between the value of the random variable and the probability of occurrence of that value can be described by the set

$$\{(0, \tfrac{1}{4}), (1, \tfrac{1}{2}), (2, \tfrac{1}{4})\}$$

This set is an example of a probability function.

The relation between the number of heads and the probability of occurrence can also be given in tabular form. The probability function is shown in Table 16.1. The table shows each possible value of the random variable together with the probability of the occurrence of the value.

Table 16.1 Probability Table for Number of Heads

$X = x$	$P(X = x)$
0	$\frac{1}{4}$
1	$\frac{1}{2}$
2	$\frac{1}{4}$

Example. Consider the experiment of tossing a pair of dice. Let the random variable X be defined as the sum of the spots on the pair of dice. Construct a probability table that describes the probability of each possible value of the random variable.

The sample space for this experiment was given on page 534. The possible values of the random variable, the number of sample points associated with each possible value of the random variable, and the probability of occurrence of each value of the random variable is shown in the table below.

Probability Table for Roll of Dice

$X = x$	No. of Sample Points	$P(X = x)$
2	1	$\frac{1}{36}$
3	2	$\frac{2}{36}$
4	3	$\frac{3}{36}$
5	4	$\frac{4}{36}$
6	5	$\frac{5}{36}$
7	6	$\frac{6}{36}$
8	5	$\frac{5}{36}$
9	4	$\frac{4}{36}$
10	3	$\frac{3}{36}$
11	2	$\frac{2}{36}$
12	1	$\frac{1}{36}$
	36	$\frac{36}{36} = 1.0$

16.2.1 DISCRETE PROBABILITY FUNCTIONS

The preceding examples illustrate that a probability function describes the relationship between the value of a random variable and the probability of the occurrence of the different possible values of the random variable. More formally, we can state that if X is a discrete random variable with values $x_1, x_2, \ldots, x_n$, then the probability function $p(x)$ is given by

$$p(x) = P(X = x_i) \qquad \text{for } i = 1, 2, \ldots, n \qquad (16.1)$$

A probability function for a discrete random variable is called a *discrete probability function* or, alternatively, a *probability mass function*. From the axioms of probability discussed in Chapter 15, we recognize that a discrete probability function has the following properties:

1. $p(x) \geq 0$ for all values of X; that is, $p(x)$ cannot be negative.
2. $\sum_{\text{all } x} p(x) = 1$; that is, the sum of the probabilities of each separate value of the random variable must be 1.

To illustrate a discrete probability function, consider an example involving testing a certain electronic part. From past experience, it has been determined that the probability of a randomly selected part's being defective is 0.3. Five parts are selected for testing. The probability mass function that describes the probabilities of various numbers of defectives has been determined and is

$$p(x) = \binom{5}{x}(0.3)^x(0.7)^{5-x} \qquad \text{for } x = 0, 1, 2, \ldots, 5$$

The probability of no defectives in the lot of five parts is calculated by evaluating the probability mass function for $x = 0$. The probability is

$$P(X = 0) = p(0) = \frac{5!}{0!5!}(0.3)^0(0.7)^5 = 0.1681$$

Similarly, the probability of exactly one defective is

$$P(X = 1) = p(1) = \frac{5!}{1!4!}(0.3)^1(0.7)^4 = 0.3601$$

The probabilities of two, three, four, and five defectives are found in the same manner. The probabilities are shown in Table 16.2.

Table 16.2

Number of defectives x	0	1	2	3	4	5
Probability $p(x)$	0.1681	0.3602	0.3087	0.1323	0.0283	0.0024

Notice in Table 16.2 that $p(x) \geq 0$ for all possible values of the random variable X. Furthermore, the sum of the probabilities of the random variables is equal to

$$\sum_{x=0}^{x=5} p(x) = 0.1681 + 0.3602 + 0.3087 + 0.1323 + 0.0283 + 0.0024 = 1.0000$$

These two properties illustrate the requirements that the probability of a random variable be greater than or equal to zero and that the sum of the probabilities of all possible values of the random variable equal one.

Example. Western Properties, Inc., has large holdings of land in northern California. This land has been subdivided and is being sold in small parcels to individuals for vacation and retirement homes. The sales manager of Western Properties has found that 40 percent of the prospective customers who actually visit the property purchase a parcel of land. A probability mass function that describes the probability of the number of purchases, X, in a group of n customers has been determined and is

$$p(x) = \binom{n}{x}(0.40)^x(0.60)^{n-x}$$

Suppose that four customers visit the property on a certain weekend. The probability of purchases by zero, one, two, three, or all four of the customers can be calculated from the probability mass function. The probabilities are

$$p(0) = \frac{4!}{0!4!}(0.40)^0(0.60)^4 = 0.1296$$

$$p(1) = \frac{4!}{1!3!}(0.40)^1(0.60)^3 = 0.3456$$

$$p(2) = \frac{4!}{2!2!}(0.40)^2(0.60)^2 = 0.3456$$

$$p(3) = \frac{4!}{3!1!}(0.40)^3(0.60)^1 = 0.1536$$

$$p(4) = \frac{4!}{4!0!}(0.40)^4(0.60)^0 = 0.0256$$

$$\sum p(x) = \overline{1.0000}$$

The reader will notice that the probability mass function used in this example is quite similar to the one used in the example on the preceding page. This function, termed the *binomial distribution*, is widely applied and will be discussed later in this chapter. Based on the function, we see that the probability of land purchases by zero, one, two, three, or all four customers are, respectively, 0.1296, 0.3456, 0.3456, 0.1536, and 0.0256. Again, the probabilities are nonnegatives for all values of the random variable, and the sum of the probabilities is one,

16.2.2 CONTINUOUS PROBABILITY FUNCTIONS

One of the important uses of integral calculus is in the evaluation of probabilities of continuous random variables. Continuous random variables, the reader will remember, can have any value along a continuum between specific limits. If, for example, an experiment involved weighing individuals, then the weights of the individuals would be a continuous random variable. In this example, the probability of a randomly selected individual's weight being between a lower and an upper limit is given by evaluating a definite integral. The function that is integrated is called a *continuous probability function*, or, alternatively, a *probability density function*.

If we let $f(x)$ represent the probability density function, a the lower limit of integration, and b the upper limit of integration, then the probability that the experiment results in the value of the random variable X being between a and b is given by

$$P(a \leq X \leq b) = \int_a^b f(x)\,dx \qquad\qquad (16.2)$$

From the axioms of probability discussed in Chapter 15, it also follows that:

1. $f(x) \geq 0$ for all values of x; that is, the probability density function cannot be negative.

2. $\displaystyle\int_{\text{all } x} f(x)\,dx = 1$; that is, the definite integral of the probability density function over all values of x must be 1.

We can illustrate these properties by referring to Fig. 16.1. This figure illustrates a hypothetical probability density function for the weights of

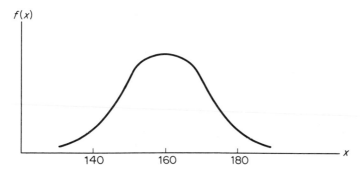

$f(x)$

140   160   180 x

Figure 16.1

adult men. The random variable in the figure is weight, and the density function is denoted by $f(x)$. The density function describes the likelihood of occurrence of weights of adult men. Since a man's weight must fall somewhere on the diagram, the integral of the density function over all possible weights must equal 1. Similarly, the density function must be greater than or equal to zero for all weights, i.e., $f(x) \geq 0$. Finally, the probability of a randomly selected individual's weight falling in some interval is given by the integral of the density function over that interval. For instance, the probability of an individual's weight being between 140 and 200 pounds would be given by

$$P(140 \leq X \leq 200) = \int_{140}^{200} f(x)\, dx$$

Example. On a certain assembly line, the time required in minutes to complete an assembly is described by the probability density function

$$f(t) = \tfrac{3}{64}t^2 \qquad \text{for } 0 \leq t \leq 4$$

Verify that $f(t)$ has the properties of a probability density function and calculate the probability that the time required to complete an assembly will be between 2 and 4 minutes.

The requirements for a probability density function are that $f(x) \geq 0$ for all x and that $\int_{\text{all } x} f(x)\, dx = 1$. In the example, $f(t) \geq 0$ for $0 \leq t \leq 4$. Furthermore,

$$\int_0^4 \tfrac{3}{64}t^2\, dt = \tfrac{1}{64}t^3 \Big|_0^4 = 1$$

Thus, $f(t)$ meets the requirements for a probability density function.

The probability that the time required to complete an assembly will be between 2 and 4 minutes is

$$P(2 \leq T \leq 4) = \int_2^4 \tfrac{3}{64}t^2\, dt = \tfrac{1}{64}t^3 \Big|_2^4 = 0.825$$

Example. The time in minutes between arrivals of customers at a certain service facility can be described by the probability density function

$$f(t) = 4e^{-4t} \qquad \text{for } t \geq 0$$

Show that $f(t)$ has the properties of a probability density function and determine the probability that a customer will arrive at the facility within 0.5 minutes.

We note that $f(t) \geq 0$ for $t \geq 0$ and that

$$\int_0^\infty 4e^{-4t}\, dt = -e^{-4t} \Big|_0^\infty = -e^{-\infty} - (-e^0)$$

Since $e^{-\infty}$ is equal to 0 and e^0 is equal to 1,

$$-e^{-\infty} - (-e^0) = 1$$

and the function has the properties required for a probability density function.

The probability that a customer will arrive within 0.5 minutes is

$$P(0 \leq T \leq 0.5) = \int_0^{0.5} 4e^{-4t}\, dt = -e^{-4t} \Big|_0^{0.5}$$

$$= -e^{-2} - (-e^0) = -0.135 + 1.000 = 0.865$$

16.3   Measures of Central Tendency and Variability

There are several important measures of central tendency and variability of a random variable that are widely used in managerial decisions. Among the important measures of central tendency are the mean, or expected value, the median, and the mode. The more important measures of variability of a random variable are the variance and the standard deviation. Our objective in this section is to introduce these measures of central tendency and variability and to illustrate their importance as parameters that describe the behavior of a random variable.

16.3.1   MEAN OR EXPECTED VALUE

Perhaps the most widely used measure of central tendency is the *mean*, or *expected value*, of a random variable. The expected value of a random variable can be interpreted as the average value of the random variable. Thus, if the random variable X represents the number of defective parts in a lot of five parts, then the expected value of the random variable is the average number of defectives in the lot. Similarly, if the random variable represents the time between arrivals of customers at a service facility, then the expected value of the random variable is the average time between arrivals.

The mean, or expected value, of a random variable is denoted by the Greek letter μ (mu) or, alternatively, by $E(X)$. The expected value of a discrete random variable is defined as

$$\mu = E(X) = \sum_{\text{all } x} xp(x) \tag{16.3}$$

where $p(x)$ is the probability mass function and x is the random variable. The expected value of a continuous random variable is similarly defined as

$$\mu = E(X) = \int_{\text{all } x} xf(x)\, dx \tag{16.4}$$

where $f(x)$ is the probability density function and x is again the random variable.

To illustrate the procedure for determining the expected value of a discrete random variable, consider the example on p. 578 of testing electronic parts. From a group of parts, five were selected for testing. The probabilities of zero, one, two, three, four, or five defectives in the lot of five were given in Table 16.2 and are repeated in Table 16.3.

Table 16.3

Number of defectives x	0	1	2	3	4	5
Probability $p(x)$	0.1681	0.3602	0.3087	0.1323	0.0283	0.0024

The expected value of the random variable is calculated from Eq. (16.3) and is

$$E(X) = \sum_{\text{all } x} xp(x) = 0(0.1681) + 1(0.3602) + 2(0.3087) + 3(0.1323)$$
$$+ 4(0.0283) + 5(0.0024) = 1.500$$

Thus, if this experiment were repeated a large number of times, we would expect an average of 1.500 defectives in the lots of five items. This does not mean that we would expect 1.500 defectives in any one lot. Quite obviously, this is impossible. Instead, after numerous repetitions of this experiment the average number of defectives would be 1.500 per lot.

Example. Jerry Clark, the western representative of Mountain Rose Wine, calls on large markets and liquor stores in an attempt to obtain orders for Mountain Rose. He has found that the probability of obtaining orders for Mountain Rose can be described by the following discrete probability function.

Number of cases x	0	1	2	3	4	5
Probability $p(x)$	0.30	0.20	0.20	0.10	0.10	0.10

Determine the expected value of the random variable.

The random variable X represents the number of cases ordered. The expected value of X is

$$E(X) = \sum_{\text{all } x} xp(x) = 0(0.30) + 1(0.20) + 2(0.20) + 3(0.10)$$
$$+ 4(0.10) + 5(0.10) = 1.8$$

Thus, the long-run average number of cases ordered per store is 1.8.

The expected value of a continuous random variable is found by using Eq. (16.4). To illustrate, consider the example concerning the time required to complete a certain assembly. This example was introduced on p. 581 and the density function was

$$f(t) = \tfrac{3}{64}t^2 \qquad \text{for } 0 \leq t \leq 4$$

The expected value of the random variable is given by

$$E(T) = \int_{\text{all } t} tf(t)\, dt$$

and is

$$E(T) = \int_0^4 \tfrac{3}{64} t^3\, dt = \tfrac{3}{256} t^4 \Big|_0^4 = 3$$

Thus, if the assembly were completed a large number of times, the average time required to complete the assembly would be 3 minutes.

Example. The useful life of a certain critical component can be described by the probability density function

$$f(x) = 0.50x \qquad \text{for } 0 \le x \le 2$$

where X is a continuous random variable that represents the life of the component in hours. Determine the expected value of the random variable. The expected value of X is

$$E(X) = \int_0^2 0.50x^2\, dx = \frac{0.50x^3}{3} \Big|_0^2 = 1.33$$

16.3.2 MEDIAN

A second important measure of central tendency in a probability distribution is the *median* value of the random variable. The median is that value of the random variable x_m such that the probability is at least 0.50 that X is greater than or equal to x_m and is at least 0.50 that X is less than or equal to x_m. For a discrete random variable, the median is that value of X such that

$$P(X \le x_m) \ge 0.50 \quad \text{and} \quad P(X \ge x_m) \ge 0.50 \qquad (16.5)$$

To illustrate determining the median of a discrete probability distribution, consider the distribution shown in Table 16.4. This distribution was first

Table 16.4

Number of defectives x	0	1	2	3	4	5
Probability $p(x)$	0.1681	0.3602	0.3087	0.1323	0.0283	0.0024

given in Table 16.2 and was repeated in Table 16.3. Notice from the table that

$$P(X \le 1) = 0.5283 \quad \text{and} \quad P(X \ge 1) = 0.8319$$

Since both requirements in Eq. (16.5) are satisfied, the median of the discrete probability distribution is $x_m = 1$.

The median of a continuous probability distribution must also satisfy the requirements of Eq. (16.5). Since a continuous random variable can take on any value in the continuum between the lower and upper limits of the random variable, the median value of the random variable is merely that value of the random variable that divides the area under the distribution in half. Expressed mathematically, the median value of the random variable is that value x_m such that

$$\int_a^{x_m} f(x)\, dx = 0.50 \qquad\qquad (16.6)$$

where $a \leq x \leq b$ is the domain of the random variable. From Eq. (16.6) we see that the probability is exactly 0.50 that $x \leq x_m$. On the basis of the properties of a continuous probability function, it also follows that

$$\int_{x_m}^b f(x)\, dx = 0.50 \qquad\qquad (16.7)$$

To illustrate determining the median of a continuous probability distribution, consider the probability density function that describes the time required to complete a certain assembly. The density function was

$$f(t) = \tfrac{3}{64}t^2 \qquad \text{for } 0 \leq t \leq 4$$

The median value of the random variable is found by evaluating the expression

$$\int_0^{t_m} \tfrac{3}{64}t^2\, dt = 0.50$$

for t_m. The median is $t_m = 3.178$. The calculations are

$$\int_0^{t_m} \tfrac{3}{64}t^2\, dt = 0.50$$

$$\left. \tfrac{1}{64}t^3 \right|_0^{t_m} = 0.50$$

$$t_m^3 = 32$$

$$t_m = 3.178$$

The median time to complete the assembly is 3.178 minutes. This means that 50 percent of the assemblies will be completed in 3.178 minutes or less, and 50 percent of the assemblies will require 3.178 minutes or more. In contrast, the mean time to complete the assembly was 3 minutes.

Example. A probability density function is described by

$$f(x) = \frac{x}{50} \qquad \text{for } 0 \leq x \leq 10$$

Determine the median value of the random variable X.

$$\int_0^{x_m} \frac{x}{50}\,dx = 0.50$$

$$\frac{x^2}{100}\Big|_0^{x_m} = 0.50$$

$$x_m^2 = 50$$

$$x_m = 7.07$$

16.3.3   MODE

A third measure of central tendency in a probability distribution is the mode. The *mode* is roughly defined as the value of the random variable most likely to occur. For a discrete probability distribution, the mode is that value of the random variable with the highest probability of occurrence. To illustrate, consider the probability distribution given in Table 16.5. The value of the

Table 16.5

Number of defectives x	0	1	2	3	4	5
Probability $p(x)$	0.1681	0.3602	0.3087	0.1323	0.0283	0.0024

random variable that has the greatest probability of occurrence is $x = 1$. Consequently, the modal value of the random variable in Table 16.5 is $x = 1$.

In certain cases, it is possible to have more than one modal value of the random variable. To illustrate, consider the plot of the discrete probability distribution in Fig. 16.2. In this distribution the values $x = 2$ and $x = 5$ are

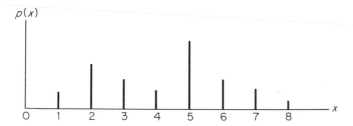

Figure 16.2

both relatively likely to occur. In cases such as this, both values of the random variable are considered to be modes, and the distribution is said to be *bimodal*.

To take into account the possibility of more than one modal value in a probability distribution, an alternative and more precise definition of the mode must be given. More formally, a mode of a probability distribution is defined as a value of the random variable at which the probability function reaches a local maximum. For a discrete probability function, a modal value is any value of the random variable that has a greater probability of occurrence than adjacent values of the random variable. Thus in Fig. 16.2, the probability that $x = 2$ is greater than that of $x = 1$ or $x = 3$ and $x = 2$ is, therefore, a mode. Similarly, $x = 5$ has a greater probability of occurring than $x = 4$ or $x = 6$, and $x = 5$ is also a mode.

The modal values of a continuous probability function are found by determining the values of the random variable for which the density function is a local maximum. In Chapter 10 we stated that a continuous function $f(x)$ reaches a local maximum at $x^* = a$ when

$$f'(x^* = a) = 0$$

and

$$f''(x = a^*) < 0$$

To illustrate the mode for a continuous random variable, consider the probability density function

$$f(x) = 4x - 3x^2 \qquad \text{for } 0 \leq x \leq 1$$

The critical point is found by equating the first derivative of the density function with zero. Thus,

$$f'(x) = 4 - 6x$$
$$4 - 6x = 0$$

and

$$x^* = \tfrac{2}{3}$$

Since the second derivative, $f''(x) = -6$, is negative, we conclude that $x^* = \tfrac{2}{3}$ is a local maximum and is, therefore, the mode.

Example. A continuous random variable has the probability density function

$$f(x) = 6x - 6x^2 \qquad \text{for } 0 \leq x \leq 1$$

Determine the modal value of X.

Equating the derivative of $f(x)$ with zero gives

$$6 - 12x = 0$$

Therefore,

$$x^* = \tfrac{1}{2}$$

Since $f''(x) = -12$, the critical point $x^* = \tfrac{1}{2}$ is a local maximum and is the mode.

16.3.4 VARIANCE AND STANDARD DEVIATION

The variance of a random variable provides a measure of the magnitude of the variation of the values of the random variable about the expected value or mean of the random variable. Figure 16.3 illustrates the concept of variance

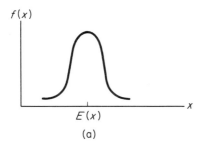 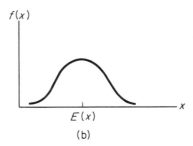

Figure 16.3

for a normally distributed random variable. The distributions in Figs. 16.3(a) and 16.3(b) are normal, and both have the same expected value. The values of the random variable in Fig. 16.3(a) are, however, more closely distributed about the expected value than those in Fig. 16.3(b). Since the variance provides a measure of the magnitude of the variation of the values of the random variable about the expected value, the variance of the distribution in Fig. 16.3(a) is less than that of the distribution in Fig. 16.3(b).

The formula for the variance of a discrete random variable is

$$\text{Var}\,(X) = \sum_{\text{all } x} (x - E(X))^2 p(x) \tag{16.8}$$

and that for a continuous random variable is

$$\text{Var}\,(X) = \int_{\text{all } x} (x - E(X))^2 f(x)\, dx \tag{16.9}$$

From the formulas we note that the variance provides a measure of the squared deviation of the values of the random variable from the expected value of the random variable. Quite obviously, the variance of a random variable that is widely distributed about the expected value will be larger than that of a random variable that is closely distributed about the expected value.

The formula for the variance of a random variable can also be written as

$$\text{Var}\,(X) = E(X^2) - E(X)^2 \tag{16.10}$$

The formula shows that the variance can be calculated in two steps. The expected value of X^2, i.e., $E(X^2)$, is first determined. The square of the expected value, $E(X)^2$, is then subtracted from $E(X^2)$. For a discrete distribution, the expected value of X^2 is given by

$$E(X^2) = \sum_{\text{all } x} x^2 p(x) \qquad (16.11)$$

and for a continuous distribution the expected value of X^2 is given by

$$E(X^2) = \int_{\text{all } x} x^2 f(x)\, dx \qquad (16.12)$$

The calculations for the variance of a discrete random variable using Eq. (16.8) are shown in Table 16.6. The expected value of the random variable

Table 16.6

x	$p(x)$	$xp(x)$	$x - E(x)$	$(x - E(x))^2$	$(x - E(x))^2 p(x)$
0	0.1	0	-2.3	5.29	0.529
1	0.2	0.2	-1.3	1.69	0.338
2	0.3	0.6	-0.3	0.09	0.027
3	0.2	0.6	0.7	0.49	0.098
4	0.1	0.4	1.7	2.89	0.289
5	0.1	0.5	2.7	7.29	0.729
		$E(X) = 2.3$			Var $(X) = 2.010$

is calculated in the third column of the table and is $E(X) = 2.3$. The deviation of the random variable from the expected value is given in the fourth column, and the squared deviation is given in the fifth column. The product of the squared deviation and the probability mass function is given in the sixth column. The summation of the product of the squared deviation and the probability mass function gives the variance of the random variable. In this example the variance is Var $(X) = 2.01$.

In order to compare the two formulas for the variance, the variance of the discrete probability function given in Table 16.6 is calculated by using Eq (16.10) in Table 16.7. The expected value of X is calculated in the third column of the table, and the expected value of X^2 is calculated in the fourth column. Using Eq. (16.10), we find that the variance is again Var $(X) = 2.01$.

To illustrate the procedure for determining the variance for a continuous random variable, consider a random variable that is distributed according to the probability density function

$$f(x) = 4x - 3x^2 \qquad \text{for } 0 \le x \le 1$$

The expected value of the random variable is

Table 16.7

x	$p(x)$	$xp(x)$	$x^2p(x)$
0	0.1	0	0
1	0.2	0.2	0.2
2	0.3	0.6	1.2
3	0.2	0.6	1.8
4	0.1	0.4	1.6
5	0.1	0.5	2.5
		$E(X) = \overline{2.3}$	$E(X^2) = \overline{7.3}$

$$\text{Var }(X) = E(X^2) - E(X)^2 = 7.3 - (2.3)^2 = 2.01$$

$$E(X) = \int_0^1 xf(x)\,dx = \int_0^1 (4x^2 - 3x^3)\,dx$$

$$= \frac{4x^3}{3} - \frac{3x^4}{4}\bigg|_0^1 = 0.583$$

The expected value of X^2 is

$$E(X^2) = \int_0^1 x^2 f(x)\,dx = \int_0^1 (4x^3 - 3x^4)\,dx$$

$$= x^4 - \frac{3x^5}{5}\bigg|_0^1 = 0.40$$

Using Eq. (16.10), we find that the variance is

$$\text{Var }(X) = E(X^2) - E(X)^2$$
$$= 0.40 - (0.583)^2$$
$$= 0.06$$

Although the calculations would be more time-consuming, the same result would be obtained by using Eq. (16.9).

Example. The time required to complete a certain assembly was described by the probability density function

$$f(t) = \tfrac{3}{64}t^2 \qquad \text{for } 0 \le t \le 4$$

Determine the variance of the random variable T.

The expected value of T was found earlier and was $E(T) = 3.0$. The expected value of T^2 is

$$E(T^2) = \int_0^4 t^2 f(t)\,dt = \int_0^4 \tfrac{3}{64}t^4\,dt$$

$$= \tfrac{3}{320}t^5\bigg|_0^4 = 9.6$$

The variance of the random variable is

$$\text{Var}\,(T) = E(T^2) - E(T)^2$$
$$= 9.6 - (3.0)^2$$
$$= 0.6$$

The *standard deviation* is a second important measure of the variability of a random variable. The standard deviation, usually denoted by the Greek letter σ (sigma), is merely the square root of the variance. That is,

$$\text{standard deviation} = \sigma_x = \sqrt{\text{Var}\,(X)} \qquad (16.13)$$

Thus, if the variance of a random variable is Var $(X) = 4$, then the standard deviation is $\sigma_x = 2$. Similarly, if the variance of a random variable is Var $(X) = 0.49$, then the standard deviation is $\sigma_x = 0.7$.

Example. The variance of the random variable T is Var $(T) = 0.6$. Determine the standard deviation.

$$\sigma_t = \sqrt{\text{Var}\,(t)} = \sqrt{0.6}$$
$$= 0.77$$

Since the standard deviation is simply the square root of the variance, the reader may question the need for both the standard deviation and the variance as measures of the variability of a random variable. The variance has certain properties that are important in statistical analysis. These properties are beyond the scope of this text but are included in any of the referenced texts in statistics. Because the variance is calculated by squaring the deviations of the random variable from the expected value, it is difficult to interpret the meaning of the variance. For this reason, the standard deviation is important as a descriptive measure of the variability of a random variable.

16.4 Commonly Used Discrete Probability Functions

This section introduces three of the more commonly used discrete probability functions. These are the hypergeometric distribution, the binomial distribution, and the Poisson distribution.

16.4.1 THE HYPERGEOMETRIC PROBABILITY DISTRIBUTION

Many statistical experiments involve sampling from a finite population. If the finite population is *dichotomous*, i.e., consists of elements that may be classified into two *mutually exclusive categories*, and the sampling is performed *without replacement*, the *hypergeometric* probability distribution applies. To illustrate populations that are both finite and dichotomous, consider the following:

1. A lot of ten parts, of which eight are good and two are defective.
2. A group of fifteen students, of which ten are men and five are women.
3. A company of 1000 employees, of which 200 are nonunion and 800 are union members.

When one is sampling without replacement from a dichotomous population, the actual number of elements in each of the two mutually exclusive classes is normally unknown. In fact, the reason for sampling from the finite population is to obtain an estimate of the number of elements on each of the two categories. This estimate can then be used in making decisions regarding the population.

In introducing the hypergeometric distribution, it is helpful to assume that the number of elements in each of the two mutually exclusive categories is known. This assumption permits one to determine the probability of obtaining a specified number of elements in the sample from each of the two mutually exclusive categories in the population. In other words, we shall limit our discussion to the hypergeometric probability distribution rather than the related problem of using the hypergeometric distribution in statistical inference. The use of a distribution, such as the hypergeometric distribution, in statistical inference is one of the major topics in a course in statistics.

To illustrate the hypergeometric distribution, consider a lot of ten electronic parts. Suppose that seven of the parts are good and three are defective. A sample of four parts is randomly selected from the lot of ten parts. Since the population is finite (i.e., the population contains a total of ten parts) and dichotomous (i.e., the elements in the population can be classified as good or defective), the hypergeometric distribution applies. Suppose we let X represent the number of defective parts in the sample. The possible values of X are $x = 0$, $x = 1$, $x = 2$, and $x = 3$. Our task is to determine the probability that $x = 0$, $x = 1$, $x = 2$, and $x = 3$.

The probabilities of the events zero defective, one defective, two defectives, and three defectives are found by using Eq. (15.10),

$$P(E) = \frac{n(E)}{n(S)} \tag{15.10}$$

In this equation, $n(E)$ represents the number of ways in which X defectives can occur, and $n(S)$ represents the total number of ways in which the sample can be selected from the population. To determine the probability of zero defectives in the sample of size four, we first determine the number of ways in which four good parts can be selected from the seven good parts in the population. This is

$$_7C_4 = \binom{7}{4} = \frac{7!}{4!3!} = 35$$

The total number of ways in which the sample of size four can be selected is

$$_{10}C_4 = \binom{10}{4} = \frac{10!}{6!4!} = 210$$

Since the event zero defectives can occur in 35 different ways and the sample can be selected in 210 different ways, the probability of the event $X = 0$ defectives is

$$P(X = 0) = \frac{\binom{7}{4}}{\binom{10}{4}} = \frac{35}{210}$$

The probability of one defective in the sample of size four is calculated in a similar fashion. The number of ways of selecting one defective part from from the three defective parts in the population is $_3C_1$, or, alternatively, $\binom{3}{1}$. The number of ways of selecting three good parts from the seven good parts in the population is $\binom{7}{3}$. Thus, the total number of ways of selecting one defective part and three good parts is

$$\binom{3}{1}\binom{7}{3} = \frac{3!}{1!2!} \cdot \frac{7!}{3!4!} = 105$$

The total number of ways of selecting a sample of size four from a population of size ten is again $\binom{10}{4}$, or 210. The probability of the event $X = 1$ is thus

$$P(X = 1) = \frac{\binom{3}{1}\binom{7}{3}}{\binom{10}{4}} = \frac{105}{210}$$

The probabilities of the events two defectives and three defectives are calculated in the same manner. The probability of two defectives is

$$P(X = 2) = \frac{\binom{3}{2}\binom{7}{2}}{\binom{10}{4}} = \frac{63}{210}$$

and the probability of three defectives is

$$P(X = 3) = \frac{\binom{3}{3}\binom{7}{1}}{\binom{10}{4}} = \frac{7}{210}$$

Since 0, 1, 2, and 3 account for all possible values of the random variable X, we would expect the probabilities associated with the values of the random variables to sum to 1.0. By summing the probabilities we see that this is indeed true, i.e.,

$$\sum_{x=0}^{x=3} p(x) = \tfrac{35}{210} + \tfrac{105}{210} + \tfrac{63}{210} + \tfrac{7}{210} = \tfrac{210}{210} = 1.0$$

The preceding example illustrates the hypergeometric distribution. To formalize the probability model, suppose we let

$N =$ the total number of elements in a finite, dichotomous population.
$n =$ the size of the random sample selected from the population.
$D =$ the number of elements in the population in one of the two mutually exclusive categories.
$X =$ the number of elements in the sample with the same characteristic as the elements in D.

The hypergeometric mass function is

$$h(x \mid N, n, D) = \frac{\binom{D}{x}\binom{N-D}{n-x}}{\binom{N}{n}} \qquad \text{for } x = 0, 1, 2, \ldots, k \quad (16.4)$$

Notice in the formula that $\binom{D}{x}$ is the number of ways of selecting x elements from the D elements in the population, $\binom{N-D}{n-x}$ is the number of ways of selecting the remaining $n - x$ elements in the sample from the $N - D$ elements in the population, and $\binom{N}{n}$ is the total number of ways a sample of size n can be selected. The number k in the formula is equal to either n or D, whichever is smaller.

Example. In a group of twelve council members, five are known to be Republicans and seven are known to be Democrats. If a committee of four is selected by lottery, what is the probability of the committee's consisting of three Democrats and one Republican?

In this problem, $N = 12$, $n = 4$, $D = 7$, and $X = 3$. The probability is

$$P(X = 3 \mid 12, 4, 7) = \frac{\binom{7}{3}\binom{5}{1}}{\binom{12}{4}} = \frac{\frac{7!}{3!4!} \frac{5!}{1!4!}}{\frac{12!}{4!8!}} = \frac{175}{495}$$

Example. A particular manufacturer orders a certain electronic part in lots of ten. From the lot of ten parts, three are selected for testing. The lot is accepted if none of the three parts tested is found to be defective. If the lot of ten parts actually contains four defective parts, what is the probability of the lot's being accepted?

In this problem, $N = 10$, $n = 3$, $D = 4$, and $X = 0$. The probability is

$$P(X = 0 \mid 10, 3, 4) = \frac{\binom{4}{0}\binom{6}{3}}{\binom{10}{3}} = \frac{\frac{6!}{3!3!}}{\frac{10!}{7!3!}} = \frac{1}{6}$$

The expected value of a random variable that is distributed according to the hypergeometric distribution is

$$\mu = E(X) = n\left(\frac{D}{N}\right) \qquad (16.15)$$

and the variance is

$$\text{Var}(X) = n\left(\frac{D}{N}\right)\left(\frac{N - D}{N}\right)\left(\frac{N - n}{N - 1}\right) \qquad (16.16)$$

These formulas were derived by substituting the hypergeometric mass function for $f(x)$ in Formulas (16.3) and (16.8). The use of these formulas makes it possible to determine the mean and the variance directly from the parameters of the distribution. This, of course, has the advantage of eliminating the calculations required by Formulas (16.3) and (16.8).

Example. Determine the expected value and variance of a random variable that is distributed according to the hypergeometric distribution with parameters $N = 10$, $n = 3$, and $D = 4$.

$$E(X) = n\left(\frac{D}{N}\right) = 3\left(\frac{4}{10}\right) = 1.20$$

$$\text{Var}(X) = n\left(\frac{D}{N}\right)\left(\frac{N - D}{N}\right)\left(\frac{N - n}{N - 1}\right) = 3\left(\frac{4}{10}\right)\left(\frac{6}{10}\right)\left(\frac{7}{9}\right)$$

$$= 0.55$$

16.4.2 THE BINOMIAL PROBABILITY DISTRIBUTION

In the preceding section we explained that the hypergeometric distribution applies when one is sampling from a finite, dichotomous population. If, as is true in many business problems, the population is either infinite or so large that it can effectively be considered as infinite, the hypergeometric distribution no longer applies. Instead, the random variable is said to be distributed according to the binomial probability distribution.

To illustrate, consider the population of registered voters in the State of California. Suppose that 40 percent of these voters favor candidate A. If a sample of size 20 is randomly selected from this population, the binomial probability distribution can be used to determine the probability of $x = 0$, $x = 1, x = 2, \ldots, x = 20$ voters in the sample favoring candidate A.

Notice in this example that the population, although finite, is large enough to essentially be considered infinite. Furthermore, the population is dichotomous; i.e., each person interviewed may be classified as favoring or not favoring candidate A. Although it is not stated in the example, we assume that the interviews of the voters are conducted so as to be independent, i.e., the results of the previous interviews have no effect on the outcome of subsequent interviews. Finally, the probability of a person's favoring candidate A is $p = 0.40$ and the probability of the person's not favoring candidate A is $(1 - p) = 0.60$. Since the results of each interview are independent, these probabilities remain constant throughout the experiment.

A sampling experiment that is conducted under conditions such as those described above is termed a Bernoulli process (named after Jacques Bernoulli, 1654–1705). The requirements for a Bernoulli process can be summarized as follows:

1. Each trial of the experiment results in one of two possible outcomes. These outcomes are often termed "success" or "failure," "defective" or "nondefective," "yes" or "no," etc.
2. Each trial is independent of all other trials. This means that the probability of a "success" on any given trial is the same, regardless of the outcomes of the preceding trials.
3. The probability of an outcome, p, remains constant from trial to trial.

Given that the requirements for a Bernoulli process are met, the random variable X that represents the number of "successes" in a sample of size n taken from the population is distributed according to the binomial probability distribution.

To develop the binomial mass function, consider a simple experiment that involves randomly selecting a sample of size five from an infinite population. Assume that the elements in the population can be classified as "success" or "failure" and that the probability of a success on any given trial is 0.20. Let the random variable X represent the number of successes in the sample of size 5. The possible values of X are 0, 1, 2, 3, 4, and 5.

The event zero successes can occur only if all five trials result in failures. Since the trials are independent and the probability of a failure on any single trial is 0.80, the probability of five successive failures is

$$P(X = 0) = (0.80)^5 = 0.3277$$

The event one success occurs when one trial results in a success and four trials result in failures. Since the success can occur on any of the five trials, there are $\binom{5}{1}$ or 5 different sequences in which the success can occur. These are

$$\{(s,f,f,f,f), (f, s,f,f,f), (f,f, s,f,f), (f,f,f, s,f), (f,f,f,f, s)\}$$

The probability of any one of these sequences is $(0.20)^1(0.80)^4$. Consequently, the probability of exactly one success is

$$P(X = 1) = \binom{5}{1}(0.20)^1(0.80)^4 = \frac{5!}{1!4!}(0.20)^1(0.80)^4 = 0.4096$$

The event two successes occurs when two trials result in successes and three trials result in failures. The number of sequences in which two successes can occur is $\binom{5}{2}$, or 10 different sequences. The probability of two successes and three failures on any sequence is $(0.20)^2(0.80)^3$. The probability of the event two successes is thus

$$P(X = 2) = \binom{5}{2}(0.20)^2(0.80)^3 = \frac{5!}{2!3!}(0.20)^2(0.80)^3 = 0.2048$$

The probabilities of the events three successes, four successes, and five successes are found in the same manner. The probabilities are

$$P(X = 3) = \binom{5}{3}(0.20)^3(0.80)^2 = \frac{5!}{3!2!}(0.20)^3(0.80)^2 = 0.0512$$

$$P(X = 4) = \binom{5}{4}(0.20)^4(0.80)^1 = \frac{5!}{4!1!}(0.20)^4(0.80)^1 = 0.0064$$

$$P(X = 5) = \binom{5}{5}(0.20)^5(0.80)^0 = \frac{5!}{5!0!}(0.20)^5(0.80)^0 = 0.0003$$

On the basis of the preceding example, we can offer a probability mass function for a random variable that is distributed according to the binomial distribution. Suppose we let

p = probability of "success" on a single trial.
n = the number of trials (i.e., the sample size).
X = the number of "successes" in the n trials.

The probability mass function for the random variable X is

$$b(x \mid n, p) = \binom{n}{x} p^x (1 - p)^{n-x} \qquad \text{for } x = 0, 1, \ldots, n \qquad (16.17)$$

This formula gives the probability of x successes in a series of n Bernoulli trials. The resulting probability distribution is called the binomial probability distribution.

Example. The probability that an item produced by a certain manufacturing process is defective is $p = 0.10$. If a sample of four items is selected for testing, what is the probability of no defectives? Of one defective? Of two defectives?

The probability of no defectives is

$$P(X = 0 \mid 4, 0.10) = \binom{4}{0}(0.10)^0(0.90)^4 = 0.6561$$

the probability of one defective is

$$P(X = 1 \mid 4, 0.10) = \binom{4}{1}(0.10)^1(0.90)^3 = 0.2916$$

and the probability of two defectives is

$$P(X = 2 \mid 4, 0.10) = \binom{4}{2}(0.10)^2(0.90)^2 = 0.0486$$

Formula (16.17) provides the mathematical definition of the binomial mass function. The formula is, however, seldom used for actual computation of binomial probabilities. Instead, tables of the binomial distribution are used. Table VIII in the Appendix contains the binomial probabilities for selected values of n and p. The use of this table is illustrated by the following examples.

Example. In a nationwide survey of business executives, 40 percent indicated that they believed that corporate profits would increase during the coming year. If a sample of size 12 is selected from this population, determine the probability of exactly four executives stating that profits will increase.

For $X = 4$, $n = 12$, and $p = 0.40$, the binomial probability is

$$P(X = 4 \mid 12, 0.40) = \binom{12}{4}(0.40)^4(0.60)^8$$

Rather than calculating the probability, we can read the probability directly from Table VIII. The probability is 0.2128.

Example. In the preceding example, determine the probability of between three and five executives stating that they believe profits will increase.

The parameters of the distribution are again $n = 12$ and $p = 0.40$. The probability of between three and five executives favoring the increase is

$$P(3 \leq X \leq 5) = P(X = 3) + P(X = 4) + P(X = 5)$$
$$= 0.1419 + 0.2128 + 0.2270$$
$$= 0.5817$$

Example. A survey indicated that 70 percent of the television viewers watched the past year's Super Bowl game. If five television viewers are selected at random, what is the probability that exactly four of the five watched the game?

For $X = 4$, $n = 5$, and $p = 0.70$, the binomial probability is

$$P(X = 4 \mid 5, 0.70) = \binom{5}{4}(0.70)^4(0.30)^1$$

Since Table VIII includes values of p only up to $p = 0.50$, we must recognize that the probability of four "successes" in five trials with a probability of 0.70 of success is equivalent to the probability of one failure in five trials with a probability of 0.30 of failure. The binomial probability is thus

$$P(X = 1 \mid 5, 0.30) = \binom{5}{1}(0.30)^1(0.70)^4$$

From Table VIII, the answer is 0.3602.

The mean or expected value of a random variable X that is distributed according to the binomial distribution is

$$\mu = E(X) = np \qquad (16.18)$$

and the variance is

$$\text{Var}(X) = np(1 - p) \qquad (16.19)$$

These formulas were derived by substituting $b(x \mid n, p)$ for $p(x)$ in Formulas (16.3) and (16.8).

Example. Determine the expected value and variance of a random variable that is distributed according to the binomial distribution with $n = 12$ and $p = 0.40$.

From Formula (16.18) the expected value is

$$E(X) = np = 12(0.40) = 4.8$$

and from Formula (16.19) the variance is

$$\text{Var}(X) = np(1 - p) = 12(0.40)(0.60) = 2.88$$

16.4.3 THE POISSON PROBABILITY DISTRIBUTION

The final discrete probability distribution discussed in this section is the *Poisson* probability distribution. This distribution, named after the French mathematician Simeon Denis Poisson (1781–1840), is often useful in describing the probability of a certain number of "successes" in some continuous unit of measurement. Unlike the binomial distribution, in which the number of trials was fixed and each trial resulted in a success or a failure, the Poisson distribution describes a process that operates continuously over some unit of measurement and generates only "successes" or "occurrences." This process is referred to as a Poisson process.

To illustrate a Poisson process, consider a manufacturing process that produces a continuous roll of magnetic tape. Flaws will occur in the tape at random points. Suppose the random variable X represents the number of flaws in the tape in a given unit measurement. The possible values of the random variable are described by the set

$$\{x \mid x = 0, 1, 2, 3, 4, \ldots\}$$

Although we can count the number of flaws that occur, we cannot count the number of flaws that do not occur. Furthermore, there is no upper limit to the possible number of flaws. Since the manufacturing process operates continuously over some unit of measurement and generates only successes (i.e., flaws), the Poisson distribution would be a likely candidate for describing the probability of different possible numbers of flaws.

As another example, consider the occurrence of equipment failure at a certain generating plant. Over some continuous unit of time, such as a month, equipment failure may occur at random points in time at the generating plant. Again, it is possible to count the number of breakdowns that occur, but it is impossible to specify an upper limit on the number of breakdowns or to ask how many breakdowns did not occur. Still other examples of a Poisson process might include the number of telephone calls received at an office switchboard during an hour, the number of automobiles arriving at a toll booth during a day, the number of defects occurring in 50 feet of telephone cable, and the number of blemishes in a square yard of wall covering.

The Poisson mass function is

$$p(x \mid \lambda) = \frac{\lambda^x e^{-\lambda}}{x!} \qquad \text{for } x = 0, 1, 2, 3, \ldots \qquad (16.20)$$

where e represents the number 2.71828, λ represents the average number of occurrences in the unit of measurement under consideration, and x represents the number of occurrences of an event that is generated by a Poisson process.†

Example. The number of cars sold at a used car lot can be described as a Poisson process with an average rate of sales of $\lambda = 2.0$ cars per day. Determine the probability of zero, one, two, three, four, etc., cars being sold in a day.

---

† Tables of the Poisson mass function for alternative values of λ are available in the statistics texts listed in the suggested references of Chapter 15.

If we let X represent the number of cars sold, the probabilities are

$$P(X = 0 \mid 2) = \frac{2^0 e^{-2}}{0!} = \frac{1(0.135)}{1} = 0.135$$

$$P(X = 1 \mid 2) = \frac{2^1 e^{-2}}{1!} = \frac{2(0.135)}{1} = 0.270$$

$$P(X = 2 \mid 2) = \frac{2^2 e^{-2}}{2!} = \frac{4(0.135)}{2} = 0.270$$

$$P(X = 3 \mid 2) = \frac{2^3 e^{-2}}{3!} = \frac{8(0.135)}{6} = 0.180$$

$$P(X = 4 \mid 2) = \frac{2^4 e^{-2}}{4!} = \frac{16(0.135)}{24} = 0.090$$

$$P(X = 5 \mid 2) = \frac{2^5 e^{-2}}{5!} = \frac{32(0.135)}{120} = 0.036$$

$$P(X = 6 \mid 2) = \frac{2^6 e^{-2}}{6!} = \frac{64(0.135)}{720} = 0.012$$

$$P(X = 7 \mid 2) = \frac{2^7 e^{-2}}{7!} - \frac{128(0.135)}{5040} = 0.004$$

$$P(X = 8 \mid 2) = \frac{2^8 e^{-2}}{8!} = \frac{256(0.135)}{40,320} = 0.001$$

Notice that the probabilities become quite small for values of X that are large relative to the value of λ. In fact, by summing the above probabilities we see that $P(X \leq 8) = 0.998$. Conversely, the $P(X \geq 9) = 0.002$. This indicates that although there is a possibility that the Poisson process will produce an indefinitely large number of occurrences in the unit measurement, the probability of this occurring is quite small.

One of the characteristics of a Poisson process is that the average number of occurrences in the unit of measurement is proportional to the size of the unit. For instance, suppose that an average of 1.5 calls per minute are received at a telephone switchboard. Since the average number of calls is proportional to the length of time, this means that an average of 3.0 calls would be expected in two minutes, 4.5 calls in three minutes, etc.

Example. Flaws occur in the manufacture of copper wire at an average rate of 0.05 flaws per foot. Assuming a Poisson process, determine the probability of no flaws, one flaw, and two flaws in ten feet of wire.

The average number of flaws in ten feet of wire is $\lambda = 0.05(10) = 0.50$. The probability of no flaws is

$$P(X = 0 \mid 0.50) = \frac{(0.50)^0 e^{-0.50}}{0!} = 1(0.607) = 0.607$$

the probability of one flaw is

$$P(X = 1 \mid 0.50) = \frac{(0.50)^1 e^{-0.50}}{1!} = \frac{0.50(0.607)}{1} = 0.304$$

and the probability of two flaws is

$$P(X = 2 \mid 0.50) = \frac{(0.50)^2 e^{-0.50}}{2!} = \frac{0.25(0.607)}{2} = 0.076$$

Example. In the preceding example, determine the probability of exactly one flaw in the first ten feet of wire and one flaw in the second ten feet of wire.

The probability of one flaw in the first ten feet of wire was 0.304. Similarly, the probability of one flaw in the second ten feet is also 0.304. Since the events are independent, the probability of one flaw in the first ten feet and one flaw in the second ten feet is

$$P(X = 1 \mid \lambda = 0.50)P(X = 1 \mid \lambda = 0.50) = (0.304)(0.304) = 0.092$$

By way of contrast, the probability of two flaws in twenty feet of wire is

$$P(X = 2 \mid \lambda = 1.0) = \frac{(1.0)^2 e^{-1}}{2!} = \frac{1(0.368)}{2} = 0.184$$

The mean or expected value of a random variable that is distributed according to the Poisson distribution is

$$\mu = E(X) = \lambda \tag{16.21}$$

and the variance is

$$\text{Var}(X) = \lambda \tag{16.22}$$

The fact that the mean and the variance are equal is a unique feature of the Poisson distribution.

16.5 Commonly Used Continuous Probability Distributions

In this final section we introduce three commonly used continuous probability distributions. These are the uniform distribution, the normal distribution and the exponential distribution.

16.5.1 THE UNIFORM DISTRIBUTION

The *uniform* distribution describes an experiment in which the random variable assumes some value in the domain $a \leq x \leq b$. The distribution is termed uniform because the probability density function has a constant value for all values of the random variable in the domain. The uniform density function is

$$f(x) = \frac{1}{b-a} \qquad \text{for } a \le x \le b \qquad (16.23)$$

The distribution is shown in Fig. 16.4.

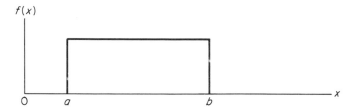

Figure 16.4

The definite integral of the uniform density function with limits of integration a and b is equal to 1, as shown below.

$$\int_a^b \frac{1}{b-a}\, dx = \frac{x}{b-a}\Big|_a^b = \frac{b-a}{b-a} = 1$$

Examples of the uniform probability distributions are shown below.

Example. Buses on a certain route arrive every 30 minutes. What is the probability that a man arriving at a random time will have to wait at least 20 minutes for a bus?

The random variable, $T = $ time to wait until the next bus, is uniformly distributed for $0 \le T \le 30$ with probability density $f(t) = \frac{1}{30}$. The probability that the man must wait at least 20 minutes is

$$P(T \ge 20) = \int_{20}^{30} \frac{1}{30}\, dt = \frac{t}{30}\Big|_{20}^{30} = \frac{1}{3}$$

Example. A four-digit random number table contains a selection of random numbers varying from 0000 to 9999. If these numbers are described by a uniform distribution, determine the probability of selecting a random number whose value is between 400 and 500, inclusive.

Since the probability of selecting any number in the table is the same, the probability density function for the selection of a random number X is

$$f(x) = \frac{1}{999.5 - (-0.5)} = \frac{1}{10,000} \qquad \text{for } 0 \le x \le 9999$$

This function is derived from the fact that we represent a discrete random number, $x = a$, as the midpoint of the continuous interval $x = a \pm 0.5$. The probability of selecting a random number that falls in the interval from 400 to 500, inclusive, is thus

$$P(400 \leq X \leq 500) = \int_{399.5}^{500.5} \frac{1}{10,000}\, dx = \frac{x}{10,000}\Big|_{399.5}^{500.5} = \frac{101}{10,000}$$

The mean or expected value of a random variable that is distributed according to the uniform distribution is determined from Eq. (16.4). The mean or expected value is

$$\mu = E(X) = \int_{\text{all } x} xf(x)\, dx$$

$$= \int_{a}^{b} \frac{x}{b-a}\, dx = \frac{x^2}{2(b-a)}\Big|_{a}^{b} = \frac{b^2 - a^2}{2(b-a)}$$

$$= \frac{(b-a)(b+a)}{2(b-a)} = \frac{1}{2}(b+a). \tag{16.24}$$

The variance is derived from Eq. (16.10). The $E(X^2)$ is

$$E(X^2) = \int_{\text{all } x} x^2 f(x)\, dx$$

$$= \int_{a}^{b} \frac{x^2}{b-a}\, dx = \frac{x^3}{3(b-a)}\Big|_{a}^{b} = \frac{b^3 - a^3}{3(b-a)}$$

Subtracting the square of the mean $E(X)^2$ from $E(X^2)$ gives

$$\text{Var}(X) = \frac{b^3 - a^3}{3(b-a)} - \frac{(b+a)^2}{4} \tag{16.25}$$

Example. Determine the mean and variance of the uniform distribution in the example on p. 603 concerning the waiting time of a man for a bus.

The density function was $f(t) = \frac{1}{30}$ for $0 \leq t \leq 30$. From Eq. (16.24) the expected waiting time is

$$E(T) = \tfrac{1}{2}(b+a) = \tfrac{1}{2}(30+0) = 15$$

Using Eq. (16.25), we find that the variance is

$$\text{Var}(T) = \frac{b^3 - a^3}{3(b-a)} - \frac{(b+a)^2}{4}$$

$$= \frac{27,000}{90} - \frac{900}{4} = 75$$

16.5.2 THE NORMAL PROBABILITY DISTRIBUTION

The most widely used continuous distribution is the normal distribution. This distribution has been found to model accurately many continuous random variables in business as well as the social, physical, and biological sciences. In addition, as those readers will discover who study statistics, it forms the basis for many problems in statistical analysis.

The density function for the normal distribution is

$$f(x) = \frac{1}{\sigma\sqrt{2\pi}} e^{-\frac{1}{2}[(x-\mu)/\sigma]^2} \qquad \text{for } -\infty < x < \infty \qquad (16.26)$$

where σ and μ are parameters of the distribution representing the standard deviation and the mean, or expected value, respectively. The distribution is symmetrical about the mean μ and is asymptotic to the horizontal axis. These features lead to a description of the normal distribution as bell-shaped, as shown by Fig. 16.5.

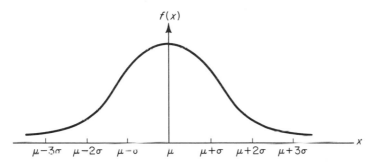

Figure 16.5

As in the case of other continuous distributions, the probability of a random variable X having a value in the interval $a \leq x \leq b$ is given by the definite integral of the normal density function with limits a and b; i.e.,

$$P(a \leq X \leq b) = \int_a^b \frac{1}{\sigma\sqrt{2\pi}} e^{-\frac{1}{2}[(x-\mu)/\sigma]^2} dx$$

This integral cannot, however, be evaluated by elementary methods. Instead, the integral is evaluated by the use of a *standard normal* probability table.

A standard normal probability table contains cumulative probabilities for a normally distributed random variable with $\mu = 0$ and $\sigma = 1$. A standard normal probability table is included as Table IX in this text. To illustrate the use of the table, assume that a random variable Z is normally distributed with $\mu = 0$ and $\sigma = 1$. The probability that $a \leq Z \leq b$ is found by determining the cumulative probabilities that $Z \leq a$ and $Z \leq b$. These probabilities are read directly from the table. Since the table gives cumulative probabilities, the $P(a \leq Z \leq b)$ is

$$P(a \leq Z \leq b) = P(Z \leq b) - P(Z \leq a) \qquad (16.27)$$

As an example, assume that $a = -2$ and $b = 1$. From Table IX,

$$P(-2 \leq Z \leq 1) = P(Z \leq 1) - P(Z \leq -2)$$

$$= 0.8413 - 0.0228 = 0.8185$$

Example. A random variable Z is normally distributed with $\mu = 0$ and $\sigma = 1$. Determine the probability that $-0.52 \leq Z \leq 1.07$.

$$P(-0.52 \leq Z \leq 1.07) = P(Z \leq 1.07) - P(Z \leq -0.52)$$

$$= 0.8577 - 0.3015 = 0.5562$$

Example. A random variable Z is normally distributed with $\mu = 0$ and $\sigma = 1$. Determine the probability that $-1.65 \leq Z \leq -0.22$.

$$P(-1.65 \leq Z \leq -0.22) = P(Z \leq -0.22) - P(Z \leq -1.65)$$

$$= 0.4129 - 0.0495 = 0.3634$$

To use a standard normal probability table to determine the probability that a normally distributed random variable X assumes a value in the interval $a \leq X \leq b$, the values of a and b must be *normalized*. This is done by converting a and b into z values. The normalized values of a and b are

$$z_a = \frac{a - \mu}{\sigma} \quad \text{and} \quad z_b = \frac{b - \mu}{\sigma} \qquad (16.28)$$

The probability that $a \leq X \leq b$ is then equal to the probability that $z_a \leq Z \leq z_b$; i.e.,

$$P(a \leq X \leq b) = P\left(\frac{a - \mu}{\sigma} \leq Z \leq \frac{b - \mu}{\sigma}\right)$$

Example. A random variable X is normally distributed with $\mu = 100$ and $\sigma = 10$. Determine the probability that $80 \leq X \leq 110$.

$$P(80 \leq X \leq 110) = P\left(\frac{80 - 100}{10} \leq Z \leq \frac{110 - 100}{10}\right)$$

$$= P(-2 \leq Z \leq 1)$$

$$= P(Z \leq 1) - P(Z \leq -2)$$

$$= 0.8413 - 0.0228 = 0.8185$$

Example. A random variable X is normally distributed with mean 60 and standard deviation 12. What is the probability that $X \geq 75$?

The probability that $X \geq 75$ is

$$P(X \geq 75) = 1 - P(X \leq 75)$$
$$= 1 - P\left(Z \leq \frac{75 - 60}{12}\right)$$
$$= 1 - P(Z \leq 1.25)$$
$$= 1 - 0.8943 = 0.1057$$

The area representing this probability is shown in the figure below.

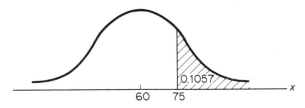

Example. A manufacture of a certain breakfast cereal states that a box of cereal contains 18.0 ounces. If the weight of the cereal is actually normally distributed with $\mu = 18.0$ and $\sigma = 0.2$, determine the probability that the contents of the box weigh more than 17.7 ounces.

$$P(X \geq 17.7) = P\left(Z \geq \frac{17.7 - 18.0}{0.2}\right)$$
$$= P(Z \geq -1.5)$$
$$= 1 - P(Z \leq -1.5)$$
$$= 1 - 0.0668 = 0.9332$$

Example. Suppose that a manufacture of fishing line claims that there is a 0.1587 probability that the line will have a breaking strength of less than 9 pounds and a 0.9772 probability that the line will have a breaking strength of less than 14 pounds. If the breaking strength of the line is normally distributed, determine the mean and standard deviation of the distribution.

To solve for the two parameters of the distribution μ and σ, we first determine the z values for the probabilities 0.1587 and 0.9772. From Table IX these are -1.00 and 2.00, respectively. We next recognize that

$$-1.00 = \frac{9 - \mu}{\sigma}$$

and

$$2.00 = \frac{14 - \mu}{\sigma}$$

Solving these two equations simultaneously gives $\mu = 10.67$ and $\sigma = 1.67$.

16.5.3 THE EXPONENTIAL PROBABILITY DISTRIBUTION

Another useful continuous probability distribution is the *exponential* probability distribution. This distribution is used in describing the time lapse between occurrences in a Poisson process. For example, if individuals arrive at a service facility according to a Poisson process, then the time lapse between the arrival of any two customers can be described by the exponential probability distribution.

The probability density function for the exponential distribution is

$$f(t) = \lambda e^{-\lambda t} \qquad \text{for } 0 \le t \le \infty \qquad (16.29)$$

In the density function, T represents the time between occurrences (e.g., arrivals at a service facility), $e = 2.71828$ is a constant, and λ is the mean number of occurrences per period. The exponential distribution is shown in Fig. 16.6.

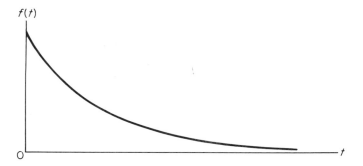

Figure 16.6

The probability of an occurrence within a time interval $T \le t$ is

$$P(T \le t) = \int_0^t \lambda e^{-\lambda t}\, dt$$

The exponential distribution satisfies the requirements for a density function, since $f(t) \ge 0$, and

$$\int_0^\infty \lambda e^{-\lambda t}\, dt = -e^{-\lambda t}\Big|_0^\infty = 0 - (-1) = 1$$

The mean or expected value of the exponential distribution can be derived by using Eq. (16.4) and is

$$\mu = E(T) = \frac{1}{\lambda} \qquad (16.30)$$

The variance is derived by using Eq. (16.10) and is

$$\text{Var}(T) = \frac{1}{\lambda^2} \qquad (16.31)$$

Several applications of the exponential distribution are illustrated in the following examples.

Example. Customers are known to arrive at a service station as a Poisson process. If an average of $\lambda = 10$ customers arrive per hour, determine the probability of an arrival within 0.2 hour. Also determine the mean and variance of the probability distribution.

The probability of an arrival within 0.2 hour is

$$P(T \leq 0.2) = \int_0^{0.2} 10e^{-10t}\, dt = -e^{-10t}\Big|_0^{0.2} = -e^{-2} + 1$$

$$= -0.135 + 1 = 0.865$$

The mean time between arrivals of customers is

$$\mu = \frac{1}{\lambda} = \frac{1}{10} = 0.1 \text{ hour}$$

and the variance is

$$\text{Var}(T) = \frac{1}{\lambda^2} = \frac{1}{100} = 0.01$$

Example. During a certain time of the day, aircraft have been observed to arrive at an airport as a Poisson process with an average of $\lambda = 20$ arrivals per hour. Determine the probability of an arrival of an aircraft within a three-minute period (i.e., $\frac{1}{20}$ of an hour).

$$P(T \leq \tfrac{1}{20}) = \int_0^{1/20} 20e^{-20t}\, dt$$

$$= -e^{-20t}\Big|_0^{1/20} = e^{-1} + 1$$

$$= -0.368 + 1 = 0.632$$

Example. In the manufacture of a certain type of steel wire, flaws have been observed to occur as a Poisson process with an average of $\lambda = 3$ flaws per 50 feet. Determine the probability of a flaw within a 10-foot length of wire (i.e., one-fifth of the 50-foot interval).

$$P(L \leq \tfrac{1}{5}) = \int_0^{1/5} 3e^{-3t}\, dt$$

$$= -e^{-3t}\Big|_0^{1/5} = -e^{-0.6} + 1$$

$$= -0.549 + 1 = 0.451$$

Example. During a certain time period, telephone calls occur at a switch-board as a Poisson process with an average of $\lambda = 30$ per hour. Determine the probability of a call's occurring within one minute (i.e., 1/60 hour).

$$P(T \le \tfrac{1}{60}) = \int_0^{1/60} 30e^{-30t} \, dt$$

$$= -e^{-30t} \Big|_0^{1/60} = -e^{-0.5} + 1$$

$$= -0.607 + 1 = 0.393$$

PROBLEMS

1. Specify the set of possible values of the random variable in the following experiments. State whether the random variable is discrete or continuous.
   (a) Four individuals are interviewed regarding their intent to vote for candidate A in an upcoming election.
   (b) In a time and motion study, an efficiency expert records the time required to complete a certain assembly.
   (c) An inspection process involves measuring the diameter of a certain steel shaft.
   (d) The number of defective parts in a lot of ten parts that are selected from a shipment of 100 parts is recorded.

2. A coin is tossed three times and the outcome of each toss is recorded as a head H or a tail T. Based on this experiment, give the set of possible values of the random variables described by the following statements.
   (a) The random variable X represents the number of heads.
   (b) The random variable X represents the square of the number of heads.
   (c) The random variable X represents the number of heads minus the number of tails.

3. Develop probability functions for the random variables in Problem 2.

4. A salesman schedules four calls per day. The probability mass function that gives the probability of 0, 1, 2, 3, and 4 sales is

$$p(x) = \frac{(5 - x)}{15} \qquad \text{for } x = 0, 1, 2, 3, 4$$

   Verify that $p(x)$ satisfies the requirements for a probability mass function and determine the probabilities, $P(X = x)$.

5. A discrete random variable X has the probability mass function

$$p(x) = \frac{(x^2 + 4)}{50} \qquad \text{for } x = 0, 1, 2, 3, 4$$

Determine the following probabilities:
(a) $P(X = 2)$
(b) $P(X \geq 3)$
(c) $P(X < 2)$
(d) $P(X > 2)$

6. A discrete random variable X has the probability mass function

$$p(x) = \frac{(2x + 1)}{25} \qquad \text{for } x = 0, 1, 2, 3, 4$$

Determine the cumulative probabilities, $P(X \leq x)$, for $x = 0, 1, 2, 3, 4$.

7. The useful life of a certain highly sensitive component can be described by the probability density function

$$f(x) = \frac{x}{18} \qquad \text{for } 0 \leq x \leq 6$$

where the random variable X represents useful life in hours.
(a) Verify that $f(x)$ satisfies the requirements for a density function.
(b) Determine the probability that the useful life of the component will be between 2 and 4 hours.
(c) Determine the probability that the useful life of the component will be less than 2 hours.
(d) Determine the probability that the useful life of the component will be more than 4 hours.

8. The time required to complete an assembly can be described by the probability density function

$$f(t) = \frac{t}{4} \qquad \text{for } 1 \leq t \leq 3$$

where the random variable T represents time in minutes.
(a) Verify that $f(t)$ satisfies the requirements for a density function.
(b) Determine the probability that the time required to complete an assembly will be greater than 2 minutes.
(c) Determine the probability that the time required will be between 1 and 2 minutes.

9. Determine the value of k such that the following functions are probability density functions.
(a) $f(x) = k(3 - x)$ for $0 \leq x \leq 3$
(b) $f(x) = k(2 + x)$ for $0 \leq x \leq 4$
(c) $f(x) = kx^2$ for $0 \leq x \leq 3$

10. The probability distribution for a discrete random variable X is given below.

x	0	1	2	3	4
$p(x)$	0.10	0.20	0.20	0.30	0.20

Determine the mean and variance of the random variable.

11. Determine the mean and variance for the discrete probability distribution in Problem 4.

12. In Problem 8, the time required to complete an assembly was described by the probability density function

$$f(t) = \frac{t}{4} \quad \text{for } 1 \leq t \leq 3$$

where T was a continuous random variable representing time. Determine the mean and variance of the random variable.

13. A continuous random variable X has the probability density function

$$f(x) = 6x - 6x^2 \quad \text{for } 0 \leq x \leq 1$$

Determine the mean and variance of X.

14. A discrete random variable X has the probability distribution given below.

x	0	1	2	3	4	5	6
$p(x)$	0.02	0.07	0.12	0.17	0.22	0.25	0.15

(a) Determine the median value of X.
(b) Determine the modal value of X.

15. A discrete random variable X has the probability mass function

$$p(x) = \begin{cases} 0.15 & \text{for } x = 0 \\ 0.05x^2 & \text{for } x = 1, 2 \\ 0.10(6 - x) & \text{for } x = 3, 4, 5 \\ 0 & \text{otherwise} \end{cases}$$

(a) Determine the median value of X.
(b) Determine the modal value of X.

16. A continuous random variable X has the probability density function

$$f(x) = 6x - 6x^2 \quad \text{for } 0 \leq x \leq 1$$

(a) Determine the median value of X.
(b) Determine the modal value of X.

17. A continuous random variable X has the density function

$$f(x) = \tfrac{3}{4}(2x - x^2) \qquad \text{for } 0 \le x \le 2$$

(a) Determine the median value of X.
(b) Determine the modal value of X.

18. A manufacturer purchases a certain component in lots of twenty. Before accepting the lot, a random sample of five components is tested. If the lot contains three defective components, what is the probability that the sample will contain one defective?

19. The Bryant Corporation has developed an acceptance sampling plan in which three components from each twelve received by the company are tested. If no more than one component of the three tested is found to be defective, the lot of twelve is accepted. If the lot of twelve components contains four defectives, what is the probability of accepting the lot?

20. The manager of a small firm has decided to create an employee council to handle employee grievances. The council will consist of the plant manager, the personnel manager, and four employee representatives. The four employees are to be selected at random from a list of fifteen nominees. If this list contains nine men and six women, what is the probability of two men and two women's being selected as the employee representatives?

21. A recent public opinion poll showed that 30 percent of the population favored an isolationist policy in foreign relations. If a sample of size 10 is selected from this population, what is the expected number that favors an isolation policy? What is the probability that exactly three individuals will favor the policy?

22. Five identical transistors are used in the assembly of a sophisticated navigational radio. All five transistors must work for the radio to function normally. If the probability of this type of transistor's failing is 0.01, what is the probability that the radio will function normally?

23. The probability that a certain type of air-to-air missile will hit a target is 0.40. How many missiles must be fired if it is desired that the probability of at least one hit be greater than or equal to 0.95?

24. Five people are chosen at random and asked if they favor a certain proposal. If only 40 percent of the population favors the proposal, what is the probability that the majority of the five people chosen will favor the proposal?

25. Customers arrive at a service facility at an average rate of six per hour.

What is the probability of exactly four customers arriving during an hour?

26. Calls arrive at a certain switchboard an average of once every two minutes. What is the probability of two calls arriving during one minute?

27. A publisher has found that an average of one type-setting error occurs in every ten pages of a text. What is the probability of no more than one error in five pages?

28. A random variable X is uniformly distributed with density

$$f(x) = \tfrac{1}{10} \qquad \text{for } 10 \leq x \leq 20$$

   (a) Determine the $P(X \geq 4)$.
   (b) Determine the $P(2 \leq X \leq 7)$.

29. An automated machine completes an operation every two minutes. What is the probability that an inspector randomly arriving will have to wait thirty seconds or less for the part?

30. A random variable X is distributed according to the normal distribution with $\mu = 20$ and $\sigma = 4$. Determine the following probabilities.
   (a) $P(X \geq 22)$ (b) $P(16 \leq X \leq 22)$
   (c) $P(X \leq 15)$ (d) $P(21 \leq X \leq 23)$

31. The life of a certain component is distributed according to the normal distribution with a mean life of fifteen months and a standard deviation of two months. Determine the probability of the component failing between the twelfth and seventeenth month.

32. A gasoline company has two million credit card holders. During the preceding month the average billing to the card holders was $\mu = \$22.00$ and the standard deviation was $\sigma = \$6.00$. If it is assumed that the billings are normally distributed, determine the number of customers who received bills exceeding \$26.

33. An investigation of the life of a particular electronic component disclosed that there was a 0.10 probability that the component would last twenty-five months or longer and a 0.67 probability that it would last seventeen months or longer. Assume that the life of the component is a normally distributed random variable. What is the mean and the standard deviation of the life of this component?

34. A television repair firm has found that calls for repair service occur as a Poisson process at a mean rate of four per hour. Determine the probability of a call within thirty minutes.

35. Ships have been observed to arrive at a certain port as a Poisson process at a mean rate of once every two days. Determine the probability of a ship's arriving during the next day.

36. Breakdowns at a computer facility occur as a Poisson process at a mean

rate of two per eight-hour day. Determine the probability of a breakdown in a two-hour interval.

SUGGESTED REFERENCES

The references for this chapter are listed in Chapter 15.

Establishing
Functions

There are several methods in which functions can be established so that the function passes through data points. The most straightforward and easily understandable method of establishing these functions is through the solution of simultaneous equations.

A.1   Linear Functions

As a first illustration, consider the problem of establishing a linear function. Two data points are required to define a linear function. We shall consider the data points ($x = 1$, $y = 4$) and ($x = 4$, $y = 2$). It is customary to enclose the data points in parentheses with the independent or x variable preceding the dependent or y variable and the two variables separated by commas. Thus, the two data points are customarily written as (1, 4) and (4, 2). The general form of the linear function is $y = a + bx$. To determine the values of a and b we substitute the two data points for x and y and solve the resulting two equations simultaneously. Thus,

$$4 = a + b(1)$$

$$2 = a + b(4)$$

must be solved simultaneously for a and b.

Subtracting the second equation from the first gives

$$2 = -3b$$

and $b = -\frac{2}{3}$. Substituting for b in either equation and solving for a gives $a = 4\frac{2}{3}$. This function is shown in Fig. A.1. A second sample follows.

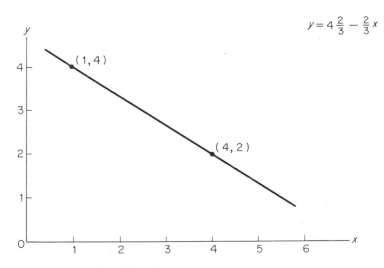

$$y = 4\tfrac{2}{3} - \tfrac{2}{3}x$$

Figure A.1. Linear Function

Example. Determine the linear function that passes through the data points $(2, -3)$ and $(10, 17)$. The two equations are

$$-3 = a + b(2)$$

$$17 = a + b(10)$$

which when solved simultaneously give

$$y = -8 + 2\tfrac{1}{2}x$$

A.2 Quadratic Functions

A quadratic, or second degree, function is of the form $y = a + bx + cx^2$. Three data points are necessary to define a quadratic function. The three data points are substituted into the general form of the quadratic function, and the resulting three equations are solved simultaneously for the parameters a, b, and c. The procedure is illustrated by the following example.

Example. Determine the quadratic function that passes through the data points $(1, 6)$, $(3, 4)$, and $(8, 7)$. These data points are shown in Fig. A.2.

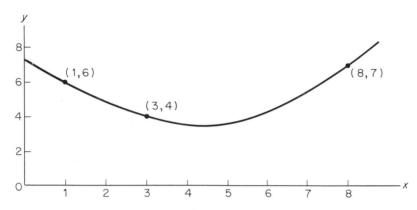

Figure A.2. Quadratic Function

The three equations are

$$6 = a + b(1) + c(1)^2$$
$$4 = a + b(3) + c(3)^2$$
$$7 = a + b(8) + c(8)^2$$

or, alternatively,

$$6 = a + b + c$$
$$4 = a + 3b + 9c$$
$$7 = a + 8b + 64c$$

These equations are solved simultaneously to give the coefficients a, b, and c; thus,

$$
\begin{array}{rl}
6 = & a + b + c \\
-4 = & -a - 3b - 9c \\
\hline
2 = & -2b - 8c
\end{array}
\qquad
\begin{array}{rl}
7 = & a + 8b + 64c \\
-4 = & -a - 3b - 9c \\
\hline
3 = & 5b + 55c
\end{array}
$$

By multiplying the equations by 5 and 2, we obtain

$$10 = -10b - 40c \quad \text{and} \quad 6 = 10b + 110c$$

These two equations can be added to give

$$10 = -10b - 40c$$
$$6 = +10b + 110c$$
$$\overline{16 = + 70c}$$

and

$$c = \frac{16}{70} = 0.228, \quad b = \frac{-2 - 8c}{2} = \frac{-3.824}{2} = -1.912$$

$$a = 6 - b - c = 6 + 1.912 - 0.228 = 7.684$$

and the quadratic function is

$$y = 7.684 - 1.912x + 0.228x^2$$

This function is shown in Fig. A.2.

A.3 Polynomial Functions

The general form of the polynomial function is

$$y = a_0 + a_1x + a_2x^2 + \ldots + a_nX^n$$

The method of establishing a function through data points, discussed in Secs. A.1 and A.2, also applies to the polynomial function of degree n. The procedure is to substitute the $(n + 1)$ data points into the general form of the equation and solve the resulting $(n + 1)$ equations simultaneously for the $(n + 1)$ parameters $a_0, a_1, a_2, \ldots, a_n$. The procedure for a cubic function is illustrated in the following example.

Example. Determine the cubic function that passes through the points $(1, 2)$, $(3, 6)$, $(5, 7)$, and $(7, 10)$. The four equations are

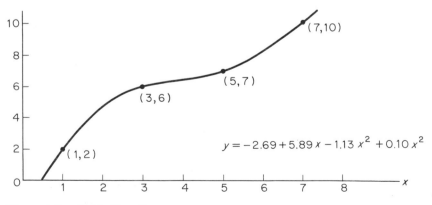

Figure A.3. Cubic Function

$$2 = a_0 + a_1(1) + a_2(1)^2 + a_3(1)^3$$
$$6 = a_0 + a_1(3) + a_2(3)^2 + a_3(3)^3$$
$$7 = a_0 + a_1(5) + a_2(5)^2 + a_3(5)^3$$
$$10 = a_0 + a_1(7) + a_2(7)^2 + a_3(7)^3$$

The four equations are solved simultaneously to give $a_0 = -2.69$, $a_1 = 5.89$, $a_2 = -1.31$, and $a_3 = 0.10$. The resulting function is shown in Fig. A.3.

A.4   Exponential Function

The exponential function has the general form $y = k(a)^{cx}$, where k and c are parameters and a is the base. An exponential function can be established through any two data points with the exception of $(x = 0,\ y = 0)$. The method of establishing the function is to write the exponential function as a logarithmic function, and solve the resulting two equations simultaneously for the parameters k and c. The procedure is illustrated by the following example.

Example. Determine the exponential function that passes through the data points $(1, 2)$ and $(4, 6)$. Use the base e. The general form of the function is

$$y = ke^{cx}$$

which when written in terms of logarithms becomes

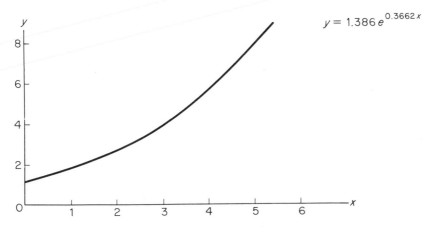

$$y = 1.386\,e^{0.3662\,x}$$

Figure A.4.   Exponential Function

$$\ln(y) = \ln(k) + cx$$

Substituting the two data points gives the two equations

$$\ln(2) = \ln(k) + c(1)$$
$$\ln(6) = \ln(k) + c(4)$$

or, alternatively,

$$0.69315 = \ln(k) + \quad c$$

$$1.79176 = \ln(k) + 4c$$

Solving these two equations simultaneously gives $\ln(k) = 0.32695$ and $c = 0.36620$. The antilog of $\ln(k)$ is $k = 1.386$. The exponential function is thus

$$y = 1.386e^{0.36620x}$$

This function is shown in Fig. A.4.

A.5 Logarithmic Functions

The general form of the logarithmic function is

$$y = a \ln(bx),$$

where x is the independent variable, y is the dependent variable, a and b are parameters, and the base of the logarithmic function is e. To determine the value of the parameters a and b, we make the following substitutions.
Let

$$y = \frac{1}{c} \ln\left(\frac{1}{k} x\right)$$

or

$$cy = \ln\left(\frac{1}{k} x\right)$$

The antilog of both sides (see Appendix B) gives

$$e^{cy} = \frac{1}{k} x$$

or

$$x = ke^{cy}$$

We can now solve for the parameters k and c, using the method discussed in Sec. A.4. We must recognize that x is the dependent variable and y is the independent variable in the above transformation. The procedure is illustrated in the following example.

Example. Determine the logarithmic function that passes through the two data points (3, 1) and (6, 3).

The two equations are

$$\ln(3) = \ln(k) + c(1)$$
$$\ln(6) = \ln(k) + c(3)$$

Solving simultaneously gives

$$2c = \ln 6 - \ln(3), \quad \text{or} \quad c = 0.34658$$

and

$$\ln(k) = 0.75203 \quad \text{and} \quad k = 2.121$$

Since $a = 1/c$ and $b = 1/k$, the logarithmic function is

$$y = 2.885 \ln(0.4715x)$$

This function is shown in Fig. A.5.

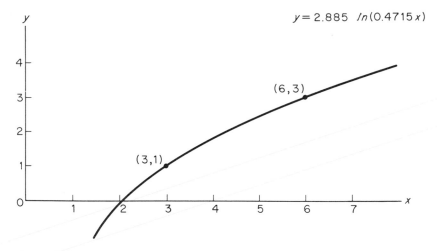

Figure A.5.   Logarithmic Function

A.6   Establishing Functions through Sets of Data Points

The method of least squares provides a technique for establishing a function through sets of data points. The technique was illustrated in Sec. 5.2 for linear, exponential, and power functions.

The method of least squares can also be used to establish polynomial functions through data points. We shall give the general formulation for establishing a pth-degree polynomial through a set of n data points. The procedure is illustrated for the second-degree polynomial.

Consider the polynomial of the form

$$y = B_0 + B_1x + B_2x^2 + \ldots + B_px^p$$

The method of least squares provides a technique for establishing a pth-degree polynomial through n data points ($n > p$). To determine the values of the parameters $B_0, B_1, B_2, \ldots, B_p$, we solve the following $p + 1$ equations simultaneously for the parameters.

$$\sum y = nB_0 + B_1 \sum x + \ldots + B_P \sum x^p$$
$$\sum xy = B_0 \sum x + B_1 \sum x^2 + \ldots + B_P \sum x^{p+1}$$
$$\sum x^2y = B_0 \sum x^2 + B_1 \sum x^3 + \ldots + B_P \sum x^{p+2}$$
$$\vdots \qquad \vdots$$
$$\sum x^py = B_0 \sum x^p + B_1 \sum x^{p+1} + \ldots + B_P \sum x^{2p}$$

This procedure is illustrated by establishing a quadratic function through a set of data points in the following example.

Example. Establish a quadratic function through the data points: (1.0, 50), (1.5, 30), (2.0, 25), (2.5, 20), (3.0, 20), (4.0, 25), and (5.0, 50). These data points are plotted in Fig. A.6.

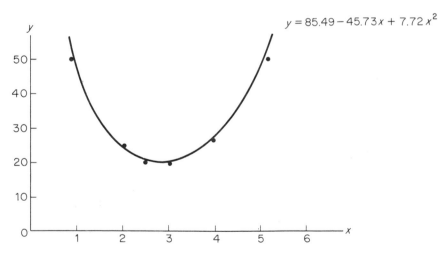

$$y = 85.49 - 45.73x + 7.72x^2$$

Figure A.6. Quadratic Function Established by Least Squares

A quadratic function is of degree 2. To determine the parameters B_0, B_1, and B_2 of the quadratic function

$$y = B_0 + B_1 x + B_2 x^2$$

we solve the following three equations simultaneously:

$$\sum y = nB_0 + B_1 \sum x + B_2 \sum x^2$$

$$\sum xy = B_0 \sum x + B_1 \sum x^2 + B_2 \sum x^3$$

$$\sum x^2 y = B_0 \sum x^2 + B_1 \sum x^3 + B_2 \sum x^4$$

where

x	y	xy	x^2y	x^2	x^3	x^4
1.0	50	50	50.0	1.0	1.00	1.00
1.5	30	45	67.5	2.25	3.38	5.06
2.0	25	50	100.0	4.00	8.00	16.00
2.5	20	50	125.0	6.25	15.62	39.06
3.0	20	60	180.0	9.00	27.00	81,00
4.0	25	100	400.0	16.00	64.00	256.00
5.0	50	250	1250.0	25.00	125.00	625.00
$\sum$ 19.0	220	605	2172.5	63.50	244.00	1023.12

Inserting these summations into the three equations gives

$$220.0 = 7.0B_0 + 19.0B_1 + 63.5B_2$$

$$605.0 = 19.0B_0 + 63.5B_1 + 244.0B_2$$

$$2172.5 = 63.5B_0 + 244.0B_1 + 1023.1B_2$$

Solving the equations simultaneously gives $B_0 = 85.49$, $B_1 = -45.73$, and $B_2 = 7.72$. The quadratic function is

$$y = 85.49 - 45.73x + 7.72x^2$$

Logarithms:
Laws of Exponents

B.1   Logarithms as Exponents

Consider the exponential function $y = a^x$, where y represents the value of the base number a raised to the exponent x. The definition of the logarithm of y is as follows: The logarithm of y to the base a is x. This is written as $\log_a (y) = x$. A *logarithm is an exponent*. To understand logarithms, it is important to remember the simple fact that a logarithm is an exponent. In fact, it might be helpful for the student to consider the terms logarithm and exponent as synonymous. To illustrate the relationship between a number and its logarithm, consider the following examples.

Example. Determine $\log_{10} (100)$. The logarithm of 100 to the base 10 is the exponent to which 10 must be raised to give 100.
Since $100 = 10^2$, we know that $\log_{10} (100) = 2$.

Example. Determine $\log_{10} (10,000)$.
Since $10,000 = 10^4$, we conclude that $\log_{10} (10,000) = 4$.

Example. Determine $\log_2 (8)$.
Since $8 = 2^3$, we conclude that $\log_2 (8) = 3$.

Example. Determine $\log_6 (36)$.

Since $36 = 6^2$, we conclude that $\log_6 (36) = 2$.

Logarithms are used for two purposes in this text. These are (1) in the calculation of the product of two numbers, the quotient of two numbers, or a number raised to a power, and (2) the algebraic manipulation of exponential functions for the purpose of expressing the function in a linear form. To understand the use of logarithms, it will be useful to review the laws of exponents.

B.2 Laws of Exponents

Let a represent the base and m and n represent exponents. The three laws of exponents are

(i) $a^m \cdot a^n = a^{m+n}$

$$\text{(ii)} \quad \frac{a^m}{a^n} = a^{m-n} = \begin{cases} a^{m-n} & \text{if } m > n \\ 1 & \text{if } m = n \\ \dfrac{1}{a^{n-m}} & \text{if } n > m \end{cases}$$

(iii) $(a^m)^n = a^{mn}$

The usefulness of logarithms in calculations of products, quotients, and powers is based upon the laws of exponents. Numbers can be expressed with a common base. The product of two numbers is then equal to the sum of the exponents, the quotient of two numbers is equal to the difference between the exponents, and a number raised to a power is equal to the product of the exponents. These properties of exponents are illustrated by the following examples.

Example. Determine the product of 100 times 1000, using the laws of exponents.

Since $100 = 10^2$ and $1000 = 10^3$, we know that

$$(100)(1000) = (10^2)(10^3) = 10^{2+3} = 10^5 = 100,000$$

Example. Determine the product of 16 times 64, using the laws of exponents.

Since $16 = 2^4$ and $64 = 2^6$, the product of 16 and 64 is

$$(2^4)(2^6) = 2^{4+6} = 2^{10} = 1024$$

Example. Determine the quotient of 1,000,000 divided by 100.

Since $1,000,000 = 10^6$ and $100 = 10^2$, $1,000,000/100 = 10^6/10^2 = 10^{6-2} = 10^4 = 10,000$.

Example. Determine the value of 100 raised to the third power.
Since $100 = 10^2$, we use the third law of exponents to show that $(10^2)^3 = 10^6 = 1,000,000$.

B.3 Laws of Logarithms

The laws of exponents can be rewritten in terms of logarithms. If we represent any two nonzero, nonnegative numbers as x and y, then the laws of logarithms are

$$\text{(i)} \quad \log_a (xy) = \log_a (x) + \log_a (y)$$

$$\text{(ii)} \quad \log_a (x/y) = \log_a (x) - \log_a (y)$$

$$\text{(iii)} \quad \log_a (x^y) = y \log_a (x)$$

Logarithms are exponents of the base. To determine the value of the product, quotient, or power we must raise the base to the logarithm (or exponent). The process of raising the base to the logarithm is termed finding the antilog of the logarithm. In symbolic form we write

$$(x)(y) = \text{antilog}_a [\log_a (x) + \log_a (y)] = a^{\log_a(x) + \log_a(y)}$$

The laws of logarithms are illustrated for the examples given in Sec. B.2.

Example. Determine the product of 100 times 1000, using logarithms.

$$\log_{10} [(100)(1000)] = \log_{10} (100) + \log_{10} (1000) = 2 + 3 = 5$$

$$\text{antilog}_{10} (5) = 10^5 = 100,000$$

Example. Determine the product of 16 times 64, using logarithms.

$$\log_2 [(16)(64)] = \log_2 (16) + \log_2 (64) = 4 + 6 = 10$$

$$\text{antilog}_2 (10) = 2^{10} = 1024$$

Example. Determine the quotient of 1,000,000 divided by 100, using logarithms.

$$\log_{10} (1,000,000/100) = \log_{10} (1,000,000) - \log_{10} (100) = 6 - 2 = 4$$

$$\text{antilog}_{10} (4) = 10^4 = 10,000$$

Example. Determine the value of 100 raised to the third power, using logarithms.

$$\log_{10}(100)^3 = 3\log_{10}(100) = 3(2) = 6$$
$$\text{antilog}_{10}(6) = 10^6 = 1,000,000$$

B.4   Determining Logarithms

To determine the logarithm of a number, it is necessary to have a table that gives the logarithms of numbers using a specific base. Two bases are commonly used, the base 10 and the base e. Logarithms using the base 10 are termed common logarithms and are written as $\log(x)$. The absence of a subscript indicates that the base is 10. Logarithms using the base e are termed Naperian or natural logarithms and are written as $\ln(x)$. The symbol "ln" refers to the natural or Naperian logarithm with base e. Any base other than 10 or e is indicated by $\log_a(x)$. We shall first illustrate the common or base 10 logarithms.

A logarithm is composed of two parts, the characteristic and the mantissa. The method of determining the characteristic and the mantissa requires that the number be written in the form: $x.xxxx \cdot 10^n$. That is, the number for which the logarithm is to be obtained must be written in a form such that a simple integer is to the left of the decimal point with the remaining significant digits to the right of the decimal point. The number x.xxxx is termed the *argument*. The magnitude of the number is expressed as the argument multiplied by a power of 10. The characteristic of the logarithm is n, the power to which 10 is raised. We can illustrate the determination of the characteristic by the following examples.

Example. Determine the characteristic of 121.
Since $121 = 1.21 \times 10^2$, the characteristic is 2. Note also that the argument is 1.21.

Example. Determine the characteristic of 1640.
Since $1640 = 1.640 \times 10^3$, the characteristic is 3.

Example. Determine the characteristic of 0.0653.
Since $0.0653 = 6.53 \times 10^{-2}$, the characteristic is -2.

Example. Determine the characteristic of 7.54.
Since $7.54 = 7.54 \times 10^0$, the characteristic is 0.

The mantissa of the number is the exponent to which the base 10 must be raised to obtain the argument x.xxxx. Since the argument has a value between 1.0000 and 9.9999, the *mantissa must have a value between 0 and 1*. The relationship, again, between the base, the mantissa, and the argument is

$$(\text{base})^{\text{mantissa}} = \text{argument}$$

If we include the characteristic, we see that

$$(\text{base})^{\text{mantissa}} \cdot (\text{base})^{\text{characteristic}} = \text{number}$$

The logarithm of the number is thus the sum of the mantissa and the characteristic. Mantissas for common logarithms are given in Table III. As an example, we shall determine the logarithm of 297. We first note that 297 = 2.97 × 10². The characteristic of the logarithm is thus 2. The mantissa of the argument 2.97 is 0.4728. In terms of exponential notation, 297 = $10^{0.4728}$ × 10². The logarithm of the number is the exponent to which 10 must be raised to obtain 297. The logarithm of 297 is, therefore, 2.4728; i.e., log (297) = 2.4728. The method of determining logarithms is further illustrated by the following examples.

Example. log (121) = 2.0828, since 121 = $10^{0.0828}$ × 10².

Example. log (1640) = 3.2148, since 1640 = $10^{0.2148}$ × 10³.

Example. log (0.0653) = -1.1851, since 0.0653 = $10^{0.8149}$ × 10^{-2}.

Example. log (7.54) = 0.8774, since 7.54 = $10^{0.8874}$ × 10^{0}.

B.5 Using Logarithms in Calculations

The first law of logarithms specifies that the product of two numbers is given by the sum of their logarithms.

Example. Multiply 78 by 194.

$$\log 78 = 1.8921$$
$$\log 194 = 2.2878$$
$$\text{Sum of logarithms} = \overline{4.1799}$$
$$\text{antilog} (4.1799) = 15{,}130$$

The fourth digit from the left, 3, was obtained by linear interpolation.

The second law of logarithms states that the quotient of two numbers is given by the difference between two logarithms.

Example. Divide 657 by 132.

$$\log 657 = 2.8041$$
$$\log 132 = 2.1206$$
$$\text{Difference of logarithms} = \overline{0.6835}$$
$$\text{antilog} (0.6835) = 4.83$$

Example. Divide 456 by 3648.

$$\begin{aligned}
\log 456 &= 2.6590 \\
\log 3648\dagger &= \underline{3.5620} \\
\text{Difference of logarithms} &= -0.9030
\end{aligned}$$

This problem requires a slight modification. Negative exponents such as -0.9030 are not included in the table of mantissas. We can, however, rewrite the logarithm as

$$-0.9030 = +0.0970 - 1.0000$$

Since logarithms are exponents, we actually have converted $10^{-0.9030}$ to $10^{0.0970} \cdot 10^{-1.0000}$. Thus, antilog $(0.0970) = 1.25$, and the quotient is $1.25 \cdot 10^{-1} = 0.125$.

The third law of logarithms states that to raise a number to a given power, the logarithm of the number is multiplied by the power.

Example. Raise 6.52 to the 3.25 power.

$$\begin{aligned}
\log (6.52) &= 0.8142 \\
&\underline{\times 3.25} \\
\text{Product} &= 2.6462
\end{aligned}$$

$$\text{antilog}\,(2.6462) = 442.8$$

The third law of logarithms also applies in determining roots of a number.

Example. Evaluate $(174)^{0.21}$.

$$\begin{aligned}
\log (174) &= 2.2405 \\
&\underline{\times 0.21} \\
\text{Product} &= 0.4705
\end{aligned}$$

$$\text{antilog}\,(0.4705) = 2.953$$

Table IV contains natural logarithms, i.e., logarithms with base e. The natural logarithms for numbers from 0 to 999 are given in this table. Both the characteristic and the mantissa are included in the table. Logarithms of numbers that exceed 999 can be determined by writing the number as a power of 10. Thus, the natural logarithm of 1230 is found by expressing 1230 as 123×10^1. The natural logarithm of 1230 is

$$\ln (1230) = \ln (123) + \ln (10)$$

$$\ln (1230) = 4.81218 + 2.30259$$

$$\ln (1230) = 7.11477$$

The laws of logarithms directly apply to natural logarithms.

---

† By interpolation.

B.6 Manipulation of Logarithmic and Exponential Functions

An exponential function may be expressed in terms of logarithms by expressing the terms of both sides of the equal sign in terms of logarithms. Thus,

$$y = kb^{f(x)}$$

can be expressed as

$$\log_a y = \log_a k + f(x) \log_a b$$

Similarly, given the function

$$y = k \log_a f(x)$$

the function can be expressed in its exponential form as

$$a^y = f(x)^k$$

A logarithm may be expressed with a different base by using the following relationship

$$\log_a x = (\log_b x)(\log_a b)$$

Example. Convert $\log_{10} (25)$ to $\log_e (25)$.

$$\log_e (25) = \log_{10} (25) \log_e (10)$$
$$\log_e (25) = (1.3979)(2.30259) = 3.21888$$

It is also helpful to recognize that

$$\log_a (a) = 1$$

and

$$\log_a (1) = 0$$

Tables

TABLE I. VALUES OF $(1 + i)^n$

n	i = 0.005	i = 0.0075	i = 0.01	i = 0.0125	i = 0.015	i = 0.02	i = 0.03	i = 0.04	i = 0.05	i = 0.06	i = 0.07	i = 0.08
1	1.005000	1.007500	1.010000	1.012500	1.015000	1.020000	1.030000	1.040000	1.050000	1.060000	1.070000	1.080000
2	1.010025	1.015056	1.020100	1.025156	1.030225	1.040400	1.060900	1.081600	1.102500	1.123600	1.144900	1.166400
3	1.015075	1.022669	1.030301	1.037971	1.045678	1.061208	1.092727	1.124864	1.157625	1.191016	1.225043	1.259712
4	1.020150	1.030339	1.040604	1.050945	1.061364	1.082432	1.125509	1.169859	1.215506	1.262477	1.310796	1.360489
5	1.025251	1.038067	1.050110	1.064082	1.077284	1.104081	1.159274	1.216653	1.276282	1.338226	1.402552	1.469328
6	1.030378	1.045852	1.061520	1.077383	1.093443	1.126162	1.194052	1.265319	1.340096	1.418519	1.500730	1.586874
7	1.035529	1.053696	1.072135	1.090850	1.109845	1.148686	1.229874	1.315932	1.407100	1.503630	1.605781	1.713824
8	1.040707	1.061599	1.082857	1.104486	1.126493	1.171659	1.266770	1.368569	1.477455	1.593848	1.718186	1.850930
9	1.045911	1.069561	1.093685	1.118292	1.143390	1.195093	1.304773	1.423312	1.551328	1.689479	1.838459	1.999004
10	1.051140	1.077583	1.104622	1.132271	1.160541	1.218994	1.343916	1.480244	1.628895	1.790848	1.967151	2.158925
11	1.056396	1.085664	1.115668	1.146424	1.177949	1.243374	1.384234	1.539454	1.710339	1.898299	2.104852	2.331629
12	1.061678	1.093807	1.126825	1.160755	1.195618	1.268242	1.425761	1.601032	1.795856	2.012196	2.252192	2.518170
13	1.066986	1.102010	1.138093	1.175264	1.213552	1.293607	1.468534	1.665074	1.885649	2.132928	2.404845	2.719624
14	1.072321	1.110276	1.149474	1.189955	1.231756	1.319479	1.512590	1.731676	1.979932	2.260904	2.578534	2.937194
15	1.077683	1.118603	1.160969	1.204829	1.250232	1.345868	1.557967	1.800944	2.078928	2.396558	2.759032	3.172169
16	1.083071	1.126992	1.172579	1.219890	1.268986	1.372786	1.604706	1.872981	2.182875	2.540352	2.952164	3.425943
17	1.088487	1.135445	1.184304	1.235138	1.288020	1.400241	1.652848	1.947900	2.292018	2.692773	3.158815	3.700018
18	1.093929	1.143960	1.196147	1.250577	1.307341	1.428246	1.702433	2.025817	2.406619	2.854339	3.379932	3.996019
19	1.099399	1.152540	1.208109	1.266210	1.326951	1.456811	1.753506	2.106849	2.526950	3.025600	3.616527	4.315701
20	1.104896	1.161184	1.220190	1.282037	1.346855	1.485947	1.806111	2.191123	2.653298	3.207135	3.869684	4.660957
21	1.110420	1.169893	1.232392	1.298063	1.367058	1.515666	1.860295	2.278768	2.785963	3.399564	4.140562	5.033834
22	1.115972	1.178667	1.244716	1.314288	1.387564	1.545980	1.916103	2.369919	2.925261	3.603537	4.430402	5.436540
23	1.121552	1.187507	1.257163	1.330717	1.408377	1.576899	1.973587	2.464716	3.071524	3.819750	4.740530	5.871464
24	1.127160	1.196414	1.269735	1.347351	1.429503	1.608437	2.032794	2.563304	3.225100	4.048935	5.072367	6.341181
25	1.132796	1.205387	1.282432	1.364193	1.450945	1.640606	2.093778	2.665836	3.386355	4.291871	5.427433	6.848475

n												
26	1.138460	1.214427	1.295256	1.381245	1.472710	1.673418	2.156591	2.772470	3.555673	4.549383	5.807353	7.396353
27	1.144152	1.223535	1.308209	1.398511	1.494800	1.706886	2.221289	2.883369	3.733456	4.822346	6.213868	7.988061
28	1.149873	1.232712	1.321291	1.415992	1.517222	1.741024	2.287928	2.998703	3.920129	5.111687	6.648838	8.627106
29	1.155622	1.241957	1.334504	1.433692	1.539981	1.775845	2.356566	3.118651	4.116136	5.418388	7.114257	9.317275
30	1.161400	1.251272	1.347849	1.451613	1.563080	1.811362	2.427263	3.243398	4.321942	5.743491	7.612255	10.062657
31	1.167207	1.260656	1.361327	1.469759	1.586526	1.847589	2.500080	3.373133	4.538039	6.088101	8.145113	10.867669
32	1.173043	1.270111	1.374941	1.488131	1.610324	1.884541	2.575083	3.508059	4.764941	6.453387	8.715271	11.737083
33	1.178908	1.279637	1.388690	1.506732	1.634479	1.922231	2.652335	3.648381	5.003186	6.840590	9.325340	12.676050
34	1.184803	1.289234	1.402577	1.525566	1.658996	1.960676	2.731905	3.794316	5.253348	7.251025	9.978114	13.690134
35	1.190727	1.298904	1.416603	1.544636	1.683881	1.999890	2.813862	3.946089	5.516015	7.686087	10.676581	14.785344
36	1.196681	1.308645	1.430769	1.563944	1.709140	2.039887	2.898278	4.103933	5.791816	8.147252	11.423942	15.968172
37	1.202664	1.318460	1.445076	1.583493	1.734777	2.080685	2.985227	4.268090	6.081407	8.636087	12.223618	17.245626
38	1.208677	1.328349	1.459527	1.603287	1.760798	2.122299	3.074783	4.438813	6.385477	9.154252	13.079271	18.625276
39	1.214721	1.338311	1.474123	1.623328	1.787210	2.164745	3.167027	4.616366	6.704751	9.703507	13.994820	20.115298
40	1.220794	1.348349	1.488864	1.643619	1.814018	2.208040	3.262038	4.801021	7.039989	10.285718	14.974458	21.724522
41	1.226898	1.358461	1.503752	1.664165	1.841229	2.252200	3.359899	4.993061	7.391988	10.902861	16.022670	23.462483
42	1.233033	1.368650	1.518790	1.684967	1.868847	2.297244	3.460696	5.192784	7.761588	11.557033	17.144257	25.339482
43	1.239198	1.378915	1.533978	1.706029	1.896880	2.343189	3.564517	5.400495	8.149667	12.250455	18.344355	27.366640
44	1.245394	1.389256	1.549318	1.727354	1.925333	2.390053	3.671452	5.616515	8.557150	12.985482	19.628460	29.555972
45	1.251621	1.399676	1.564811	1.748946	1.954213	2.437854	3.781596	5.841176	8.985008	13.764611	21.002452	31.920449
46	1.257879	1.410173	1.580459	1.770808	1.983526	2.486611	3.895044	6.074823	9.434258	14.590487	22.472623	34.474085
47	1.264168	1.420750	1.596263	1.792943	2.013279	2.536344	4.011895	6.317816	9.905971	15.465917	24.045707	37.232012
48	1.270489	1.431405	1.612226	1.815355	2.043478	2.587070	4.132252	6.570528	10.401270	16.393872	25.728907	40.210573
49	1.276842	1.442141	1.628348	1.838047	2.074130	2.638812	4.256219	6.833349	10.921333	17.377504	27.529930	43.427420
50	1.283226	1.452957	1.644632	1.861022	2.105242	2.691588	4.383906	7.106683	11.467400	18.420154	29.457025	46.901613
51	1.289642	1.463854	1.661078	1.884285	2.136821	2.745420	4.515423	7.390951	12.040770	19.525364	31.519017	50.653742
52	1.296090	1.474833	1.677689	1.907839	2.168873	2.800328	4.650886	7.686589	12.642808	20.696885	33.725348	54.706041
53	1.302571	1.485894	1.694466	1.931687	2.201406	2.856335	4.790412	7.994052	13.274949	21.938699	36.086122	59.082524
54	1.309083	1.497038	1.711410	1.955833	2.234428	2.913461	4.934125	8.313814	13.938696	23.255020	38.612151	63.809126
55	1.315629	1.508266	1.728525	1.980281	2.267944	2.971731	5.082149	8.646367	14.635631	24.650322	41.315002	68.913856
56	1.322207	1.519578	1.745810	2.005034	2.301963	3.031165	5.234613	8.992222	15.367413	26.129341	44.207052	74.426965
57	1.328818	1.530975	1.763268	2.030097	2.336493	3.091789	5.391651	9.351911	16.133783	27.697101	47.301545	80.381122
58	1.335462	1.542457	1.780901	2.055473	2.371540	3.153624	5.553401	9.725987	16.942572	29.358927	50.612653	86.811612
59	1.342139	1.554026	1.798710	2.081167	2.407113	3.216697	5.720003	10.115026	17.789701	31.120463	54.155539	93.756540
60	1.348850	1.565681	1.816697	2.107181	2.443220	3.281031	5.891603	10.519627	18.679186	32.987691	57.946427	101.257064

TABLE II. VALUES OF $(1 + i)^{-n}$

n	$i = 0.005$	$i = 0.0075$	$i = 0.01$	$i = 0.0125$	$i = 0.015$	$i = 0.02$	$i = 0.03$	$i = 0.04$	$i = 0.05$	$i = 0.06$	$i = 0.07$	$i = 0.08$
1	0.995025	0.992556	0.990099	0.987654	0.985222	0.980392	0.970874	0.961538	0.952381	0.943396	0.934779	0.925926
2	0.990074	0.985167	0.980296	0.975461	0.970662	0.961169	0.942596	0.924556	0.907029	0.889996	0.873439	0.857339
3	0.985149	0.977833	0.970590	0.963418	0.956317	0.942322	0.915142	0.888996	0.863838	0.839619	0.816298	0.793832
4	0.980248	0.970554	0.960980	0.951524	0.942184	0.923845	0.888487	0.854804	0.822702	0.792094	0.762895	0.735030
5	0.975371	0.963329	0.951466	0.939777	0.928260	0.905731	0.862609	0.821927	0.783526	0.747258	0.712986	0.680583
6	0.970518	0.956158	0.942045	0.928175	0.914542	0.887971	0.837484	0.790315	0.746215	0.704961	0.666342	0.630170
7	0.965690	0.949040	0.932718	0.916716	0.901027	0.870560	0.813092	0.759918	0.710681	0.665057	0.622750	0.583490
8	0.960885	0.941975	0.923483	0.905398	0.887711	0.853490	0.789409	0.730690	0.676839	0.627412	0.582009	0.540269
9	0.956105	0.934963	0.914340	0.894221	0.874592	0.836755	0.766417	0.702587	0.644609	0.591898	0.543934	0.500249
10	0.951348	0.928003	0.905287	0.883181	0.861667	0.820348	0.744094	0.675564	0.613913	0.558395	0.508349	0.463193
11	0.946615	0.921095	0.896324	0.872277	0.848933	0.804263	0.722421	0.649581	0.584679	0.526788	0.475093	0.428883
12	0.941905	0.914238	0.887449	0.861509	0.836387	0.778493	0.701380	0.624597	0.556837	0.496969	0.444012	0.397114
13	0.937219	0.907432	0.878663	0.850873	0.824027	0.773033	0.680951	0.600574	0.530321	0.468839	0.414964	0.367698
14	0.932556	0.900677	0.869963	0.840368	0.811849	0.757875	0.661118	0.577475	0.505068	0.442301	0.387817	0.340461
15	0.927917	0.893973	0.861349	0.829993	0.799852	0.743015	0.641862	0.555264	0.481017	0.417265	0.362446	0.315243
16	0.923300	0.887318	0.852821	0.819746	0.788031	0.728446	0.623167	0.533908	0.458112	0.393646	0.338735	0.291890
17	0.918707	0.880712	0.844377	0.809626	0.776385	0.714163	0.605016	0.513373	0.436297	0.371364	0.316574	0.270269
18	0.914136	0.874156	0.836017	0.799631	0.764912	0.700159	0.587395	0.493628	0.415521	0.350344	0.295864	0.250249
19	0.909588	0.867649	0.827740	0.789759	0.753607	0.686431	0.570286	0.474642	0.395734	0.330513	0.276508	0.231712
20	0.905063	0.861190	0.819544	0.780009	0.742470	0.672971	0.553676	0.456387	0.376889	0.311805	0.258419	0.214548
21	0.900560	0.854779	0.811430	0.770379	0.731498	0.659776	0.537549	0.438834	0.358942	0.294155	0.241513	0.198656
22	0.896080	0.848416	0.803396	0.760868	0.720688	0.646839	0.521892	0.421955	0.341850	0.277505	0.225713	0.183941
23	0.891622	0.842100	0.795442	0.751475	0.710037	0.634156	0.506692	0.405726	0.325571	0.261797	0.210947	0.170315
24	0.887186	0.835831	0.787566	0.742197	0.699544	0.621721	0.491934	0.390121	0.310068	0.246979	0.197147	0.157699
25	0.882772	0.829609	0.779768	0.733034	0.689206	0.609531	0.477606	0.375117	0.295303	0.232999	0.184249	0.146018

26	0.135202	0.172195	0.219810	0.281241	0.360689	0.463695	0.597579	0.679021	0.723984	0.772048	0.823434	0.878380
27	0.125187	0.160930	0.207368	0.267848	0.346817	0.450189	0.585862	0.668986	0.715046	0.764404	0.817304	0.874010
28	0.115914	0.150402	0.195630	0.255094	0.333477	0.437077	0.574375	0.659099	0.706219	0.756836	0.811220	0.869662
29	0.107328	0.140563	0.184557	0.242946	0.320651	0.424346	0.563112	0.649359	0.697500	0.749342	0.805181	0.865335
30	0.099377	0.131367	0.174110	0.231377	0.308319	0.411987	0.552071	0.639762	0.688889	0.741923	0.799187	0.861030
31	0.092016	0.122773	0.164255	0.220359	0.296460	0.399987	0.541246	0.630308	0.680384	0.734577	0.793238	0.856746
32	0.085200	0.114741	0.154957	0.209866	0.285058	0.388337	0.530633	0.620993	0.671984	0.727304	0.787333	0.852484
33	0.078889	0.107235	0.146186	0.199873	0.274094	0.377026	0.520229	0.611816	0.663688	0.720103	0.781472	0.848242
34	0.073045	0.100219	0.137912	0.190355	0.263552	0.366045	0.510028	0.602774	0.655494	0.712973	0.775654	0.844022
35	0.067635	0.093663	0.130105	0.181290	0.253415	0.355383	0.500028	0.593866	0.647402	0.705914	0.769880	0.839823
36	0.062635	0.087535	0.122741	0.172657	0.243669	0.345032	0.490223	0.585090	0.639409	0.698925	0.764149	0.835645
37	0.057986	0.081809	0.115793	0.164436	0.234297	0.334983	0.480611	0.576443	0.631515	0.692005	0.758461	0.831487
38	0.053690	0.076457	0.109239	0.156605	0.225285	0.325226	0.471187	0.567924	0.623719	0.685153	0.752814	0.827351
39	0.049713	0.071455	0.103056	0.149148	0.216621	0.315754	0.461948	0.559531	0.616018	0.678370	0.747210	0.823235
40	0.046031	0.066780	0.097222	0.142046	0.208289	0.306557	0.452890	0.551262	0.608413	0.671653	0.741648	0.819139
41	0.042621	0.062412	0.091719	0.135282	0.200278	0.297628	0.444010	0.543116	0.600902	0.665003	0.736127	0.815064
42	0.039464	0.058329	0.086527	0.128840	0.192575	0.288959	0.435304	0.535089	0.593484	0.658419	0.730647	0.811008
43	0.036541	0.054513	0.081630	0.122704	0.185168	0.280543	0.426769	0.527182	0.586157	0.651900	0.725208	0.806974
44	0.033834	0.050946	0.077009	0.116861	0.178046	0.272372	0.418401	0.519391	0.578920	0.645445	0.719810	0.802959
45	0.031329	0.047613	0.072650	0.111297	0.171198	0.264439	0.410197	0.511715	0.571773	0.639055	0.714451	0.798964
46	0.029007	0.044499	0.068538	0.105997	0.164614	0.256737	0.402154	0.504153	0.564714	0.632728	0.709133	0.794989
47	0.026859	0.041587	0.064658	0.100949	0.158283	0.249259	0.394268	0.496702	0.557742	0.626463	0.703854	0.791034
48	0.024869	0.038867	0.060998	0.096142	0.152195	0.241999	0.386538	0.489362	0.550856	0.620260	0.698614	0.787098
49	0.023027	0.036324	0.057546	0.091564	0.146341	0.234950	0.378958	0.482130	0.544056	0.614119	0.693414	0.783182
50	0.021321	0.033948	0.054288	0.087204	0.140713	0.228107	0.371528	0.475005	0.537339	0.608039	0.688252	0.779286
51	0.019742	0.031727	0.051215	0.083051	0.135301	0.221463	0.364243	0.467985	0.530705	0.602019	0.683128	0.775409
52	0.018280	0.029651	0.048316	0.079096	0.130097	0.215013	0.357101	0.461069	0.524153	0.596058	0.678043	0.771551
53	0.016926	0.027712	0.045582	0.075330	0.125093	0.208750	0.350100	0.454255	0.517682	0.590156	0.672995	0.767713
54	0.015672	0.025899	0.043002	0.071743	0.120282	0.202670	0.343234	0.447542	0.511291	0.584313	0.667986	0.763893
55	0.014511	0.024204	0.040567	0.068326	0.115656	0.196767	0.336504	0.440928	0.504979	0.578528	0.663013	0.760093
56	0.013436	0.022621	0.038271	0.065073	0.111207	0.191036	0.329906	0.434412	0.498745	0.572800	0.658077	0.756311
57	0.012441	0.021141	0.036105	0.061974	0.106930	0.185472	0.323437	0.427992	0.492587	0.567129	0.653178	0.752548
58	0.011519	0.019758	0.034061	0.059023	0.102817	0.180070	0.317095	0.421667	0.486506	0.561514	0.648316	0.748804
59	0.010666	0.018465	0.032133	0.056213	0.098863	0.174826	0.310878	0.415435	0.480500	0.555954	0.643490	0.745079
60	0.009876	0.017257	0.030314	0.053536	0.095060	0.169733	0.304782	0.409296	0.474568	0.550450	0.638700	0.741372

TABLE III. COMMON LOGARITHMS

N	0	1	2	3	4	5	6	7	8	9
10	0000	0043	0086	0128	0170	0212	0253	0294	0334	0374
11	0414	0453	0492	0531	0569	0607	0645	0682	0719	0755
12	0792	0828	0864	0899	0934	0969	1004	1038	1072	1106
13	1139	1173	1206	1239	1271	1303	1335	1367	1399	1430
14	1461	1492	1523	1553	1584	1614	1644	1673	1703	1732
15	1761	1790	1818	1847	1875	1903	1931	1959	1987	2014
16	2041	2068	2095	2122	2148	2175	2201	2227	2253	2279
17	2304	2330	2355	2380	2405	2430	2455	2480	2504	2529
18	2553	2577	2601	2625	2648	2672	2695	2718	2742	2765
19	2788	2810	2833	2856	2878	2900	2923	2945	2967	2989
20	3010	3032	3054	3075	3096	3118	3139	3160	3181	3201
21	3222	3243	3263	3284	3304	3324	3345	3365	3385	3404
22	3424	3444	3464	3483	3502	3522	3541	3560	3579	3598
23	3617	3636	3655	3674	3692	3711	3729	3747	3766	3784
24	3802	3820	3838	3856	3874	3892	3909	3927	3945	3962
25	3979	3997	4014	4031	4048	4065	4082	4099	4116	4133
26	4150	4166	4183	4200	4216	4232	4249	4265	4281	4298
27	4314	4330	4346	4362	4378	4393	4409	4425	4440	4456
28	4472	4487	4502	4518	4533	4548	4564	4579	4594	4609
29	4624	4639	4654	4669	4683	4698	4713	4728	4742	4757
30	4771	4786	4800	4814	4829	4843	4857	4871	4886	4900
31	4914	4928	4942	4955	4969	4983	4997	5011	5024	5038
32	5051	5065	5079	5092	5105	5119	5132	5145	5159	5172
33	5185	5198	5211	5224	5237	5250	5263	5276	5289	5302
34	5315	5328	5340	5353	5366	5378	5391	5403	5416	5428
35	5441	5453	5465	5478	5490	5502	5514	5527	5539	5551
36	5563	5575	5587	5599	5611	5623	5635	5647	5658	5670
37	5682	5694	5705	5717	5729	5740	5752	5763	5775	5786
38	5798	5809	5821	5832	5843	5855	5866	5877	5888	5899
39	5911	5922	5933	5944	5955	5966	5977	5988	5999	6010
40	6021	6031	6042	6053	6064	6075	6085	6096	6107	6117
41	6128	6138	6149	6160	6170	6180	6191	6201	6212	6222
42	6232	6243	6253	6263	6274	6284	6294	6304	6314	6325
43	6335	6345	6355	6365	6375	6385	6395	6405	6415	6425
44	6435	6444	6454	6464	6474	6484	6493	6503	6513	6522
45	6532	6542	6551	6561	6571	6580	6590	6599	6609	6618
46	6628	6637	6646	6656	6665	6675	6684	6693	6702	6712
47	6721	6730	6739	6749	6758	6767	6776	6785	6794	6803
48	6812	6821	6830	6839	6848	6857	6866	6875	6884	6893
49	6902	6911	6920	6928	6937	6946	6955	6964	6972	6981
50	6990	6998	7007	7016	7024	7033	7042	7050	7059	7067
51	7076	7084	7093	7101	7110	7118	7126	7135	7143	7152
52	7160	7168	7177	7185	7193	7202	7210	7218	7226	7235
53	7243	7251	7259	7267	7275	7284	7292	7300	7308	7316
54	7324	7332	7340	7348	7356	7364	7372	7380	7388	7396

TABLE III. COMMON LOGARITHMS (Continued)

N	0	1	2	3	4	5	6	7	8	9
55	7404	7412	7419	7427	7435	7443	7451	7459	7466	7474
56	7482	7490	7497	7505	7513	7520	7528	7536	7543	7551
57	7559	7566	7574	7582	7589	7597	7604	7612	7619	7627
58	7634	7642	7649	7657	7664	7672	7679	7686	7694	7701
59	7709	7716	7723	7731	7738	7745	7752	7760	7767	7774
60	7782	7789	7796	7803	7810	7818	7825	7832	7839	7846
61	7853	7860	7868	7875	7882	7889	7896	7903	7910	7917
62	7924	7931	7938	7945	7952	7959	7966	7973	7980	7987
63	7993	8000	8007	8014	8021	8028	8035	8041	8048	8055
64	8062	8069	8075	8082	8089	8096	8102	8109	8116	8122
65	8129	8136	8142	8149	8156	8162	8169	8176	8182	8189
66	8195	8202	8209	8215	8222	8228	8235	8241	8248	8254
67	8261	8267	8274	8280	8287	8293	8299	8306	8312	8319
68	8325	8331	8338	8344	8351	8357	8363	8370	8376	8382
69	8388	8395	8401	8407	8414	8420	8426	8432	8439	8445
70	8451	8457	8463	8470	8476	8482	8488	8494	8500	8506
71	8513	8519	8525	8531	8537	8543	8549	8555	8561	8567
72	8573	8579	8585	8591	8597	8603	8609	8615	8621	8627
73	8633	8639	8645	8651	8657	8663	8669	8675	8681	8686
74	8692	8698	8704	8710	8716	8722	8727	8733	8739	8745
75	8751	8756	8762	8768	8774	8779	8785	8791	8797	8802
76	8808	8814	8820	8825	8831	8837	8842	8848	8854	8859
77	8865	8871	8876	8882	8887	8893	8899	8904	8910	8915
78	8921	8927	8932	8938	8943	8949	8954	8960	8965	8971
79	8976	8982	8987	8993	8998	9004	9009	9015	9020	9025
80	9031	9036	9042	9047	9053	9058	9063	9069	9074	9079
81	9085	9090	9096	9101	9106	9112	9117	9122	9128	9133
82	9138	9143	9149	9154	9159	9165	9170	9175	9180	9186
83	9191	9196	9201	9206	9212	9217	9222	9227	9232	9238
84	9243	9248	9253	9258	9263	9269	9274	9279	9284	9289
85	9294	9299	9304	9309	9315	9320	9325	9330	9335	9340
86	9345	9350	9355	9360	9365	9370	9375	9380	9385	9390
87	9395	9400	9405	9410	9415	9420	9425	9430	9435	9440
88	9445	9450	9455	9460	9465	9469	9474	9479	9484	9489
89	9494	9499	9504	9509	9513	9518	9523	9528	9533	9538
90	9542	9547	9552	9557	9562	9566	9571	9576	9581	9586
91	9590	9595	9600	9605	9609	9614	9619	9624	9628	9633
92	9638	9643	9647	9652	9657	9661	9666	9671	9675	9680
93	9685	9689	9694	9699	9703	9708	9713	9717	9722	9727
94	9731	9736	9741	9745	9750	9754	9759	9763	9768	9773
95	9777	9782	9786	9791	9795	9800	9805	9809	9814	9818
96	9823	9827	9832	9836	9841	9845	9850	9854	9859	9863
97	9868	9872	9877	9881	9886	9890	9894	9899	9903	9908
98	9912	9917	9921	9926	9930	9934	9939	9943	9948	9952
99	9956	9961	9965	9969	9974	9978	9983	9987	9991	9996

TABLE IV. NATURAL OR NAPERIAN LOGARITHMS
0.000–0.499

N	0	1	2	3	4	5	6	7	8	9
0.00	− ∞	− 6⁺ .90776	−6 .21461	−5 .80914	−5 .52146	−5 .29832	−5 .11000	−4 96185	−4 .82831	−4 .71053
.01	−4.60517	.50986	.42285	.34281	.26870	.19971	.13517	.07454	.01738	*.96332
.02	−3.91202	.86323	.81671	.77226	.72970	.68888	.64966	.61192	.57555	.54046
.03	.50656	.47377	.44202	.41125	.38139	.35241	.32424	.29684	.27017	.24419
.04	.21888	.19418	.17009	.14656	.12357	.10109	.07911	.05761	.03655	.01593
.05	−2.99573	.97593	.95651	.93746	.91877	.90042	.88240	.86470	.84731	.83022
.06	.81341	.79688	.78062	.76462	.74887	.73337	.71810	.70306	.68825	.67365
.07	.65926	.64508	.63109	.61730	.60369	.59027	.57702	.56395	.55105	.53831
.08	.52573	.51331	.50104	.48891	.47694	.46510	.45341	.44185	.43042	.41912
.09	.40795	.39690	.38597	.37516	.36446	.35388	.34341	.33304	.32279	.31264
0.10	−2.30259	.29263	.28278	.27303	.26336	.25379	.24432	.23493	.22562	.21641
.11	.20727	.19823	.18926	.18037	.17156	.16282	.15417	.14558	.13707	.12863
.12	.12026	.11196	.10373	.09557	.08747	.07944	.07147	.06357	.05573	.04794
.13	.04022	.03256	.02495	.01741	.00992	.00248	*.99510	*.98777	*.98050	*.97328
.14	−1.96611	.95900	.95193	.94491	.93794	.93102	.92415	.91732	.91054	.90381
.15	.89712	.89048	.88387	.87732	.87080	.86433	.85790	.85151	.84516	.83885
.16	.83258	.82635	.82016	.81401	.80789	.80181	.79577	.78976	.78379	.77786
.17	.77196	.76609	.76026	.75446	.74870	.74297	.73727	.73161	.72597	.72037
.18	.71480	.70926	.70375	.69827	.69282	.68740	.68201	.67665	.67131	.66601
.19	.66073	.65548	.65026	.64507	.63990	.63476	.62964	.62455	.61949	.61445
0.20	−1.60944	.60445	.59949	.59455	.58964	.58475	.57988	.57504	.57022	.56542
.21	.56065	.55590	.55117	.54646	.54178	.53712	.53248	.52786	.52326	.51868
.22	.51413	.50959	.50508	.50058	.49611	.49165	.48722	.48281	.47841	.47403
.23	.46968	.46534	.46102	.45672	.45243	.44817	.44392	.43970	.43548	.43129
.24	.42712	.42296	.41882	.41469	.41059	.40650	.40242	.39837	.39433	.39030
.25	.38629	.38230	.37833	.37437	.37042	.36649	.36258	.35868	.35480	.35093
.26	.34707	.34323	.33941	.33560	.33181	.32803	.32426	.32051	.31677	.31304
.27	.30933	.30564	.30195	.29828	.29463	.29098	.28735	.28374	.28013	.27654
.28	.27297	.26940	.26585	.26231	.25878	.25527	.25176	.24827	.24479	.24133
.29	.23787	.23443	.23100	.22758	.22418	.22078	.21740	.21402	.21066	.20731
0.30	−1.20397	.20065	.19733	.19402	.19073	.18744	.18417	.18091	.17766	.17441
.31	.17118	.16796	.16475	.16155	.15836	.15518	.15201	.14885	.14570	.14256
.32	.13943	.13631	.13320	.13010	.12701	.12393	.12086	.11780	.11474	.11170
.33	.10866	.10564	.10262	.09961	.09661	.09362	.09064	.08767	.08471	.08176
.34	.07881	.07587	.07294	.07002	.06711	.06421	.06132	.05843	.05555	.05268
.35	−1.04982	.04697	.04412	.04129	.03846	.03564	.03282	.03002	.02722	.02443
.36	.02165	.01888	.01611	.01335	.01060	.00786	.00512	.00239	*.99967	*.99696
.37	−0.99425	.99155	.98886	.98618	.98350	.98083	.97817	.97551	.97286	.97022
.38	.96758	.96496	.96233	.95972	.95711	.95451	.95192	.94933	.94675	.94418
.39	.94161	.93905	.93649	.93395	.93140	.92887	.92634	.92382	.92130	.91879
0.40	−0.91629	.91379	.91130	.90882	.90634	.90387	.90140	.89894	.89649	.89404
.41	.89160	.88916	.88673	.88431	.88189	.87948	.87707	.87467	.87227	.86988
.42	.86750	.86512	.86275	.86038	.85802	.85567	.85332	.85097	.84863	.84630
.43	.84397	.84165	.83933	.83702	.83471	.83241	.83011	.82782	.82554	.82326
.44	.82098	.81871	.81645	.81419	.81193	.80968	.80744	.80520	.80296	.80073
.45	.79851	.79629	.79407	.79186	.78966	.78746	.78526	.78307	.78089	.77871
.46	.77653	.77436	.77219	.77003	.76787	.76572	.76357	.76143	.75929	.75715
.47	.75502	.75290	.75078	.74866	.74655	.74444	.74234	.74024	.73814	.73605
.48	.73397	.73189	.72981	.72774	.72567	.72361	.72155	.71949	.71744	.71539
.49	.71335	.71131	.70928	.70725	.70522	.70320	.70118	.69917	.69716	.69515

From *Standard Mathematical Tables*, 21st ed., Samuel Selby, © The Chemical Rubber Co., 1973. Used by permission of The Chemical Rubber Co.

TABLE IV. NATURAL OR NAPERIAN LOGARITHMS (Continued)
0.500–0.999

N	0	1	2	3	4	5	6	7	8	9
0.50	−0.69315	.69115	.68916	.68717	.68518	.68320	.68122	.67924	.67727	.67531
.51	.67334	.67139	.66934	.66748	.66553	.66359	.66165	.65971	.65778	.65585
.52	.65393	.65201	.65009	.64817	.64626	.64436	.64245	.64055	.63866	.63677
.53	.63488	.63299	.63111	.62923	.62736	.62549	.62362	.62176	.61990	.61804
.54	.61619	.61434	.61249	.61065	.60881	.60697	.60514	.60331	.60148	.59966
.55	.59784	.59602	.59421	.59240	.59059	.58879	.58699	.58519	.58340	.58161
.56	.57982	.57803	.57625	.57448	.57270	.57093	.56916	.56740	.56563	.56387
.57	.56212	.56037	.55862	.55687	.55513	.55339	.55165	.54991	.54818	.54645
.58	.54473	.54300	.54128	.53957	.53785	.53614	.53444	.53273	.53103	.52933
.59	.52763	.52594	.52425	.52256	.52088	.51919	.51751	.51584	.51416	.51249
0.60	−0.51083	.50916	.50750	.50584	.50418	.50253	.50088	.49923	.49758	.49594
.61	.49430	.49266	.49102	.48939	.48776	.48613	.48451	.48289	.48127	.47965
.62	.47804	.47642	.47482	.47321	.47160	.47000	.46840	.46681	.46522	.46362
.63	.46204	.46045	.45887	.45728	.45571	.45413	.45256	.45099	.44942	.44785
.64	.44629	.44473	.44317	.44161	.44006	.43850	.43696	.43541	.43386	.43232
.65	.43078	.42925	.42771	.42618	.42465	.42312	.42159	.42007	.41855	.41703
.66	.41552	.41400	.41249	.41098	.40947	.40797	.40647	.40497	.40347	.40197
.67	.40048	.39899	.39750	.39601	.39453	.39304	.39156	.39008	.38861	.38713
.68	.38566	.38419	.38273	.38126	.37980	.37834	.37688	.37542	.37397	.37251
.69	.37106	.36962	.36817	.36673	.36528	.36384	.36241	.36097	.35954	.35810
0.70	−0.35667	.35525	.35382	.35240	.35098	.34956	.34814	.34672	.34531	.34390
.71	.34249	.34108	.33968	.33827	.33687	.33547	.33408	.33268	.33129	.32989
.72	.32850	.32712	.32573	.32435	.32296	.32158	.32021	.31883	.31745	.31608
.73	.31471	.31334	.31197	.31061	.30925	.30788	.30653	.30517	.30381	.30246
.74	.30111	.29975	.29841	.29706	.29571	.29437	.29303	.29169	.29035	.28902
.75	.28768	.28635	.28502	.28369	.28236	.28104	.27971	.27839	.27707	.27575
.76	.27444	.27312	.27181	.27050	.26919	.26788	.26657	.26527	.26397	.26266
.77	.26136	.26007	.25877	.25748	.25618	.25489	.25360	.25231	.25103	.24974
.78	.24846	.24718	.24590	.24462	.24335	.24207	.24080	.23953	.23826	.23699
.79	.23572	.23446	.23319	.23193	.23067	.22941	.22816	.22690	.22565	.22439
0.80	−0.22314	.22189	.22065	.21940	.21816	.21691	.21567	.21433	.21319	.21196
.81	.21072	.20949	.20825	.20702	.20579	.20457	.20334	.20212	.20089	.19967
.82	.19845	.19723	.19601	.19480	.19358	.19237	.19116	.18995	.18874	.18754
.83	.18633	.18513	.18392	.18272	.18152	.18032	.17913	.17793	.17674	.17554
.84	.17435	.17316	.17198	.17079	.16960	.16842	.16724	.16605	.16487	.16370
.85	−0.16252	.16134	.16017	.15900	.15782	.15665	.15548	.15432	.15315	.15199
.86	.15032	.14966	.14850	.14734	.14618	.14503	.14387	.14272	.14156	.14041
.87	.13926	.13811	.13697	.13582	.13467	.13353	.13239	.13125	.13011	.12897
.88	.12783	.12670	.12556	.12443	.12330	.12217	.12104	.11991	.11878	.11766
.89	.11653	.11541	.11429	.11317	.11205	.11093	.10981	.10870	.10759	.10647
0.90	−0.10536	.10425	.10314	.10203	.10093	.09982	.09872	.09761	.09651	.09541
.91	.09431	.09321	.09212	.09102	.08992	.08883	.08744	.08665	.08556	.08447
.92	.08338	.08230	.08121	.08013	.07904	.07796	.07688	.07580	.07472	.07365
.93	.07257	.07150	.07042	.06935	.06828	.06721	.06614	.06507	.06401	.06294
.94	.06188	.06081	.05975	.05869	.05763	.05657	.05551	.05446	.05340	.05235
.95	.05129	.05024	.04919	.04814	.04709	.04604	.04500	.04395	.04291	.04186
.96	.04082	.03978	.03874	.03770	.03666	.03563	.03459	.03356	.03252	.03149
.97	.03046	.02943	.02840	.02737	.02634	.02532	.02429	.02327	.02225	.02122
.98	.02020	.01918	.01816	.01715	.01613	.01511	.01410	.01309	.01207	.01106
.99	.01005	.00904	.00803	.00702	.00602	.00501	.00401	.00300	.00200	.00100

TABLE IV. NATURAL OR NAPERIAN LOGARITHMS (Continued)
1.00–5.49

N	0	1	2	3	4	5	6	7	8	9
1.0	0.00000	.00995	.01980	.02956	.03922	.04879	.05827	.06766	.07696	.08618
.1	.09531	.10436	.11333	.12222	.13103	.13976	.14842	.15700	.16551	17395
.2	.18232	.19062	.19885	.20701	.21511	.22314	.23111	.23902	.24686	.25464
.3	.26236	.27003	.27763	.28518	.29267	.30010	.30748	.31481	.32208	.32930
.4	.33647	.34359	.35066	.35767	.36464	.37156	.37844	.38526	.39204	.39878
.5	.40547	.41211	.41871	.42527	.43178	.43825	.44469	.45108	.45742	.46373
.6	.47000	.47623	.48243	.48858	.49470	.50078	.50682	.51282	.51879	.52473
.7	.53063	.53649	.54232	.54812	.55389	.55962	.56531	.57098	.57661	.58222
.8	.58779	.59333	.59884	.60432	.60977	.61519	.62058	.62594	.63127	.63658
.9	.64185	.64710	.65233	.65752	.66269	.66783	.67294	.67803	.68310	.68813
2.0	0.69315	.69813	.70310	.70804	.71295	.71784	.72271	.72755	.73237	.73716
.1	.74194	.74669	.75142	.75612	.76081	.76547	.77011	.77473	.77932	.78390
.2	.78846	.79299	.79751	.80200	.80648	.81093	.81536	.81978	.82418	.82855
.3	.83291	.83725	.84157	.84587	.85015	.85442	.85866	.86289	.86710	.87129
.4	.87547	.87963	.88377	.88789	.89200	.89609	.90016	.90422	.90826	.91228
.5	.91629	.92028	.92426	.92822	.93216	.93609	.94001	.94391	.94779	.95166
.6	.95551	.95935	.96317	.96698	.97078	.97456	.97833	.98208	.98582	.98954
.7	.99325	.99695	*.00063	*.00430	*.00796	*.01160	*.01523	*.01885	*.02245	*.02604
.8	1.02962	.03318	.03674	.04028	.04380	.04732	.05082	.05431	.05779	.06126
.9	.06471	.06815	.07158	.07500	.07841	.08181	.08519	.08856	.09192	.09527
3.0	1.09861	.10194	.10526	.10856	.11186	.11514	.11841	.12168	.12493	.12817
.1	.13140	.13462	.13783	.14103	.14422	.14740	.15057	.15373	.15688	.16002
.2	.16315	.16627	.16938	.17248	.17557	.17865	.18173	.18479	.18784	.19089
.3	.19392	.19695	.19996	.20297	.20597	.20896	.21194	.21491	.21788	.22083
.4	.22378	.22671	.22964	.23256	.23547	.23837	.24127	.24415	.24703	.24990
.5	.25276	.25562	.25846	.26130	.26413	.26695	.26976	.27257	.27536	.27815
.6	.28093	.28371	.28647	.28923	.29198	.29473	.29746	.30019	.30291	.30563
.7	.30833	.31103	.31372	.31641	.31909	.32176	.32442	.32708	.32972	.33237
.8	.33500	.33763	.34025	.34286	.34547	.34807	.35067	.35325	.35584	.35841
.9	.36098	.36354	.36609	.36864	.37118	.37372	.37624	.37877	.38128	.38379
4.0	1.38629	.38879	.39128	.39377	.39624	.39872	.40118	.40364	.40610	.40854
.1	.41099	.41342	.41585	.41828	.42070	.42311	.42552	.42792	.43031	.43270
.2	.43508	.43756	.43984	.44220	.44456	.44692	.44927	.45161	.45395	.45629
.3	.45862	.46094	.46326	.46557	.46787	.47018	.47247	.47476	.47705	.47933
.4	.48160	.48387	.48614	.48840	.49065	.49290	.49515	.49739	.49962	.50185
.5	.50408	.50630	.50851	.51072	.51293	.51513	.51732	.51951	.52170	.52388
.6	.52606	.52823	.53039	.53256	.53471	.53687	.53902	.54116	.54330	.54543
.7	.54756	.54969	.55181	.55393	.55604	.55814	.56025	.56235	.56444	.56653
.8	.56862	.57070	.57277	.57485	.57691	.57898	.58104	.58309	.58515	.58719
.9	.58924	.59127	.59331	.59534	.59737	.59939	.60141	.60342	.60543	.60744
5.0	1.60944	.61144	.61343	.61542	.61741	.61939	.62137	.62334	.62531	.62728
.1	.62924	.63120	.63315	.63511	.63705	.63900	.64094	.64287	.64481	.64673
.2	.64866	.65058	.65250	.65441	.65632	.65823	.66013	.66203	.66393	.66582
.3	.66771	.66959	.67147	.67335	.67523	.67710	.67896	.68083	.68269	.68455
.4	.68640	.68825	.69010	.69194	.69378	.69562	.69745	.69928	.70111	.7029~

TABLE IV. NATURAL OR NAPERIAN LOGARITHMS (Continued)

5.50–9.99

N	0	1	2	3	4	5	6	7	8	9
.5	.70475	.70656	.70838	.71019	.71199	.71380	.71560	.71740	.71919	.72098
.6	.72277	.72455	.72633	.72811	.72988	.73166	.73342	.73519	.73695	.73871
.7	.74047	.74222	.74397	.74572	.74746	.74920	.75094	.75267	.75440	.75613
.8	.75786	.75958	.76130	.76302	.76473	.76644	.76815	.76985	.77156	.77326
.9	.77495	.77665	.77834	.78002	.78171	.78339	.78507	.78675	.78842	.79009
6.0	1.79176	.79342	.79509	.79675	.79840	.80006	.80171	.80336	.80500	.80665
.1	.80829	.80993	.81156	.81319	.81482	.81645	.81808	.81970	.82132	.82294
.2	.82455	.82616	.82777	.82938	.83098	.83258	.83418	.83578	.83737	.83896
.3	.84055	.84214	.84372	.84530	.84688	.84845	.85003	.85160	.85317	.85473
.4	.85630	.85786	.85942	.86097	.86253	.86408	.86563	.86718	.86872	.87026
.5	.87180	.87334	.87487	.87641	.87794	.87947	.88099	.88251	.88403	.88555
.6	.88707	.88858	.89010	.89160	.89311	.89462	.89612	.89762	.89912	.90061
.7	.90211	.90360	.90509	.90658	.90806	.90954	.91102	.91250	.91398	.91545
.8	.91692	.91839	.91986	.92132	.92279	.92425	.92571	.92716	.92862	.93007
.9	.93152	.93297	.93442	.93586	.93730	.93874	.94018	.94162	.94305	.94448
7.0	1.94591	.94734	.94876	.95019	.95161	.95303	.95445	.95586	.95727	.95869
.1	.96009	.96150	.96291	.96431	.96571	.96711	.96851	.96991	.97130	.97269
.2	.97408	.97547	.97685	.97824	.97962	.98100	.98238	.98376	.98513	.98650
.3	.98787	.98924	.99061	.99198	.99334	.99470	.99606	.99742	.99877	*.00013
.4	2.00148	.00283	.00418	.00553	.00687	.00821	.00956	.01089	.01223	.01357
.5	.01490	.01624	.01757	.01890	.02022	.02155	.02287	.02419	.02551	.02683
.6	.02815	.02946	.03078	.03209	.03340	.03471	.03601	.03732	.03862	.03992
.7	.04122	.04252	.04381	.04511	.04640	.04769	.04898	.05027	.05156	.05284
.8	.05412	.05540	.05668	.05796	.05924	.06051	.06179	.06306	.06433	.06560
.9	.06686	.06813	.06939	.07065	.07191	.07317	.07443	.07568	.07694	.07819
8.0	2.07944	.08069	.08194	.08318	.08443	.08567	.08691	.08815	.08939	.09063
.1	.09186	.09310	.09433	.09556	.09679	.09802	.09924	.10047	.10169	.10291
.2	.10413	.10535	.10657	.10779	.10900	.11021	.11142	.11263	.11384	.11505
.3	.11626	.11746	.11866	.11986	.12106	.12226	.12346	.12465	.12585	.12704
.4	.12823	.12942	.13061	.13180	.13298	.13417	.13535	.13653	.13771	.13889
.5	.14007	.14124	.14242	.14359	.14476	.14593	.14710	.14827	.14943	.15060
.6	.15176	.15292	.15409	.15524	.15640	.15756	.15871	.15987	.16102	.16217
.7	.16332	.16447	.16562	.16677	.16791	.16905	.17020	.17134	.17248	.17361
.8	.17475	.17589	.17702	.17816	.17929	.18042	.18155	.18267	.18380	.18493
.9	.18605	.18717	.18830	.18942	.19054	.19165	.19277	.19389	.19500	.19611
9.0	2.19722	.19834	.19944	.20055	.20166	.20276	.20387	.20497	.20607	.20717
.1	.20827	.20937	.21047	.21157	.21266	.21375	.21485	.21594	.21703	.21812
.2	.21920	.22029	.22138	.22246	.22354	.22462	.22570	.22678	.22786	.22894
.3	.23001	.23109	.23216	.23324	.23431	.23538	.23645	.23751	.23858	.23965
.4	.24071	.24177	.24284	.24390	.24496	.24601	.24707	.24813	.24918	.25024
.5	.25129	.25234	.25339	.25444	.25549	.25654	.25759	.25863	.25968	.26072
.6	.26176	.26280	.26384	.26488	.26592	.26696	.26799	.26903	.27006	.27109
.7	.27213	.27316	.27419	.27521	.27624	.27727	.27829	.27932	.28034	.28136
.8	.28238	.28340	.28442	.28544	.28646	.28747	.28849	.28950	.29051	.29152
.9	.29253	.29354	.29455	.29556	.29657	.29757	.29858	.29958	.30058	.30158

TABLE IV. NATURAL OR NAPERIAN LOGARITHMS (Continued)
0–499

N	0	1	2	3	4	5	6	7	8	9
0	$-\infty$	0.00000	0.69315	1.09861	.38629	60944	.79176	.94591	*.07944	*.19722
1	2.30259	.39790	.48491	.56495	.63906	.70805	.77259	.83321	.89037	.94444
2	.99573	*.04452	*.09104	*.13549	*.17805	*.21888	*.25810	*.29584	*.33220	*.36730
3	3.40120	.43399	.46574	.49651	.52636	.55535	.58352	.61092	.63759	.66356
4	.68888	.71357	.73767	.76120	.78419	.80666	.82864	.85015	.87120	.89182
5	.91202	.93183	.95124	.97029	.98898	*.00733	*.02535	*.04305	*.06044	*.07754
6	4.09434	.11087	.12713	.14313	.15888	.17439	.18965	.20469	.21951	.23411
7	.24850	.26268	.27667	.29046	.30407	.31749	.33073	.34381	.35671	.36945
8	.38203	.39445	.40672	.41884	.43082	.44265	.45435	.46591	.47734	.48864
9	.49981	.51086	.52179	.53260	.54329	.55388	.56435	.57471	.58497	.59512
10	4.60517	.61512	.62497	.63473	.64439	.65396	.66344	.67283	.68213	.69135
11	.70048	.70953	.71850	.72739	.73620	.74493	.75359	.76217	.77068	.77912
12	.78749	.79579	.80402	.81218	.82028	.82831	.83628	.84419	.85203	.85981
13	.86753	.87520	.88280	.89035	.89784	.90527	.91265	.91998	.92725	.93447
14	.94164	.94876	.95583	.96284	.96981	.97673	.98361	.99043	.99721	*.00395
15	5.01064	.01728	.02388	.03044	.03695	.04343	.04986	.05625	.06260	.06890
16	.07517	.08140	.08760	.09375	.09987	.10595	.11199	.11799	.12396	.12990
17	.13580	.14166	.14749	.15329	.15906	.16479	.17048	.17615	.18178	.18739
18	.19296	.19850	.20401	.20949	.21494	.22036	.22575	.23111	.23644	.24175
19	.24702	.25227	.25750	.26269	.26786	.27300	.27811	.28320	.28827	.29330
20	5.29832	.30330	.30827	.31321	.31812	.32301	.32788	.33272	.33754	.34233
21	.34711	.35186	.35659	.36129	.36598	.37064	.37528	.37990	.38450	.38907
22	.39363	.39816	.40268	.40717	.41165	.41610	.42053	.42495	.42935	.43372
23	.43808	.44242	.44674	.45104	.45532	.45959	.46383	.46806	.47227	.47646
24	.48064	.48480	.48894	.49306	.49717	.50126	.50533	.50939	.51343	.51745
25	.52146	.52545	.52943	.53339	.53733	.54126	.54518	.54908	.55296	.55683
26	.56068	.56452	.56834	.57215	.57595	.57973	.58350	.58725	.59099	.59471
27	.59842	.60212	.60580	.60947	.61313	.61677	.62040	.62402	.62762	.63121
28	.63479	.63835	.64191	.64545	.64897	.65249	.65599	.65948	.66296	.66643
29	.66988	.67332	.67675	.68017	.68358	.68698	.69036	.69373	.69709	.70044
30	5.70378	.70711	.71043	.71373	.71703	.72031	.72359	.72685	.73010	.73334
31	.73657	.73979	.74300	.74620	.74939	.75257	.75574	.75890	.76205	.76519
32	.76832	.77144	.77455	.77765	.78074	.78383	.78690	.78996	.79301	.79606
33	.79909	.80212	.80513	.80814	.81114	.81413	.81711	.82008	.82305	.82600
34	.82895	.83188	.83481	.83773	.84064	.84354	.84644	.84932	.85220	.85507
35	.85793	.86079	.86363	.86647	.86930	.87212	.87493	.87774	.88053	.88332
36	.88610	.88888	.89164	.89440	.89715	.89990	.90263	.90536	.90808	.91080
37	.91350	.91620	.91889	.92158	.92426	.92693	.92959	.93225	.93489	.93754
38	.94017	.94280	.94542	.94803	.95064	.95324	.95584	.95842	.96101	.96358
39	.96615	.96871	.97126	.97381	.97635	.97889	.98141	.98394	.98645	.98896
40	5.99146	.99396	.99645	.99894	*.00141	*.00389	*.00635	*.00881	*.01127	*.01372
41	6.01616	.01859	.02102	.02345	.02587	.02828	.03069	.03309	.03548	.03787
42	.04025	.04263	.04501	.04737	.04973	.05209	.05444	.05678	.05192	.06146
43	.06379	.06611	.06843	.07074	.07304	.07535	.07764	.07993	.08222	.08450
44	.08677	.08904	.09131	.09357	.09582	.09807	.10032	.10256	.10479	.10702
45	.10925	.11147	.11368	.11589	.11810	.12030	.12249	.12468	.12687	.12905
46	.13123	.13340	.13556	.13773	.13988	.14204	.14419	.14633	.14847	.15060
47	.15273	.15486	.15698	.15910	.16121	.16331	.16542	.16752	.16961	.17170
48	.17379	.17587	.17794	.18002	.18208	.18415	.18621	.18826	.19032	.19236
49	.19441	.19644	.19848	.20051	.20254	.20456	.20658	.20859	.21060	.21261

TABLE IV. NATURAL OR NAPERIAN LOGARITHMS (Continued)
500–999

N	0	1	2	3	4	5	6	7	8	9
50	6.21461	.21661	.21860	.22059	.22258	.22456	.22654	.22851	.23048	.23245
51	.23441	.23637	.23832	.24028	.24222	.24417	.24611	.24804	.24998	.25190
52	.25383	.25575	.25767	.25958	.26149	.26340	.26530	.26720	.26910	.27099
53	.27288	.27476	.27664	.27852	.28040	.28227	.28413	.28600	.28786	.28972
54	.29157	.29342	.29527	.29711	.29895	.30079	.30262	.30445	.30628	.30810
55	.30992	.31173	.31355	.31536	.31716	.31897	.32077	.32257	.32436	.32615
56	.32794	.32972	.33150	.33328	.33505	.33683	.33859	.34036	.34212	.34388
57	.34564	.34739	.34914	.35089	.35263	.35437	.35611	.35784	.35957	.36130
58	.36303	.36475	.36647	.36819	.36990	.37161	.37332	.37502	.37673	.37843
59	.38012	.38182	.38351	.38519	.38688	.38856	.39024	.39192	.39359	.39526
60	6.30693	.39859	.40026	.40192	.40357	.40523	.40688	.40853	.41017	.41182
61	.41346	.41510	.41673	.41836	.41999	.42162	.42325	.42487	.42649	.42811
62	.42972	.43133	.43294	.43455	.43615	.43775	.43935	.44095	.44254	.44413
63	.44572	.44731	.44889	.45047	.45205	.45362	.45520	.45677	.45834	.45990
64	.46147	.46303	.46459	.46614	.46770	.46925	.47080	.47235	.47389	.47543
65	.47697	.47851	.48004	.48158	.48311	.48464	.48616	.48768	.48920	.49072
66	.49224	.49375	.49527	.49677	.49828	.49979	.50129	.50279	.50429	.50578
67	.50728	.50877	.51026	.51175	.51323	.51471	.51619	.51767	.51915	.52062
68	.52209	.52356	.52503	.52649	.52796	.52942	.53088	.53233	.53379	.53524
69	.53669	.53814	.53959	.54103	.54247	.54391	.54535	.54679	.54822	.54965
70	6.55108	.55251	.55393	.55536	.55678	.55820	.55962	.56103	.56244	.56386
71	.56526	.56667	.56808	.56948	.57088	.57228	.57368	.57508	.57647	.57786
72	.57925	.58064	.58203	.58341	.58479	.58617	.58755	.58893	.59030	.59167
73	.59304	.59441	.59578	.59715	.59851	.59987	.60123	.60259	.60394	.60530
74	.60665	.60800	.60935	.61070	.61204	.61338	.61473	.61607	.61740	.61874
75	.62007	.62141	.62274	.62407	.62539	.62672	.62804	.62936	.63068	.63200
76	.63332	.63463	.63595	.63726	.63857	.63988	.64118	.64249	.64379	.64509
77	.64639	.64769	.64898	.65028	.65157	.65286	.65415	.65544	.65673	.65801
78	.65929	.66058	.66185	.66313	.66441	.66568	.66696	.66823	.66950	.67077
79	.67203	.67330	.67456	.67582	.67708	.67834	.67960	.68085	.68211	.68336
80	6.68461	.68586	.68711	.68835	.68960	.69084	.69208	.69332	.69456	.69580
81	.69703	.69827	.69950	.70073	.70196	.70319	.70441	.70564	.70686	.70808
82	.70930	.71052	.71174	.71296	.71417	.71538	.71659	.71780	.71901	.72022
83	.72143	.72263	.72383	.72503	.72623	.72743	.72863	.72982	.73102	.73221
84	.73340	.73459	.73578	.73697	.73815	.73934	.74052	.74170	.74288	.74406
85	.74524	.74641	.74759	.74876	.74993	.75110	.75227	.75344	.75460	.75577
86	.75693	.75809	.75926	.76041	.76157	.76273	.76388	.76504	.76619	.76734
87	.76849	.76964	.77079	.77194	.77308	.77422	.77537	.77651	.77765	.77878
88	.77992	.78106	.78219	.78333	.78446	.78559	.78672	.78784	.78897	.79010
89	.79122	.79234	.79347	.79459	.79571	.79682	.79794	.79906	.80017	.80128
90	6.80239	.80351	.80461	.80572	.80683	.80793	.80904	.81014	.81124	.81235
91	.81344	.81454	.81564	.81674	.81783	.81892	.82002	.82111	.82220	.82329
92	.82437	.82546	.82655	.82763	.82871	.82979	.83087	.83195	.83303	.83411
93	.83518	.83626	.83733	.83841	.83948	.84055	.84162	.84268	.84375	.84482
94	.84588	.84694	.84801	.84907	.85013	.85118	.85224	.85330	.85435	.85541
95	.85646	.85751	.85857	.85961	.86066	.86171	.86276	.86380	.86485	.86589
96	.86693	.86797	.86901	.87005	.87109	.87213	.87316	.87420	.87523	.87626
97	.87730	.87833	.87936	.88038	.88141	.88244	.88346	.88449	.88551	.88653
98	.88755	.88857	.88959	.89061	.89163	.89264	.89366	.89467	.89568	.89669
99	.89770	.89871	.89972	.90073	.90174	.90274	.90375	.90475	.90575	.90675

TABLE V. EXPONENTIAL FUNCTIONS

x	e^x	e^{-x}	x	e^x	e^{-x}
0.00	1.0000	1.000000	0.50	1.6487	0.606531
0.01	1.0101	0.990050	0.51	1.6653	.600496
0.02	1.0202	.980199	0.52	1.6820	.594521
0.03	1.0305	.970446	0.53	1.6989	.588605
0.04	1.0408	.960789	0.54	1.7160	.582748
0.05	1.0513	0.951229	0.55	1.7333	0.576950
0.06	1.0618	.941765	0.56	1.7507	.571209
0.07	1.0725	.932394	0.57	1.7683	.565525
0.08	1.0833	.923116	0.58	1.7860	.559898
0.09	1.0942	.913931	0.59	1.8040	.554327
0.10	1.1052	0.904837	0.60	1.8221	0.548812
0.11	1.1163	.895834	0.61	1.8404	.543351
0.12	1.1275	.886920	0.62	1.8589	.537944
0.13	1.1388	.878095	0.63	1.8776	.532592
0.14	1.1503	.869358	0.64	1.8965	.527292
0.15	1.1618	0.860708	0.65	1.9155	0.522046
0.16	1.1735	.852114	0.66	1.9348	.516851
0.17	1.1853	.843665	0.67	1.9542	.511709
0.18	1.1972	.835270	0.68	1.9739	.506617
0.19	1.2092	.826959	0.69	1.9937	.501576
0.20	1.2214	0.818731	0.70	2.0138	0.496585
0.21	1.2337	.810584	0.71	2.0340	.491644
0.22	1.2461	.802519	0.72	2.0544	.486752
0.23	1.2586	.794534	0.73	2.0751	.481909
0.24	1.2712	.786628	0.74	2.0959	.477114
0.25	1.2840	0.778801	0.75	2.1170	0.472367
0.26	1.2969	.771052	0.76	2.1383	.467666
0.27	1.3100	.763379	0.77	2.1598	.463013
0.28	1.3231	.755784	0.78	2.1815	.458406
0.29	1.3364	.748264	0.79	2.2034	.453845
0.30	1.3499	0.740818	0.80	2.2255	0.449329
0.31	1.3634	.733447	0.81	2.2479	.444858
0.32	1.3771	.726149	0.82	2.2705	.440432
0.33	1.3910	.718924	0.83	2.2933	.436049
0.34	1.4049	.711770	0.84	2.3164	.431711
0.35	1.4191	0.704688	0.85	2.3396	0.427415
0.36	1.4333	.697676	0.86	2.3632	.423162
0.37	1.4477	.690734	0.87	2.3869	.418952
0.38	1.4623	.683861	0.88	2.4109	.414783
0.39	1.4770	.677057	0.89	2.4351	.410656
0.40	1.4918	0.670320	0.90	2.4596	0.406570
0.41	1.5068	.663650	0.91	2.4843	.402524
0.42	1.5220	.657047	0.92	2.5093	.398519
0.43	1.5373	.650509	0.93	2.5345	.394554
0.44	1.5527	.644036	0.94	2.5600	.390628
0.45	1.5683	0.637628	0.95	2.5857	0.386741
0.46	1.5841	.631284	0.96	2.6117	.382893
0.47	1.6000	.625002	0.97	2.6379	.379083
0.48	1.6161	.618783	0.98	2.6645	.375311
0.49	1.6323	.612626	0.99	2.6912	.371577
0.50	1.6487	0.606531	1.00	2.7183	0.367870

From *Standard Mathematical Tables*, 21st ed., Samuel Selby, © The Chemical Rub... Co., 1973. Used by permission of The Chemical Rubber Co.

TABLE V. EXPONENTIAL FUNCTIONS (Continued)

x	e^x	e^{-x}	x	e^x	e^{-x}
1.0	2.718	0.368	5.5	244.7	0.0041
1.1	3.004	0.333	5.6	270.4	0.0037
1.2	3.320	0.301	5.7	298.9	0.0033
1.3	3.669	0.273	5.8	330.3	0.0030
1.4	4.055	0.247	5.9	365.0	0.0027
1.5	4.482	0.223	6.0	403.4	0.0025
1.6	4.953	0.202	6.1	445.9	0.0022
1.7	5.474	0.183	6.2	492.8	0.0020
1.8	6.050	0.165	6.3	544.6	0.0018
1.9	6.686	0.150	6.4	601.8	0.0017
2.0	7.389	0.135	6.5	665.1	0.0015
2.1	8.166	0.122	6.6	735.1	0.0014
2.2	9.025	0.111	6.7	812.4	0.0012
2.3	9.974	0.100	6.8	897.8	0.0011
2.4	11.023	0.091	6.9	992.3	0.0010
2.5	12.18	0.082	7.0	1,096.6	0.0009
2.6	13.46	0.074	7.1	1,212.0	0.0008
2.7	14.88	0.067	7.2	1,339.4	0.0007
2.8	16.44	0.061	7.3	1,480.3	0.0007
2.9	18.17	0.055	7.4	1,636.0	0.0006
3.0	20.09	0.050	7.5	1,808.0	0.00055
3.1	22.20	0.045	7.6	1,998.2	0.00050
3.2	24.53	0.041	7.7	2,208.3	0.00045
3.3	27.11	0.037	7.8	2,440.6	0.00041
3.4	29.96	0.033	7.9	2,697.3	0.00037
3.5	33.12	0.030	8.0	2,981.0	0.00034
3.6	36.60	0.027	8.1	3,294.5	0.00030
3.7	40.45	0.025	8.2	3,641.0	0.00027
3.8	44.70	0.022	8.3	4,023.9	0.00025
3.9	49.40	0.020	8.4	4,447.1	0.00022
4.0	54.60	0.018	8.5	4,914.8	0.00020
4.1	60.34	0.017	8.6	5,431.7	0.00018
4.2	66.69	0.015	8.7	6,002.9	0.00017
4.3	73.70	0.014	8.8	6,634.2	0.00015
4.4	81.45	0.012	8.9	7,332.0	0.00014
4.5	90.02	0.011	9.0	8,103.1	0.00012
4.6	99.48	0.010	9.1	8,955.3	0.00011
4.7	109.95	0.009	9.2	9,897.1	0.00010
4.8	121.51	0.008	9.3	10,938	0.00009
4.9	134.29	0.007	9.4	12,088	0.00008
5.0	148.4	0.0067	9.5	13,360	0.00007
5.1	164.0	0.0061	9.6	14,765	0.00007
5.2	181.3	0.0055	9.7	16,318	0.00006
5.3	200.3	0.0050	9.8	18,034	0.00006
5.4	221.4	0.0045	9.9	19,930	0.00005

TABLE VI. TABLE OF INTEGRALS

Elementary Forms

1. $\int [f(x) + g(x) - h(x)]\, dx = \int f(x)\, dx + \int g(x)\, dx - \int h(x)\, dx$

2. $\int kf(x)\, dx = k \int f(x)\, dx$

3. $\int dx = x + c$

4. $\int k\, dx = kx + c$

5. $\int x^n\, dx = \dfrac{x^{n+1}}{n+1} + c \qquad$ if $n \neq -1$

6. $\int x^{-1}\, dx = \ln x + c$

Forms Containing $a + bx$

7. $\int (a + bx)^n\, dx = \dfrac{1}{b(n+1)}(a + bx)^{n+1} + c \qquad$ if $n \neq -1$

$\qquad\qquad\qquad = \dfrac{1}{b} \ln (a + bx) + c \qquad\qquad$ if $n = -1$

8. $\int x(a + bx)^n\, dx = \dfrac{(a + bx)^{n+2}}{b^2(n+2)} - \dfrac{a(a + bx)^{n+1}}{b^2(n+1)} + c \qquad$ if $n \neq -1, -2$

$\qquad\qquad\qquad = \dfrac{1}{b^2}[a + bx - a \ln (a + bx)] + c \qquad$ if $n = -1$

$\qquad\qquad\qquad = \dfrac{1}{b^2}\left[\ln (a + bx) + \dfrac{a}{a + bx}\right] + c \qquad$ if $n = -2$

Forms Containing $x^2 \pm a^2$, or $a^2 \pm x^2$

9. $\int \dfrac{dx}{a^2 - x^2} = \dfrac{1}{2a} \ln \left(\dfrac{a + x}{a - x}\right) + c$

10. $\int \dfrac{dx}{x^2 - a^2} = \dfrac{1}{2a} \ln \left(\dfrac{x - a}{x + a}\right) + c$

11. $\int (x^2 \pm a^2)^{1/2}\, dx = \tfrac{1}{2}[x(x^2 \pm a^2)^{1/2} \pm a^2 \ln (x + (x^2 \pm a^2)^{1/2}] + c$

12. $\int (x^2 \pm a^2)^{-1/2}\, dx = \ln [x + (x^2 \pm a^2)^{1/2}] + c$

13. $\int x(x^2 \pm a^2)^{1/2}\, dx = \tfrac{1}{3}(x^2 \pm a^2)^{3/2} + c$

14. $\int x(x^2 \pm a^2)^{-1/2} \, dx = (x^2 \pm a^2)^{1/2} + c$

15. $\int x(a^2 - x^2)^{1/2} \, dx = -\frac{1}{3}(a^2 - x^2)^{3/2} + c$

16. $\int x(a^2 - x^2)^{-1/2} \, dx = -(a^2 - x^2)^{1/2} + c$

Exponential Forms

17. $\int e^x \, dx = e^x + c$

18. $\int a^{kx} \, dx = \frac{a^{kx}}{k \ln a} + c$

19. $\int xe^{kx} \, dx = \frac{e^{kx}}{k^2} (kx - 1) + c$

20. $\int e^{kx} \, dx = \frac{e^{kx}}{k} + c$

Logarithmic Forms

21. $\int \ln x \, dx = x \ln x - x + c$

22. $\int \log_a x \, dx = x \log_a x - x \log_a e + c$

23. $\int \frac{dx}{x \, (\ln x)} = \ln (\ln x) + c$

24. $\int (\ln x)^2 \, dx = x(\ln x)^2 - 2x(\ln x) + 2x + c$

25. $\int x \ln x \, dx = \frac{x^2 \ln x}{2} - \frac{x^2}{4} + c$

TABLE VII. SQUARES AND SQUARE ROOTS
1-1000

N	N²	√N	N	N²	√N	N	N²	√N
			50	2 500	7.071 068	100	10 000	10.00000
1	1	1.000 000	51	2 601	7.141 428	101	10 201	10.04988
2	4	1.414 214	52	2 704	7.211 103	102	10 404	10.09950
3	9	1.732 051	53	2 809	7.280 110	103	10 609	10.14889
4	16	2.000 000	54	2 916	7.348 469	104	10 816	10.19804
5	25	2.236 068	55	3 025	7.416 198	105	11 025	10.24695
6	36	2.449 490	56	3 136	7.483 315	106	11 236	10.29563
7	49	2.645 751	57	3 249	7.549 834	107	11 449	10.34408
8	64	2.828 427	58	3 364	7.615 773	108	11 664	10.39230
9	81	3.000 000	59	3 481	7.681 146	109	11 881	10.44031
10	100	3.162 278	60	3 600	7.745 967	110	12 100	10.48809
11	121	3.316 625	61	3 721	7.810 250	111	12 321	10.53565
12	144	3.464 102	62	3 844	7.874 008	112	12 544	10.58301
13	169	3.605 551	63	3 969	7.937 254	113	12 769	10.63015
14	196	3.741 657	64	4 096	8.000 000	114	12 906	10.67708
15	225	3.872 983	65	4 225	8.062 258	115	13 225	10.72381
16	256	4.000 000	66	4 356	8.124 038	116	13 456	10.77033
17	289	4.123 106	67	4 489	8.185 353	117	13 689	10.81665
18	324	4.242 641	68	4 624	8.246 211	118	13 924	10.86278
19	361	4.358 899	69	4 761	8.306 624	119	14 161	10.90871
20	400	4.472 136	70	4 900	8.366 600	120	14 400	10.95445
21	441	4.582 576	71	5 041	8.426 150	121	14 641	11.00000
22	484	4.690 416	72	5 184	8.485 281	122	14 884	11.04536
23	529	4.795 832	73	5 329	8.544 004	123	15 129	11.09054
24	576	4.898 979	74	5 476	8.602 325	124	15 376	11.13553
25	625	5.000 000	75	5 625	8.660 254	125	15 625	11.18034
26	676	5.099 020	76	5 776	8.717 798	126	15 876	11.22497
27	729	5.196 152	77	5 929	8.774 964	127	16 129	11.26943
28	784	5.291 503	78	6 084	8.831 761	128	16 384	11.31371
29	841	5.385 165	79	6 241	8.888 194	129	16 641	11.35782
30	900	5.477 226	80	6 400	8.944 272	130	16 900	11.40175
31	961	5.567 764	81	6 561	9.000 000	131	17 161	11.44552
32	1 024	5.656 854	82	6 724	9.055 385	132	17 424	11.48913
33	1 089	5.744 563	83	6 889	9.110 434	133	17 689	11.53256
34	1 156	5.830 952	84	7 056	9.165 151	134	17 956	11.57584
35	1 225	5.916 080	85	7 225	9.219 544	135	18 225	11.61895
36	1 296	6.000 000	86	7 396	9.273 618	136	18 496	11.66190
37	1 369	6.082 763	87	7 569	9.327 379	137	18 769	11.70470
38	1 444	6.164 414	88	7 744	9.380 832	138	19 044	11.74734
39	1 521	6.244 998	89	7 921	9.433 981	139	19 321	11.78983
40	1 600	6.324 555	90	8 100	9.486 833	140	19 600	11.83216
41	1 681	6.403 124	91	8 281	9.539 392	141	19 881	11.87434
42	1 764	6.480 741	92	8 464	9.591 663	142	20 164	11.91638
43	1 849	6.557 439	93	8 649	9.643 651	143	20 449	11.95826
44	1 936	6.633 250	94	8 836	9.695 360	144	20 736	12.00000
45	2 025	6.708 204	95	9 025	9.746 794	145	21 025	12.04159
46	2 116	6.782 330	96	9 216	9.797 959	146	21 316	12.08305
47	2 209	6.855 655	97	9 409	9.848 858	147	21 609	12.12436
48	2 304	6.928 203	98	9 604	9.899 405	148	21 904	12.1655
49	2 401	7.000 000	99	9 801	9.949 874	149	22 201	12.206

TABLE VII. SQUARES AND SQUARE ROOTS (Continued)
1–1000

N	N²	√N	N	N²	√N	N	N²	√N
150	22 500	12.24745	200	40 000	14.14214	250	62 500	15.81139
151	22.801	12.28821	201	40 401	14.17745	251	63 001	15.84298
152	23 104	12.32883	202	40 804	14.21267	252	63 504	15.87451
153	23 409	12.36932	203	41 209	14.24781	253	64 009	15.90597
154	23 716	12.40967	204	41 616	14.28286	254	64 516	15.93738
155	24 025	12.44990	205	42 025	14.31782	255	65 025	15.96872
156	24 336	12.49000	206	42 436	14.35270	256	65 536	16.00000
157	24 649	12.52996	207	42 849	14.38749	257	66 049	16.03122
158	24 964	12.56981	208	43 264	14.42221	258	66 564	16.06238
159	25 281	12.60952	209	43 681	14.45683	259	67 081	16.09348
160	25 600	12.64911	210	44 100	14.49138	260	67 600	16.12452
161	25 921	12.68858	211	44 521	14.52584	261	68 121	16.15549
162	26 244	12.72792	212	44 944	14.56022	262	68 644	16.18641
163	26 569	12.76715	213	45 369	14.59452	263	69 169	16.21727
164	26 896	12.80625	214	45 796	14.62874	264	69 696	16.24808
165	27 225	12.84523	215	46 225	14.66288	265	70 225	16.27882
166	27 556	12.88410	216	46 656	14.69694	266	70 756	16.30951
167	27 889	12.92285	217	47 089	14.73092	267	71 289	16.34013
168	28 224	12.96148	218	47 524	14.76482	268	71 824	16.37071
169	28 561	13.00000	219	47 961	14.79865	269	72 361	16.40122
170	28 900	13.03840	220	48 400	14.83240	270	72 900	16.43168
171	29 241	13.07670	221	48 841	14.86607	271	73 441	16.46208
172	29 584	13.11488	222	49 284	14.89966	272	73 984	16.49242
173	29 929	13.15295	223	49 729	14.93318	273	74 529	16.52271
174	30 276	13.19091	224	50 176	14.96663	274	75 076	16.55295
175	30 625	13.22876	225	50 625	15.00000	275	75 625	16.58312
176	30 976	13.26650	226	51 076	15.03330	276	76 176	16.61325
177	31 329	13.30413	227	51 529	15.06652	277	76 729	16.64332
178	31 684	13.34166	228	51 984	15.09967	278	77 284	16.67333
179	32 041	13.37909	229	52 441	15.13275	279	77 841	16.70329
180	32 400	13.41641	230	52 900	15.16575	280	78 400	16.73320
181	32 761	13.45362	231	53 361	15.19868	281	78 961	16.76305
182	33 124	13.49074	232	53 824	15.23155	282	79 524	16.79286
183	33 489	13.52775	233	54 289	15.26434	283	80 089	16.82260
184	33 856	13.56466	234	54 756	15.29706	284	80 656	16.85230
185	34 225	13.60147	235	55 225	15.32971	285	81 225	16.88194
186	34 596	13.63818	236	55 696	15.36229	286	81 796	16.91153
187	34 969	13.67479	237	56 169	15.39480	287	82 369	16.94107
188	35 344	13.71131	238	56 644	15.42725	288	82 944	16.97056
189	35 721	13.74773	239	57 121	15.45962	289	83 521	17.00000
190	36 100	13.78405	240	57 600	15.49193	290	84 100	17.02939
191	36 481	13.82027	241	58 081	15.52417	291	84 681	17.05872
192	36 864	13.85641	242	58 564	15.55635	292	85 264	17.08801
193	37 249	13.89244	243	59 049	15.58846	293	85 849	17.11724
194	37 636	13.92839	244	59 536	15.62050	294	86 436	17.14643
195	38 025	13.96424	245	60 025	15.65248	295	87 025	17.17556
196	38 416	14.00000	246	60 516	15.68439	296	87 616	17.20465
197	38 809	14.03567	247	61 009	15.71623	297	88 209	17.23369
198	39 204	14.07125	248	61 504	15.74802	298	88 804	17.26268
199	39 601	14.10674	249	62 001	15.77973	299	89 401	17.29162

TABLE VII. SQUARES AND SQUARE ROOTS (Continued)
1–1000

N	N²	√N	N	N²	√N	N	N²	√N
300	90 000	17.32051	350	122 500	18.70829	400	160 000	20.00000
301	90 601	17.34935	351	123 201	18.73499	401	160 801	20.02498
302	91 204	17.37815	352	123 904	18.76166	402	161 604	20.04994
303	91 809	17.40690	353	124 609	18.78829	403	162 409	20.07486
304	92 416	17.43560	354	125 316	18.81489	404	163 216	20.09975
305	93 025	17.46425	355	126 025	18.84144	405	164 025	20.12461
306	93 636	17.49286	356	126 736	18.86796	406	164 836	20.14944
307	94 249	17.52142	357	127 449	18.89444	407	165 649	20.17424
308	94 864	17.54993	358	128 164	18.92089	408	166 464	20.19901
309	95 481	17.57840	359	128 881	18.94730	409	167 231	20.22375
310	96 100	17.60682	360	129 600	18.97367	410	168 100	20.24846
311	96 721	17.63519	361	130 321	19.00000	411	168 921	20.27313
312	97 344	17.66352	362	131 044	19.02630	412	169 744	20.29778
313	97 969	17.69181	363	131 769	19.05256	413	170 569	20.32249
314	98 596	17.72005	364	132 496	19.07878	414	171 396	20.34699
315	99 225	17.74824	365	133 225	19.10497	415	172 225	20.37155
316	99 856	17.77639	366	133 956	19.13113	416	173 056	20.39608
317	100 489	17.80449	367	134 689	19.15724	417	173 889	20.42058
318	101 124	17.83255	368	135 424	19.18333	418	174 724	20.44505
319	101 761	17.86057	369	136 161	19.20937	419	175 561	20.46949
320	102 400	17.88854	370	136 900	19.23538	420	176 400	20.49390
321	103 041	17.91647	371	137 641	19.26136	421	177 241	20.51828
322	103 684	17.94436	372	138 384	19.28730	422	178 084	20.54264
323	104 329	17.97220	373	139 129	19.31321	423	178 929	20.56696
324	104 976	18.00000	374	139 876	19.33908	424	179 776	20.59126
325	105 625	18.02776	375	140 625	19.36492	425	180 625	20.61553
326	106 276	18.05547	376	141 376	19.39072	426	181 476	20.63977
327	106 929	18.08314	377	142 129	19.41649	427	182 329	20.66398
328	107 584	18.11077	378	142 884	19.44222	428	183 184	20.68816
329	108 241	18.13836	379	143 641	19.46792	429	184 041	20.71232
330	108 900	18.16590	380	144 400	19.49359	430	184 900	20.73644
331	109 561	18.19341	381	145 161	19.51922	431	185 761	20.76054
332	110 224	18.22087	382	145 924	19.54483	432	186 624	20.78461
333	110 889	18.24829	383	146 689	19.57039	433	187 489	20.80865
334	111 556	18.27567	384	147 456	19.59592	434	188 356	20.83267
335	112 225	18.30301	385	148 225	19.62142	435	189 225	20.85665
336	112 896	18.33030	386	148 996	19.64688	436	190 096	20.88061
337	113 569	18.35756	387	149 769	19.67232	437	190 969	20.90454
338	114 244	18.38478	388	150 544	19.69772	438	191 844	20.92845
339	114 921	18.41195	389	151 321	19.72308	439	192 721	20.95233
340	115 600	18.43909	390	152 100	19.74842	440	193 600	20.97618
341	116 281	18.46619	391	152 881	19.77372	441	194 481	21.00000
342	116 964	18.49324	392	153 664	19.79899	442	195 364	21.02380
343	117 649	18.52026	393	154 449	19.82423	443	196 249	21.04757
344	118 336	18.54724	394	155 236	19.84943	444	197 136	21.07131
345	119 025	18.57418	395	156 025	19.87461	445	198 025	21.09502
346	119 716	18.60108	396	156 816	19.89975	446	198 916	21.11871
347	120 409	18.62794	397	157 609	19.92486	447	199 809	21.14237
348	121 104	18.65476	398	158 404	19.94994	448	200 704	21.16601
349	121 801	18.68154	399	159 201	19.97498	449	201 601	21.18962

TABLE VII. SQUARES AND SQUARE ROOTS (Continued)
1–1000

N	N²	√N	N	N²	√N	N	N²	√N
450	202 500	21.21320	500	250 000	22.36068	550	302 500	23.45208
451	203 401	21.23676	501	251 001	22.38303	551	303 601	23.47339
452	204 304	21.26029	502	252 004	22.40536	552	304 704	23.49468
453	205 209	21.28380	503	253 009	22.42766	553	305 809	23.51595
454	206 116	21.30728	504	254 016	22.44994	554	306 916	23.53720
455	207 025	21.33073	505	255 025	22.47221	555	308 025	23.55844
456	207 936	21.35416	506	256 036	22.49444	556	309 136	23.57965
457	208 849	21.37756	507	257 049	22.51666	557	310 249	23.60085
458	209 764	21.40093	508	258 064	22.53886	558	311 364	23.62202
459	210 681	21.42429	509	259 081	22.56103	559	312 481	23.64318
460	211 600	21.44761	510	260 100	22.58318	560	313 600	23.66432
461	212 521	21.47091	511	261 121	22.60531	561	314 721	23.68544
462	213 444	21.49419	512	262 144	22.62742	562	315 844	23.70654
463	214 369	21.51743	513	263 169	22.64950	563	316 969	23.72762
464	215 296	21.54066	514	264 196	22.67157	564	318 096	23.74868
465	216 225	21.56386	515	265 225	22.69361	565	319 225	23.76973
466	217 156	21.58703	516	266 256	22.71563	566	320 356	23.79075
467	218 089	21.61018	517	267 289	22.73763	567	321 489	23.81176
468	219 024	21.63331	518	268 324	22.75961	568	322 624	23.83275
469	219 961	21.65641	519	269 361	22.78157	569	323 761	23.85372
470	220 900	21.67948	520	270 400	22.80351	570	324 900	23.87467
471	221 841	21.70253	521	271 441	22.82542	571	326 041	23.89561
472	222 781	21.72556	522	272 484	22.84732	572	327 184	23.91652
473	223 729	21.74856	523	273 529	22.86919	573	328 329	23.93742
474	224 676	21.77154	524	274 576	22.89105	574	329 476	23.95830
475	225 625	21.79449	525	275 625	22.91288	575	330 625	23.97916
476	226 576	21.81742	526	276 676	22.93469	576	331 776	24.00000
477	227 529	21.84033	527	277 729	22.95648	577	332 929	24.02082
478	228 484	21.86321	528	278 784	22.97825	578	334 084	24.04163
479	229 441	21.88607	529	279 841	23.00000	579	335 241	24.06242
480	230 400	21.90890	530	280 900	23.02173	580	336 400	24.08319
481	231 361	21.93171	531	281 961	23.04344	581	337 561	24.10394
482	232 324	21.95450	532	283 024	23.06513	582	338 724	24.12468
483	233 289	21.97726	533	284 089	23.08679	583	339 889	24.14539
484	234 256	22.00000	534	285 156	23.10844	584	341 056	24.16609
485	235 225	22.02272	535	286 225	23.13007	585	342 225	24.18677
486	236 196	22.04541	536	287 296	23.15167	586	343 396	24.20744
487	237 169	22.06808	537	288 369	23.17326	587	344 569	24.22808
488	238 144	22.09072	538	289 444	23.19483	588	345 744	24.24871
489	239 121	22.11334	539	290 521	23.21637	589	346 921	24.26932
490	240 100	22.13594	540	291 600	23.23790	590	348 100	24.28992
491	241 081	22.15852	541	292 681	23.25941	591	349 281	24.31049
492	242 064	22.18107	542	293 764	23.28089	592	350 464	24.33105
493	243 049	22.20360	543	294 849	23.30236	593	351 649	24.35159
494	244 036	22.22611	544	295 936	23.32381	594	352 836	24.37212
495	245 025	22.24860	545	297 025	23.34524	595	354 025	24.39262
496	246 016	22.27106	546	298 116	23.36664	596	355 216	24.41311
497	247 009	22.29350	547	299 209	23.38803	597	356 409	24.43358
498	248 004	22.31591	548	300 304	23.40940	598	357 604	24.45404
499	249 001	22.33831	549	301 401	23.43075	599	358 801	24.47448

TABLE VII. SQUARES AND SQUARE ROOTS (Continued)
1–1000

N	N²	√N	N	N²	√N	N	N²	√N
600	360 000	24.49490	650	422 500	25.49510	700	490 000	26.45751
601	361 201	24.51530	651	423 801	25.51470	701	491 401	26.47640
602	362 404	24.53569	652	425 104	25.53429	702	492 804	26.49528
603	363 609	24.55606	653	426 409	25.56386	703	494 209	26.51415
604	364 816	24.57641	654	427 716	25.57342	704	495 616	26.53200
605	366 025	24.59675	655	429 025	25.59297	705	497 025	26.55184
606	367 236	24.61707	656	430 336	25.61250	706	498 436	26.57066
607	368 449	24.63737	657	431 649	25.63201	707	499 849	26.58947
608	369 664	24.65766	658	432 964	25.65151	708	501 264	26.60827
609	370 881	24.67793	659	434 281	25.67100	709	502 681	26.62705
610	372 100	24.69818	660	435 600	25.69047	710	504 100	26.64583
611	373 321	24.71841	661	436 921	25.70992	711	505 521	26.66458
612	374 544	24.73863	662	438 244	25.72936	712	506 944	26.68383
613	375 769	24.75884	663	439 569	25.74829	713	508 369	26.70206
614	376 996	24.77902	664	440 896	25.76820	714	509 796	26.72078
615	378 225	24.79919	665	442 225	25.78759	715	511 225	26.73948
616	379 456	24.81935	666	443 556	25.80698	716	512 656	26.75818
617	380 689	24.83948	667	444 889	25.82634	717	514 089	26.77686
618	381 924	24.85961	668	446 224	25.84570	718	515 524	26.79552
619	383 161	24.87971	669	447 561	25.86503	719	516 961	26.81418
620	384 400	24.89980	670	448 900	25.88436	720	518 400	26.83282
621	385 641	24.91987	671	450 241	25.90367	721	519 841	26.85144
622	386 884	24.93993	672	451 584	25.92296	722	521 284	26.87006
623	388 129	24.95997	673	452 929	25.94224	723	522 729	26.88966
624	389 376	24.97999	674	454 276	25.96151	724	524 176	26.90725
625	390 625	25.00000	675	455 625	25.98076	725	525 625	26.92582
626	391 876	25.01999	676	456 976	26.00000	726	527 076	26.94439
627	393 129	25.03997	677	458 329	26.01922	727	528 529	26.96294
628	394 384	25.05993	678	459 684	26.03843	728	529 984	26.98148
629	395 641	25.07987	679	461 041	26.05763	729	531 441	27.00000
630	396 900	25.09980	680	462 400	26.07681	730	532 900	27.01851
631	398 161	25.11971	681	463 761	26.09598	731	534 361	27.03701
632	399 424	25.13961	682	465 124	26.11513	732	535 824	27.03550
633	400 689	25.15949	683	466 489	26.13427	733	537 289	27.02397
634	401 956	25.17936	684	467 856	26.15339	734	538 756	27.09243
635	403 225	25.19921	685	469 225	26.17250	735	540 225	27.11088
636	404 496	25.21904	686	470 596	26.19160	736	541 696	27.12932
637	405 769	25.23886	687	471 969	26.21068	737	543 169	27.14774
638	407 044	25.25866	688	473 344	26.22975	738	544 644	27.16616
639	408 321	25.27845	689	474 721	26.24881	739	546 121	27.18455
640	409 600	25.29822	690	476 100	26.26785	740	547 600	27.20294
641	410 881	25.31798	691	477 481	26.28688	741	549 081	27.22132
642	412 164	25.33772	692	478 864	26.30589	742	550 564	27.23968
643	413 449	25.35744	693	480 249	26.32489	743	552 049	27.25803
644	414 736	25.37716	694	481 636	26.34388	744	553 536	27.27636
645	416 025	25.39685	695	483 025	26.36285	745	555 025	27.29469
646	417 319	25.41653	696	484 416	26.38181	746	556 516	27.31300
647	418 609	25.43619	697	485 809	26.40076	747	558 009	27.33130
648	419 904	25.45584	698	487 204	26.41969	748	559 504	27.3495
649	421 201	25.47510	699	488 601	26.43861	749	561 001	27.367

TABLE VII. SQUARES AND SQUARE ROOTS (Continued)
1–1000

N	N²	√N	N	N²	√N	N	N²	√N
750	562 500	27.38613	800	640 000	28.28427	850	722 500	29.15476
751	564 001	27.40438	801	641 601	28.30194	851	724 201	29.17190
752	565 504	27.42262	802	643 204	28.31960	852	725 904	29.18904
753	567 009	27.44085	803	644 809	28.33725	853	727 609	29.20616
754	568 516	27.45906	804	646 416	28.35489	854	729 316	29.22328
755	570 025	27.47726	805	648 025	28.37252	855	731 025	29.24038
756	571 536	27.49545	806	649 636	28.39014	856	732 736	29.25748
757	573 049	27.51363	807	651 249	28.40775	857	734 449	29.27456
758	574 564	27.53180	808	652 864	28.42534	858	736 164	29.29164
759	576 081	27.54995	809	654 481	28.44293	859	737 881	29.30870
760	577 600	27.56810	810	656 100	28.46050	860	739 600	29.32576
761	579 121	27.58623	811	657 721	28.47806	861	741 321	29.34280
762	580 644	27.60435	812	659 344	28.49561	862	743 044	29.35984
763	582 169	27.62245	813	660 969	28.51315	863	744 769	29.37686
764	583 696	27.64055	814	662 596	28.53069	864	746 496	29.39388
765	585 225	27.65863	815	664 225	28.54820	865	748 225	29.41088
766	586 756	27.67671	816	665 856	28.56571	866	749 956	29.42788
767	588 289	27.69476	817	667 489	28.58321	867	751 689	29.44486
768	589 824	27.71281	818	669 124	28.60070	868	753 424	29.46184
769	591 361	27.73085	819	670 761	28.61818	869	755 161	29.47881
770	592 900	27.74887	820	672 400	28.63564	870	756 900	29.49576
771	594 441	27.76689	821	674 041	28.65310	871	758 641	29.51271
772	595 984	27.78489	822	675 684	28.67054	872	760 384	29.52965
773	597 529	27.80288	823	677 329	28.68798	873	762 129	29.54657
774	599 076	27.82086	824	678 976	28.70540	874	763 876	29.56349
775	600 625	27.83882	825	680 625	28.72281	875	765 625	29.58040
776	602 176	27.85678	826	682 276	28.74022	876	767 376	29.59730
777	603 729	27.87472	827	683 929	28.75761	877	769 129	29.61419
778	605 284	27.89265	828	685 584	28.77499	878	770 884	29.63106
779	606 341	27.91057	829	687 241	28.79236	879	772 641	29.64793
780	608 400	27.92848	830	688 900	28.80972	880	774 400	29.66479
781	609 961	27.94638	831	690 561	28.82707	881	776 161	29.68164
782	611 524	27.96426	832	692 224	28.84441	882	777 924	29.69848
783	613 089	27.98214	833	693 889	28.86174	883	779 689	29.71532
784	614 656	28.00000	834	695 556	28.87906	884	781 456	29.73214
785	616 225	28.01785	835	697 225	28.89637	885	783 225	29.74895
786	617 796	28.03569	836	698 896	28.91366	886	784 996	29.76575
787	619 369	28.05352	837	700 569	28.93095	887	786 769	29.78255
788	620 944	28.07134	838	702 244	28.94823	888	788 544	29.79933
789	622 521	28.08914	839	703 921	28.96550	889	790 321	29.81610
790	624 100	28.10694	840	705 600	28.98275	890	792 100	29.83287
791	625 681	28.12472	841	707 281	29.00000	891	793 881	29.84962
792	627 264	28.14249	842	708 964	29.01724	892	795 664	29.86637
793	628 849	28.16026	843	710 649	29.03446	893	797 449	29.88341
794	630 436	28.17801	844	712 336	29.05168	894	799 236	29.89983
795	632 025	28.19574	845	714 025	29.06888	895	801 025	29.91655
796	633 616	28.21347	846	715 716	29.08608	896	802 816	29.93326
797	635 209	28.23119	847	717 409	29.10326	897	804 609	29.94996
98	636 804	28.24889	848	719 104	29.12044	898	806 404	29.96665
9	638 401	28.26659	849	720 801	29.13760	899	808 201	29.98333

TABLE VII. SQUARES AND SQUARE ROOTS (Continued)
1–1000

N	N²	√N	N	N²	√N
900	810 000	30.00000	950	902 500	30.82207
901	811 801	30.01666	951	904 401	30.83829
902	813 604	30.03331	952	906 304	30.85450
903	815 409	30.04996	953	908 209	30.87070
904	817 216	30.06659	954	910 116	30.88689
905	819 025	30.08322	955	912 025	30.90307
906	820 836	30.09983	956	913 936	30.91925
907	822 649	30.11644	957	915 849	30.93542
908	824 464	30.13304	958	917 764	30.95158
909	826 281	30.14963	959	919 681	30.96773
910	828 100	30.16621	960	921 600	30.98387
911	829 921	30.18278	961	923 521	31.00000
912	831 744	30.19934	962	925 444	31.01612
913	833 569	30.21589	963	927 369	31.03224
914	835 396	30.23243	964	929 296	31.04835
915	837 225	30.24897	965	931 225	31.06445
916	839 056	30.26549	966	933 156	31.08054
917	840 889	30.28201	967	935 089	31.09662
918	842 724	30.29851	968	937 024	31.11270
919	844 561	30.31501	969	938 961	31.12876
920	846 400	30.33150	970	940 900	31.14482
921	848 241	30.34798	971	942 841	31.16087
922	850 084	30.36445	972	944 784	31.17691
923	851 929	30.38092	973	946 729	31.19295
924	853 776	30.39737	974	948 676	31.20897
925	855 625	30.41381	975	950 625	31.22499
926	857 476	30.43025	976	952 576	31.24100
927	859 329	30.44667	977	954 529	31.25700
928	861 184	30.46309	978	956 484	31.27299
929	863 041	30.47950	979	958 441	31.28898
930	864 900	30.49590	980	960 400	31.30495
931	866 761	30.51229	981	962 361	31.32092
932	868 624	30.52868	982	964 324	31.33688
933	870 489	30.54505	983	966 289	31.35283
934	872 356	30.56141	984	968 256	31.36877
935	874 225	30.57777	985	970 225	31.38471
936	876 096	30.59412	986	972 196	31.40064
937	877 969	30.61046	987	974 169	31.41656
938	879 844	30.62679	988	976 144	31.43247
939	881 721	30.64311	989	978 121	31.44837
940	883 600	30.65942	990	980 100	31.46427
941	885 481	30.67572	991	982 081	31.48015
942	887 364	30.69202	992	984 064	31.49603
943	889 249	30.70831	993	986 049	31.51190
944	891 136	30.72458	994	988 036	31.52777
945	893 025	30.74085	995	990 025	31.54362
946	894 916	30.75711	996	992 016	31.55947
947	896 809	30.77337	997	994 009	31.57531
948	898 704	30.78961	998	996 004	31.59114
949	900 601	30.80584	999	998 001	31.60696
			1000	1 000 000	31.62278

TABLE VIII. BINOMIAL DISTRIBUTION

$$b(x \mid n, p) = \binom{n}{x} p^x(1 - p)^{n-x}$$

N	X							P					
		.01	.02	.03	.04	.05	.10	.15	.20	.25	.30	.40	.50
2	0	.9801	.9404	.9409	.9216	.9025	.8100	.7225	.6400	.5625	.4900	.3600	.2500
	1	.0198	.0392	.0582	.0768	.0950	.1800	.2550	.3200	.3750	.4200	.4800	.5000
	2	.0001	.0004	.0009	.0016	.0025	.0100	.0225	.0400	.0625	.0900	.1600	.2500
3	0	.9703	.9412	.9127	.8847	.8574	.7290	.6141	.5120	.4219	.3430	.2160	.1250
	1	.0294	.0576	.0847	.1106	.1354	.2430	.3251	.3840	.4219	.4410	.4320	.3750
	2	.0003	.0012	.0026	.0046	.0071	.0270	.0574	.0960	.1406	.1890	.2880	.3750
	3	.0000	.0000	.0000	.0001	.0001	.0010	.0034	.0080	.0156	.0270	.0640	.1250
4	0	.9606	.9224	.8853	.8493	.8145	.6561	.5220	.4096	.3164	.2401	.1296	.0625
	1	.0388	.0753	.1095	.1416	.1715	.2916	.3685	.4096	.4219	.4116	.3456	.2500
	2	.0006	.0023	.0051	.0088	.0135	.0486	.0975	.1536	.2109	.2646	.3456	.3750
	3	.0000	.0000	.0001	.0002	.0005	.0036	.0115	.0256	.0469	.0756	.1536	.2500
	4	.0000	.0000	.0000	.0000	.0000	.0001	.0005	.0016	.0039	.0081	.0256	.0625
5	0	.9510	.9039	.8587	.8154	.7738	.5905	.4437	.3277	.2373	.1681	.0778	.0313
	1	.0480	.0922	.1328	.1699	.2036	.3280	.3915	.4096	.3955	.3602	.2592	.1563
	2	.0010	.0038	.0082	.0142	.0214	.0729	.1382	.2048	.2637	.3087	.3456	.3125
	3	.0000	.0001	.0003	.0006	.0011	.0081	.0244	.0512	.0879	.1323	.2304	.3125
	4	.0000	.0000	.0000	.0000	.0000	.0004	.0022	.0064	.0146	.0283	.0768	.1563
	5	.0000	.0000	.0000	.0000	.0000	.0000	.0001	.0003	.0010	.0024	.0102	.0313
6	0	.9415	.8858	.8330	.7828	.7351	.5314	.3771	.2621	.1780	.1176	.0467	.0156
	1	.0571	.1085	.1546	.1957	.2321	.3543	.3993	.3932	.3560	.3025	.1866	.0938
	2	.0014	.0055	.0120	.0204	.0305	.0984	.1762	.2458	.2966	.3241	.3110	.2344
	3	.0000	.0002	.0005	.0011	.0021	.0146	.0415	.0819	.1318	.1852	.2765	.3125
	4	.0000	.0000	.0000	.0000	.0001	.0012	.0055	.0154	.0330	.0595	.1382	.2344
	5	.0000	.0000	.0000	.0000	.0000	.0001	.0004	.0015	.0044	.0102	.0369	.0938
	6	.0000	.0000	.0000	.0000	.0000	.0000	.0000	.0001	.0002	.0007	.0041	.0156
7	0	.9321	.8681	.8080	.7514	.6983	.4783	.3206	.2097	.1335	.0824	.0280	.0078
	1	.0659	.1240	.1749	.2192	.2573	.3720	.3960	.3670	.3115	.2471	.1306	.0547
	2	.0020	.0076	.0162	.0274	.0406	.1240	.2097	.2753	.3115	.3177	.2613	.1641
	3	.0000	.0003	.0008	.0019	.0036	.0230	.0617	.1147	.1730	.2269	.2903	.2734
	4	.0000	.0000	.0000	.0001	.0002	.0026	.0109	.0287	.0577	.0972	.1935	.2734
	5	.0000	.0000	.0000	.0000	.0000	.0002	.0012	.0043	.0115	.0250	.0774	.1641
	6	.0000	.0000	.0000	.0000	.0000	.0000	.0001	.0004	.0013	.0036	.0172	.0547
	7	.0000	.0000	.0000	.0000	.0000	.0000	.0000	.0000	.0001	.0002	.0016	.0078
8	0	.9227	.8508	.7837	.7214	.6634	.4305	.2725	.1678	.1001	.0576	.0168	.0039
	1	.0746	.1389	.1939	.2405	.2793	.3826	.3847	.3355	.2670	.1977	.0896	.0313
	2	.0026	.0099	.0210	.0351	.0515	.1488	.2376	.2936	.3115	.2965	.2090	.1096
	3	.0001	.0004	.0013	.0029	.0054	.0331	.0839	.1468	.2076	.2541	.2787	.2188

TABLE VIII.   BINOMIAL DISTRIBUTION (Continued)

N	X	.01	.02	.03	.04	.05	.10	.15	.20	.25	.30	.40	.50
8	4	.0000	.0000	.0001	.0002	.0004	.0046	.0185	.0459	.0865	.1361	.2322	.2734
	5	.0000	.0000	.0000	.0000	.0000	.0004	.0026	.0092	.0231	.0467	.1239	.2188
	6	.0000	.0000	.0000	.0000	.0000	.0000	.0002	.0011	.0038	.0100	.0413	.1094
	7	.0000	.0000	.0000	.0000	.0000	.0000	.0000	.0001	.0004	.0012	.0079	.0313
	8	.0000	.0000	.0000	.0000	.0000	.0000	.0000	.0000	.0000	.0001	.0007	.0039
9	0	.9135	.8337	.7602	.6925	.6302	.3874	.2316	.1342	.0751	.0404	.0101	.0020
	1	.0830	.1531	.2116	.2597	.2985	.3874	.3679	.3020	.2253	.1556	.0605	.0176
	2	.0034	.0125	.0262	.0433	.0629	.1722	.2597	.3020	.3003	.2668	.1612	.0703
	3	.0001	.0006	.0019	.0042	.0077	.0446	.1069	.1762	.2336	.2668	.2508	.1641
	4	.0000	.0000	.0001	.0003	.0006	.0074	.0283	.0661	.1168	.1715	.2508	.2461
	5	.0000	.0000	.0000	.0000	.0000	.0008	.0050	.0165	.0389	.0735	.1672	.2461
	6	.0000	.0000	.0000	.0000	.0000	.0001	.0006	.0028	.0087	.0210	.0743	.1641
	7	.0000	.0000	.0000	.0000	.0000	.0000	.0000	.0003	.0012	.0039	.0212	.0703
	8	.0000	.0000	.0000	.0000	.0000	.0000	.0000	.0000	.0001	.0004	.0035	.0176
	9	.0000	.0000	.0000	.0000	.0000	.0000	.0000	.0000	.0000	.0000	.0003	.0020
10	0	.9044	.8171	.7374	.6648	.5987	.3487	.1969	.1074	.0563	.0282	.0060	.0010
	1	.0914	.1667	.2281	.2770	.3151	.3874	.3474	.2684	.1877	.1211	.0403	.0098
	2	.0042	.0153	.0317	.0519	.0746	.1937	.2759	.3020	.2816	.2335	.1209	.0439
	3	.0001	.0008	.0026	.0058	.0105	.0574	.1298	.2013	.2503	.2668	.2150	.1172
	4	.0000	.0000	.0001	.0004	.0010	.0112	.0401	.0881	.1460	.2001	.2508	.2051
	5	.0000	.0000	.0000	.0000	.0001	.0015	.0085	.0264	.0584	.1029	.2007	.2461
	5	.0000	.0000	.0000	.0000	.0000	.0001	.0012	.0055	.0162	.0368	.1115	.2051
	7	.0000	.0000	.0000	.0000	.0000	.0000	.0001	.0008	.0031	.0090	.0425	.1172
	8	.0000	.0000	.0000	.0000	.0000	.0000	.0000	.0001	.0004	.0014	.0106	.0439
	9	.0000	.0000	.0000	.0000	.0000	.0000	.0000	.0000	.0000	.0001	.0016	.0098
	10	.0000	.0000	.0000	.0000	.0000	.0000	.0000	.0000	.0000	.0000	.0001	.0010
11	0	.8953	.8007	.7153	.6282	.5688	.3138	.1673	.0859	.0422	.0198	.0036	.0005
	1	.0995	.1798	.2433	.2925	.2793	.3835	.3248	.2362	.1549	.0932	.0266	.0054
	2	.0050	.0183	.0376	.0609	.0867	.2131	.2866	.2953	.2581	.1998	.0887	.0269
	3	.0002	.0011	.0035	.0076	.0137	.0710	.1517	.2215	.2581	.2568	.1774	.0806
	4	.0000	.0000	.0002	.0006	.0014	.0158	.0536	.1107	.1721	.2201	.2365	.1611
	5	.0000	.0000	.0000	.0000	.0001	.0025	.0132	.0388	.0803	.1321	.2207	.2256
	6	.0000	.0000	.0000	.0000	.0000	.0003	.0023	.0097	.0268	.0566	.1471	.2256
	7	.0000	.0000	.0000	.0000	.0000	.0000	.0003	.0017	.0064	.0173	.0701	.1611
	8	.0000	.0000	.0000	.0000	.0000	.0000	.0000	.0002	.0011	.0037	.0234	.0806
	9	.0000	.0000	.0000	.0000	.0000	.0000	.0000	.0000	.0001	.0005	.0052	.0269
	10	.0000	.0000	.0000	.0000	.0000	.0000	.0000	.0000	.0000	.0000	.0007	.0054
	11	.0000	.0000	.0000	.0000	.0000	.0000	.0000	.0000	.0000	.0000	.0000	.0005
12	0	.8864	.7847	.6938	.6127	.5404	.2824	.1422	.0687	.0317	.0138	.0022	.0002
	1	.1074	.1922	.2575	.3064	.3413	.3766	.3012	.2062	.1267	.0712	.0174	.0029
	2	.0060	.0216	.0438	.0702	.0988	.2301	.2924	.2835	.2323	.1678	.0639	.0161
	3	.0002	.0015	.0045	.0098	.0173	.0852	.1720	.2362	.2581	.2397	.1419	.0537

TABLE VIII. BINOMIAL DISTRIBUTION (Continued)

N	X	.01	.02	.03	.04	.05	.10	.15	.20	.25	.30	.40	.50
12	4	.0000	.0001	.0003	.0009	.0021	.0213	.0683	.1329	.1936	.2311	.2128	.1208
	5	.0000	.0000	.0000	.0001	.0002	.0038	.0193	.0532	.1032	.1585	.2270	.1934
	6	.0000	.0000	.0000	.0000	.0000	.0005	.0040	.0155	.0401	.0792	.1766	.2256
	7	.0000	.0000	.0000	.0000	.0000	.0000	.0006	.0033	.0115	.0291	.1009	.1934
	8	.0000	.0000	.0000	.0000	.0000	.0000	.0001	.0005	.0024	.0078	.0420	.1208
	9	.0000	.0000	.0000	.0000	.0000	.0000	.0000	.0001	.0004	.0015	.0125	.0537
	10	.0000	.0000	.0000	.0000	.0000	.0000	.0000	.0000	.0000	.0002	.0025	.0161
	11	.0000	.0000	.0000	.0000	.0000	.0000	.0000	.0000	.0000	.0000	.0003	.0029
	12	.0000	.0000	.0000	.0000	.0000	.0000	.0000	.0000	.0000	.0000	.0000	.0002
13	0	.8775	.7690	.6730	.5882	.5133	.2542	.1209	.0550	.0238	.0097	.0013	.0001
	1	.1152	.2040	.2706	.3186	.3512	.3672	.2774	.1787	.1029	.0540	.0113	.0016
	2	.0070	.0250	.0502	.0797	.1109	.2448	.2937	.2680	.2059	.1388	.0453	.0095
	3	.0003	.0019	.0057	.0122	.0214	.0997	.1900	.2457	.2517	.2181	.1107	.0349
	4	.0000	.0001	.0004	.0013	.0028	.0277	.0838	.1535	.2097	.2337	.1845	.0873
	5	.0000	.0000	.0000	.0001	.0003	.0055	.0266	.0691	.1258	.1803	.2214	.1571
	6	.0000	.0000	.0000	.0000	.0000	.0008	.0063	.0230	.0559	.1030	.1968	.2095
	7	.0000	.0000	.0000	.0000	.0000	.0001	.0011	.0058	.0186	.0442	.1312	.2095
	8	.0000	.0000	.0000	.0000	.0000	.0000	.0001	.0011	.0047	.0142	.0656	.1571
	9	.0000	.0000	.0000	.0000	.0000	.0000	.0000	.0001	.0009	.0034	.0243	.0873
	10	.0000	.0000	.0000	.0000	.0000	.0000	.0000	.0000	.0001	.0006	.0065	.0349
	11	.0000	.0000	.0000	.0000	.0000	.0000	.0000	.0000	.0000	.0001	.0012	.0095
	12	.0000	.0000	.0000	.0000	.0000	.0000	.0000	.0000	.0000	.0000	.0001	.0016
	13	.0000	.0000	.0000	.0000	.0000	.0000	.0000	.0000	.0000	.0000	.0000	.0001
14	0	.8687	.7536	.6528	.5647	.4877	.2288	.1028	.0440	.0178	.0068	.0008	.0001
	1	.1229	.2153	.2827	.3294	.3593	.3559	.2539	.1539	.0832	.0407	.0073	.0009
	2	.0081	.0286	.0568	.0892	.1229	.2570	.2912	.2501	.1802	.1134	.0317	.0056
	3	.0003	.0023	.0070	.0149	.0259	.1142	.2056	.2501	.2402	.1943	.0845	.0222
	4	.0000	.0001	.0006	.0017	.0037	.0349	.0998	.1720	.2202	.2290	.1549	.0611
	5	.0000	.0000	.0000	.0001	.0004	.0078	.0352	.0860	.1468	.1963	.2066	.1222
	6	.0000	.0000	.0000	.0000	.0000	.0013	.0093	.0322	.0734	.1262	.2066	.1833
	7	.0000	.0000	.0000	.0000	.0000	.0002	.0019	.0092	.0280	.0618	.1574	.2095
	8	.0000	.0000	.0000	.0000	.0000	.0000	.0003	.0020	.0082	.0232	.0918	.1833
	9	.0000	.0000	.0000	.0000	.0000	.0000	.0000	.0003	.0018	.0066	.0408	.1222
	10	.0000	.0000	.0000	.0000	.0000	.0000	.0000	.0000	.0003	.0014	.0136	.0611
	11	.0000	.0000	.0000	.0000	.0000	.0000	.0000	.0000	.0000	.0002	.0033	.0222
	12	.0000	.0000	.0000	.0000	.0000	.0000	.0000	.0000	.0000	.0000	.0005	.0056
	13	.0000	.0000	.0000	.0000	.0000	.0000	.0000	.0000	.0000	.0000	.0001	.0009
	14	.0000	.0000	.0000	.0000	.0000	.0000	.0000	.0000	.0000	.0000	.0000	.0001
15	0	.8601	.7386	.6333	.5421	.4633	.2059	.0874	.0352	.0134	.0047	.0005	.0000
	1	.1303	.2261	.2938	.3388	.3658	.3432	.2312	.1319	.0668	.0305	.0047	.0005
	2	.0092	.0323	.0636	.0988	.1348	.2669	.2856	.2309	.1559	.0916	.0219	.0032
	3	.0004	.0029	.0085	.0178	.0307	.1285	.2184	.2501	.2252	.1700	.0634	.0139

TABLE VIII.   BINOMIAL DISTRIBUTION (Continued)

							P						
N	X	.01	.02	.03	.04	.05	.10	.15	.20	.25	.30	.40	.50
15	4	.0000	.0002	.0008	.0022	.0049	.0428	.1156	.1876	.2252	.2186	.1268	.0417
	5	.0000	.0000	.0001	.0002	.0006	.0105	.0449	.1032	.1651	.2061	.1859	.0916
	6	.0000	.0000	.0000	.0000	.0000	.0019	.0132	.0430	.0917	.1472	.2066	.1527
	7	.0000	.0000	.0000	.0000	.0000	.0003	.0030	.0138	.0393	.0811	.1771	.1964
	8	.0000	.0000	.0000	.0000	.0000	.0000	.0005	.0035	.0131	.0348	.1181	.1964
	9	.0000	.0000	.0000	.0000	.0000	.0000	.0001	.0007	.0034	.0116	.0612	.1527
	10	.0000	.0000	.0000	.0000	.0000	.0000	.0000	.0001	.0007	.0030	.0245	.0916
	11	.0000	.0000	.0000	.0000	.0000	.0000	.0000	.0000	.0001	.0006	.0074	.0417
	12	.0000	.0000	.0000	.0000	.0000	.0000	.0000	.0000	.0000	.0001	.0016	.0139
	13	.0000	.0000	.0000	.0000	.0000	.0000	.0000	.0000	.0000	.0000	.0003	.0032
	14	.0000	.0000	.0000	.0000	.0000	.0000	.0000	.0000	.0000	.0000	.0000	.0005
16	0	.8515	.7238	.6143	.5204	.4401	.1853	.0743	.0281	.0100	.0033	.0003	.0000
	1	.1376	.2363	.3040	.3469	.3706	.3294	.2096	.1126	.0535	.0228	.0030	.0002
	2	.0104	.0362	.0705	.1084	.1463	.2745	.2775	.2111	.1336	.0732	.0150	.0018
	3	.0005	.0034	.0102	.0211	.0359	.1423	.2285	.2463	.2079	.1465	.0468	.0085
	4	.0000	.0002	.0010	.0029	.0061	.0514	.1311	.2001	.2252	.2040	.1014	.0278
	5	.0000	.0000	.0001	.0003	.0008	.0137	.0555	.1201	.1802	.2099	.1623	.0667
	6	.0000	.0000	.0000	.0000	.0001	.0028	.0180	.0550	.1101	.1649	.1983	.1222
	7	.0000	.0000	.0000	.0000	.0000	.0004	.0045	.0197	.0524	.1010	.1889	.1746
	8	.0000	.0000	.0000	.0000	.0000	.0001	.0009	.0055	.0197	.0487	.1417	.1964
	9	.0000	.0000	.0000	.0000	.0000	.0000	.0001	.0012	.0058	.0185	.0840	.1746
	10	.0000	.0000	.0000	.0000	.0000	.0000	.0000	.0002	.0014	.0056	.0392	.1222
	11	.0000	.0000	.0000	.0000	.0000	.0000	.0000	.0000	.0002	.0013	.0142	.0667
	12	.0000	.0000	.0000	.0000	.0000	.0000	.0000	.0000	.0000	.0002	.0040	.0278
	13	.0000	.0000	.0000	.0000	.0000	.0000	.0000	.0000	.0000	.0000	.0008	.0085
	14	.0000	.0000	.0000	.0000	.0000	.0000	.0000	.0000	.0000	.0000	.0001	.0018
	15	.0000	.0000	.0000	.0000	.0000	.0000	.0000	.0000	.0000	.0000	.0000	.0002
17	0	.8429	.7093	.5958	.4996	.4181	.1668	.0631	.0225	.0075	.0023	.0002	.0000
	1	.1447	.2461	.3133	.3539	.3741	.3150	.1893	.0957	.0426	.0169	.0019	.0001
	2	.0117	.0402	.0775	.1180	.1575	.2800	.2673	.1914	.1136	.0581	.0102	.0010
	3	.0006	.0041	.0120	.0246	.0415	.1556	.2359	.2393	.1893	.1245	.0341	.0052
	4	.0000	.0003	.0013	.0036	.0076	.0605	.1457	.2093	.2209	.1868	.0796	.0182
	5	.0000	.0000	.0001	.0004	.0010	.0175	.0668	.1361	.1914	.2081	.1379	.0472
	6	.0000	.0000	.0000	.0000	.0001	.0039	.0236	.0680	.1276	.1784	.1839	.0944
	7	.0000	.0000	.0000	.0000	.0000	.0007	.0065	.0267	.0668	.1201	.1927	.1484
	8	.0000	.0000	.0000	.0000	.0000	.0001	.0014	.0084	.0279	.0644	.1606	.1855
	9	.0000	.0000	.0000	.0000	.0000	.0000	.0003	.0021	.0093	.0276	.1070	.1855
	10	.0000	.0000	.0000	.0000	.0000	.0000	.0000	.0004	.0025	.0095	.0571	.1484
	11	.0000	.0000	.0000	.0000	.0000	.0000	.0000	.0001	.0005	.0026	.0242	.0944
	12	.0000	.0000	.0000	.0000	.0000	.0000	.0000	.0000	.0001	.0006	.0081	.0472
	13	.0000	.0000	.0000	.0000	.0000	.0000	.0000	.0000	.0000	.0001	.0021	.0182

TABLE VIII. BINOMIAL DISTRIBUTION (Continued)

N	X	.01	.02	.03	.04	.05	.10	.15	.20	.25	.30	.40	.50
17	14	.0000	.0000	.0000	.0000	.0000	.0000	.0000	.0000	.0000	.0000	.0004	.0052
	15	.0000	.0000	.0000	.0000	.0000	.0000	.0000	.0000	.0000	.0000	.0001	.0010
	16	.0000	.0000	.0000	.0000	.0000	.0000	.0000	.0000	.0000	.0000	.0000	.0001
18	0	.8345	.6951	.5780	.4796	.3972	.1501	.0536	.0180	.0056	.0016	.0001	.0000
	1	.1517	.2554	.3217	.3597	.3763	.3002	.1704	.0811	.0338	.0126	.0012	.0001
	2	.0130	.0443	.0846	.1274	.1683	.2835	.2556	.1723	.0958	.0458	.0069	.0006
	3	.0007	.0048	.0140	.0283	.0473	.1680	.2406	.2297	.1704	.1046	.0246	.0031
	4	.0000	.0004	.0016	.0044	.0093	.0700	.1592	.2153	.2130	.1681	.0614	.0117
	5	.0000	.0000	.0001	.0005	.0014	.0218	.0787	.1507	.1988	.2017	.1146	.0327
	6	.0000	.0000	.0000	.0000	.0002	.0052	.0301	.0816	.1436	.1873	.1655	.0708
	7	.0000	.0000	.0000	.0000	.0000	.0010	.0091	.0350	.0820	.1376	.1892	.1214
	8	.0000	.0000	.0000	.0000	.0000	.0002	.0022	.0120	.0376	.0811	.1734	.1669
	9	.0000	.0000	.0000	.0000	.0000	.0000	.0004	.0033	.0139	.0386	.1284	.1855
	10	.0000	.0000	.0000	.0000	.0000	.0000	.0001	.0008	.0042	.0149	.0771	.1669
	11	.0000	.0000	.0000	.0000	.0000	.0000	.0000	.0001	.0010	.0046	.0374	.1214
	12	.0000	.0000	.0000	.0000	.0000	.0000	.0000	.0000	.0002	.0012	.0145	.0708
	13	.0000	.0000	.0000	.0000	.0000	.0000	.0000	.0000	.0000	.0002	.0045	.0327
	14	.0000	.0000	.0000	.0000	.0000	.0000	.0000	.0000	.0000	.0000	.0011	.0117
	15	.0000	.0000	.0000	.0000	.0000	.0000	.0000	.0000	.0000	.0000	.0002	.0031
	16	.0000	.0000	.0000	.0000	.0000	.0000	.0000	.0000	.0000	.0000	.0000	.0006
	17	.0000	.0000	.0000	.0000	.0000	.0000	.0000	.0000	.0000	.0000	.0000	.0001
19	0	.8262	.6812	.5606	.4604	.3774	.1351	.0456	.0144	.0042	.0011	.0001	.0000
	1	.1586	.2642	.3294	.3645	.3774	.2852	.1529	.0685	.0268	.0093	.0008	.0000
	2	.0144	.0485	.0917	.1367	.1787	.2852	.2428	.1540	.0803	.0358	.0046	.0003
	3	.0008	.0056	.0161	.0323	.0533	.1796	.2428	.2182	.1517	.0869	.0175	.0018
	4	.0000	.0005	.0020	.0054	.0112	.0798	.1714	.2182	.2023	.1491	.0467	.0074
	5	.0000	.0000	.0002	.0007	.0018	.0266	.0907	.1636	.2023	.1916	.0933	.0222
	6	.0000	.0000	.0000	.0001	.0002	.0069	.0374	.0955	.1574	.1916	.1451	.0518
	7	.0000	.0000	.0000	.0000	.0000	.0014	.0122	.0443	.0974	.1525	.1797	.0961
	8	.0000	.0000	.0000	.0000	.0000	.0002	.0032	.0166	.0487	.0981	.1797	.1442
	9	.0000	.0000	.0000	.0000	.0000	.0000	.0007	.0051	.0198	.0514	.1464	.1762
	10	.0000	.0000	.0000	.0000	.0000	.0000	.0001	.0013	.0066	.0220	.0976	.1762
	11	.0000	.0000	.0000	.0000	.0000	.0000	.0000	.0003	.0018	.0077	.0532	.1442
	12	.0000	.0000	.0000	.0000	.0000	.0000	.0000	.0000	.0004	.0022	.0237	.0961
	13	.0000	.0000	.0000	.0000	.0000	.0000	.0000	.0000	.0001	.0005	.0085	.0518
	14	.0000	.0000	.0000	.0000	.0000	.0000	.0000	.0000	.0000	.0001	.0024	.0222
	15	.0000	.0000	.0000	.0000	.0000	.0000	.0000	.0000	.0000	.0000	.0005	.0074
	16	.0000	.0000	.0000	.0000	.0000	.0000	.0000	.0000	.0000	.0000	.0001	.0018
	17	.0000	.0000	.0000	.0000	.0000	.0000	.0000	.0000	.0000	.0000	.0000	.0003
20	0	.8179	.6676	.5438	.4420	.3585	.1216	.0388	.0115	.0032	.0008	.0000	.0000
	1	.1652	.2725	.3364	.3683	.3774	.2702	.1368	.0576	.0211	.0068	.0005	.0000
	2	.0159	.0528	.0988	.1458	.1887	.2852	.2293	.1369	.0669	.0278	.0031	.0002
	3	.0010	.0065	.0183	.0364	.0596	.1901	.2428	.2054	.1339	.0716	.0123	.0011

TABLE VIII. BINOMIAL DISTRIBUTION (Continued)

N	X	.01	.02	.03	.04	.05	P .10	.15	.20	.25	.30	.40	.50
20	4	.0000	.0006	.0024	.0065	.0133	.0898	.1821	.2182	.1897	.1304	.0350	.0046
	5	.0000	.0000	.0002	.0009	.0022	.0319	.1028	.1746	.2023	.1789	.0746	.0148
	6	.0000	.0000	.0000	.0001	.0003	.0089	.0454	.1091	.1686	.1916	.1244	.0370
	7	.0000	.0000	.0000	.0000	.0000	.0020	.0160	.0545	.1124	.1643	.1659	.0739
	8	.0000	.0000	.0000	.0000	.0000	.0004	.0046	.0222	.0609	.1144	.1797	.1201
	9	.0000	.0000	.0000	.0000	.0000	.0001	.0011	.0074	.0271	.0654	.1597	.1602
	10	.0000	.0000	.0000	.0000	.0000	.0000	.0002	.0020	.0099	.0308	.1171	.1762
	11	.0000	.0000	.0000	.0000	.0000	.0000	.0000	.0005	.0030	.0120	.0710	.1602
	12	.0000	.0000	.0000	.0000	.0000	.0000	.0000	.0001	.0008	.0039	.0355	.1201
	13	.0000	.0000	.0000	.0000	.0000	.0000	.0000	.0000	.0002	.0010	.0146	.0739
	14	.0000	.0000	.0000	.0000	.0000	.0000	.0000	.0000	.0000	.0002	.0049	.0370
	15	.0000	.0000	.0000	.0000	.0000	.0000	.0000	.0000	.0000	.0000	.0013	.0148
	16	.0000	.0000	.0000	.0000	.0000	.0000	.0000	.0000	.0000	.0000	.0003	.0046
	17	.0000	.0000	.0000	.0000	.0000	.0000	.0000	.0000	.0000	.0000	.0000	.0011
	18	.0000	.0000	.0000	.0000	.0000	.0000	.0000	.0000	.0000	.0000	.0000	.0002
25	0	.7778	.6035	.4670	.3604	.2774	.0718	.0172	.0038	.0008	.0001	.0000	.0000
	1	.1964	.3079	.3611	.3754	.3650	.1994	.0759	.0236	.0063	.0014	.0000	.0000
	2	.0238	.0754	.1340	.1877	.2305	.2659	.1607	.0708	.0251	.0074	.0004	.0000
	3	.0018	.0118	.0318	.0600	.0930	.2265	.2174	.1358	.0641	.0243	.0019	.0001
	4	.0001	.0013	.0054	.0137	.0269	.1384	.2110	.1867	.1175	.0572	.0071	.0004
	5	.0000	.0001	.0007	.0024	.0060	.0646	.1564	.1960	.1645	.1030	.0199	.0016
	6	.0000	.0000	.0001	.0003	.0010	.0239	.0920	.1633	.1828	.1472	.0442	.0053
	7	.0000	.0000	.0000	.0000	.0001	.0072	.0441	.1108	.1654	.1712	.0800	.0143
	8	.0000	.0000	.0000	.0000	.0000	.0018	.0175	.0623	.1241	.1651	.1200	.0322
	9	.0000	.0000	.0000	.0000	.0000	.0004	.0058	.0294	.0781	.1336	.1511	.0609
	10	.0000	.0000	.0000	.0000	.0000	.0001	.0016	.0118	.0417	.0916	.1612	.0974
	11	.0000	.0000	.0000	.0000	.0000	.0000	.0004	.0040	.0189	.0536	.1465	.1328
	12	.0000	.0000	.0000	.0000	.0000	.0000	.0001	.0012	.0074	.0268	.1139	.1550
	13	.0000	.0000	.0000	.0000	.0000	.0000	.0000	.0003	.0025	.0115	.0760	.1550
	14	.0000	.0000	.0000	.0000	.0000	.0000	.0000	.0001	.0007	.0042	.0434	.1328
	15	.0000	.0000	.0000	.0000	.0000	.0000	.0000	.0000	.0002	.0013	.0212	.0974
	16	.0000	.0000	.0000	.0000	.0000	.0000	.0000	.0000	.0000	.0004	.0088	.0609
	17	.0000	.0000	.0000	.0000	.0000	.0000	.0000	.0000	.0000	.0001	.0031	.0322
	18	.0000	.0000	.0000	.0000	.0000	.0000	.0000	.0000	.0000	.0000	.0009	.0143
	19	.0000	.0000	.0000	.0000	.0000	.0000	.0000	.0000	.0000	.0000	.0002	.0053
	20	.0000	.0000	.0000	.0000	.0000	.0000	.0000	.0000	.0000	.0000	.0000	.0016
	21	.0000	.0000	.0000	.0000	.0000	.0000	.0000	.0000	.0000	.0000	.0000	.0004
	22	.0000	.0000	.0000	.0000	.0000	.0000	.0000	.0000	.0000	.0000	.0000	.0001
50	0	.6050	.3642	.2181	.1299	.0769	.0052	.0003	.0000	.0000	.0000	.0000	.0000
	1	.3056	.3716	.3372	.2706	.2025	.0286	.0026	.0002	.0000	.0000	.0000	.0000
	2	.0756	.1858	.2555	.2762	.2611	.0779	.0113	.0011	.0001	.0000	.0000	.0000
	3	.0122	.0607	.1264	.1842	.2199	.1386	.0319	.0044	.0004	.0000	.0000	.0000

TABLE VIII. BINOMIAL DISTRIBUTION (Continued)

N	X	.01	.02	.03	.04	.05	.10	.15	.20	.25	.30	.40	.50
50	4	.0015	.0145	.0459	.0902	.1360	.1809	.0661	.0128	.0016	.0001	.0000	.0000
	5	.0001	.0027	.0131	.0346	.0658	.1849	.1072	.0295	.0049	.0006	.0000	.0000
	6	.0000	.0004	.0030	.0108	.0260	.1541	.1419	.0554	.0123	.0018	.0000	.0000
	7	.0000	.0001	.0006	.0028	.0086	.1076	.1575	.0870	.0259	.0048	.0000	.0000
	8	.0000	.0000	.0001	.0006	.0024	.0643	.1493	.1169	.0463	.0110	.0002	.0000
	9	.0000	.0000	.0000	.0001	.0006	.0333	.1230	.1364	.0721	.0220	.0005	.0000
	10	.0000	.0000	.0000	.0000	.0001	.0152	.0890	.1398	.0985	.0386	.0014	.0000
	11	.0000	.0000	.0000	.0000	.0000	.0661	.0571	.1271	.1194	.0602	.0035	.0000
	12	.0000	.0000	.0000	.0000	.0000	.0022	.0328	.1033	.1294	.0838	.0076	.0001
	13	.0000	.0000	.0000	.0000	.0000	.0007	.0169	.0755	.1261	.1050	.0147	.0003
	14	.0000	.0000	.0000	.0000	.0000	.0002	.0079	.0499	.1110	.1189	.0260	.0008
	15	.0000	.0000	.0000	.0000	.0000	.0001	.0033	.0299	.0888	.1223	.0415	.0020
	16	.0000	.0000	.0000	.0000	.0000	.0000	.0013	.0164	.0648	.1147	.0606	.0044
	17	.0000	.0000	.0000	.0000	.0000	.0000	.0005	.0082	.0432	.0983	.0808	.0087
	18	.0000	.0000	.0000	.0000	.0000	.0000	.0001	.0037	.0264	.0772	.0987	.0160
	19	.0000	.0000	.0000	.0000	.0000	.0000	.0000	.0016	.0148	.0558	.1109	.0270
	20	.0000	.0000	.0000	.0000	.0000	.0000	.0000	.0006	.0077	.0370	.1146	.0419
	21	.0000	.0000	.0000	.0000	.0000	.0000	.0000	.0002	.0036	.0227	.1091	.0598
	22	.0000	.0000	.0000	.0000	.0000	.0000	.0000	.0001	.0016	.0128	.0959	.0788
	23	.0000	.0000	.0000	.0000	.0000	.0000	.0000	.0000	.0006	.0067	.0778	.0960
	24	.0000	.0000	.0000	.0000	.0000	.0000	.0000	.0000	.0002	.0032	.0584	.1080
	25	.0000	.0000	.0000	.0000	.0000	.0000	.0000	.0000	.0001	.0014	.0405	.1123
	26	.0000	.0000	.0000	.0000	.0000	.0000	.0000	.0000	.0000	.0006	.0259	.1080
	27	.0000	.0000	.0000	.0000	.0000	.0000	.0000	.0000	.0000	.0002	.0154	.0960
	28	.0000	.0000	.0000	.0000	.0000	.0000	.0000	.0000	.0000	.0001	.0084	.0788
	29	.0000	.0000	.0000	.0000	.0000	.0000	.0000	.0000	.0000	.0000	.0043	.0598
	30	.0000	.0000	.0000	.0000	.0000	.0000	.0000	.0000	.0000	.0000	.0020	.0419
	31	.0000	.0000	.0000	.0000	.0000	.0000	.0000	.0000	.0000	.0000	.0009	.0270
	32	.0000	.0000	.0000	.0000	.0000	.0000	.0000	.0000	.0000	.0000	.0003	.0160
	33	.0000	.0000	.0000	.0000	.0000	.0000	.0000	.0000	.0000	.0000	.0001	.0087
	34	.0000	.0000	.0000	.0000	.0000	.0000	.0000	.0000	.0000	.0000	.0000	.0044
	35	.0000	.0000	.0000	.0000	.0000	.0000	.0000	.0000	.0000	.0000	.0000	.0020
	36	.0000	.0000	.0000	.0000	.0000	.0000	.0000	.0000	.0000	.0000	.0000	.0008
	37	.0000	.0000	.0000	.0000	.0000	.0000	.0000	.0000	.0000	.0000	.0000	.0003
	38	.0000	.0000	.0000	.0000	.0000	.0000	.0000	.0000	.0000	.0000	.0000	.0001

TABLE IX. STANDARD NORMAL CUMULATIVE DISTRIBUTION

Z	.00	.01	.02	.03	.04	.05	.06	.07	.08	.09
−3.8	.0001	.0001	.0001	.0001	.0001	.0001	.0001	.0001	.0001	.0001
−3.7	.0001	.0001	.0001	.0001	.0001	.0001	.0001	.0001	.0001	.0001
−3.6	.0002	.0002	.0001	.0001	.0001	.0001	.0001	.0001	.0001	.0001
−3.5	.0002	.0002	.0002	.0002	.0002	.0002	.0002	.0002	.0002	.0002
−3.4	.0003	.0003	.0003	.0003	.0003	.0003	.0003	.0003	.0003	.0002
−3.3	.0005	.0005	.0005	.0004	.0004	.0004	.0004	.0004	.0004	.0003
−3.2	.0007	.0007	.0006	.0006	.0006	.0006	.0006	.0005	.0005	.0005
−3.1	.0010	.0009	.0009	.0009	.0008	.0008	.0008	.0008	.0007	.0007
−3.0	.0014	.0013	.0013	.0012	.0012	.0011	.0011	.0011	.0010	.0010
−2.9	.0019	.0018	.0018	.0017	.0016	.0016	.0015	.0015	.0014	.0014
−2.8	.0026	.0025	.0024	.0023	.0023	.0022	.0021	.0021	.0020	.0019
−2.7	.0035	.0034	.0033	.0032	.0031	.0030	.0029	.0028	.0027	.0026
−2.6	.0047	.0045	.0044	.0043	.0041	.0040	.0039	.0038	.0037	.0036
−2.5	.0062	.0060	.0059	.0057	.0055	.0054	.0052	.0051	.0049	.0048
−2.4	.0082	.0080	.0078	.0076	.0073	.0071	.0069	.0068	.0066	.0064
−2.3	.0107	.0104	.0102	.0099	.0096	.0094	.0091	.0089	.0087	.0084
−2.2	.0139	.0136	.0132	.0129	.0125	.0122	.0119	.0116	.0113	.0110
−2.1	.0179	.0174	.0170	.0166	.0162	.0158	.0154	.0150	.0146	.0143
−2.0	.0228	.0222	.0217	.0212	.0207	.0202	.0197	.0192	.0188	.0183
−1.9	.0287	.0281	.0274	.0268	.0262	.0256	.0250	.0244	.0239	.0233
−1.8	.0359	.0352	.0344	.0336	.0329	.0322	.0314	.0307	.0301	.0294
−1.7	.0446	.0436	.0427	.0418	.0409	.0401	.0392	.0384	.0375	.0367
−1.6	.0548	.0537	.0526	.0516	.0505	.0495	.0485	.0475	.0465	.0455
−1.5	.0668	.0655	.0643	.0630	.0618	.0606	.0594	.0582	.0571	.0559
−1.4	.0808	.0793	.0778	.0764	.0749	.0735	.0721	.0708	.0694	.0681
−1.3	.0968	.0951	.0934	.0918	.0901	.0885	.0869	.0853	.0838	.0823
−1.2	.1151	.1131	.1112	.1094	.1075	.1057	.1038	.1020	.1003	.0985
−1.1	.1357	.1335	.1314	.1292	.1271	.1251	.1230	.1210	.1190	.1170
−1.0	.1587	.1562	.1539	.1515	.1492	.1469	.1446	.1423	.1401	.1379
−0.9	.1841	.1814	.1788	.1762	.1736	.1711	.1685	.1660	.1635	.1611
−0.8	.2119	.2090	.2061	.2033	.2005	.1977	.1949	.1922	.1894	.1867
−0.7	.2420	.2389	.2358	.2327	.2296	.2266	.2236	.2206	.2177	.2148
−0.6	.2743	.2709	.2676	.2643	.2611	.2578	.2546	.2514	.2483	.2451
−0.5	.3085	.3050	.3015	.2981	.2946	.2912	.2877	.2843	.2810	.2776
−0.4	.3446	.3409	.3372	.3336	.3300	.3264	.3228	.3192	.3156	.3121
−0.3	.3821	.3783	.3745	.3707	.3669	.3632	.3594	.3557	.3520	.3483
−0.2	.4207	.4168	.4129	.4090	.4052	.4013	.3947	.3936	.3897	.3859
−0.1	.4602	.4562	.4522	.4483	.4443	.4404	.4364	.4325	.4286	.4247
−0.0	.5000	.4960	.4920	.4880	.4840	.4801	.4761	.4721	.4681	.4641

This table is reproduced by permission from P. Ewart, J. Ford, and C. Lin, *Probability for Statistical Decision Making*, Prentice-Hall, Inc., 1974.

TABLE IX.   STANDARD NORMAL CUMULATIVE DISTRIBUTION
(Continued)

Z	.00	.01	.02	.03	.04	.05	.06	.07	.08	.09
0.0	.5000	.5040	.5080	.5120	.5160	.5199	.5239	.5279	.5319	.5359
0.1	.5398	.5438	.5478	.5517	.5557	.5596	.5636	.5675	.5714	.5753
0.2	.5793	.5832	.5871	.5910	.5948	.5987	.6026	.6064	.6103	.6141
0.3	.6179	.6217	.6255	.6293	.6331	.6368	.6406	.6443	.6480	.6517
0.4	.6554	.6591	.6628	.6664	.6700	.6736	.6772	.6808	.6844	.6879
0.5	.6915	.6950	.6985	.7019	.7054	.7088	.7123	.7157	.7190	.7224
0.6	.7257	.7291	.7324	.7357	.7389	.7422	.7454	.7486	.7517	.7549
0.7	.7580	.7611	.7642	.7673	.7704	.7734	.7764	.7794	.7823	.7852
0.8	.7881	.7910	.7939	.7967	.7995	.8023	.8015	.8078	.8106	.8133
0.9	.8159	.8186	.8212	.8238	.8264	.8289	.8315	.8340	.8365	.8389
1.0	.8413	.8438	.8461	.8485	.8508	.8531	.8554	.8577	.8599	.8621
1.1	.8643	.8665	.8686	.8708	.8729	.8749	.8770	.8790	.8810	.8830
1.2	.8849	.8869	.8888	.8906	.8925	.8943	.8962	.8980	.8997	.9015
1.3	.9032	.9049	.9066	.9082	.9099	.9115	.9131	.9147	.9162	.9177
1.4	.9192	.9207	.9222	.9236	.9251	.9265	.9279	.9292	.9306	.9319
1.5	.9332	.9345	.9357	.9370	.9382	.9394	.9406	.9418	.9429	.9441
1.6	.9452	.9463	.9474	.9484	.9495	.9505	.9515	.9525	.9535	.9545
1.7	.9554	.9564	.9573	.9582	.9591	.9599	.9608	.9616	.9625	.9633
1.8	.9641	.9648	.9656	.9664	.9671	.9678	.9686	.9693	.9699	.9706
1.9	.9713	.9719	.9726	.9732	.9738	.9744	.9750	.9756	.9761	.9767
2.0	.9772	.9778	.9783	.9788	.9793	.9798	.9803	.9808	.9812	.9817
2.1	.9821	.9826	.9830	.9834	.9838	.9842	.9846	.9850	.9854	.9857
2.2	.9861	.9864	.9868	.9871	.9875	.9878	.9881	.9884	.9887	.9890
2.3	.9893	.9896	.9898	.9901	.9904	.9906	.9909	.9911	.9913	.9916
2.4	.9918	.9920	.9922	.9924	.9927	.9929	.9931	.9932	.9934	.9936
2.5	.9938	.9940	.9941	.9943	.9945	.9946	.9948	.9949	.9951	.9952
2.6	.9953	.9955	.9956	.9957	.9959	.9960	.9961	.9962	.9963	.9964
2.7	.9965	.9966	.9967	.9968	.9969	.9970	.9971	.9972	.9973	.9974
2.8	.9974	.9975	.9976	.9977	.9977	.9978	.9979	.9979	.9980	.9981
2.9	.9981	.9982	.9982	.9983	.9984	.9984	.9985	.9985	.9986	.9986
3.0	.9986	.9987	.9987	.9988	.9988	.9989	.9989	.9989	.9990	.9990
3.1	.9990	.9991	.9991	.9991	.9992	.9992	.9992	.9992	.9993	.9993
3.2	.9993	.9993	.9994	.9994	.9994	.9994	.9994	.9995	.9995	.9995
3.3	.9995	.9995	.9995	.9996	.9996	.9996	.9996	.9996	.9996	.9997
3.4	.9997	.9997	.9997	.9997	.9997	.9997	.9997	.9997	.9997	.9998
3.5	.9998	.9998	.9998	.9998	.9998	.9998	.9998	.9998	.9998	.9998
3.6	.9998	.9998	.9999	.9999	.9999	.9999	.9999	.9999	.9999	.9999
3.7	.9999	.9999	.9999	.9999	.9999	.9999	.9999	.9999	.9999	.9999
3.8	.9999	.9999	.9999	.9999	.9999	.9999	.9999	.9999	.9999	.9999

Selected Answers

to

Odd-Numbered Questions

CHAPTER 1

1. No. The individuals that will vote in an upcoming election are not known with certainty.
3. Yes. It is not necessary to have a listing of all family units with incomes of less than $6000 to describe this group as a set.
5. $\{a, b, c\}$, $\{a, b\}$, $\{a, c\}$, $\{b, c\}$, $\{a\}$, $\{b\}$, $\{c\}$, $\{\ \}$
7. 64
9. (a) Q', (b) ϕ, (c) U
11. 60
13. (a) 20, (b) 280
17. 12
19. 630
21. Cities where revenue is less than $5000 per month.

CHAPTER 2

1. (a) 0, (c) 2, (e) -0.5, (g) -3
3. (a) $f(x) = 3000 + \frac{2}{3}x,$ (c) $f(x) = 3500 + 10x$
    (e) $P = 11.50 - 0.001q$ (g) $y = 10{,}000 + 500t$
5. $A \cap B = \{4\},\ A \cup B = \{-5, -4, -3, -2, -1, 0, 1, 2, 3, 4, 5\}$
7. Domain: $x = 0, 3, 4, 5$; Range: $f(x) = 6, 7, 8, 10$
9. (a) $2x + 3y \le 80$ for $x, y \ge 0$
    (c) $20{,}000x + 60{,}000y + 80{,}000z \le 1{,}200{,}000$ for x, y, z nonnegative integers
11. $f(x) = -45 + 9x$ for $5 \le x \le 12$; Range: $0 \le f(x) \le 63$; Breakeven: 5 hours
13. $f(x) = -10{,}000{,}000 + 6000x - 0.5x^2$
15. $f(x) = -16 + 5.5x - 0.25x^2$
17.
$$f(x) = \begin{cases} 100 & \text{for } 0 \le x \le 1000 \\ 125 - 0.025x & \text{for } 1000 < x \le 3000 \\ 20 + 0.01x & \text{for } 3000 < x \le 4000 \end{cases}$$
19. $x_A = 6,\ x_B = 3$
21. An infinite number of solutions.
23. (a) Consistent, unique solution $x = 7,\ y = 5$
    (c) Inconsistent

CHAPTER 3

1. $A - B = (-4, -2, 7)$, (c) The product of two row vectors is not defined.

3. (a) $A + B = \begin{pmatrix} 14 \\ 2 \\ 8 \end{pmatrix},$ (c) $3A - 2B = \begin{pmatrix} 2 \\ 16 \\ -6 \end{pmatrix}$

5. $A(B + C) = AB + AC = 192$

7. (a) $A + B = \begin{pmatrix} 5 & 0 \\ 5 & 12 \\ -4 & 9 \end{pmatrix},$ (c) $3A - B = \begin{pmatrix} 7 & 20 \\ 3 & 12 \\ 8 & 7 \end{pmatrix}$

$$A(B + C) = AB + AC = \begin{pmatrix} 72 & 84 \\ -12 & -60 \\ 60 & 96 \end{pmatrix}$$

11. $VW = \begin{pmatrix} 10 & 16 & 6 \\ -10 & -16 & -6 \\ 20 & 32 & 12 \end{pmatrix}$

13. $AI = IA = \begin{pmatrix} 6 & 3 & 8 \\ 4 & 2 & 5 \\ 3 & 1 & 7 \end{pmatrix}$

15. (a) $x_1 = 4,\ x_2 = 6,$ (c) $x_1 = 3,\ x_2 = 5,\ x_3 = 4$

17. (a) $A^{-1} = \begin{pmatrix} 3.5 & -1.5 \\ -2 & 1 \end{pmatrix}$,   (c) $A^{-1} = \begin{pmatrix} 2 & -3 & 1 \\ 1 & 3 & -2 \\ -1 & -1 & 1 \end{pmatrix}$

19. $(I - A)^{-1}D = \begin{pmatrix} 124.32 \\ 143.30 \\ 159.76 \\ 159.19 \end{pmatrix}$

CHAPTER 4

1. (a) $A^t = \begin{pmatrix} 3 & 4 \\ 1 & 3 \end{pmatrix}$,   (c) $A' = \begin{pmatrix} 2 & 3 \\ 1 & -2 \\ -3 & 4 \end{pmatrix}$

3. $AB = \begin{pmatrix} 30 & 54 \\ 45 & 78 \\ 31 & 83 \\ 40 & 87 \end{pmatrix}$

5. minor $a_{11} = 9$,   minor $a_{12} = -17$, minor $a_{13} = 29$
   minor $a_{21} = -21$, minor $a_{22} = -34$, minor $a_{23} = 6$
   minor $a_{31} = 26$,   minor $a_{32} = 0$,   minor $a_{33} = -39$

7. $|A| = 221$

9. (a) $|A| = 16$,   (b) $|A| = 191$

11. adj $A = (\text{cof } A)^t = \begin{pmatrix} 9 & 21 & 26 \\ 17 & -34 & 0 \\ 29 & -6 & -39 \end{pmatrix}$

13. (a) $A^{-1} = \begin{pmatrix} \frac{1}{2} & \frac{1}{4} & -1 \\ 0 & -\frac{1}{2} & 1 \\ -\frac{1}{2} & -\frac{1}{4} & 2 \end{pmatrix}$,   (c) $A^{-1} = -\frac{1}{64}\begin{pmatrix} 16 & -16 & 0 \\ -12 & -4 & 16 \\ -6 & 14 & -24 \end{pmatrix}$

15. (a) $x_1 = 6,\ x_2 = -4,\ x_3 = 2$,   (c) $x_1 = 40,\ x_2 = 42,\ x_3 = -3$

17. (a) $r = 2$,   (c) $r = 2$,   (e) $r = 1$

CHAPTER 5

1. (a) $x_1 = 6$, $x_2 = 7$, $Z = 72$
   (c) $x_1 = 12.27$, $x_2 = 3.83$, $C = 88.16$
3. $x_1 = 4.4$, $x_2 = 5.5$, Max $Z = 9.9$, Min $-Z = -9.9$
5. $x_1 = 461.5$, $x_2 = 615.4$, Profit $= \$16,923$
7. $x_1 = 200$, $x_2 = 600$, Profit $= \$520$
9. $x_1 = 5$ oz., $x_2 = 1$ oz., Cost $= \$0.19$
11. Let x_j = number of units manufactured, for $j = 1, 2, 3$

$$\text{Maximize:} \quad Z = 0.15x_1 + 0.12x_2 + 0.09x_3$$
$$\text{Subject to:} \quad 0.03x_1 + 0.02x_2 + 0.03x_3 \leq 1000$$
$$0.11x_1 + 0.14x_2 + 0.20x_3 \leq 4000$$
$$0.30x_1 + 0.20x_2 + 0.26x_3 \leq 9000$$
$$0.08x_1 + 0.07x_2 + 0.08x_3 \leq 3000$$
$$x_1 \geq 9000, \; x_2 \geq 9000, \; x_3 \geq 6000$$

13. Let x_{ijk} represent the amount of grain i in feed j sold at price k, for $i = $ 1(Wheat) and 2(Barley), $j = $ 1(Fertilex) and 2(Multiplex), and $k = $ 1(regular price) and 2(discount price).

$$\text{Maximize:} \quad Z = 1.10x_{111} + 0.94x_{112} + 1.23x_{211} + 1.07x_{212}$$
$$+ 0.86x_{121} + 0.69x_{122} + 0.99x_{221} + 0.82x_{222}$$

$$\text{Subject to:} \quad x_{111} + \quad x_{112} + \quad x_{121} + \quad x_{122} \leq 1000$$
$$x_{211} + \quad x_{212} + \quad x_{221} + \quad x_{222} \leq 1200$$
$$x_{111} + \quad x_{211} \leq 99$$
$$x_{121} + \quad x_{221} \leq 99$$
$$x_{111} \geq \quad 2x_{211}$$
$$x_{112} \geq \quad 2x_{212}$$
$$x_{221} \geq \quad 2x_{121}$$
$$x_{222} \geq \quad 2x_{122}$$
$$x_{ijk} \geq 0$$

15. Let x_{ij} represent the amount of blending component i in brand j for $i = 1, 2$ and $j = 1, 2, 3$.

$$\text{Maximize:} \quad Z = 0.23x_{11} + 0.28x_{12} + 0.33x_{13} + 0.13x_{21}$$
$$+ 0.18x_{22} + 0.23x_{23}$$
$$\text{Subject to:} \quad x_{11} + \quad x_{12} + \quad x_{13} \leq 4000$$
$$x_{21} + \quad x_{22} + \quad x_{23} \leq 2000$$
$$10x_{11} + \quad 50x_{21} \geq \quad 20(x_{11} + x_{21})$$
$$10x_{12} + \quad 50x_{22} \geq \quad 30(x_{12} + x_{22})$$
$$10x_{13} + \quad 50x_{23} \geq \quad 40(x_{13} + x_{23})$$
$$x_{ij} \geq 0$$

17. Let x_{ij} represent job i scheduled during period j, for $i = 1, 2$ and $j = 1, 2$.

Assumptions: (1) Jobs scheduled during the day can be completed during the day or night.
(2) Jobs scheduled during the night are constrained by the night capacity and the overflow of day jobs.

Maximize: $Z = 275(x_{11} + x_{12}) + 125(x_{21} + x_{22}) + 225(x_{31} + x_{32})$

Subject to: $1200(x_{11} + x_{12}) + 1400(x_{21} + x_{22}) + 800(x_{31} + x_{32})$
$$\leq 13{,}400$$

$20(x_{11} + x_{12}) + 15(x_{21} + x_{22}) + 35(x_{31} + x_{32})$
$$\leq 400$$

$100(x_{11} + x_{12}) + 60(x_{21} + x_{22}) + 80(x_{31} + x_{32})$
$$\leq 1050$$

$$1200\,x_{12} + 1400x_{22} + 800x_{32} \leq 9200$$
$$20\,x_{12} + 15x_{22} + 35x_{32} \leq 250$$
$$100\,x_{12} + 60x_{22} + 80x_{32} \leq 650$$
$$x_{ij} \geq 0$$

19. Let x_j represent the number of advertisements in media j, for $j = 1, 2, 3$.

Media	Effectiveness Coefficient
1	$0.80(0.4) + 0.70(0.2) + 0.15(0.4) = 0.52$
2	$0.70(0.4) + 0.80(0.2) + 0.20(0.4) = 0.52$
3	$0.20(0.4) + 0.60(0.2) + 0.40(0.4) = 0.36$

Maximize: $Z = 0.52(600{,}000)x_1 + 0.52(800{,}000)x_2 + 0.36(300{,}000)x_3$

Subject to. $600x_1 + 800x_2 + 450x_3 \leq 20{,}000$
$x_1 \leq 12, \quad x_2 \leq 24, \quad x_3 \leq 12$
$x_1 \geq 3. \quad x_2 \geq 6, \quad x_3 \geq 2$

CHAPTER 6

1. (a) Multiple optimal solutions: $x_1 = 8, x_2 = 0, x_3 = 0, x_4 = 28, Z = 24$;
   and $x_1 = 5.45, x_2 = 3.82, x_3 = 0, x_4 = 0, Z = 24$
   (c) $x_1 = 2.8, x_2 = 4.4, x_3 = 0, x_4 = 0, x_5 = 0, Z = 54.8$

3. (a) $x_1 = 13.75, x_2 = 15, x_3 = 20, x_4 = 0, x_5 = 0, x_6 = 0, Z = 525$

5. (a) $x_1 = 20, x_2 = 6, x_3 = 96, x_4 = 0, Z = 152$

7. (a) No feasible solution,   (b) Unbounded solution,
   (c) No feasible solution,   (d) Unbounded solution

9. 533.3 type 1200 lamps, 133.3 type 1201 lamps, profit = $893

marginal value of assembly labor is $8.67 and the marginal value of wiring labor is $1.67.

11. 61.5 Toots, 149.25 Wheets, 0 Honks, Profit $109.76.

13. 99 pounds of Fertilex at regular price, 301 pounds of Fertilex at discount price, 99 pounds of Multiplex at regular price, 1701 pounds of Multiplex at discount price. Total profit is $2116.67.

15. Total production is 5000 quarts of Brand S and 1000 quarts of Brand SSS. Profit is $1280.

17. 1.05 payrolls jobs scheduled during the day, 1.45 payroll jobs scheduled during the night, 3.68 inventory jobs scheduled during the day, 6.32 inventory jobs scheduled during the night. Total profit is $2937.50.

CHAPTER 7

1. (a) $\quad$ Maximize: $\quad P = 234w_1 + 214w_2$
$\qquad$ Subject to: $\qquad 13w_1 + 16w_2 \leq 2$
$\qquad\qquad\qquad\qquad 18w_1 + 14w_2 \leq 3$
$\qquad\qquad\qquad\qquad\qquad w_1, w_2 \geq 0$

(c) $\quad$ Maximize: $\quad P = 42w_1 + 60w_2 + 63w_3$
$\qquad$ Subject to: $\qquad 12w_1 + 5w_2 + 3w_3 \leq 2$
$\qquad\qquad\qquad\qquad 7w_1 + 12w_2 + 21w_3 \leq 1$
$\qquad\qquad\qquad\qquad\qquad w_1, w_2, w_3 \geq 0$

3. (a) $x_1 = 20$, $x_2 = 6$, $x_3 = 0$, $x_4 = 96$, $C = 1520$
$\qquad w_1 = 14.5$, $w_2 = 0$, $w_3 = 18.2$, $w_4 = 0$, $w_5 = 0$, $P = 1520$

5. (a) $x_1 = 0$, $x_2 = 25.33$, $x_3 = 0$, $x_4 = 20.67$, $C = 152$

7. $x_1 = 0$, $x_2 = 8$, $x_3 = 9$, $x_4 = 5$, $C = \$786$

9. $x_1 = 4.8$, $x_2 = 12.6$, $x_3 = 0$, $x_4 = 4.8$, $x_5 = 0$, $C = \$0.444$

CHAPTER 8

1. (a)

	A_1	A_2	A_3	Available
E_1	300	200		500
E_2		100	400	500
Scheduled	300	300	400	1000

(b)

	A_1	A_2	A_3	Available
E_1		100	400	500
E_2	300	200		500
Scheduled	300	300	400	1000

The total shipping cost is $20,500.

3. (a)

	1	2	3	4	5	6	Capacity
1	0.14	0.16	0.20	0.23	0.23	0	75
2	0.16	0.13	0.18	0.19	0.20	0	100
3	0.17	0.16	0.17	0.16	0.20	0	125
Req'd.	30	30	100	50	40	50	300

(b)

	1	2	3	4	5	6	Capacity
1	30		5		40		75
2		30	70				100
3			25	50		50	125
Req'd.	30	30	100	50	40	50	300

(c)

	1	2	3	4	5	6	Capacity
1	30					45	75
2		30	25		40	5	100
3			75	50			125
Req'd.	30	30	100	50	40	50	300

The total cost is $41,350.

5. (a)

Plant	Gas.	Kero.	Diesel	Jet	Asph.	Slack	Capacity
1	0.165	$-M$	0.140	0.125	0.138	0	70,000
2	0.140	0.146	0.126	$-M$	0.133	0	90,000
3	$-M$	0.139	0.134	0.134	0.130	0	40,000
Req'd.	60,000	15,000	40,000	25,000	20,000	40,000	200,000

(b)

Plant	Gas.	Kero.	Diesel	Jet	Asph.	Slack	Capacity
1	60,000		10,000				70,000
2		15,000	15,000		20,000	40,000	90,000
3			15,000	25,000			40,000
Req'd.	60,000	15,000	40,000	25,000	20,000	40,000	200,000

The initial solution is optimal. Profit is $24,400.

9.

	1	2	3	4	5	Supply
1		240	260	600		1100
2	220	160	540			920
3	620					620
4			200		820	1020
Req'd.	840	400	1000	600	820	3660

The total transportation cost is $8,110,000.

11.

	1	2	3	4	5	6
1	20	0	20	⓪	0	10
2	40	⓪	40	30	30	20
3	⓪	20	20	50	0	20
4	20	50	⓪	20	50	20
5	50	20	30	20	⓪	0
6	20	10	70	0	50	⓪

An optimal assignment is shown by the circled elements. The total time is 970 minutes.

13. (a) Since the problem involves maximizing the number of viewers, it is necessary to reverse the magnitudes of the entries. This reversed magnitude tableau is

	1	2	3	4	5	6	7	8
1	22	28	15	24	10	35	25	32
2	26	26	10	22	8	34	22	34
3	28	25	18	22	4	21	23	36
4	30	20	24	26	0	27	24	36
5	20	29	20	25	12	33	26	33
6	24	28	14	23	16	29	23	35
7	26	26	20	15	0	30	24	37
8	28	24	26	26	10	34	27	38

(b)

	1	2	3	4	5	6	7	8
1	3	2	2	0	1	8	1	⓪
2	9	2	⓪	0	1	9	0	4
3	15	5	12	4	1	⓪	5	10
4	20	3	21	11	②	9	9	13
5	⓪	2	7	0	2	5	1	0
6	6	3	3	0	8	3	⓪	4
7	16	9	17	⓪	0	12	9	4
8	11	⓪	16	4	3	9	5	7

CHAPTER 9

1. (a) $2x$, (c) $5x^4$, (e) $-4x^{-5}$, (g) $\frac{1}{2}x^{-1/2}$, (i) $\frac{1}{4}x^{-3/4}$, (k) $-2/x^3$
   (m) nx^{n-1}, (o) 0, (q) $\frac{1}{2}x^{-1/2}$

3. (a) $6x + 2$, (c) $75x^2 + 32x + 6$, (e) $\frac{3}{4}x^{-3/4} + x^{-1/2}$
   (g) $4ax^3 + 3bx^2 + 2cx + d$

5. (a) $5/(x + 3)^2$, (c) $[(x^2 + 2)(4x - 5) - (2x^2 - 5x)(2x)]/(x^2 + 2)^2$
   (e) $[(x^4 + 3x^2)(-3x^{-4} - 2x^{-3}) - (x^{-3} + x^{-2})(4x^3 + 6x)]/(x^4 + 3x^2)^2$
7. (a) 8, (c) 1, (e) 1.5, (g) 1.5, (i) 72

CHAPTER 10

1. (a) $x = 3$ minimum, (c) $x = 3$ minimum, (e) $x = 3$ minimum,
   $x = 2$ maximum, (g) $x = -1$ maximum, $x = 1$ minimum
   (i) $x = -1.37$ maximum, $x = 4.37$ minimum, (k) $x = 2/3$ minimum,
   $x = -3$ maximum

3.

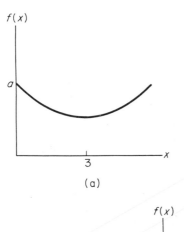

(a)

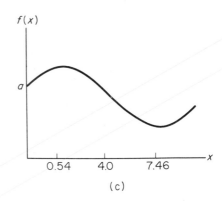

(c)

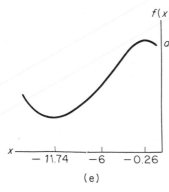

(e)

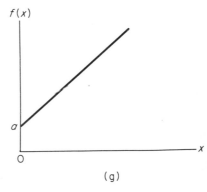

(g)

$(x) = 590 - 1.45x + 0.001x^2$, $x^* = 725.$
$) = 533 - 3.30x + 0.0067x^2$, $x^* = 246$
$= -3910 + 105.5x - 0.534x^2$, $x^* = 99$

CHAPTER 11

1. $TR = 5.00q - 0.75q^2$
   $MR = 5.00 - 1.50q$
   $AR = 5.00 - 0.75q$
   $E = \dfrac{1}{-0.75} \cdot \dfrac{5.00 - 0.75q}{q}$
   Revenue is maximum when $q = 3.33$.

3. $TR = 2500q - 60q^2 - 0.5q^3$
   $MR = 2500 - 120q - 1.5q^2$
   $AR = 2500 - 60q - 0.5q^2$
   $E = \dfrac{1}{-60 - q} \cdot \dfrac{(2500 - 60q - 0.5q^2)}{q}$
   $q = 17.2$ for maximum revenue
   $TR(q = 17.2) = 22{,}750$
   $AR = 1320$, $E = -1$, or $|E| = 1$

5. $p = 10 - 0.005q \qquad AR = 10 - 0.005q$
   $TR = 10q - 0.005q^2 \qquad E = \dfrac{1}{-0.005} \cdot \dfrac{(10 - 0.005q)}{q}$
   $MR = 10 - 0.01q \qquad q = 1000$ and $p = \$5$ for maximum revenue

7. $E = -1.20$; fare increase should not be made since $|E| > 1$.

9. $TC = 1000 + 20q - q^2 + 0.05q^3 \qquad MC = 20 - 2q + 0.15q^2$
   $AC = \dfrac{1000}{q} + 20 - q + 0.05q^2$
   $AVC = 20 - q + 0.05q^2 \qquad AFC = \dfrac{1000}{q}$

   AC is a minimum when $\dfrac{d(AC)}{dq} = 0$, $q = 25.5$.

   AVC is a minimum when $\dfrac{d(AVC)}{dq} = 0$, $q = 10.0$.

   MC is a minimum when $\dfrac{d(MC)}{dq} = 0$, $q = 6.67$.

   AFC is a minimum at the end point of the domain, $q = 30$.

11. $\pi = 600q - 0.5q^2 - 1500 - 150q + 4q^2 - 0.5q^3$
    Profits are maximum for $q = 19.8$.
    Revenue is maximum for $q = 600.0$.
    AC is minimum for $q = 13$.

13. $TR = 200q - q^2 \qquad TC = 5000 + 0.50q$
   $\pi = 200q - q^2 - 5000 - 0.50q$
   $q = 99.75$

15. $q = 1414$

17. $q_1 = 3830, q_2 = 8250$

CHAPTER 12

1. (a) $\dfrac{\partial f}{\partial x} = 2 + 3y \qquad \dfrac{\partial f}{\partial y} = 3x + 3 \qquad \dfrac{\partial^2 f}{\partial x^2} = 0$

   $\dfrac{\partial^2 f}{\partial y^2} = 0 \qquad \dfrac{\partial^2 f}{\partial x\, \partial y} = 3$

   (c) $\dfrac{\partial f}{\partial x} = 6x(x^2 + y^2)^2 \qquad \dfrac{\partial f}{\partial y} = 6y(x^2 + y^2)^2$

   $\dfrac{\partial^2 f}{\partial x^2} = 6(x^2 + y^2)^2 + 24x^2(x^2 + y^2)$

   $\dfrac{\partial^2 f}{\partial y^2} = 6(x^2 + y^2)^2 + 24y^2(x^2 + y^2) \qquad \dfrac{\partial^2 f}{\partial x\, \partial y} = 24xy(x^2 + y^2)$

   (e) $\dfrac{\partial f}{\partial x} = 3(x + y)^2 \qquad \dfrac{\partial f}{\partial y} = 3(x + y)^2 \qquad \dfrac{\partial^2 f}{\partial x^2} = 6(x + y)$

   $\dfrac{\partial^2 f}{\partial y^2} = 6(x + y) \qquad \dfrac{\partial^2 f}{\partial x\, \partial y} = 6(x + y)$

   (g) $\dfrac{\partial f}{\partial x} = \dfrac{1}{y} \qquad \dfrac{\partial f}{\partial y} = \dfrac{-x}{y^2} \qquad \dfrac{\partial^2 f}{\partial x^2} = 0 \qquad \dfrac{\partial^2 f}{\partial y^2} = \dfrac{2x}{y^3} \qquad \dfrac{\partial^2 f}{\partial x\, \partial y} = -\dfrac{1}{y^2}$

   (i) $\dfrac{\partial f}{\partial x} = \dfrac{4}{2x + y} \qquad \dfrac{\partial f}{\partial y} = \dfrac{2}{2x + y} \qquad \dfrac{\partial^2 f}{\partial x^2} = \dfrac{-8}{(2x + y)^2}$

   $\dfrac{\partial^2 f}{\partial y^2} = \dfrac{-2}{(2x + y)^2} \qquad \dfrac{\partial^2 f}{\partial x\, \partial y} = \dfrac{-4}{(2x + y)^2}$

   (k) $\dfrac{\partial f}{\partial x} = \dfrac{x + 2y}{x + 3y} + \ln(x + 3y) \qquad \dfrac{\partial f}{\partial y} = \dfrac{3x + 6y}{x + 3y} + 2\ln(x + 3y)$

   $\dfrac{\partial^2 f}{\partial x^2} = \dfrac{y}{(x + 3y)^2} + \dfrac{1}{(x + 3y)} \qquad \dfrac{\partial^2 f}{\partial y^2} = \dfrac{-3x}{(x + 3y)^2} + \dfrac{6}{(x + 3y)}$

   $\dfrac{\partial^2 f}{\partial x\, \partial y} = \dfrac{-x}{(x + 3y)^2} + \dfrac{3}{x + 3y}$

   (m) $\dfrac{\partial f}{\partial x} = 2xe^{x^2 + y^2} \qquad \dfrac{\partial f}{\partial y} = 2ye^{x^2 + y^2} \qquad \dfrac{\partial^2 f}{\partial x^2} = 2e^{x^2 + y^2}(2x^2 + 1)$

   $\dfrac{\partial^2 f}{\partial y^2} = 2e^{x^2 + y^2}(2y^2 + 1) \qquad \dfrac{\partial^2 f}{\partial x\, \partial y} = 4xye^{x^2 + y^2}$

   (o) All derivatives are zero.

3. (a) $x^* = 1$, $y^* = 2$, minimum
   (c) $x^* = 5$, $y^* = 7.5$, saddle point
   (e) $x^* = 16$, $y^* = 7$, minimum
   (g) $x^* = 7.5$, $y^* = 15.6$, saddle point
   (i) $x^* = 2$, $y^* = 4$, maximum
   (k) $x^* = -4$, $y^* = -2$ minimum
   (m) $x^* = 18$, $y^* = 8$, minimum

4. (a) $x = -43.2$, $y = 33.6$, $\lambda = 230.4$
   $f(-43.2, 33.6) = -2754.8$, $f(-42.2, 33.1) = -2753.5$
   $f(-44.2, 34.1) = -2753.5$. Constrained minimum.
   (c) $x = 5$, $y = 3$, $\lambda = -7$
   $f(5, 3) = 28$, $f(4, 4) = 32$, $f(6, 2) = 32$. Constrained minimum.
   (e) $x = 5$, $y = 4$, $\lambda = 2$
   $f(5, 4) = -19$, $f(5\frac{1}{3}, 3) = -25\frac{1}{3}$, $f(4\frac{2}{3}, 5) = -25\frac{1}{3}$
   Constrained maximum.
   (g) $x = 7.45$, $y = 4.55$, $\lambda = -10.8$
   $f(7.45, 4.55) = 214.5$, $f(7, 5) = 210$, $f(8, 4) = 208$
   Constrained maximum.

5. (a) Increases sales by approximately $70.
   (c) Increases sales by approximately $93,400.

7. (a) $x = 164$, $y = 63.2$, $p = \$77{,}252$

9. $x = 10$, $y = 50$, $\lambda = 180$

CHAPTER 13

1. $q = 121$, $s = 66$, $t = 0.242$

3. $q = 667$, $t_1 = 0.0333$, $t_2 = 0.0500$,
   $t_3 = 0.1500$, $t_4 = 0.1000$, $t = 0.3333$
   Peak inventory $= 300$

5. $q = 20.25 \sim 20$, $s = 14$,
   $t = \frac{2}{3}$, $t_2 = \frac{1}{5}$ month

9. $i = 1.12\%$

11. $i = 7.6\%$

13. $i_{daily} = 5.13\%$, $i_{qtrly.} = 5.09\%$

15. 8% for alternative 1
    8.10% for alternative 2

17. $p = \$2665$

CHAPTER 14

1. (a) $6x + c$ (c) $\dfrac{x}{3} + c$ (e) $ax + bx + c$

 (g) $x + c$ (i) $\dfrac{x^2}{2} + c$ (k) $-x^{-1} + c$

 (m) $-\tfrac{1}{2}x^{-2} + c$ (o) $\dfrac{2x^{3/2}}{3} + c$ (q) $\tfrac{8}{3}$

 (s) 5   (u) $7499\tfrac{1}{4}$

2. (a) $\dfrac{3x^2}{2} + 4x + c$ (c) $\dfrac{x^4}{2} + \dfrac{x^3}{3} + c$

 (e) $\dfrac{x^3}{3} + x^2 + x + c$ (g) 15

 (i) $e^{3x} + c$ (k) $\dfrac{e^{2x}}{4} + c$

 (m) 1   (o) $x \ln(x) - x + c$
 (q) $x \log_a(3x) - x \log_a e + c$ (s) 14.0260

3. (a) $\tfrac{2}{7}(x - 1)^{7/2} + c$ (c) $\dfrac{(4x^3 - 2)^{10}}{120} + c$

 (e) $\dfrac{e^{x^3}}{3} + c$ (g) $\tfrac{1}{5} \ln(5x - 6) + c$

 (i) 2.4423,   (k) $\tfrac{1}{2}(x^2 - 6) \ln(x^2 - 6) - \tfrac{1}{2}(x^2 - 6) + c$
 (m) 0.6932

5. Using $n = 5$, $\Delta x = 1.0$, $A = 41.25$
 Using $n = 10$, $\Delta x = 0.5$, $A = 41.56$
 Using 14.4, $A = 41.67$

7. (a) 36,   (c) 132,   (e) 128,   (g) 21.25,
 (i) -7.5,   (k) 30,   (m) $-41\tfrac{2}{3}$,   (o) 54,
 (q) 2893,   (s) $-87\tfrac{1}{12}$,   (u) $\tfrac{2}{3}(e^{60} - 1)$,   (w) 0.393,
 (y) 0.234

CHAPTER 15

(a) $\{(mm), (mf), (fm), (ff)\}$
) $\{(x, y, z), (x, z, y), (y, x, z), (y, z, x), (z, x, y), (z, y, x)\}$

  $C), (S, O), (S, A), (J, C), (J, O), (J, A), (C, C), (C, O), (C, A)\}$

5.

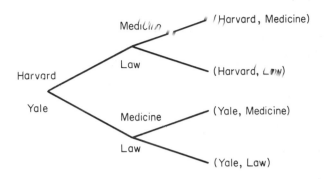

7. (a)

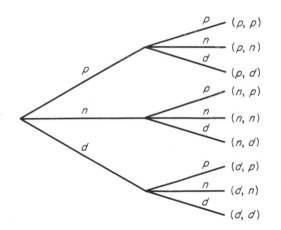

9. $S = \{0, 1, 2, 3, \ldots\}$, Countably infinite

11. (a) $\{(ggg)\}$, Simple
    (c) $\{(ggd), (gdg), (dgg), (gdd), (dgd), (ddg), (ddd)\}$, Composite
    (e) $\{(dgg), (dgd), (ddg), (ddd)\}$, Composite

13. (a) 8,   (c) 68,   (e) 100,   (g) 17

15. 24

17. 3,628,800

19. 11,232,000

21. 3003

23. 66
25. (a) Objective,   (c) 0.30,   (e) 0.1344
27. $P(L) = 0.40, P(B) = 0.70, P(L \cap B) = 0.28$
29. $P(B) = 0.46, P(A' \cap B') = 0.36$
31. (a) $P(P|p) = 0.435, P(C|p) = 0.435, P(A|p) = 0.130$

CHAPTER 16

1. (a) $\{x|x = 0, 1, 2, 3, 4\}$,   Discrete
   (c) $\{d|d \geq 0\}$,   Continuous

3. (a)

x	0	1	2	3
$P(X = x)$	$\frac{1}{8}$	$\frac{3}{8}$	$\frac{3}{8}$	$\frac{1}{8}$

5. (a) $\frac{8}{50}$,   (c) $\frac{9}{50}$
7. (b) $\frac{12}{36}$,   (c) $\frac{4}{36}$
9. (a) 4.5,   (c) $\frac{1}{9}$
11. $E(X) = 1\frac{1}{3}, \text{Var}(x) = 1\frac{5}{9}$
13. $E(X) = 0.5, \text{Var}(X) = 0.05$
15. Median $= 3$, Mode $= 3$
17. $x_m = 1$
19. $P(X = 0) = 0.255, P(X = 1) = 0.509, P(X \leq 1) = 0.764$
21. $E(X) = 3, P(X = 3) = 0.2668$
23. $n = 6$
25. 0.135
27. 0.909
29. 0.25
31. 0.7745
33. $\mu = 19.05, \sigma = 4.65$
35. 0.393

Index